AF614758

NEUROMETHODS

Series Editor
Wolfgang Walz
University of Saskatchewan
Saskatoon, SK, Canada

For further volumes:
http://www.springer.com/series/7657

Protein Kinase Technologies

Edited by

Hideyuki Mukai

Biosignal Research Center, Kobe University, Kobe, Japan

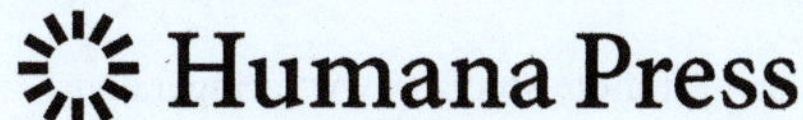

Editor
Hideyuki Mukai
Biosignal Research Center
Kobe University
Kobe, Japan

Please note that additional material for this book can be downloaded from
http://extras.springer.com

ISSN 0893-2336 e-ISSN 1940-6045
ISBN 978-1-61779-823-8 e-ISBN 978-1-61779-824-5
DOI 10.1007/978-1-61779-824-5
Springer New York Dordrecht Heidelberg London

Library of Congress Control Number: 2012938260

Printed on acid-free paper

Humana Press is part of Springer Science+Business Media (www.springer.com)

Preface to the Series

Under the guidance of its founders Alan Boulton and Glen Baker, the Neuromethods series by Humana Press has been very successful since the first volume appeared in 1985. In about 17 years, 37 volumes have been published. In 2006, Springer Science + Business Media made a renewed commitment to this series. The new program will focus on methods that either are unique to the nervous system and excitable cells or need special consideration to be applied to the neurosciences. The program will strike a balance between recent and exciting developments like those concerning new animal models of disease, imaging, in vivo methods, and more established techniques. These include immunocytochemistry and electrophysiological technologies. New trainees in neurosciences still need a sound footing in these older methods in order to apply a critical approach to their results. The careful application of methods is probably the most important step in the process of scientific inquiry. In the past, new methodologies led the way in developing new disciplines in the biological and medical sciences. For example, Physiology emerged out of Anatomy in the nineteenth century by harnessing new methods based on the newly discovered phenomenon of electricity. Nowadays, the relationships between disciplines and methods are more complex. Methods are now widely shared between disciplines and research areas. New developments in electronic publishing also make it possible for scientists to download chapters or protocols selectively within a very short time of encountering them. This new approach has been taken into account in the design of individual volumes and chapters in this series.

Wolfgang Walz

Preface

Part I of this book contains recent methodologies and techniques generally applicable to protein kinase research. Part II focuses on individual protein kinases which require special attention in neuroscience. These chapters will not only be practical instructions useful for readers' daily work in setting up and performing research but also be thought-provoking and enjoyable reviews of recent advancements of individual protein kinases in neuroscience, provided by prominent investigators. It is important to emphasize that while each of these chapters in Part II deals with different protein kinases, they will contain a wealth of common methodologies. As such, the techniques and approaches pioneered with respect to one class of protein kinases are likely to be equally applicable to other classes. In addition to fostering cross-talk among investigators who study different protein kinases, these chapters will also be beneficial for the entry of new investigators into the field.

Kobe, Japan *Hideyuki Mukai*

Contents

Preface to the Series v
Preface vii
Contributors xi

PART I RECENT GENERAL METHODOLOGIES

1 Visualization of Protein Kinase Activities in Living Cells 3
Kazuhiro Aoki, Naoki Komatsu, Akihiro Goto, and Michiyuki Matsuda

2 Phos-tag Affinity Electrophoresis for Protein Kinase Profiling 13
Eiji Kinoshita, Emiko Kinoshita-Kikuta, and Tohru Koike

3 Large-Scale Protein Phosphorylation Analysis by Mass Spectrometry-Based Phosphoproteomics 35
Wei-Chi Ku, Naoyuki Sugiyama, and Yasushi Ishihama

PART II INDIVIDUAL PROTEIN KINASES

4 Ca^{2+}/Calmodulin-Dependent Protein Kinase II (CaMKII) 49
Steven J. Coultrap and K. Ulrich Bayer

5 CASK: A Specialized Neuronal Kinase 73
Konark Mukherjee

6 Cyclin-Dependent Kinase 5 (Cdk5): Preparation and Measurement of Kinase Activity 87
Seiji Minegishi, Taro Saito, and Shin-ichi Hisanaga

7 ErbB Membrane Tyrosine Kinase Receptors: Analyzing Migration in a Highly Complex Signaling System 105
Nicole M. Brossier, Stephanie J. Byer, Lafe T. Peavler, and Steven L. Carroll

8 The Extracellular Signal-Regulated Kinase (ERK) Cascade in Neuronal Cell Signaling 133
Daniel Orellana, Ilaria Morella, Marzia Indrigo, Alessandro Papale, and Riccardo Brambilla

9 Glycogen Synthase Kinase-3 in Neurological Diseases 153
Oksana Kaidanovich-Beilin and James Robert Woodgett

10 Cortical Neurons Culture to Study c-Jun N-Terminal Kinase Signaling Pathway 189
Alessandra Sclip, Xanthi Antoniou, and Tiziana Borsello

11 MuSK: A Kinase Critical for the Formation and Maintenance of the Neuromuscular Junction 203
Arnab Barik, Wen-cheng Xiong, and Lin Mei

12 Assays for Pten-Induced Novel Kinase 1 (PINK1) and Leucine-Rich Repeat Kinase 2 (LRRK2), Kinases Associated with Parkinson's Disease 219
Alexandra Beilina and Mark R. Cookson

13 Imaging PKA Activation Inside Neurons in Brain Slice Preparations 237
Marina Brito, Elvire Guiot, and Pierre Vincent

14 Imaging Oscillations of Protein Kinase C Activity in Cells 251
Maya T. Kunkel and Alexandra C. Newton

15 Diacylglycerol Kinase (DGK) as a Regulator of PKC 259
Yasuhito Shirai and Naoaki Saito

16 Trafficking of Trk Receptors 273
Daniel Bodmer and Rejji Kuruvilla

17 Mammalian Target of Rapamycin 291
Lukasz J. Swiech, Malgorzata Urbanska, Matylda Macias, Agnieszka Skalecka, and Jacek Jaworski

18 Protein Kinase G (PKG): Involvement in Promoting Neural Cell Survival, Proliferation, Synaptogenesis, and Synaptic Plasticity and the Use of New Ultrasensitive Capillary-Electrophoresis-Based Methodologies for Measuring PKG Expression and Molecular Actions 319
Ronald R. Fiscus and Mary G. Johlfs

19 Electrophysiological Technique for Analysis of Synaptic Function of PKN1 in Hippocampus. 349
Hiroki Yasuda and Hideyuki Mukai

Index. *361*

Contributors

XANTHI ANTONIOU • *Istituto di Ricerche Farmacologiche "Mario Negri", Milano, Italy*

KAZUHIRO AOKI • *Department of Pathology and Biology of Diseases, Graduate School of Medicine, Kyoto University, Kyoto, Japan; PRESTO, Japan Science and Technology Agency (JST), Saitama, Japan*

ARNAB BARIK • *Department of Neurology, Institute of Molecular Medicine and Genetics, Georgia Health Sciences University, Augusta, GA, USA*

K. ULRICH BAYER • *Department of Pharmacology, School of Medicine, University of Colorado, Denver, Aurora, CO, USA*

ALEXANDRA BEILINA • *Cell Biology and Gene Expression Section, Laboratory of Neurogenetics, National Institute on Aging, National Institutes of Health, Bethesda, MD, USA*

DANIEL BODMER • *Department of Biology, Johns Hopkins University, Baltimore, MD, USA*

TIZIANA BORSELLO • *Istituto di Ricerche Farmacologiche "Mario Negri", Milano, Italy*

RICCARDO BRAMBILLA • *San Raffaele Scientific Institute and University, Milano, Italy*

MARINA BRITO • *Centre National de la Recherche Scientifique, Unité Mixe de Recherche (UMR) 7102, Université Pierre et Marie Curie, UPMC, Paris, France*

NICOLE M. BROSSIER • *Departments of Pathology, Cell Biology, and the Medical Scientist Training Program, University of Alabama at Birmingham, Birmingham, AL, USA*

STEPHANIE J. BYER • *Department of Pathology, University of Alabama at Birmingham, Birmingham, AL, USA*

STEVEN L. CARROLL • *Departments of Pathology, Cell Biology and Neurobiology, University of Alabama at Birmingham, Birmingham, AL, USA*

MARK R. COOKSON • *Cell Biology and Gene Expression Section, Laboratory of Neurogenetics, National Institute on Aging, National Institutes of Health, Bethesda, MD, USA*

STEVEN J. COULTRAP • *Department of Pharmacology, School of Medicine, University of Colorado, Denver, Aurora, CO, USA*

RONALD R. FISCUS • *Center for Diabetes and Obesity Prevention, Treatment, Research, and Education, College of Pharmacy, Roseman University of Health Sciences (Formerly the University of Southern Nevada), Henderson, NV, USA*

AKIHIRO GOTO • *Department of Pathology and Biology of Diseases, Graduate School of Medicine, Kyoto University, Kyoto, Japan*

ELVIRE GUIOT • *Centre National de la Recherche Scientifique, Unité Mixe de Recherche (UMR) 7102, Université Pierre et Marie Curie, UPMC, Paris, France*

SHIN-ICHI HISANAGA • *Department of Biological Sciences, Graduate School of Science, Tokyo Metropolitan University, Tokyo, Japan*

MARZIA INDRIGO • *San Raffaele Scientific Institute and University, Milano, Italy*
YASUSHI ISHIHAMA • *Graduate School of Pharmaceutical Sciences, Kyoto University, Kyoto, Japan; Institute for Advanced Biosciences, Keio University, Yamagata, Japan*
JACEK JAWORSKI • *International Institute of Molecular and Cell Biology, Warsaw, Poland*
MARY G. JOHLFS • *Center for Diabetes and Obesity Prevention, Treatment, Research, and Education, The College of Pharmacy, Roseman University of Health Sciences (Formerly the University of Southern Nevada), Henderson, NV, USA*
OKSANA KAIDANOVICH-BEILIN • *Samuel Lunenfeld Research Institute, Mount Sinai Hospital, Toronto, ON, Canada*
EIJI KINOSHITA • *Department of Functional Molecular Science, Institute of Biomedical and Health Sciences, Hiroshima University, Hiroshima, Japan*
EMIKO KINOSHITA-KIKUTA • *Department of Functional Molecular Science, Institute of Biomedical and Health Sciences, Hiroshima University, Hiroshima, Japan*
TOHRU KOIKE • *Department of Functional Molecular Science, Institute of Biomedical and Health Sciences, Hiroshima University, Hiroshima, Japan*
NAOKI KOMATSU • *Department of Pathology and Biology of Diseases, Graduate School of Medicine, Kyoto University, Kyoto, Japan*
WEI-CHI KU • *Graduate School of Pharmaceutical Sciences, Kyoto University, Kyoto, Japan*
MAYA T. KUNKEL • *Department of Pharmacology, University of California at San Diego, La Jolla, CA, USA*
REJJI KURUVILLA • *Department of Biology, Johns Hopkins University, Baltimore, MD, USA*
MATYLDA MACIAS • *International Institute of Molecular and Cell Biology, Warsaw, Poland*
MICHIYUKI MATSUDA • *Department of Pathology and Biology of Diseases, Graduate School of Medicine, Kyoto University, Kyoto, Japan*
LIN MEI • *Department of Neurology, Institute of Molecular Medicine and Genetics, Georgia Health Sciences University, Augusta, GA, USA*
SEIJI MINEGISHI • *Department of Biological Sciences, Graduate School of Science, Tokyo Metropolitan University, Tokyo, Japan*
ILARIA MORELLA • *San Raffaele Scientific Institute and University, Milano, Italy*
HIDEYUKI MUKAI • *Biosignal Research Center, Kobe University, Kobe, Japan*
KONARK MUKHERJEE • *Virginia Tech Carilion Research Institute, Roanoke, VA, USA*
ALEXANDRA C. NEWTON • *Department of Pharmacology, University of California at San Diego, La Jolla, CA, USA*
DANIEL ORELLANA • *San Raffaele Scientific Institute and University, Milano, Italy*
ALESSANDRO PAPALE • *San Raffaele Scientific Institute and University, Milano, Italy*
LAFE T. PEAVLER • *Department of Pathology, University of Alabama at Birmingham, Birmingham, AL, USA*
NAOAKI SAITO • *Laboratory of Molecular Pharmacology, Biosignal Research Center, Kobe, Japan*
TARO SAITO • *Department of Biological Sciences, Graduate School of Science, Tokyo Metropolitan University, Tokyo, Japan*

AGNIESZKA SKALECKA • *International Institute of Molecular and Cell Biology, Warsaw, Poland*
ALESSANDRA SCLIP • *Istituto di Ricerche Farmacologiche "Mario Negri", Milano, Italy*
YASUHITO SHIRAI • *Laboratory of Chemistry and Utilization of Animal Production Resources, Applied Chemistry in Bioscience Division, Graduate School of Agricultural Science, Biosignal Research Center, Kobe, Japan*
NAOYUKI SUGIYAMA • *Institute for Advanced Biosciences, Keio University, Yamagata, Japan*
LUKASZ J. SWIECH • *International Institute of Molecular and Cell Biology, Warsaw, Poland*
MALGORZATA URBANSKA • *International Institute of Molecular and Cell Biology, Warsaw, Poland*
PIERRE VINCENT • *Centre National de la Recherche Scientifique, Unité Mixe de Recherche (UMR) 7102, Université Pierre et Marie Curie, UPMC, Paris, France*
JAMES ROBERT WOODGETT • *Samuel Lunenfeld Research Institute, Mount Sinai Hospital, Toronto, ON, Canada; Department of Medical Biophysics, University of Toronto, Toronto, ON, Canada*
Wen-cheng Xiong • *Department of Neurology, Institute of Molecular Medicine and Genetics, Georgia Health Sciences University, Augusta, GA, USA*
HIROKI YASUDA • *Education and Research Support Center, Gunma University Graduate School of Medicine, Maebashi, Japan*

Part I

Recent General Methodologies

Chapter 1

Visualization of Protein Kinase Activities in Living Cells

Kazuhiro Aoki, Naoki Komatsu, Akihiro Goto, and Michiyuki Matsuda

Abstract

Protein kinases are key modulators of intracellular signal transduction cascades, which determine various events in neuronal cells such as replication and differentiation. For many years, protein kinases were analyzed mostly by biochemical methods, which could handle the cells only en masse. For a better understanding of the role of kinases in neuronal cells, one would like to know the subcellular distribution of kinase activities and to follow a particular kinase activity for a specific period in a single cell. Genetically encoded biosensors based on the principle of Förster (or fluorescence) resonance energy transfer (FRET) and fluorescent proteins have been developed to accommodate such requirements. The method involves expression of the FRET biosensors in neuronal cells, time-lapse imaging under fluorescence microscopes, image processing, and quantification of FRET. This technique could be applicable to living organisms ranging from *Caenorhabditis elegans* to mouse, permitting visualization of spatio-temporal regulation of kinase activities and systemic understanding of the signaling networks in living animals.

Key words: FRET, Kinase, Time-lapse imaging

1. Introduction

Protein kinases constitute one of the largest gene families, covering ~2% of the human genome (1). It is estimated that approximately 30% of all cellular proteins are phosphorylated on at least one residue (2). Thus, protein kinases have key roles in many fundamental processes of not only neurons but also other cell types (1). Biochemical methods have been widely used to investigate whether or not a protein kinase of interest is active. Although biochemical methods are robust in vitro, they generally do not provide information about protein kinase activities at specific subcellular compartments; nor do they provide information about activity changes at the single-cell level.

Hideyuki Mukai (ed.), *Protein Kinase Technologies*, Neuromethods, vol. 68,
DOI 10.1007/978-1-61779-824-5_1, © Springer Science+Business Media, LLC 2012

The recent advent of green fluorescent proteins (GFP) and sophisticated microscopes has opened a new window onto the dynamics of spatio-temporal regulation of protein kinase activities (3). Förster (or fluorescence) resonance energy transfer (FRET) is a non-radiative energy transfer process that is observed when an excited donor fluorophore is placed in close proximity to an acceptor fluorophore (see reviews (3–5)). This phenomenon depends on the spectral overlap of the donor emission and acceptor excitation, the distance between them, and the relative orientation of the fluorophore's transition dipole moments (6). Biosensors based on the principle of FRET and GFP derivatives enabled us to visualize the spatio-temporal dynamics of intracellular signaling cascades involving protein kinases, small GTPases, phospholipids, and so on (7–10).

There are two types of FRET biosensors: intramolecular (or unimolecular) or intermolecular (or bimolecular). Most of the hitherto reported FRET biosensors having a high signal-to-noise ratio belong to the intramolecular type (6, 11). The intramolecular FRET biosensors for kinase activity can be further sub-classified into a substrate type and an enzyme type. In the former design, the FRET biosensor comprises a substrate peptide specific to the kinase of interest and a phosphopeptide-binding domain, sandwiched between donor and acceptor fluorophores (Fig. 1a). Upon activation of the kinase of interest, the substrate peptide is phosphorylated and bound to the phosphopeptide-binding domain, thereby inducing a global change of the biosensor conformation and concomitant increase (or decrease in some cases) of the FRET efficiency from the donor to the acceptor. Meanwhile, in the latter design, a full-length kinase of interest, or a region of that kinase, is included in the biosensor. The FRET efficiency of the biosensor depends on the conformation of the kinase, which will change by binding to (12–14) or phosphorylation by upstream regulators (15) (Fig. 1b). These FRET biosensors have been utilized to observe kinase activities in a single neuron or at specific subcellular regions such as spines and growth cones (16–18).

Here, as an example, we provide a method for visualizing the spatio-temporal activity of protein kinase A (PKA) beneath the plasma membrane in live PC12D cells by pm-AKAR3, a substrate-type intramolecular FRET biosensor (19). PKA activation is required for neurite outgrowth of PC12D cells induced by dibutyryl cAMP (dbcAMP) (20). An outline of this procedure is as follows: (1) PC12D cells are seeded onto glass-based dishes for FRET imaging, (2) an expression plasmid for pm-AKAR3 is introduced into PC12D cells by lipofection, (3) FRET images are acquired by an epifluorescence microscope, and (4) the acquired imaging data are processed and quantified by image analysis software offline.

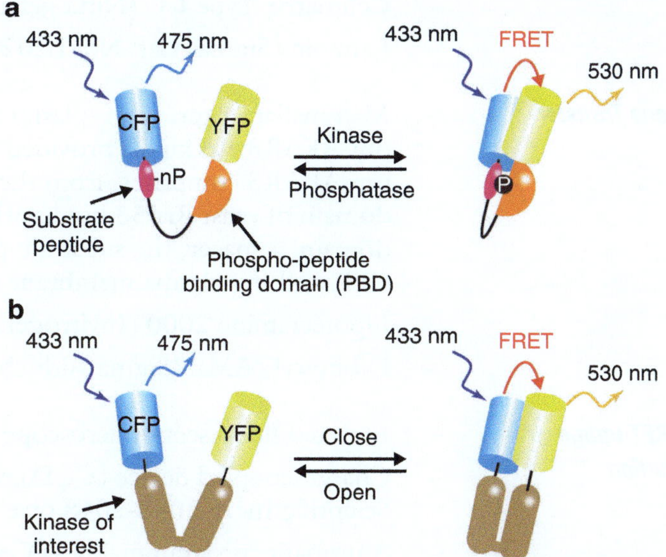

Fig. 1. Schematic representation of an intramolecular FRET biosensor of kinases. (**a**) Structure of a substrate-type intramolecular FRET biosensor, which consists of a yellow-emitting mutant of green fluorescent protein (YFP), a phospho-binding domain, a substrate peptide, and a cyan-emitting mutant of green fluorescent protein (CFP). When the substrate peptide is not phosphorylated, cyan fluorescence of 475 nm emanates from CFP upon excitation at 433 nm. When the substrate peptide is phosphorylated by the kinase of interest, the intramolecular interaction of the phosphorylated peptide and the binding domain brings YFP into close proximity with CFP to induce FRET, evoking 525 nm fluorescence from YFP. (**b**) Structure of an enzyme-type intramolecular FRET biosensor, which includes a full or part of a kinase of interest within the probe. When the kinase perceives an input signal such as phosphorylation, the biosensor changes its conformation, leading to a corresponding increase or decrease in FRET efficiency.

2. Materials

2.1. Cell Culture

- PC12D cells kindly provided by Dr. Sano (20)
- Bovine serum albumin (BSA) (Sigma-Aldrich Co., Cat. No. A2153-50G)
- DMEM (Sigma-Aldrich, St. Louis, MO) supplemented with 10% fetal bovine serum (FBS)
- Phenol-red-free DMEM/F12 medium (Invitrogen, Cat. No. 11039-021)
- Collagen coated-tissue culture plastic dishes (Asahi Techno Glass Co., Cat. No. 4000-010)
- Plastic petri dishes (Nunc, Cat. No. 150350)
- Cover-glass bottom dishes for imaging (Asahi Techno Glass Co., Cat. No. 3911-035)

- Cellmatrix Type I-C (Nitta gelatin)
- Laminin (Sigma, Cat. No. L2020-1MG)

2.2. Gene Transfer

- Mammalian expression plasmids for the FRET biosensor: pm-AKAR3 was kindly provided by Dr. Jin Zhang (19). Breifly, pm-AKAR3 comprises, from the amino terminus, YFP, FHA1 domain of yeast Rad53 used as the phosphothreonine-binding domain, a spacer, the substrate peptide against PKA, a spacer, CFP, and the plasma membrane targeting signal of KRas
- Lipofectamine 2000 (Invitrogen, Cat. No. 11668-019)
- Dibutyryl cAMP (Sigma-Aldrich Co., Cat. No. D026)

2.3. FRET Image Acquisition

- Inverted fluorescent microscope (IX81; Olympus) (Fig. 2)
- Charge-coupled device (CCD) camera (CoolSNAP K4; Roper Scientific Inc.): 2048-2048 pixel
- Automatic programmable XY stage (MD-XY30100T-Meta, SIGMA KOKI)
- Laser-based autofocus system (IX2-ZDC, Olympus)
- Optical filters used for the dual-emission imaging: an XF1071 (440AF21) excitation filter, an XF2034 (455DRLP) dichroic mirror, and two emission filters (XF3075 [480AF30] for CFP and XF3079 [535AF26] for FRET) (all from Omega Optical) and neutral density filters (Olympus). Filters with similar optical properties can also be purchased from Chroma Technology Co., Asahi Spectra Co., and Semrock Co.
- LED illumination system, CoolLED precisExcite (Molecular device), or 75-W Xenon arc lamp and power supply (Olympus)

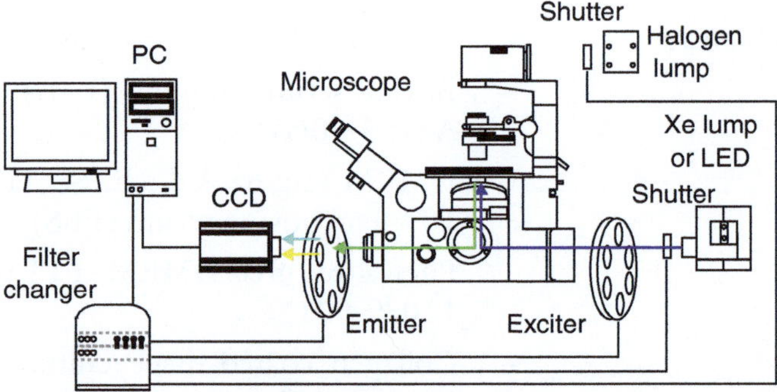

Fig. 2. Layout of the microscopic setup for FRET imaging. Illustration of an inverted fluorescence microscope equipped with an XY stage, a Xenon lamp or LED illumination system, excitation and emission filter wheels, a CCD camera, and a thermal incubation box. All of these devices are controlled by MetaMorph software.

- A thermal incubation box, IBMU (Olympus) or thermally controlled chamber, to maintain 37°C
- 60x oil-immersion objective lens (UPlanSApo 60x/1.42, Olympus)
- MetaMorph software (Universal Imaging): version 7.0 or higher
- MetaMorph software (Universal Imaging): version 7.0 or higher
- Excel (Microsoft)

3. Methods

3.1. Cell Culture

1. PC12D cells maintained in 10 cm tissue culture dishes are trypsinized and seeded onto 35 mm cover-glass bottom dishes (Note 1).

3.2. Gene Transfer

1. One day after seeding, an expression plasmid for pm-AKAR3 (Fig. 3a) is introduced into the cells by lipofection with Lipofectamine2000 according to the manufacturer's protocol (Note 2).
2. Three hours after the addition of transfection reagents, the culture medium is replaced with fresh medium to reduce the toxicity.

3.3. Time-Lapse Imaging

1. One day or two days after transfection, the culture medium is further replaced with 2 ml phenol-red-free DMEM/F12 with 0.1% BSA (Note 3) for serum starvation (Note 4). Then, cells are cultured for 3–12 h.
2. Turn on all electronic devices of the epifluorescent microscope at least 1 h before imaging (Note 5).
3. Transfer 500 μl of medium from the culture dish into a microtube. Add dbcAMP (final concentration, 1 mM) to this conditioned medium. Keep warm at 37°C.
4. Set the 35 mm cover-glass bottom dish on the stage of the microscope.
5. Set imaging conditions. Typical conditions are listed as follows: illumination setting, CFP (ex. 440 nm, em. 475 nm), FRET (ex. 440 nm, em. 510 nm), differential interference contrast (DIC) or phase contrast; 6–25% neutral density filter for 75-W Xenon lamp or 5–15% laser power for CoolLED precisExcite illumination; interval time, 30 s to 2 min; exposure time, 100 ms to 1 s; CCD camera binning, 3 × 3 to 8 × 8.
6. Focus and select the cell of interest (Note 6).

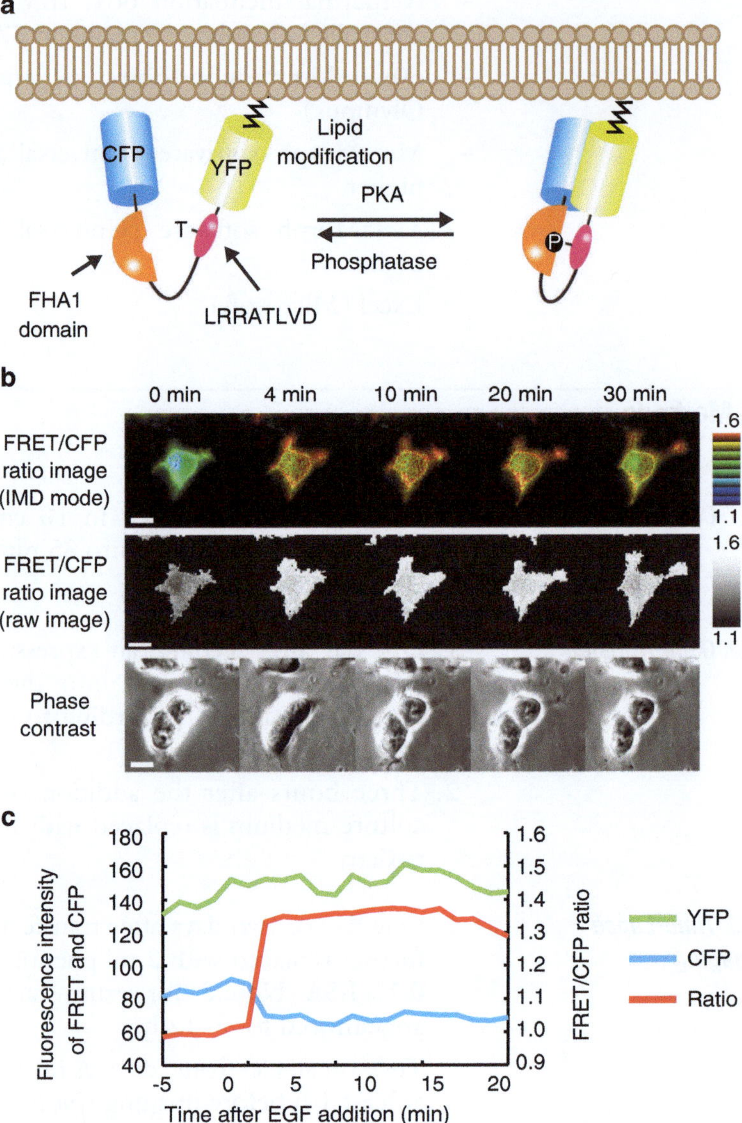

Fig. 3. PKA activation at the plasma membrane visualized by pm-AKAR3 in PC12D cells. PC12D cells expressing pm-AKAR3 were starved for 3 h and stimulated with 1 mM dbcAMP. Images were obtained every 2 min for 40 min. (**a**) Schematic representation of mode of action of pm-AKAR3 is shown. (**b**) Time-lapse images of FRET/CFP ratios are represented in the intensity modulated display (IMD) mode (*upper row*) or the raw values (*middle row*). Phase contrast images (*lower row*) are also shown. Scale bars, 10 μm. (**c**) Representative time courses of FRET, CFP and FRET/CFP ratios after EGF addition (min). The ratio data is normalized to its average values before EGF stimulation. Note the inverse correlation between the fluorescence intensities of FRET and CFP.

7. Wait at least 1 min to recover from photochromism of YFP, a reversible photochemical reaction (21, 22).
8. Start time-lapse imaging.
9. After 10 min of imaging, pause the image acquisition and add dbcAMP- or EGF-containing conditioned-medium (from step 3) dropwise around the dishes with a micropipette or dropper (Note 7). Resume data acquisition and continue for 30–60 min.

3.4. Image Processing

1. Launch MetaMorph software (Note 8).
2. Assemble time-lapse images into stack files such as FRET.stk, CFP.stk, and DIC.stk.
3. Set a region of interest (ROI) in a cell-free area and subtract background with the following functions: process → background and shading correction → statistical correction (Note 9).
4. Create a FRET/CFP ratio image in the intensity modulated display (IMD) mode with the functions: process → ratio image. Set parameters as follows: numerator, the FRET stack file; denominator, the CFP stack file; IMD display, eight ratios with 32 intensities. Thereafter, min (minimum) and max (maximum) ratio values should be set in a trial-and-error manner so that the image fully uses the eight color hues (Note 10) (Fig. 3b).
5. If required, make movie files (.avi or .mov) using the function stack → make movie in the menu of the MetaMorph program.
6. To quantify the increase in FRET/CFP ratio upon stimulation, set ROIs over the cells of interest in the background-subtracted FRET stack file. Confirm that these ROIs are large enough to keep the cells within it during the whole course of imaging. Transfer the ROIs to the background-subtracted CFP stack file.
7. Measure the fluorescence intensity of FRET and CFP in the cells of interest with the functions: measure → region measurements, and export the values in an ASCII or an Excel format.
8. Open the data file in Excel and plot the FRET/CFP value versus time after stimulation (Fig. 3c).

4. Notes

1. Coating extracellular matrix such as collagen or laminin on the surface of glass bottom dishes helps attachment of cells to the dish, which enhances gene transfer with lipofection and facilitates time-lapse imaging. Of note, this procedure is essential

for primary and immortalized neuronal cells due to their weak binding affinity to the surface of cover-glass bottom dishes.

2. Alternative strategies are applicable for expressing exogenous proteins such as electroporation and virus-mediated gene transfer.
3. Phenol-red emits weak fluorescence at the wavelength of 440 nm. Therefore, it is preferable to use phenol-red-free medium during the course of time-lapse imaging. In addition, we usually add 0.1% BSA (a final concentration) as a surrogate of serum.
4. Serum, a source of growth factors and auto-fluorescence, should be excluded from the culture medium to increase the signal-to-noise ratio.
5. It is essential to warm up the thermal incubator box or imaging chamber to 37°C before imaging to avoid thermal focus drift.
6. To obtain reproducible results, one must find healthy-looking cells expressing a proper amount of FRET biosensors. Generally, higher expression levels of fluorescence proteins allow a lower intensity of excitation light, which reduces the risk of phototoxicity to the cells and photo-bleaching of the FRET biosensors. However, over-expression of FRET biosensors may perturb endogenous signaling events. The key is moderation: do not choose too bright or too dim cells.
7. Never touch the dish when reagents are applied to the cells, or cells will be bumped out of the viewfield.
8. Other image processing software like NIH ImageJ (freely downloadable at http://rsbweb.nih.gov/ij/index.html) is also applicable for the data processing of FRET imaging.
9. If the background fluorescence intensity fluctuates during the course of imaging, the background subtraction of FRET and CFP images should be performed for every image plane. For this purpose, the custom-made macro (called "Journal" in MetaMorph software) is described (23) and also available online (http://www.lif.kyoto-u.ac.jp/labs/fret/phogemon/journal/Subtract_FRET_CFP.JNL).
10. Alternatively, the min and max ratio values can be evenly determined by a custom-made journal, "Normalized_Ratio.jnl" (23) available online (http://www.lif.kyoto-u.ac.jp/labs/fret/phogemon/journal/Normalized_Ratio.JNL).

Acknowledgments

We thank Dr. J. Zhang for the plasmids. Y. Inaoka, K. Hirano, R. Sakai, and N. Nonaka are also to be thanked for their technical assistance. We are grateful to the members of the Matsuda Laboratory for their helpful discussions. K.A. was supported by a Grant-in-Aid for Scientific Research on Priority Areas and by the JST PRESTO program. M.M. was supported by the Research Program of Innovative Cell Biology by Innovative Technology (Cell Innovation) from the Ministry of Education, Culture, Sports, and Science (MEXT), Japan.

References

1. Ubersax JA, Ferrell JE Jr (2007) Mechanisms of specificity in protein phosphorylation. Nat Rev Mol Cell Biol 8:530–541
2. Cohen P (2000) The regulation of protein function by multisite phosphorylation–a 25 year update. Trends Biochem Sci 25: 596–601
3. Tsien RY, Miyawaki A (1998) Seeing the machinery of live cells. Science 280: 1954–1955
4. Pollok BA, Heim R (1999) Using GFP in FRET-based applications. Trends Cell Biol 9:57–60
5. Jares-Erijman EA, Jovin TM (2003) FRET imaging. Nat Biotechnol 21:1387–1395
6. Miyawaki A (2003) Visualization of the spatial and temporal dynamics of intracellular signaling. Dev Cell 4:295–305
7. Mochizuki N, Yamashita S, Kurokawa K, Ohba Y, Nagai T, Miyawaki A, Matsuda M (2001) Spatio-temporal images of growth-factor-induced activation of Ras and Rap1. Nature 411:1065–1068
8. Miyawaki A, Llopis J, Heim R, McCaffery JM, Adams JA, Ikura M, Tsien RY (1997) Fluorescent indicators for Ca^{2+} based on green fluorescent proteins and calmodulin. Nature 388:882–887
9. Sato M, Ueda Y, Takagi T, Umezawa Y (2003) Production of PtdInsP3 at endomembranes is triggered by receptor endocytosis. Nat Cell Biol 5:1016–1022
10. Itoh RE, Kurokawa K, Ohba Y, Yoshizaki H, Mochizuki N, Matsuda M (2002) Activation of rac and cdc42 video imaged by fluorescent resonance energy transfer-based single-molecule probes in the membrane of living cells. Mol Cell Biol 22:6582–6591
11. Kiyokawa E, Aoki K, Nakamura T, Matsuda M (2010) Spatiotemporal regulation of small GTPases as revealed by probes based on the principle of Förster Resonance Energy Transfer (FRET): implications for signaling and pharmacology. Annu Rev Pharmacol Toxicol 51:337–358
12. Terai K, Matsuda M (2006) The amino-terminal B-Raf-specific region mediates calcium-dependent homo- and hetero-dimerization of Raf. EMBO J 25:3556–3564
13. Terai K, Matsuda M (2005) Ras binding opens c-Raf to expose the docking site for mitogen-activated protein kinase kinase. EMBO Rep 6:251–255
14. Fujioka A, Terai K, Itoh RE, Aoki K, Nakamura T, Kuroda S, Nishida E, Matsuda M (2006) Dynamics of the Ras/ERK MAPK cascade as monitored by fluorescent probes. J Biol Chem 281:8917–8926
15. Yoshizaki H, Mochizuki N, Gotoh Y, Matsuda M (2007) Akt-PDK1 complex mediates epidermal growth factor-induced membrane protrusion through Ral activation. Mol Biol Cell 18:119–128
16. Harvey CD, Ehrhardt AG, Cellurale C, Zhong H, Yasuda R, Davis RJ, Svoboda K (2008) A genetically encoded fluorescent sensor of ERK activity. Proc Natl Acad Sci USA 105:19264–19269
17. Murray AJ, Tucker SJ, Shewan DA (2009) cAMP-dependent axon guidance is distinctly regulated by Epac and protein kinase A. J Neurosci 29:15434–15444
18. Shelly M, Lim BK, Cancedda L, Heilshorn SC, Gao H, Poo MM (2010) Local and long-range reciprocal regulation of cAMP and cGMP in axon/dendrite formation. Science 327:547–552

19. Allen MD, Zhang J (2006) Subcellular dynamics of protein kinase A activity visualized by FRET-based reporters. Biochem Biophys Res Commun 348:716–721
20. Katoh-Semba R, Kitajima S, Yamazaki Y, Sano M (1987) Neuritic growth from a new subline of PC12 pheochromocytoma cells: cyclic AMP mimics the action of nerve growth factor. J Neurosci Res 17:36–44
21. Sinnecker D, Voigt P, Hellwig N, Schaefer M (2005) Reversible photobleaching of enhanced green fluorescent proteins. Biochemistry 44:7085–7094
22. Henderson JN, Ai HW, Campbell RE, Remington SJ (2007) Structural basis for reversible photobleaching of a green fluorescent protein homologue. Proc Natl Acad Sci USA 104:6672–6677
23. Aoki K, Matsuda M (2009) Visualization of small GTPase activity with fluorescence resonance energy transfer-based biosensors. Nat Protoc 4:1623–1631

Chapter 2

Phos-tag Affinity Electrophoresis for Protein Kinase Profiling

Eiji Kinoshita, Emiko Kinoshita-Kikuta, and Tohru Koike

Abstract

Protein kinase profiling can provide a basis for understanding the molecular origins of diseases and, potentially, for developing tools for therapeutic intervention. It is therefore very important to develop advanced experimental procedures for convenient and accurate determination of the phosphorylation status of certain substrate proteins in the life sciences. Here, we introduce a method for protein kinase profiling by using a novel type of phosphate-affinity sodium dodecyl sulfate-polyacrylamide gel electrophoresis (SDS-PAGE). The phosphate-affinity site is a polyacrylamide-bound dinuclear metal complex of a phosphate-binding tag molecule known as Phos-tag. The Phos-tag SDS-PAGE method permits detection of changes in the mobility of phosphorylated proteins in comparison with their nonphosphorylated counterparts and thereby allows quantitative analysis of protein kinase reactions without any special apparatus, radioactive isotopes, or chemical labels. If a kinase reaction occurs at one residue of a substrate protein, the monophosphorylated and nonphosphorylated forms can be simultaneously detected as two migration bands on a Phos-tag SDS-PAGE gel. In the case of hyperphosphorylation, the phosphorylated products appear as multiple migration bands, depending on the phosphorylation status in terms of the numbers and the positions of attached phosphate groups. This article discusses applications of label-free kinase activity profiling by the Phos-tag SDS-PAGE method in the analysis of phosphorylated substrates derived from various kinase reactions. The resolving power of the affinity electrophoresis provides detailed information that leads to an overview of the kinase-dependent dynamics of various substrate proteins.

Key words: Affinity electrophoresis, Phosphoproteomics, Phosphorylation, Phos-tag, Protein kinase, SDS-PAGE, Western blotting

1. Introduction

Protein phosphorylation is among the most common post-translational modifications across biological species and it controls a number of key cellular processes by means of changes in the balance between the opposing reversible reactions of specific protein kinases and phosphatases (1). More than 500 protein kinases are predicted to occur in the human proteome alone (2), a number that clearly reflects the

Hideyuki Mukai (ed.), *Protein Kinase Technologies*, Neuromethods, vol. 68,
DOI 10.1007/978-1-61779-824-5_2, © Springer Science+Business Media, LLC 2012

importance of protein phosphorylation. In fact, abnormal protein phosphorylation is closely involved with many human diseases, including cancer and neurodegenerative diseases (3). Many mutations in protein kinase-encoding genes that cause aberrant enzyme activity are associated with various forms of cancer. There has been considerable progress in the development of selective inhibitors for certain kinases as potential drug targets, and some of these inhibitors have been approved for use in humans for the treatment of cancer. Imatinib mesylate (Glivec; Novartis, Basel, Switzerland) and gefitinib (Iressa; AstraZeneca, London, UK) are typical examples of such drugs. The former was designed to inhibit Bcr/Abl tyrosine kinase, which is generated by a chimeric gene resulting from a chromosomal translocation that is characteristic of chronic myeloid leukemia. The latter was developed as a potent inhibitor of epidermal growth factor (EGF) receptor tyrosine kinase, and it selectively inhibits EGF-stimulated tumor cell growth. In human brain neurons, on the other hand, the activities of several protein kinases can be dysregulated, leading to hyperphosphorylation of the microtubule-associated protein Tau, which is a classic hallmark of Alzheimer's disease (4). The sites and stoichiometry of phosphorylation of the Tau protein are correlated with the pathological characteristics of the disease.

Methods for determining the phosphorylation status of certain substrate proteins and for screening for novel inhibitors of certain kinases have therefore become increasingly important. A method conventionally used for defining a particular phosphorylation event is the incorporation of a radioisotope (RI) such as ^{32}P or ^{33}P into a phosphorylated protein. The phosphorylation status of the target protein can then be detected and quantified from its radioactivity. A non-RI method using polyclonal and monoclonal antibodies is also well established for the detection of protein phosphorylation. The readout from the antibody that has recognized a phosphopeptide or phosphoprotein is measured as a fluorescence, luminescence, or polarization signal, and this can be utilized in many analytical procedures, including enzyme-linked immunosorbent assay, immunoblotting, or immunocytochemistry. Because these procedures require specific antibody-based reagents, a lack of availability or specificity of an appropriate antibody can directly affect the performance of the assay in some cases. As an alternative to immunoassays, which are frequently problematic in relation to the antibody-based reagents, chemical labeling of phosphate groups has been used for phosphospecific-site mapping in conjunction with mass spectrometry (MS) (5). Affinity chromatography using the characteristics of metal ions is widely accepted as a technique for comprehensively determining phosphorylation sites. Recently, improvements in the specificity of the chromatography have been accomplished in a number of ways for MS-based studies on the phosphoproteome (6–10). To establish novel therapeutic approaches to human diseases, phosphoproteomic methods for the

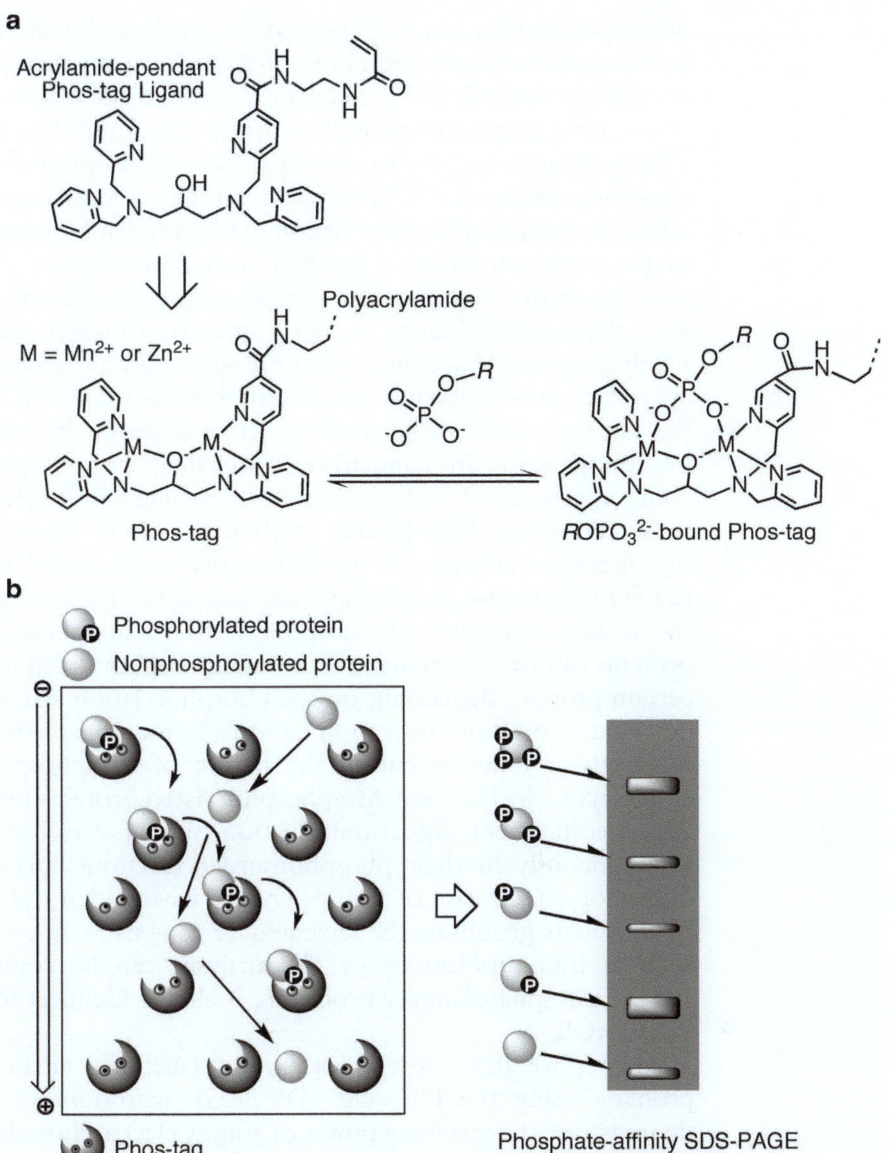

Fig. 1. Structure of acrylamide-pendant Phos-tag ligand and scheme for reversible capture of a phosphomonoester dianion ($RO PO_3^{2-}$) by Phos-tag (**a**). Schematic representation of the principle of phosphate-affinity SDS-PAGE (**b**).

determination of the phosphorylation status of proteins are subject to continual development efforts.

We have developed the Phos-tag technology as a novel approach for the analysis of protein phosphorylation (Phos-tag consortium, http://www.phos-tag.com/english/index.html). The Phos-tag technology utilizes a novel phosphate-binding tag molecule, Phos-tag {1,3-bis[bis(pyridin-2-ylmethyl)amino]propan-2-olato dizinc(II) complex}, which binds to anionic substituents, especially phosphomonoester dianions, under physiological conditions (Fig. 1a) (11). To date, the technology has contributed to the development of a range of procedures, including matrix-assisted laser desorption/ionization time-of-flight MS analysis of phosphorylated compounds (12), electrospray ionization MS analysis for online mass tagging of phosphopeptides (13), immobilized metal-affinity chromatography for the separation of phosphopeptides and phosphoproteins (14–17), surface plasmon resonance analysis for reversible peptide phosphorylation (18), Western blotting analysis of phosphoproteins on a protein-blotted membrane (19), and solid-phase-free fluorescence analysis of protein phosphorylation and dephosphorylation in an aqueous solution (20, 21). Furthermore, we have shown that phosphate-affinity sodium dodecyl sulfate-polyacrylamide gel electrophoresis (SDS-PAGE) using the tag molecule can be used to detect shifts in the mobility of phosphoproteins in comparison with their nonphosphorylated counterparts (22–27) (Fig. 1b). This Phos-tag affinity electrophoresis (Phos-tag SDS-PAGE) technique offers the following significant advantages (1) no radioactive or chemical labels are required for kinase and phosphatase assays; (2) the time course of the quantitative ratio of phosphorylated to nonphosphorylated proteins can be determined; (3) several phosphorylated forms of a certain protein, depending on the phosphorylation status, can be detected as multiple migration bands; (4) the phosphate-binding specificity is independent of the nature of the phosphorylated amino acid; (5) His- and Asp-phosphorylated proteins involved in a two-component signal-transduction system can be detected simultaneously in their phosphotransfer reactions; (6) different phosphorylated forms of a single protein having identical numbers of phosphate groups can be separated; (7) downstream procedures, such as immunoblotting or MS analysis, can be applied; and (8) the phosphate-affinity procedure is almost identical to normal SDS-PAGE.

Here, we describe applications of label-free kinase activity profiling using the Phos-tag SDS-PAGE methodology, and we demonstrate the resolving power of affinity electrophoresis in separating phosphorylated substrate proteins.

2. Materials (See Note 1)

2.1. Preparation of Tau Proteins Phosphorylated by Various Ser/Thr Kinases

1. Substrate Tau protein: His-tagged recombinant human Tau isoform F (molecular weight≈60 kDa), consisting of 441 amino acid residues, purchased from Calbiochem (La Jolla, CA, USA) is used. Store at −80°C.
2. Ser/Thr kinases: Recombinant human glycogen synthase kinase-3β (GSK-3β) and recombinant mouse protein kinase A (PKA) catalytic subunit purchased from Calbiochem are used. Recombinant human cyclin-dependent kinase 5 (cdk5)/p35, recombinant mouse mitogen-activated protein kinase 2

(MAPK), recombinant human casein kinase II (CKII), and rat forebrain calcium/calmodulin-dependent protein kinase II (CaMKII) were purchased from Millipore (Billerica, MA, USA). GSK-3β and cdk5/p35 must be stored at −80°C; the other proteins can be stored at −20°C.

3. In vitro Ser/Thr kinase reaction: The in vitro phosphorylation assay is carried out by using the recombinant Tau protein (4.1 μg) at 30°C.

4. Kinase reaction buffer for phosphorylation of Tau by GSK-3β, cdk5/p35, PKA, MAPK, and CKII: The reaction buffer (20 μL) contains 25 mM Tris–HCl (pH 7.5) (e.g., Nacalai Tesque, Kyoto, Japan), 5.0 mM β-glycerol phosphate (e.g., glycerol 2-phosphate disodium salt hydrate, Sigma, St. Luis, MO, USA), 12 mM $MgCl_2$ (e.g., Nacalai Tesque), 2.0 mM dithiothreitol [(2*S*,3*S*)-1,4-disulfanylbutane-2,3-diol; e.g., Nacalai Tesque], 0.10 mM Na_3VO_4 (e.g., Nacalai Tesque), 50 μM ATP (e.g., Sigma), 37 kBq [γ-^{32}P]ATP (e.g., GE Healthcare Bio-Sciences, Piscataway, NJ, USA), and 4.1 μg of Tau. The amounts of the kinases in the buffer are as follows: GSK-3β 2.0 μg, cdk5/p35 0.10 μg, PKA 2,500 U, MAPK 0.20 μg, and CKII 0.25 μg. The buffer is prepared immediately before use.

5. Kinase reaction buffer for phosphorylation by CaMKII: The reaction buffer (20 μL) contains 20 mM MOPS (pH 7.2) (e.g., Nacalai Tesque), 25 mM β-glycerol phosphate, 15 mM $MgCl_2$, 1.0 mM dithiothreitol, 1.0 mM Na_3VO_4, 1.0 mM $CaCl_2$, 20 μg/mL recombinant bovine calmodulin (Millipore), 50 μM ATP, 37 kBq [γ-^{32}P]ATP, 4.1 μg of Tau, and 50 ng of CaMKII. The buffer is prepared immediately before use.

6. Sample-loading buffer (3×) for stopping the kinase reaction: 195 mM Tris–HCl (pH 6.8), 3.0% (w/v) SDS (e.g., Nacalai Tesque), 15% (v/v) 2-mercaptoethanol (e.g., Nacalai Tesque) (see Note 2), 30% (v/v) glycerol (e.g., Nacalai Tesque), and 0.1% (w/v) bromophenol blue (BPB) (e.g., Nacalai Tesque). Store at −20°C. After incubation for various reaction times (0–300 min), 3.0 μL of the reaction mixture is taken out and added to the sample-loading buffer (1.5 μL).

7. Ser/Thr-phosphorylation sample of Tau: Store at −20°C. An aliquot (1.2 μL) of the resulting solution is used in the Phos-tag SDS-PAGE (164 ng of Tau/lane).

2.2. Preparation of Tau Proteins Phosphorylated by Various Tyr Kinases

1. Substrate Tau protein: see (1) under Sect. 1.

2. Tyr kinases: Recombinant Tyr kinases (ABL, ACK, AXL, EGFR, EPHA1, FES, FGFR1, FYN, INSR, JAK1, LCK, LYNa, MET, SRC, TEC, TIE2, TYK2, and YES) purchased from Carna Biosciences (Kobe, Japan) are used. Store at −80°C.

3. In vitro Tyr kinase reaction: The in vitro phosphorylation assay is carried out using the recombinant Tau protein (2.5 μg) at 30°C.

4. Kinase reaction buffer for phosphorylation of Tau with 18 types of Tyr kinase: The reaction buffer (40 μL) contains 60 mM HEPES-NaOH (pH 7.5) (e.g., Nacalai Tesque), 10 mM β-glycerol phosphate, 10 mM $MgCl_2$, 1.25 mM dithiothreitol, 0.30 mM Na_3VO_4, 0.20 mM ATP, 2.5 μg of Tau, and 0.20 μg of each kinase. Prepare immediately before use.

5. Sample-loading buffer (3×) for stopping the kinase reaction: see (6) under Sect. 1. After incubation for the appropriate time for each kinase (ABL: 5 min, ACK: 60 min, AXL: 60 min, EGFR: 5 min, EPHA1: 5 min, FES: 60 min, FGFR1: 5 min, FYN: 2 min, INSR: 2 min, JAK1: 60 min, LCK: 2 min, LYNa: 2 min, MET: 2 min, SRC: 2 min, TEC: 60 min, TIE2: 60 min, and YES: 5 min), the sample-loading buffer (20 μL) is added to the reaction mixture.

6. Tyr-phosphorylation sample of Tau: Store at −20°C. An aliquot (3.0 μL) of the resulting solution is used in the Phos-tag SDS-PAGE (125 ng of Tau/lane).

2.3. Preparation of Cell Lysate

1. Culture medium for SW480 cells: Dulbecco's modified Eagle medium (DMEM) containing 10% (v/v) fetal bovine serum (FBS), 100 U/mL penicillin, and 100 μg/mL streptomycin purchased from Invitrogen (Carlsbad, CA, USA) is used. The cells (10^7 cells) on a 90-mm culture dish (e.g., Sumitomo Bakelite, Tokyo, Japan) are incubated in the medium under a humidified atmosphere of 5% CO_2 and 95% air at 37°C.

2. Cell stimulants: A 100 mM solution of Na_3VO_4 in distilled water is used as a Tyr phosphatase inhibitor and is stored at room temperature. For stimulation with 1.0 mM pervanadate for 30 min, a mixture of 1.0 mM Na_3VO_4 and 3.0 mM H_2O_2 (final concentration) (e.g., Nacalai Tesque) is added to the culture medium. Mix immediately before use. As an adenylyl cyclase activator, a 10 mM solution of forskolin (Enzo Life Sciences, Plymouth Meeting, PA, USA) in dimethyl sulfoxide (e.g., Nacalai Tesque) is prepared and stored at −20°C. For stimulation for 30 min, 10 μM forskolin (final concentration) is added to the culture medium.

3. Washing buffer: Tris-buffered saline (TBS) containing 10 mM Tris–HCl (pH 7.5) and 0.10 M NaCl (e.g., Nacalai Tesque). Store at room temperature. Each culture is washed twice before preparation of the cell lysate.

4. Lysis buffer: a sample-loading buffer (1×) consisting of 65 mM Tris–HCl (pH 6.8), 1.0% (w/v) SDS, 5.0% (v/v) 2-mercaptoethanol, 10% (v/v) glycerol, and 0.03% (w/v) BPB is used. Store at −20°C.

5. Lysate sample: The lysate sample solution is sonicated briefly to reduce its viscosity and then boiled for 5 min. Store at –20°C. An aliquot (10 μL) of the resulting solution is used in the Phos-tag SDS-PAGE.

2.4. Mn^{2+}–Phos-tag SDS-PAGE Using an Alkaline-pH Buffer System (Laemmli's Buffer System)

1. Phos-tag solution: 5.0 mM acrylamide-pendant Phos-tag ligand, which is commercially available from Wako Pure Chemical (Osaka, Japan), in distilled water containing 3% (v/v) methanol (e.g., Nacalai Tesque) (see Note 3). Store at room temperature in the dark.
2. Manganese(II) chloride solution: 10 mM $MnCl_2 \cdot 4H_2O$ (e.g., Nacalai Tesque) in distilled water (see Note 4). Store at room temperature.
3. Thirty percent (w/v) acrylamide/bis solution (29:1 ratio of acrylamide to *N,N′*-methylenebisacrylamide) (e.g., Nacalai Tesque) (see Note 5). Store at room temperature in the dark.
4. Separating gel buffer (4×): 1.5 M Tris–HCl (pH 8.8) and 0.4% (w/v) SDS. Store at room temperature.
5. Stacking gel buffer (4×): 0.5 M Tris–HCl (pH 6.8) and 0.4% (w/v) SDS. Store at room temperature.
6. Ammonium persulfate (APS) (e.g., Nacalai Tesque) solution and *N,N,N′,N′*-tetramethylethane-1,2-diamine (TEMED) (e.g., Nacalai Tesque) (see Note 6): 10% (w/v) APS in distilled water. Prepare immediately before use.
7. Electrophoresis running buffer (see Note 7): 25 mM Tris, 192 mM glycine (e.g., Nacalai Tesque), and 0.1% (w/v) SDS. Store at room temperature.
8. Gel staining solutions: Sil-Best Stain for Protein/PAGE purchased from Nacalai Tesque and SYPRO Ruby protein gel stain purchased from Invitrogen are used.

2.5. Zn^{2+}–Phos-tag SDS-PAGE Using a Neutral-pH Buffer System

1. Phos-tag solution: see (1) under Sect. 4.
2. Zinc(II) nitrate solution: 10 mM $Zn(NO_3)_2 \cdot 6H_2O$ (e.g., Nacalai Tesque) in distilled water (see Note 8). Store at room temperature.
3. Thirty percent (w/v) acrylamide/bis solution (29:1 ratio of acrylamide to *N,N′*-methylenebisacrylamide): see (3) under Sect. 4.
4. Gel buffer (2.8×) (see Note 9): 1.0 M Bis-Tris–HCl (pH 6.8) (e.g., Nacalai Tesque). Store at room temperature.
5. APS solution and TEMED: see (6) under Sect. 4.
6. Electrophoresis running buffer (see Note 10): 100 mM Tris, 100 mM MOPS, and 0.1% (w/v) SDS. Store at room temperature. Sodium bisulfite is dissolved to a concentration of 5 mM in the buffer solution immediately before use (see Note 11).
7. Gel staining solutions: see (8) under Sect. 4.

2.6. Electroblotting

1. Blotting buffer (see Note 12): 25 mM Tris, 192 mM glycine, and 10% (v/v) methanol. Store at room temperature.
2. Disodium ethylenediaminetetraacetate (EDTA) (e.g., Katayama Chemical, Osaka, Japan) solution: 0.5 M EDTA–NaOH (pH 8). Store at room temperature.
3. Blotting buffer containing 1 mM EDTA: Add 0.2 mL of EDTA solution to 100 mL of blotting buffer.
4. TBS-T solution: 10 mM Tris–HCl (pH 7.5), 0.10 M NaCl, and 0.10% (v/v) Tween-20 (e.g., Nacalai Tesque). Store at room temperature.

2.7. Equipment Setup

1. SDS-PAGE equipment: The instructions assume the use of an Atto model AE-6500 mini-slab gel system (1-mm thick, 9-cm wide, and 9-cm long). The setup can be readily adapted to other formats, including large-type gels.
2. Electroblotting equipment: The instructions assume the use of a Nippon Eido model NA-1511C electroblotting wet-tank unit. The setup can be readily adapted to other formats but not to a semi-dry one (see Note 13).

3. Methods

3.1. Phosphate-Binding Tag Molecule, Phos-tag

To capture a phosphorylated compound effectively under physiological conditions, we developed a dinuclear metal complex {1,3-bis[bis(pyridin-2-ylmethyl)amino]propan-2-olato dizinc(II) complex}, which was synthesized to mimic the active center of an alkaline phosphatase (11). The dinuclear zinc(II) complex forms stable 1:1 complexes with phosphate monoester dianions ($ROPO_3^{2-}$) in an aqueous solution (see Fig. 1a). An X-ray crystal structure analysis of the 1:1 dinuclear zinc(II) complex of a 4-nitrophenyl phosphate dianion showed that each phosphate oxygen anion binds to a zinc(II) atom at the fifth coordination site and that the two zinc(II) ions are separated by a distance of 3.6 Å. Thus, the dinuclear zinc(II) complex, which has a vacancy on the two zinc(II) ions, is suitable for access by a phosphate monoester dianion, which forms a bridging ligand. In an aqueous solution at neutral pH, the dinuclear zinc(II) complex binds strongly with a phenyl phosphate dianion ($K_d = 2.5 \times 10^{-8}$ M). The anion selectivity indexes against SO_4^{2-}, CH_3COO^-, Cl^-, and bisphenyl phosphate monoanion at 25°C are 5.2×10^3, 1.6×10^4, 8.0×10^5, and $>2 \times 10^6$, respectively. We named this dizinc(II) complex "Phos-tag" as an abbreviation of "phosphate-binding tag molecule."

3.2. Mn^{2+}–Phos-tag SDS-PAGE Using the Laemmli's Buffer System

We synthesized a Phos-tag ligand with a pendant acrylamide moiety (see Fig. 1a) and we used this as a novel additive (comonomer) for a separating gel in Laemmli's SDS-PAGE method (28), which is a technique widely used for the separation and detection of proteins. The principle underlying the analytical procedure is the change in the mobility of phosphorylated proteins that results from reversible trapping of phosphate moieties by the Phos-tag molecules immobilized in the gel (see Fig. 1b). When we attempted to use polyacrylamide-bound Zn^{2+}–Phos-tag in SDS-PAGE, the expected phosphate-selective shift in mobility was not observed under the general SDS-PAGE conditions used in Laemmli's method. Presumably, the electrophoresis conditions, such as the abundance of SDS anions and the alkaline buffers, are not conducive to selective trapping of phosphate by the zinc(II) complex. We therefore investigated complexes of other metals (Mn^{2+}, Cu^{2+}, Co^{2+}, Fe^{2+}, and Ni^{2+}) as phosphate-trapping molecules under SDS-PAGE conditions. As a result, we found that a polyacrylamide-bound manganese(II) homologue (Mn^{2+}–Phos-tag) can act as a phosphate-affinity site in Laemmli's buffer system (19). In a separating gel containing copolymerized acrylamide-pendant Mn^{2+}–Phos-tag, the degree of migration of a phosphoprotein is less than that of its nonphosphorylated counterpart, because the tag molecules trap the phosphoprotein reversibly during electrophoresis (see Fig. 1b). On this basis, we developed a novel type of phosphate-affinity SDS-PAGE, known as Mn^{2+}–Phos-tag SDS-PAGE, for the separation of phosphoproteins from their corresponding nonphosphorylated analogues (22–26). Subsequent gel staining or Western blotting with phosphorylation-independent antibodies enables the simultaneous detection of phosphorylated and nonphosphorylated proteins. This method is therefore appropriate for quantitative analyses of protein phosphorylation reactions in vitro and in vivo.

3.3. Zn^{2+}–Phos-tag SDS-PAGE Using a Neutral-pH Buffer System

Affinity electrophoresis using a polyacrylamide-bound Mn^{2+}–Phos-tag and Laemmli's buffer system under alkaline pH conditions has been widely used in determining the phosphorylation states of many proteins (29). However, it has some limitations for separation analysis of certain phosphoproteins. For example, it has been reported that no up-shifted band of Tau protein is observed, following treatment by a certain Tyr kinase (23). An additional disadvantage is that a Mn^{2+}–Phos-tag SDS-PAGE gel cast using an alkaline buffer system is unstable and undergoes partial hydrolysis during long-term storage, resulting in changes in the pore size and a poor resolution. To overcome these limitations, we have recently developed an improved Phos-tag SDS-PAGE (Zn^{2+}–Phos-tag SDS-PAGE) method that uses a dizinc(II) complex of acrylamide-pendant Phos-tag in conjunction with a Bis-Tris-buffered neutral-pH gel system (27). Zn^{2+}–Phos-tag SDS-PAGE is performed at 40 mA/

Table 1
Comparison of the buffer system for Zn^{2+}–Phos-tag SDS-PAGE with that for Mn^{2+}–Phos-tag SDS-PAGE

	Zn^{2+}–Phos-tag SDS-PAGE	Mn^{2+}–Phos-tag SDS-PAGE
Running buffer	100 mM Tris 100 mM MOPS 0.1% (w/v) SDS 5.0 mM sodium bisulfite (pH 7.8, no adjustment)	25 mM Tris 192 mM glycine 0.1% (w/v) SDS (pH 8.3, no adjustment)
Stacking gel buffer	357 mM Bis-Tris–HCl (pH 6.8)	125 mM Tris–HCl 0.1% (w/v) SDS (pH 6.8)
Separating gel buffer	357 mM Bis-Tris–HCl (pH 6.8)	375 mM Tris–HCl 0.1% (w/v) SDS (pH 8.8)

gel and room temperature using a 1-mm-thick, 9-cm-wide, 9-cm-long gel on a mini-slab PAGE apparatus. The gel consists of 1.8 mL of a stacking gel [4.0% (w/v) polyacrylamide and 357 mM Bis-Tris–HCl, pH 6.8] and 6.3 mL of a separating gel, consisting of an appropriate percentage of polyacrylamide and 357 mM Bis-Tris–HCl buffer (pH 6.8). The acrylamide-pendant Phos-tag ligand (at an appropriate concentration) and two equivalents of $Zn(NO_3)_2$ are added to the separating gel before polymerization. The running buffer for electrophoresis consists of 100 mM Tris and 100 mM MOPS containing 0.1% (w/v) SDS and 5.0 mM sodium bisulfite. The electrophoresis is continued until the BPB dye reached the bottom of the separating gel. Table 1 shows a comparison between the buffer systems used for Zn^{2+}–Phos-tag SDS-PAGE and that used for Mn^{2+}–Phos-tag SDS-PAGE.

Here we introduce a procedure for an improved phosphate-affinity electrophoresis by Zn^{2+}–Phos-tag SDS-PAGE and we demonstrate its utility by visualization of novel up-shifted bands of the phosphorylated substrates derived from reactions involving various kinases.

3.4. Procedure for Preparation of the Zn^{2+}–Phos-tag SDS-PAGE Gel (See Note 14)

1. Clean the glass plates for casting the gels. It is vital that the glass plates (1-mm thick, 9-cm wide, and 9-cm long) are washed thoroughly with a rinsable detergent, and that they are rinsed extensively with distilled water before the gels are cast.
2. Prepare the separating gel solution. As a typical example, a 7.5% (w/v) separating gel solution (~7 mL) (see Note 15) containing 80 μM polyacrylamide-bound Zn^{2+}–Phos-tag (see Note 16) is prepared by mixing 1.75 mL of 30% (w/v)

acrylamide/bis solution, 2.5 mL of gel buffer (2.8×), 112 μL of Phos-tag acrylamide solution, 112 μL of zinc(II) nitrate solution (two equivalents on Phos-tag), 10 μL of TEMED, and 2.42 mL of distilled water in a 50-mL centrifuge tube (e.g., Sumitomo Bakelite).

3. Add 100 μL of APS solution and mix gently.
4. Transfer the separating gel solution to the gap between the glass plates, pour distilled water on top of the separating gel solution, and allow the gel solution to polymerize for about 20 min at room temperature.
5. Prepare the stacking gel solution. In a similar manner to the preparation of the separating gel solution, a 4.0% (w/v) stacking gel solution (~1.8 mL) is prepared by mixing 0.24 mL of 30% (w/v) acrylamide/bis solution, 0.64 mL of gel buffer (2.8×), 2 μL of TEMED, and 0.87 mL of distilled water in a 50-mL centrifuge tube.
6. Rinse the top of the separating gel with distilled water and remove the residual liquid with a paper towel.
7. Add 50 μL of APS solution to the stacking gel solution, mix gently, and then pour the solution onto the separating gel.
8. Insert a sample-well comb and allow the gel solution to polymerize for about 20 min.
9. Carefully remove the comb from the stacking gel, and assemble the gel plate and electrophoresis apparatus.

3.5. Procedure for Electrophoresis

1. Fill the electrode chambers with the electrophoresis running buffer (see Note 17).
2. Apply the protein samples mixed with the sample-loading buffer into the wells (see Note 18).
3. Attach the leads to the power supply (e.g., ATTO AE-8750 Power Station 1000XP). Run the gels under a constant-current condition of 40 mA/gel at room temperature until the BPB dye, which contains in the sample-loading buffer, reaches the bottom of the separating gel (80–90 min).

3.6. Procedures for Autoradiography and Gel Staining

1. For autoradiography, remove the fully run gel from the apparatus and dry it using a gel dryer (e.g., an ATTO AE-3750 dryer). Expose an X-ray film to the gel at –80°C for an appropriate time. The X-ray film is then developed to show the radioactive signals on the phosphoprotein bands.
2. For silver staining of the gel by using Sil-Best Stain for Protein/PAGE, remove the gel from the apparatus and stain it according to the manufacturer's instructions to visualize the protein bands.
3. For SYPRO Ruby gel staining, fix the gel in an aqueous solution containing 10% (v/v) methanol and 7.0% (v/v) acetic acid

for 30 min, stain it in a solution of SYPRO Ruby protein gel stain for 3–12 h, and then wash it in 10% (v/v) methanol and 7.0% (v/v) acetic acid for 30 min. Detect the SYPRO Ruby dye-bound proteins from their 575-nm emission signals on excitation at 473 nm using a laser scanner (e.g., an FLA 5000 laser scanner, Fujifilm, Tokyo, Japan).

3.7. Procedure for Electroblotting

1. When the run is complete, remove the gel from the apparatus and soak it in 100 mL of blotting buffer containing 1 mM EDTA for 10 min (see Note 19).
2. Soak the gel in 100 mL of blotting buffer without EDTA for 10 min.
3. Prepare the poly(vinylidene difluoride) (PVDF) membrane (e.g., Fluorotrans W, Nippon Pall, Tokyo, Japan) by cutting it to the same size as the gel and soak it for 30 s in 100% methanol. The membrane is then incubated for 15 min in the blotting buffer.
4. Prepare four pieces of 3MM paper (e.g., Whatman, Maidstone, UK) by cutting them to the same size as the gel.
5. To form a "blotting sandwich" on the electroblotting screen attached to the electroblotting equipment, assemble the gel, PVDF membrane, and 3MM paper as follows. Soak the blotting sponge attached to the electroblotting equipment in the blotting buffer and place it on the electroblotting screen. Then soak two pieces of 3MM paper in the blotting buffer and place them on the sponge, followed sequentially by the gel and PVDF membrane; avoid incorporating air between the various layers. Then place two more sheets of 3MM paper and one sponge soaked in the blotting buffer on the membrane and close the electroblotting screen. Insert the electroblotting screen in the chamber unit of the electroblotting equipment and fill up with the blotting buffer (see Note 20).
6. Subject the gel to constant-voltage conditions (3.5 V/cm) for 16 h (overnight).
7. After blotting, perform the immunoblotting analysis.

4. Typical/Anticipated Results

4.1. Protein Kinase Profiling Toward Tau by Using Mn^{2+}–Phos-tag SDS-PAGE

Through profiling of kinase activity by using Mn^{2+}–Phos-tag SDS-PAGE, we characterized six kinds of Ser/Thr kinases that phosphorylated recombinant human Tau protein and might, therefore, be involved in Alzheimer's disease. The products of the individual kinase reactions involving GSK-3β, cdk5/p35, PKA, MAPK, CKII, and CaMKII were analyzed by normal SDS-PAGE and by

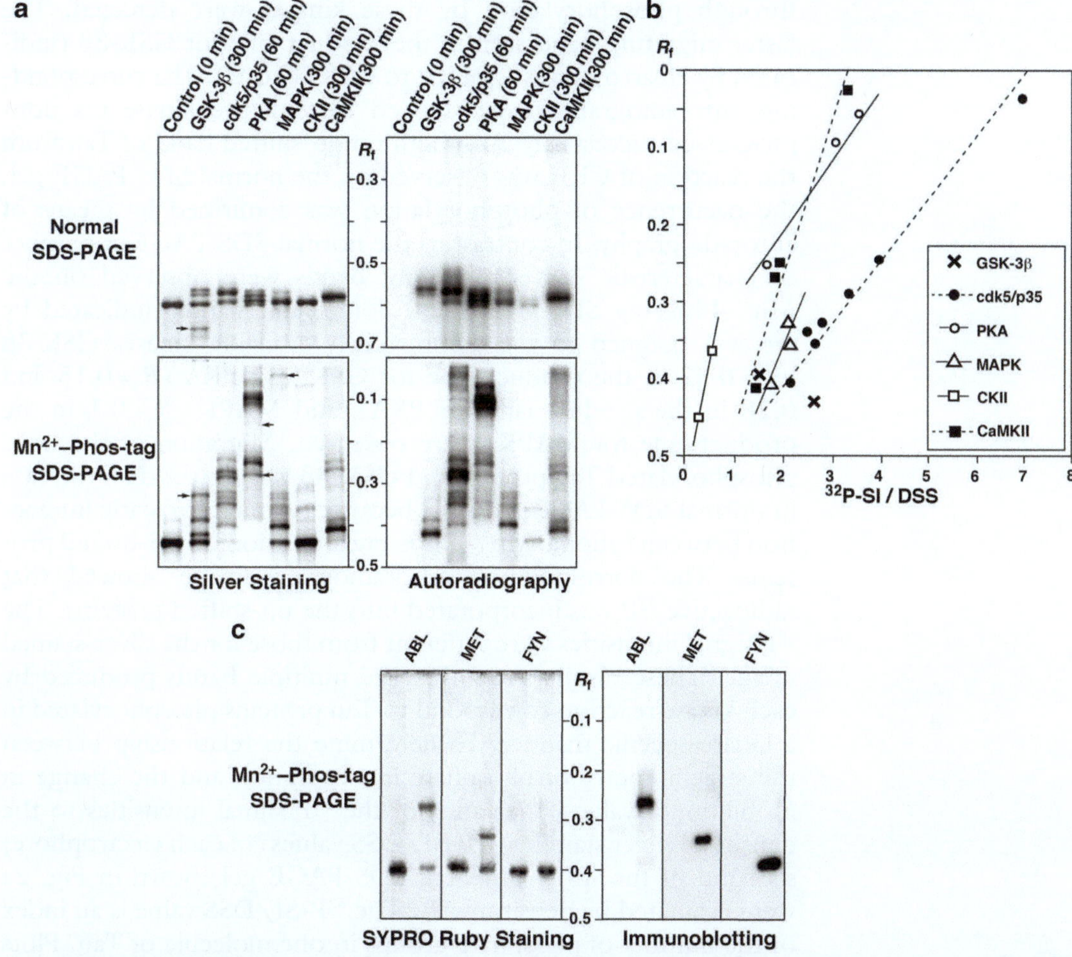

Fig. 2. In vitro Ser/Thr kinase activity profiles in the phosphorylation of Tau protein. (**a**) Normal SDS-PAGE [7.5% (w/v) polyacrylamide] and Mn^{2+}–Phos-tag SDS-PAGE [80 μM polyacrylamide-bound Mn^{2+}–Phos-tag and 7.5% (w/v) polyacrylamide] for six kinds of Ser/Thr kinase products of the Tau protein, followed by silver gel staining (*left*) or autoradiography (*right*). (**b**) Plots of the phosphate-incorporation ratios (ratios of the ^{32}P signal intensity to the density of silver staining for each electrophoresis band in Mn^{2+}–Phos-tag SDS-PAGE; ^{32}P-SI/DSS values) against the R_f values. [Reproduced, with permission, from (22), © 2007, The American Society for Biochemistry and Molecular Biology, Inc.] (**c**) Mn^{2+}–Phos-tag SDS-PAGE [80 μM polyacrylamide-bound Mn^{2+}–Phos-tag and 7.5% (w/v) polyacrylamide] for three kinds of Tyr kinase products of the Tau protein, followed by SYPRO Ruby gel staining (*left*) or immunoblotting with an anti-phosphotyrosine antibody (*right*, horseradish peroxidase-conjugated clone PY20, GE Healthcare Biosciences). In each kinase analysis, nonphosphorylated Tau (*left lane*) and the reaction product (*right lane*) were applied. [Reproduced, with permission, from (27) © (2010), Wiley-VCH Verlag GmbH & Co. KGaA, Weinheim].

Mn^{2+}–Phos-tag SDS-PAGE followed, in both cases, by silver gel staining and autoradiography (Fig. 2a); nonphosphorylated Tau was applied to the leftmost lane in each case as a control. In the normal SDS-PAGE, nonphosphorylated and phosphorylated Tau were observed as migration bands at an R_f value of ~0.6. The R_f value 1.0 is defined as the position of the BPB dye (the bottom of the separating gel). Some bands that were slightly up-shifted through phosphorylation by these kinases were detected. The faster-migrating band seen in the product lane for GSK-3β (indicated by an arrow) was assigned to GSK-3β itself. The corresponding autoradiogram image showed that all the kinase reactions progressed successfully. Although no up-shifted band of Tau from the reaction of CKII was observed on the normal SDS-PAGE gel, the occurrence of phosphorylation was confirmed by means of autoradiography. In contrast to the normal SDS-PAGE, a number of characteristic slower-migrating bands were observed on the Mn^{2+}–Phos-tag SDS-PAGE gel. Some faint bands (indicated by arrows) assigned to the commercially available kinases GSK-3β ($R_f = 0.32$ in the product lane for GSK-3β), PKA ($R_f = 0.15$ and 0.28 in the product lane for PKA), and MAPK ($R_f < 0.1$ in the product lane for MAPK) were observed. Migration of the non-phosphorylated Tau protein and GSK-3β was slower than the case in normal SDS-PAGE, possibly because of an electrostatic interaction between cationic Mn^{2+}–Phos-tag and anionic SDS-bound proteins. The corresponding autoradiogram image showed that radioactive ^{32}P was incorporated into the up-shifted proteins. The ^{32}P signal intensities were different from those for the silver-stained image. These results show that the multiple bands produced by each kinase reaction correspond to Tau proteins phosphorylated in a kinase-specific manner. To determine the relationship between the stoichiometry of phosphate incorporation and the change in mobility (R_f value), the ratios of the ^{32}P signal intensities to the density of silver staining (^{32}P-SI/DSS values) of each electrophoresis band of the Mn^{2+}–Phos-tag SDS-PAGE gel shown in Fig. 2a were evaluated by densitometry. The ^{32}P-SI/DSS value is an index of the number of phosphate groups in one molecule of Tau. Plots of the values of ^{32}P-SI/DSS against the R_f values are shown in Fig. 2b. In each kinase reaction, except for that of GSK-3β, the R_f value decreased although there was an increase in the ^{32}P-SI/DSS value. The reverse relationships between the ^{32}P-SI/DSS values and the R_f values were markedly different for these kinase reactions. These results suggest that the degree of change in mobility of a phosphoprotein might not be related exclusively to the stoichiometry of phosphate incorporation, but that other factors, such as kinase-specific phosphorylation sites, may be involved.

To show that the degree of migration of phosphorylated Tau in Mn^{2+}–Phos-tag SDS-PAGE depends on the particular kinase-specific phosphorylation sites, we characterized three distinct monophosphorylated forms of Tau by using Mn^{2+}–Phos-tag SDS-PAGE. The three monophosphorylated forms of Tau were specifically phosphorylated in vitro at the Tyr-394, Tyr-197, or Tyr-18 residues by ABL, MET, and FYN, respectively (Fig. 2c). Each monophosphorylated form of Tau was detected as three distinct migration bands. This shows that Mn^{2+}–Phos-tag SDS-PAGE is capable of separating substrate proteins that are phosphorylated at kinase-specific phosphorylation sites. Although no up-shifted band of Tau in the FYN reaction was observed, the phosphorylation of tyrosine was confirmed by immunoblotting analysis using an anti-phosphotyrosine antibody. This demonstrates a limitation of Mn^{2+}–Phos-tag SDS-PAGE in the analysis of protein phosphorylation under the experimental conditions.

4.2. Zn^{2+}–Phos-tag SDS-PAGE for Advanced Protein Kinase Profiling

Each of the three forms of Tau monophosphorylated at the Tyr-394, Tyr-197, or Tyr-18 residue, respectively, was visualized as three distinct migration bands on the Mn^{2+}–Phos-tag SDS-PAGE gel, but no shift was detected for the Tau protein phosphorylated at the Tyr-18 residue by FYN, which is a member of the Src family of kinases. We therefore examined the mobilities of the three monophosphorylated Tau proteins by using the Zn^{2+}–Phos-tag SDS-PAGE method. Figure 3a shows a typical result obtained by using Zn^{2+}–Phos-tag SDS-PAGE, followed, in each case, by staining of the gel with SYPRO Ruby and immunoblotting with an anti-phosphotyrosine antibody. The Zn^{2+}–Phos-tag SDS-PAGE method followed by gel staining permitted the detection of shifts in the mobility of Tyr-phosphorylated Tau in the reactions of all the kinases. Only the up-shifted bands were confirmed by immunoblotting to be Tyr-phosphorylated proteins. The differences in the R_f values of the various monophosphorylated Tau proteins were dependent on the kinase-specific phosphorylation site. In a major improvement, a novel up-shifted band of the Tyr-phosphorylated Tau produced by the reaction of FYN was identified. We also confirmed that the up-shifted Tau bands produced by other Src family kinases (YES, SRC, LCK, and LYNa) all showed identical R_f values to that produced by FYN in Zn^{2+}–Phos-tag SDS-PAGE (Fig. 3b), reflecting an Src family specific phosphorylation at the Tyr-18 residue. Moreover, the utility of Zn^{2+}–Phos-tag SDS-PAGE was demonstrated by visualizing up-shifted bands of Tyr-phosphorylated Tau in reactions of 11 other kinases; these bands could not be detected by using the Mn^{2+}–Phos-tag SDS-PAGE method (Fig. 3b). Improvements in the detection of shifts in the mobility might be due to an increase in affinity for the phosphorylated targets as a result of using the dizinc(II) complex of Phos-tag acrylamide under conditions of neutral pH buffered with Bis-Tris.

4.3. Advanced Resolving Power for the Analysis of Cell Lysates

Finally, we examined the advanced resolving power of Zn^{2+}–Phos-tag SDS-PAGE in the analysis of biologically crude samples. As a typical example, we profiled the hyperphosphorylation status of endogeneous β-catenin by using an untreated lysate of human colon adenocarcinoma SW480 cells (Fig. 4). β-Catenin is an 85-kDa protein that binds to the cytoplasmic tail of E-cadherin, a transmembrane adhesion molecule. The binding is regulated by phosphorylation of three critical Tyr residues (Tyr-142, Tyr-489,

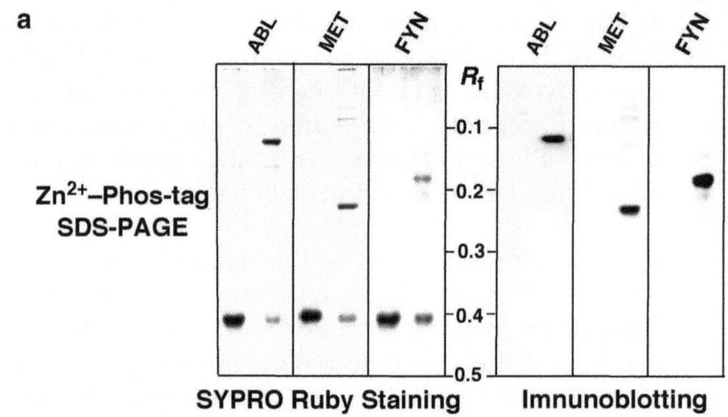

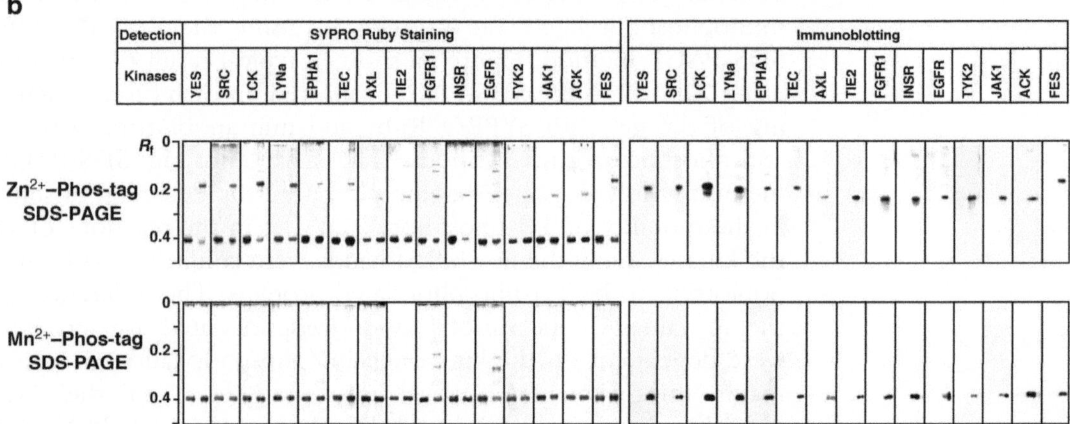

Fig. 3. In vitro Tyr kinase activity profiles in the phosphorylation of Tau protein. (**a**) Zn^{2+}–Phos-tag SDS-PAGE [80 μM polyacrylamide-bound Zn^{2+}–Phos-tag and 7.5% (w/v) polyacrylamide] of three kinds of Tyr kinase products of Tau protein, followed by SYPRO Ruby gel staining (*left*) or immunoblotting with the anti-phosphotyrosine antibody (*right*). In each kinase analysis, nonphosphorylated Tau (*left lane*) and the reaction products (*right lane*) were applied. (**b**) Comparison of mobilities of Tau phosphorylated on Tyr moieties by 15 types of Tyr kinase in Zn^{2+}–Phos-tag SDS-PAGE and in Mn^{2+}–Phos-tag SDS-PAGE. In each kinase analysis, nonphosphorylated Tau (*left lane*) and the reaction products (*right lane*) were applied to Phos-tag SDS-PAGE [7.5% (w/v) polyacrylamide containing 80 μM Phos-tag]. Proteins were stained with SYPRO Ruby or detected by immunoblotting with the anti-phosphotyrosine antibody. [Reproduced, with permission, from (27), © 2010, Wiley-VCH Verlag GmbH & Co. KGaA, Weinheim].

and Tyr-654) of β-catenin during developmental and physiological processes (30). β-Catenin is also a key component of the Wnt signaling pathway, which affects cell proliferation and differentiation in many types of cell (31). In Wnt signaling, CKIα phosphorylates β-catenin at the Ser-45 residue. The reaction enhances subsequent phosphorylation by GSK-3β at the Ser-33, Ser-37, and Tyr-41 residues. β-Catenin is also phosphorylated at the Ser-552 and Ser-675 residues by PKA and AKT, and this phosphorylation induces the transcriptional activity of β-catenin (32, 33). Thus, β-catenin is regulated by sundry protein kinases in vivo, and its various phosphorylation states are closely involved with specific cellular events.

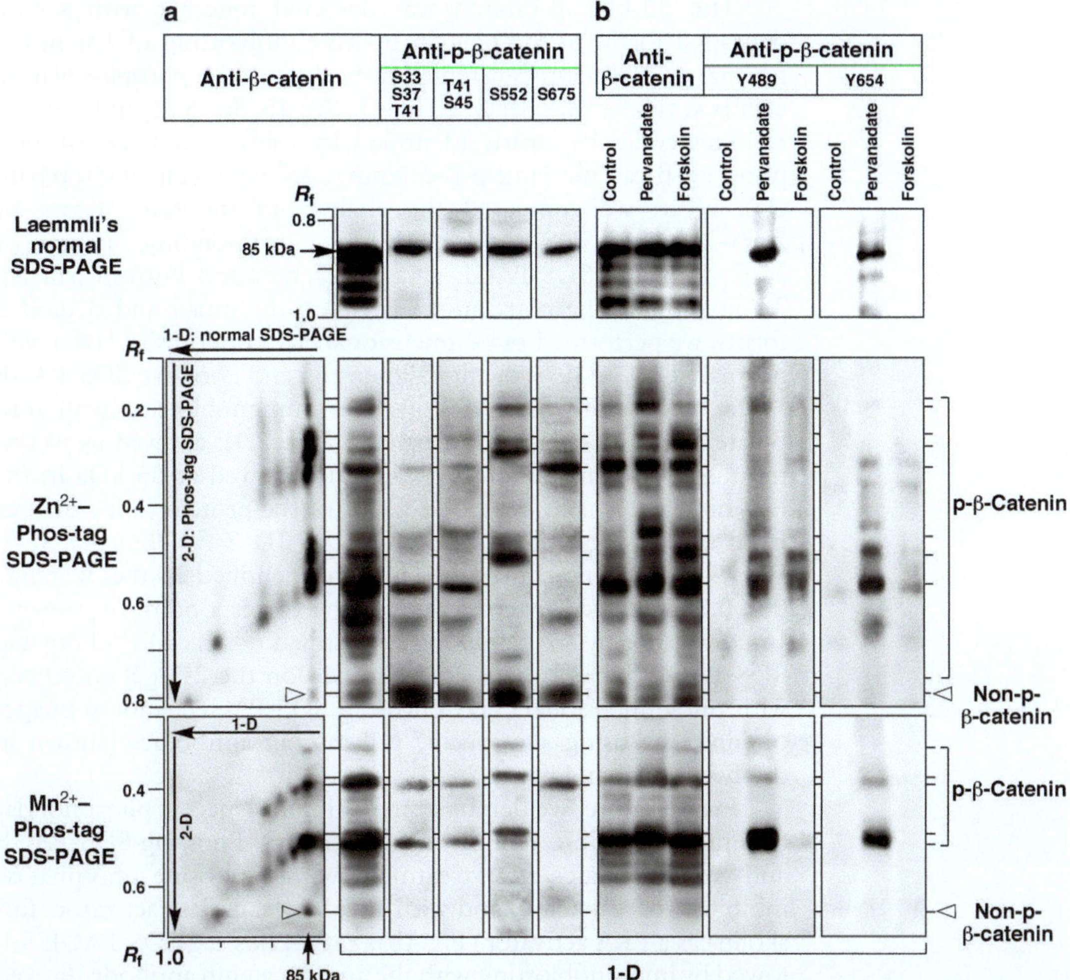

Fig. 4. Profiling of the phosphorylation status of intracellular β-catenin. Laemmli's normal gels using 1DE and 2DE consisted of 5.5% (w/v) and 6.0% (w/v) polyacrylamide, respectively. Each Phos-tag gel consisted of 5.5% (w/v) polyacrylamide containing 25 μM Phos-tag. β-Catenin was detected by immunoblotting with the antibodies shown above each panel. Anti-β-catenin antibody (against the C-terminal, clone 14) was purchased from BD Bioscience (San Jose, CA, USA). The site-specific antibodies against pS33/S37/T41, pS552, and pS675 were purchased from Cell Signaling Technology (Danvers, MA, USA). The site-specific antibodies against pY489 and pY654 were purchased from ECM Biosciences (Versailles, KY, USA) and Abcam Japan (Tokyo, Japan), respectively. The site-specific antibody against pT41/S45 (clone EP1905Y) was purchased from Millipore. (**a**) The lysate was prepared from cells treated without drug. (**b**) The lysates were prepared from cells treated without drug (control) and with pervanadate (1.0 mM) for 30 min, or with forskolin (10 μM) for 30 min. The top panels show the results of analyses by Laemmli's normal 1D SDS-PAGE. The center panels show results of analyses by Zn^{2+}–Phos-tag SDS-PAGE. The *leftmost panel* shows 2DE coupling of Laemmli's normal SDS-PAGE as the first dimension (1D, *horizontal arrow*) and Zn^{2+}–Phos-tag SDS-PAGE as the second dimension (2D, *vertical arrow*). The *bottom panels* show the results of analyses by Mn^{2+}–Phos-tag SDS-PAGE. The *leftmost panel* shows 2DE coupling of Laemmli's normal SDS-PAGE (1D, *horizontal arrow*) and Mn^{2+}–Phos-tag SDS-PAGE (2D, *vertical arrow*). [Reproduced, with permission, from (27), © 2010, Wiley-VCH Verlag GmbH & Co. KGaA, Weinheim].

The 85-kDa β-catenin was detected together with several degraded forms of the protein by immunoblotting of Laemmli's normal gel with anti-β-catenin antibody, and the phosphorylation events at the Ser-33, Ser-37, Thr-41, Ser-45, Ser-552, and Ser-675 residues could be clearly identified by using four kinds of anti-phospho-β-catenin (anti-p-β-catenin) antibodies (Fig. 4a; top panels). Next, we compared the analysis of the same lysate by Zn^{2+}–Phos-tag SDS-PAGE with the corresponding analysis by Mn^{2+}–Phos-tag SDS-PAGE (Fig. 4a; center and bottom panels). To investigate the correlations between the intact and degraded forms, we performed two-dimensional electrophoresis (2DE) with normal SDS-PAGE as the first dimension and Phos-tag SDS-PAGE as the second dimension, followed by immunoblotting with anti-β-catenin antibody (leftmost panels). The 2DE allowed us to distinguish between the intact β-catenin (arrowed at 85 kDa in the first dimension) and degraded forms of the protein (left-side area from the position of 85 kDa). On the 2D Zn^{2+}–Phos-tag SDS-PAGE gel, ten up-shifted bands were identified in the R_f range 0.2–0.8 (indicated by cross lines at the corresponding R_f values). In contrast, only six bands were identified on the Mn^{2+}–Phos-tag SDS-PAGE gel. All the spots identified on the 2D gel correlated with one-dimensional (1D) Phos-tag SDS-PAGE banding images obtained by using site-specific p-β-catenin antibodies (shown in the five right-hand panels).

Furthermore, we demonstrated stimulus-specific phosphorylation profiling of β-catenin by using 1D Zn^{2+}–Phos-tag SDS-PAGE and SW480 lysates after treatment with a tyrosine phosphatase inhibitor (pervanadate) and with an adenylyl cyclase activator (forskolin) as a PKA activator (Fig.4b). Zn^{2+}–Phos-tag SDS-PAGE followed by immunoblotting with the anti-β-catenin antibody showed characteristic migration patterns of β-catenin produced by individual stimulation. For the pervanadate-treated sample, the signal intensities of the bands at R_f values of 0.20, 0.25, and 0.45 increased, and a novel up-shifted band appeared at an R_f value of 0.43. For the forskolin-treated sample, the signal intensities of the bands at R_f values of 0.25, 0.28, 0.45, and 0.50 increased. These bands had identical R_f values to those detected by anti-p-β-catenin antibodies against the phosphorylated Ser-552 or Ser-675 residue (see Fig. 4a). On the other hand, there was almost no difference in the migration patterns of β-catenin between the lysates analyzed by Laemmli's normal SDS-PAGE and those analyzed by Mn^{2+}–Phos-tag SDS-PAGE. In immunoblotting of the Zn^{2+}–Phos-tag SDS-PAGE gel with p-β-catenin antibodies against the phosphorylated Tyr-489 and Tyr-654 residues, prominent signals of Tyr-phosphorylated β-catenin were detected on the multiple up-shifted bands for the pervanadate-treated sample. Thus, Zn^{2+}–Phos-tag SDS-PAGE permits determination of drug-specific phosphorylation events of β-catenin.

Advanced separation by using Zn^{2+}–Phos-tag SDS-PAGE should permit greater coverage for a larger number of phosphoproteins and should increase the sensitivity of the detection of hierarchical protein phosphorylation and dephosphorylation. This method can therefore assist in mapping low-abundance phosphorylation events, and it should be a useful tool for the profiling of complicated protein kinase–phosphatase networks. Furthermore, it has been demonstrated that Zn^{2+}–Phos-tag SDS-PAGE gels cast in the neutral buffer are stable during long-term storage (at least 3 months) (27). The storage of the gel does not require any special knowhow or equipment. The gel, together with the casting glass plates, is merely wrapped in a Saran wrap to protect it from drying and then stored at room temperature under normal laboratory illumination until required. Therefore, the Zn^{2+}–Phos-tag SDS-PAGE method offers a better shelf life, which makes it particularly appealing for laboratory practice. We can thus present a simple, convenient, and more reliable "in-house" gel system for phosphate-affinity SDS-PAGE.

5. Notes

1. All reagents and solvents used are purchased at the highest commercial quality available and used without further purification. All aqueous solutions are prepared by using deionized and distilled water.
2. 2-Mercaptoethanol is toxic by inhalation, ingestion, and skin contact. When handling this chemical, work in a chemical fume hood, wear gloves and a mask, and use a pipetting aid.
3. The oily product, acrylamide-pendant Phos-tag ligand (10 mg), is placed in a plastic tube and completely dissolved in methanol (0.10 mL). The solution is diluted with distilled water (3.2 mL) by pipetting. Methanol is an inhalation toxin that causes depression of the central nervous system; when handling this chemical, work in a chemical fume hood, wear gloves, and use a pipetting aid.
4. Do not use any other salts such as $Mn(NO_3)_2$ or $Mn(OCOCH_3)_2$.
5. Because acrylamide monomer is a neurotoxin and a suspected human carcinogen and teratogen, exposure to this substance should be carefully avoided. When weighing powdered acrylamide, work in a chemical fume hood, wear gloves, eye protection, and a mask. Acrylamide is unstable and can polymerize violently on heating to its melting point (84.5°C). It is incompatible with acids, bases, oxidizing agents, reducing agents, iron and its salts, copper, aluminum, brass, and free-radical initiators.

6. TEMED is stored in a desiccator at room temperature. Buy small bottles as its quality may degrade, and gels will take longer to polymerize after opening of the container.
7. Do not adjust the pH with acid or base.
8. Zinc(II) chloride solution (10 mM $ZnCl_2$ in distilled water) is a suitable substitute.
9. This buffer is used for the separating and stacking gels.
10. Do not adjust the pH with acid or base.
11. Sulfite ion (SO_3^{2-}) is a reducing reagent that diminishes O_2 levels in the electrophoresis running buffer solution and inhibits the oxidation of reduced proteins in the gel. Buffer solution in which sodium bisulfite has been dissolved should be promptly used for electrophoresis. Do not store buffer solution containing sodium bisulfite.
12. Do not adjust the pH with acid or base.
13. Although the semi-dry method is generally the most efficient method for protein blotting in terms of time and consumption of buffer reagents, it is not suitable for electroblotting from the Phos-tag SDS-PAGE gel. The transfer efficiency of proteins is lower in the semi-dry method than that in the wet-tank method.
14. Except for the buffer system, the procedure for Mn^{2+}–Phos-tag SDS-PAGE is almost identical. For details of the procedure see (26). Various contaminants (e.g., EDTA, inorganic salts, or surfactants) in the sample protein solutions often cause disruption (waving and/or tailing) of the electrophoresis bands in Phos-tag SDS-PAGE. To minimize this disruption, it is recommended that the sample is desalted before loading. For example, dialysis filtration can be used to decrease the amounts the small-molecule substances in the sample. Furthermore, to avoid distortion of bands, do not apply commercially available pre-stained molecular-weight protein markers to Phos-tag SDS-PAGE.
15. Because the optimal percentage of polyacrylamide depends on the molecular weight of the target protein, an appropriate value should be determined for each target.
16. In Phos-tag SDS-PAGE, the R_f values (the degrees of electrophoresis migration) of both phosphorylated and nonphosphorylated proteins are generally smaller than those in normal SDS-PAGE. The optimal concentration of Zn^{2+}–Phos-tag (e.g., 5–100 μM) to achieve sufficient separation between the phosphorylated and nonphosphorylated proteins should be determined for each target. It is recommended that tests would be conducted using low concentrations of 5–25 μM Zn^{2+}–Phos-tag for complex samples, such as cell lysates, that contain various phosphorylated and nonphosphorylated proteins.

17. Take care not to produce bubbles at the bottom surface of the gel set. When bubbles are observed, they should be carefully and completely removed.
18. The same sample-loading buffer is used for both the Mn^{2+}–Phos-tag SDS-PAGE and Zn^{2+}–Phos-tag SDS-PAGE methods.
19. Zn^{2+}–Phos-tag in the gel causes inefficient electroblotting. This can be ameliorated by treatment with EDTA to chelate the zinc ions.
20. The use of wet-tank equipment is strongly recommended for optimal efficiency of protein transfer from the Zn^{2+}–Phos-tag SDS-PAGE gel. The efficiency of transfer from the gel is much higher in the wet-tank method than in the semi-dry method.

Acknowledgments

We wish to thank the Research Center for Molecular Medicine and the Analysis Center of Life Science, Hiroshima University, Japan, for the use of their facilities. This work was supported in part by Grants-in-Aid for Scientific Research (B, 22390006; C, 22590037; C, 24590050) from the Japan Society for the Promotion of Science (JSPS), by a Grant-in Aid for Scientific Research on Innovative Areas (23117522) from the Ministry of Education Culture, Sports, Science, and Technology (MEXT), and by a exploratory research grant (AS232Z01251F) for Adaptable and Seamless Technology Transfer Program through Target-driven R&D (A-STEP) from the Japan Science and Technology Agency (JST). Financial support was also provided by The Takeda Science Foundation and The Ube Foundation.

References

1. Hunter T (2000) Signaling: 2000 and beyond. Cell 100:113–127
2. Johnson SA, Hunter T (2005) Kinomics: methods for deciphering the kinome. Nat Methods 2:17–25
3. Cohen P (2002) Protein kinase: the major drug targets of the twenty-first century? Nat Rev Drug Discov 1:309–315
4. Lee VMY, Goedert M, Trojanowski JQ (2001) Neurodegenerative tauopathies. Annu Rev Neurosci 24:1121–1159
5. Oda Y, Nagasu T, Chait BT (2001) Enrichment analysis of phosphorylated proteins as a tool for probing the phosphoproteome. Nat Biotechnol 19:379–382
6. Ficarro SB, McCleland ML, Stukenberg PT, Burke DJ, Ross MM, Shabanowitz J, Hunt DF, White FM (2002) Phosphoproteome analysis by mass spectrometry and its application to *Saccharomyces cerevisiae.* Nat Biotechnol 20: 301–305
7. Larsen MR, Thingholm TE, Jensen ON, Roepstorff P, Jørgensen TJD (2005) Highly selective enrichment of phosphorylated peptides from peptide mixtures using titanium dioxide microcolumns. Mol Cell Proteomics 4:873–886
8. Kokubu M, Ishihama Y, Sato T, Nagasu T, Oda Y (2005) Specificity of immobilized metal affinity-based IMAC/C18 tip enrichment of phosphopeptides for protein phosphorylation analysis. Anal Chem 77:5144–5154
9. Kweon HK, Håkansson K (2006) Selective zirconium dioxide-based enrichment of

phosphorylated peptides for mass spectrometric analysis. Anal Chem 78:1743–1749

10. Sugiyama N, Masuda T, Shinoda K, Nakamura A, Tomita M, Ishihama Y (2007) Phosphopeptide enrichment by aliphatic hydroxy acid-modified metal oxide chromatography for nano-LC-MS/MS in proteomics applications. Mol Cell Proteomics 6:1103–1109
11. Kinoshita E, Takahashi M, Takeda H, Shiro M, Koike T (2004) Recognition of phosphate monoester dianion by an alkoxide-bridged dinuclear zinc(II) complex. Dalton Trans: 1189–1193
12. Takeda H, Kawasaki A, Takahashi M, Yamada A, Koike T (2003) Matrix-assisted laser desorption/ionization time-of-flight mass spectrometry of phosphorylated compounds using a novel phosphate capture molecule. Rapid Commun Mass Spectrom 17:2075–2081
13. Prudent M, Rossier JS, Lion N, Girault HH (2008) Microfabricated dual sprayer for on-line mass tagging of phosphopeptides. Anal Chem 80:2531–2538
14. Kinoshita E, Yamada A, Takeda H, Kinoshita-Kikuta E, Koike T (2005) Novel immobilized zinc(II) affinity chromatography for phosphopeptides and phosphorylated proteins. J Sep Sci 28:155–162
15. Kinoshita-Kikuta E, Kinoshita E, Yamada A, Endo M, Koike T (2006) Enrichment of phosphorylated proteins from cell lysate using a novel phosphate-affinity chromatography at physiological pH. Proteomics 6:5088–5095
16. Kinoshita-Kikuta E, Kinoshita E, Koike T (2009) Phos-tag beads as an immunoblotting enhancer for selective detection of phosphoproteins in cell lysates. Anal Biochem 389:83–85
17. Kinoshita-Kikuta E, Yamada A, Inoue C, Kinoshita E, Koike T (2011) A novel phosphate-affinity bead with immobilized Phos-tag for separation and enrichment of phosphopeptides and phosphoproteins. J Integr OMICS 1:157–169
18. Inamori K, Kyo M, Nishiya Y, Inoue Y, Sonoda T, Kinoshita E, Koike T, Katayama Y (2005) Detection and quantification of on-chip phosphorylated peptides by surface plasmon resonance imaging techniques using a phosphate capture molecule. Anal Chem 77:3979–3985
19. Kinoshita E, Kinoshita-Kikuta E, Takiyama K, Koike T (2006) Phosphate-binding tag, a new tool to visualize phosphorylated proteins. Mol Cell Proteomics 5:749–757
20. Takiyama K, Kinoshita E, Kinoshita-Kikuta E, Fujioka Y, Kubo Y, Koike T (2009) A Phos-tag-based fluorescence resonance energy transfer system for the analysis of the dephosphorylation of phosphopeptides. Anal Biochem 388:235–241
21. Somura M, Takiyama K, Kinoshita-Kikuta E, Kinoshita E, Koike T (2011) A Phos-tag-based fluorescence resonance energy transfer system for the analysis of the kinase reaction of a substrate peptide. Anal Methods 3:1303–1309
22. Kinoshita-Kikuta E, Aoki Y, Kinoshita E, Koike T (2007) Label-free kinase profiling using phosphate affinity polyacrylamide gel electrophoresis. Mol Cell Proteomics 6: 356–366
23. Kinoshita E, Kinoshita-Kikuta E, Matsubara M, Yamada S, Nakamura H, Shiro Y, Aoki Y, Okita K, Koike T (2008) Separation of phosphoprotein isotypes having the same number of phosphate groups using phosphate-affinity SDS-PAGE. Proteomics 8:2994–3003
24. Kinoshita E, Kinoshita-Kikuta E, Matsubara M, Aoki Y, Ohie S, Mouri Y, Koike T (2009) Two-dimensional phosphate-affinity gel electrophoresis for the analysis of phosphoprotein isotypes. Electrophoresis 30:550–559
25. Kinoshita E, Kinoshita-Kikuta E, Ujihara H, Koike T (2009) Mobility shift detection of phosphorylation on large proteins using a Phos-tag SDS-PAGE gel strengthened with agarose. Proteomics 9:4098–4101
26. Kinoshita E, Kinoshita-Kikuta E, Koike T (2009) Separation and detection of large phosphoproteins using Phos-tag SDS-PAGE. Nat Protoc 4:1513–1521
27. Kinoshita E, Kinoshita-Kikuta E (2011) Improved Phos-tag SDS-PAGE under neutral pH conditions for advanced protein phosphorylation profiling. Proteomics 11:319–323
28. Laemmli UK (1970) Cleavage of structural proteins during the assembly of the head of bacteriophage T4. Nature 227:680–685
29. Kinoshita E, Kinoshita-Kikuta E, Koike T (2009) Phosphate-affinity gel electrophoresis using a Phos-tag molecule for phosphoproteome study. Curr Proteomics 6:104–121
30. Lilien J, Balsamo J (2005) The regulation of cadherin-mediated adhesion by tyrosine phosphorylation/dephosphorylation of β-catenin. Curr Opin Cell Biol 17:459–465
31. Kikuchi A (2003) Tumor formation by genetic mutations in the components of the Wnt signaling pathway. Cancer Sci 94:225–229
32. Taurin S, Sandbo N, Qin Y, Browning D, Dulin NO (2006) Phosphorylation of β-catenin by cyclic AMP-dependent protein kinase. J Biol Chem 281:9971–9976
33. Fang D, Hawke D, Zheng Y, Xia Y, Meisenhelder J, Nika H, Mills GB, Kobayashi R, Hunter T, Lu Z (2007) Phosphorylation of β-catenin by AKT promotes β-catenin transcriptional activity. J Biol Chem 282:11221–11229

Chapter 3

Large-Scale Protein Phosphorylation Analysis by Mass Spectrometry-Based Phosphoproteomics

Wei-Chi Ku, Naoyuki Sugiyama, and Yasushi Ishihama

Abstract

Protein phosphorylation is one of the key mechanisms controlling cellular signaling networks. Due to the low abundance of phosphorylated proteins and weaker ionization efficiency of phosphopeptides during mass spectrometric analyses, it is highly required to remove abundant non-phosphopeptides from complex mixtures, such as cell lysates, allowing successful detection of low abundant phosphopeptides. We recently developed an aliphatic hydroxy acid-modified metal oxide chromatography (HAMMOC) to efficiently and selectively enrich phosphopeptides prior to mass spectrometry (MS) analysis. Here we describe a detailed workflow of HAMMOC for enriching phosphopeptides from small amounts, e.g., 100 μg, of tryptic digests of whole cell lysate. We also discuss the importance of confidently assigning phosphorylation site(s) from an identified phosphopeptide after MS analyses.

Key words: Hydroxy acids, Lactic acid, Mass spectrometry, Metal oxide chromatography, Phosphopeptide enrichment, Phosphoproteomics, Protein phosphorylation, Titanium dioxide, TiO_2

1. Introduction

1.1. Detection of Protein Phosphorylation by Mass Spectrometry

Protein phosphorylation is one of the most important post-translational modifications and participates in a variety of cellular metabolism (1). To date more than one-third of all proteins in a eukaryotic cell are subjective to phosphorylation modifications (2). Protein phosphorylation in human cells is regulated by more than 500 members of kinase superfamily, which is one of the top 15 functional categories in human proteome (3, 4). Aberrant kinase activities cause uncontrolled phosphorylation signaling cascades, and eventually lead to cellular transformation such as tumorigenesis (5, 6). A systematic investigation of kinases-derived substrate

Hideyuki Mukai (ed.), *Protein Kinase Technologies*, Neuromethods, vol. 68,
DOI 10.1007/978-1-61779-824-5_3, © Springer Science+Business Media, LLC 2012

phosphorylation at a large scale, therefore, is important to understand regulatory roles of specific kinase(s) in cellular processes.

With the advance of modern mass spectrometry (MS) technologies, it becomes possible to conduct qualitative or quantitative proteomics studies in response to stimuli, including the phosphoproteomics (7–9). Since phosphorylated proteins are in lower abundance and, after proteolysis, phosphopeptides are poorly ionized during MS analyses in positive ion mode, efficient and specific enrichment of the phosphopeptides is highly required. A variety of approaches for phosphopeptide enrichment have been published in the past decades, such as immobilized metal ion affinity chromatography, strong cation exchange chromatography, metal oxide chromatography (MOC), etc. (7–9). With the ongoing development of enrichment methods, to date over 25,000 phosphorylation sites in human proteome have been identified by the MS-based phosphoproteomics approaches (10, 11), including the identification of specific kinase substrates (12–16) or kinase superfamilies (17, 18).

1.2. Hydroxy Acid-Modified MOC

Among the aforementioned MS-based phosphoproteomics technologies, MOC using titanium dioxide (TiO_2) is one of the most widely used methods to enrich phosphopeptides. TiO_2 was first introduced by Ikeguchi and Nakamura for enriching organophosphates including digested phosphopeptides (19), and later in phosphoproteomics studies by other groups (20, 21). However, TiO_2-based MOC suffers low specificity due to the competitive binding of acidic residues on non-phosphopeptides (22). In addition, TiO_2-based approach tends to identify mono-phosphorylated peptides better than multi-phosphorylated peptides (22). To overcome these problems of TiO_2-based MOC, we have recently developed an aliphatic hydroxy acid-modified metal oxide chromatography (HAMMOC) for highly selective phosphopeptide enrichment (23).

The principle of HAMMOC is illustrated in Fig. 1. As shown in Fig. 1a, aliphatic hydroxy acid, e.g., lactic acid, and phosphate anion can form cyclic chelate bridging with one and two TiO_2, respectively (24–26). Therefore, TiO_2 has higher affinity toward phosphate groups followed by lactic acid, with weaker affinity to the carboxylic groups of acidic amino acid residues, i.e., aspartic and glutamic acid (Fig. 1a). When the TiO_2 beads are blocked with lactic acid first, phosphopeptides in successive loading step compete out lactic acid for TiO_2 binding, whereas acidic non-phosphopeptides remain unbound due to the weaker affinity than lactic acid to TiO_2 (Fig. 1b). Our HAMMOC approach drastically reduces the non-specific binding of acidic non-phosphorylated peptides to TiO_2 and allows enriching phosphopeptides directly from complex cell lysates without prefractionation, when compared to other phosphopeptide enrichment methods, such as immobilized metal ion affinity chromatography and DHB/phthalate-titanium dioxide

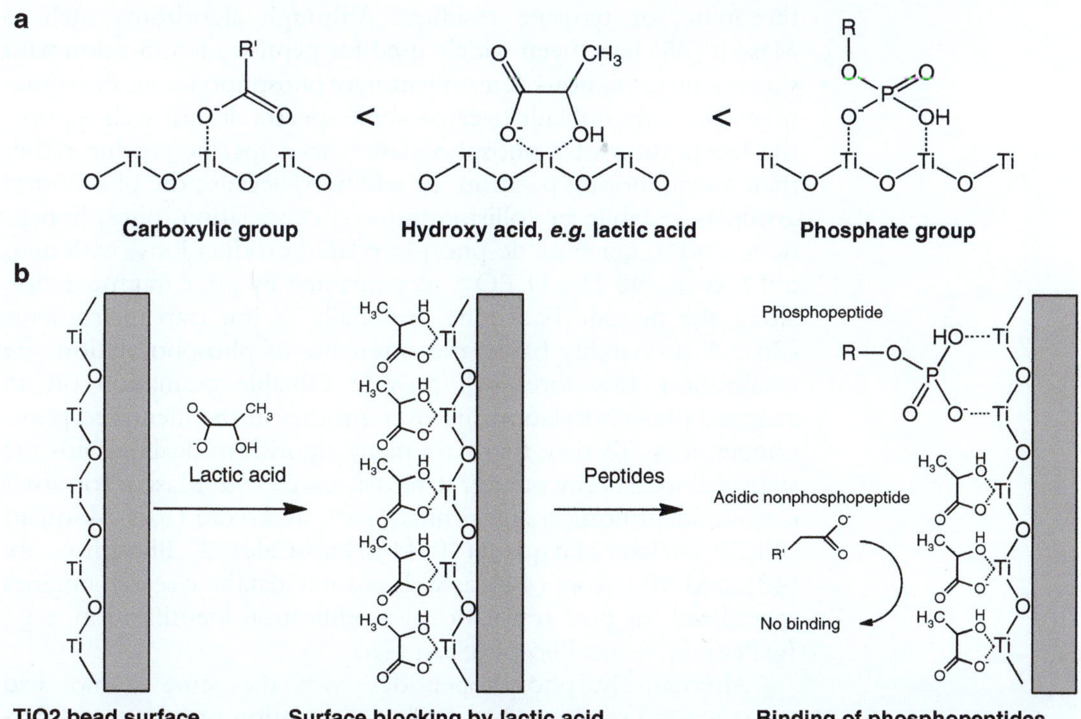

Fig. 1. Principle of HAMMOC for selective enrichment of phosphopeptides. (**a**) Preferential binding affinity of TiO_2 toward different acids; phosphate group of phosphopeptides > lactic acid > carboxylic groups of non-phosphopeptides (from strong to weak). (**b**) The phosphopeptide can bind to lactic acid-blocked TiO_2 beads by competing with lactic acid. In contrast, non-phosphopeptides cannot bind to the beads due to weaker affinity of carboxylic groups than lactic acid.

chromatography (27, 28). In addition, aliphatic hydroxy acids can be easily removed by desalting reversed phase cartridges, which is necessary for subsequent LC-MS/MS analyses. This observation was later confirmed by another group, in which glycolic acid was employed instead of their original DHB/phthalate methods (29). Furthermore, we optimized the elution conditions in HAMMOC to maximize the coverage of phosphoproteome (30), and combined chemical dephosphorylation strategy with HAMMOC to improve the efficiency of identification of multiply phosphorylated peptides (31). Using the HAMMOC strategy, we have successfully identified thousands of phosphorylation sites in *Arabidopsis* (31, 32) and rice (33), as well as more than 60,000 phosphorylation sites in 11,800 human proteins (Ishihama et al., unpublished data). In addition, the HAMMOC strategy can be also easily streamlined with stable isotope labeling by amino acids in cell culture (SILAC) for quantitative phosphoproteomics (34).

1.3. Assignment of Phosphorylation Sites

In typical spectra of phosphopeptides from LC-MS/MS experiments, protein phosphorylation can be detected as a mass shift (+79.99 Da) corresponding to the addition of HPO_3 at serine, threonine, or tyrosine residues. Although algorithms such as Mascot (35) have been widely used for peptide identification with success, unambiguous determination of phosphorylation site sometimes becomes difficult because some spectra do not well support the localization of a phosphorylation to a specific residue rather than a neighboring position. In addition, because the phosphoryl group is so labile in collision-induced dissociation, phosphopeptides tend to generate de-phosphorylated product ions (with neutral loss of −98 Da; H_3PO_4) accompanied by poor fragmentation along the peptide backbone, especially in ion trap instruments (36). A probability-based measurement of phosphorylation site localization, therefore, will provide valuable evaluation on an assigned phosphorylation site over another in the identified phosphopeptides. To date there are many algorithms dealing with the statistical measurement after database search with Mascot (or other peptide identification algorithms), such as Ascore (37), MSquant (38, 39) (or later Maxquant (40, 41)), PhosCalc (42), PhosphoScore (43), and MD-score (44), as well as some database search engines specialized for post-translational modification identification, e.g., InsPecT (45) and PhosphoScan (46).

Alternatively, phosphopeptides with the same amino acid sequence and total number of phosphorylation sites but on different residues can elute at almost the same retention time in reversed phase chromatography. In this case it is hard to accept both modification candidates by using probability-based algorithms described above, even if the MS/MS spectrum consists of fragment ions derived from both phosphopeptides. To solve this ambiguity, we have developed a non-probability-based method called site-determining ion combination, or SIDIC, to discriminate the co-existence of multiple phosphorylation variants. We have used the SIDIC in complementary with the probability-based method to successfully increase the confidence in site-determination as far as possible (33).

Here we present the workflow of phosphopeptide enrichment by HAMMOC. A generic procedure for the preparation of tryptic peptides from cell lysate is introduced, followed by the enrichment of phosphopeptides by HAMMOC. Finally, we also discuss the importance of determination of phosphorylation site(s) in the identified phosphopeptides after MS/MS analyses.

2. Materials

2.1. In-Solution Digestion of Cell Lysate

1. Lysis buffer: 20 mM HEPES·NaOH (pH 7.4) (Wako, Cat. No. 346-01373), 0.25 M sucrose (Wako, Cat. No. 196-00015), 1.5 mM $MgCl_2$ (Wako, Cat. No. 135-00165), 10 mM

KCl (Wako, Cat. No. 163-03545), 0.05% NP-40 (Calbiochem, Cat. No. 492018). Add protease inhibitor cocktails (Sigma, Cat. No. P8340) and phosphatase inhibitor cocktails before use (Sigma, Cat. No. P5726 and P0044) (see Note 1)

2. Urea (Wako, Cat. No. 217-00615)
3. 1 M Tris-HCl, pH 9.0 (Wako, Cat. No. 204-07885)
4. 1-*O*-*n*-Octyl-β-D-glucopyranoside (Nacalai Tesque, Cat. No. 22235-24)
5. Reduction buffer: 1 M dithiothreitol (Wako, Cat. No. 045-08974) in water (see Note 2)
6. *S*-Alkylation buffer: 1 M iodoacetamide (Wako, Cat. No. 095-02151) in water
7. Amicon® Ultra centrifugal filters; 10,000 MWCO (Millipore, Cat. No. UFC501096)
8. Lysyl Endopeptidase (Lys-C) (Wako, Cat. No. 129-02541): 2.5 AU/mL in water (see Note 3)
9. 50 mM ammonium bicarbonate (Wako, Cat. No. 017-02875) in water
10. Sequencing Grade Modified Trypsin, 0.5 mg/mL (Promega, Cat. No. V5113)

2.2. Peptide Desalting

1. Solution A: 5% acetonitrile (Wako, Cat. No. 018-19853) and 0.1% trifluoroacetic acid (Wako, Cat. No. 204-02743) in water
2. Solution B: 80% acetonitrile and 0.1% trifluoroacetic acid in water
3. Empore™ high-performance extraction disk cartridge (3M, Cat. No. 4340HD)

2.3. Phosphopeptide Enrichment by HAMMOC

1. Empore™ C8 disk membrane (3M, Cat. No. 2214)
2. Titansphere TiO_2 beads (GL Sciences, 10 μm, Cat. No. 5020-75010) (see Note 4)
3. Methanol, HPLC grade (Wako, Cat. No. 134-14523)
4. Loading buffer: 300 mg/mL lactic acid (Wako, Cat. No. 128-00056) in a buffer consisting of 80% acetonitrile and 0.1% trifluoroacetic acid (see Note 5)
5. Washing buffer: 80% acetonitrile and 0.1% trifluoroacetic acid
6. Elution buffer: 0.5% piperidine (Wako, Cat. No. 166-02773) in water (see Note 6)

3. Methods

3.1. In-Solution Digestion of Cell Lysate (See Note 7)

1. Harvest the cultured cells from one 15-cm culture dish by scraping. Wash the cells extensively with ice-cold PBS. Centrifuge at 1,000×*g* for 5 min at 4°C to collect the cell pellet.
2. Resuspend the cell pellet in 1-mL ice-cold lysis buffer.
3. Sonicate the resuspended mixture on ice-chilled water for 5 min.
4. Centrifuge at 1,500×*g* for 5 min at 4°C. Transfer the supernatant as cytoplasmic lysate to a new 2.0-mL pre-chilled microcentrifuge tube.
5. Add 960 mg urea, 200 μL of 1 M Tris-HCl (pH 9.0), and 80 μL of 10% 1-*O*-*n*-Octyl-β-D-glucopyranoside to the lysate. Allow the urea to be fully dissolved.
6. Adjust the volume to 2 mL with water. Determine the protein concentration by BCA protein assay.
7. Add 20 μL of 1 M dithiothreitol and react the mixture for 30 min at room temperature (see Note 8).
8. Add 100 μL of 1 M iodoacetamide and react the mixture for 30–60 min at room temperature in the dark.
9. Concentrate the lysate mixture to ~200 μL using Amicon® Ultra centrifugal filters (10,000 MWCO) at 12,000×*g* at room temperature.
10. For 1 mg of cell lysate, add 10 μL of 2.5 AU/mL Lys-C and perform digestion for 3 h at room temperature (see Note 9).
11. Dilute the lysate mixture with fourfold volume of 50 mM ammonium bicarbonate (see Note 10).
12. For 1 mg of cell lysate, add 20 μL of 0.5 mg/mL trypsin and perform digestion for overnight at room temperature.

3.2. Peptide Desalting (See Note 11)

1. Terminate the digestion by adding 10% trifluoroacetic acid to the final concentration of 0.5%. Remove any particulates by centrifugation.
2. Put an Empore™ high-performance extraction disk cartridge into a 50-mL falcon tube. Equilibrate the cartridge with 1 mL Solution A by centrifuging at 1,000×*g* for 1 min at room temperature.
3. Load the acidified peptide solution by centrifuging at 1,000×*g* for 1 min at room temperature.
4. Wash the cartridge with 1 mL Solution A by centrifuging at 1,000×*g* for 1 min at room temperature.

5. Elute the peptides with more than 500 μL Solution B by centrifuging at 1,000×*g* for 1 min at room temperature.
6. Adjust the desalted peptides to 1 mg/mL with Solution B. Proceed to HAMMOC procedure or store the desalted peptides in aliquot at −20°C for later use (see Note 12).

3.3. Phosphopeptide Enrichment by HAMMOC

1. Prepare C8 StageTip for each sample by inserting an Empore™ C8 disk membrane punched with 20-Gauge needle into a 10-μL pipette tip, as previously described (47, 48).
2. For each C8 StageTip, weigh 0.5 mg TiO_2 beads into a 1.5-mL microcentrifuge tube containing a small stir bar (see Note 13).
3. Add 20 μL methanol for 1 mg TiO_2 beads. Mix well by continuous stirring.
4. Transfer 10 μL of the TiO_2 beads/methanol mixture to each C8 StageTip (see Note 14).
5. Remove the residual methanol in the tip by centrifuging the TiO_2/C8 StageTip in an adapter microcentrifuge tube (47, 48) at 500×*g* for 5 min.
6. Equilibrate the TiO_2/C8 StageTip with 20 μL loading buffer by centrifuging at 1,000×*g* for 1 min at room temperature.
7. Mix 100 μg of the total peptide solution (1 mg/mL in Solution B) with 100 μL loading buffer.
8. Load the peptide mixture into the pre-equilibrated TiO_2/C8 StageTip by centrifuging at 1,500×*g* for 1 min at room temperature (see Note 15).
9. Wash the tip with 50 μL loading buffer by centrifuging at 1,500×*g* for 2 min at room temperature.
10. Wash the tip with 50 μL washing buffer by centrifuging at 1,500×*g* for 2 min at room temperature.
11. Elute the phosphopeptides with 50 μL elution buffer by centrifuging at 1,500×*g* for 2 min at room temperature (see Note 16).
12. Acidify the eluted phosphopeptides with 10% trifluoroacetic acid to the final concentration of 0.5%.
13. Desalt the phosphopeptides by C18 desalting StageTip (47, 48).
14. Vacuum dry the desalted phosphopeptides to ~1 to 2 μL. Do not over dry the samples.
15. Resuspend the phosphopeptides samples in 0.1% trifluoroacetic acid. The samples are now ready for LC-MS (see Note 17).

3.4. Determination of Phosphorylation Sites in Identified Phosphopeptides

After LC-MS and peptide identification using database search engine such as Mascot or SEQUEST (49), the identified phosphopeptides are filtered by some criteria to reject hits which are doubtful at primary sequence level as follows: (1) At least five b, y-ions and three successive b, y-ions should be observed; (2) short peptides less than seven amino acid residues are rejected; and (3) as an optional criterion, a dominant fragment ion comes from a precursor ion-originated neutral loss of phosphate in phosphoserine or phosphothreonine-containing peptides in some cases if the MS/MS spectrum is acquired in collision-induced dissociation mode using an ion trap MS.

To further confirm whether the matched phosphopeptides is identified as the top-one hit with unambiguously assigned phosphorylation site(s) by database search, the existence of site-determining ions, which are derived from b, y-ion series, supporting an identified phosphorylation site exclusively are checked by manual inspection (23) or a tool script (SIDIC). For more stringent site-determination considering other modification variants, PTM score (38, 39) and SIDIC-based approach are used. PTM-score-based phosphorylation site determination is performed by using PhosCalc (42). The score of SIDIC is calculated as follows:

1. Calculate a probability-based score (P) for each possible candidates in the same manner as the PTM score (38, 39) (see Note 18).
2. Reject candidates with the delta score greater than 20.
3. Count the number of theoretical SIDIC and that of observed SIDIC for the remaining candidates.
4. Accept the candidate with the highest P value as a possible hit if at least one SIDIC is observed.
5. For other candidates, accept the candidates as a possible hit if the ratio of observed SIDIC to theoretical SIDIC is over 0.3.
6. If only one possible hit is accepted from one MS/MS spectrum, the sites are regarded as an unambiguous site.
7. If more than one possible hit are accepted, the spectrum may consist of multiple modification variants.
8. If no possible hit is accepted, no phosphorylation site can be determined from the spectrum.

4. Notes

1. All buffers in this protocol should be prepared using Milli-Q deionized water with 18.2 MΩ cm. This deionized water is abbreviated into "water" throughout the protocol.

2. For the best result, all solutions should be prepared freshly before use. *S*-Alkylation buffer should be also kept in the dark, since iodoacetamide is unstable and light-sensitive.
3. The resuspended Lys-C solution should be aliquoted and stored at −80°C. Thaw the required amounts of Lys-C on ice for each digestion and discard the remaining solution after use. Do not freeze and thaw again.
4. Store the TiO_2 beads in vacuum desiccators. The binding specificity and capacity of TiO_2 beads to phosphopeptides are very sensitive to humidity (50, 51).
5. For the best result, all solutions in HAMMOC should be prepared freshly before use.
6. The stability of piperidine decreases with time once the bottle is opened. Discard the piperidine stock after 3 months upon open.
7. This is a generic protocol for extracting cytoplasmic extracts from cultured cells, e.g., HeLa cells. For other lysates in solution, consider to perform acetone precipitation before starting this protocol.
8. Do not perform the reduction and the following *S*-alkylation steps above room temperature, e.g., 25°C. Undesired carbamylation of free amino groups can occur in the presence of high concentration of urea at elevated temperature.
9. Adjust the amounts of Lys-C and trypsin for digestion according to actual amounts of the cell lysate.
10. Check the pH value by dipping a drop of mixture on a pH test paper. The pH should be above 8.0 after dilution.
11. This protocol is to desalt ~1 mg of digested peptides using commercial desalting cartridge. Alternatively, one can use in-house-made StageTip (47, 48) for peptide desalting.
12. In our experience, the peptides are stable in Solution B at −20°C for weeks. Do not attempt to vacuum dry the desalted peptide before freezing, since significant loss of peptides may occur for dried peptides.
13. The amount of TiO_2 beads for each StageTip is sufficient for enriching phosphopeptides from 100 μg of total peptide mixture. In addition, it is suggested to weigh more TiO_2 beads for easier handling. For example, use 1–2 mg TiO_2 beads for making 1 or 2 StageTip.
14. Work quickly during this step and make sure the mixture is continuously stirred, since the TiO_2 beads very easily settle down. Failure to maintain homogenous mixture solution will result in inadequate loading of TiO_2 beads into the C8 StageTip.

15. The maximal loading volume for a typical TiO_2/C8 StageTip is ~50 μL. To load all sample mixture, simply perform multiple loadings for each tip. Therefore, four loadings are usually required for each sample.
16. The residual detergent, i.e., NP-40 and octylglycoside, in the peptide mixture is effectively removed during the HAMMOC procedure, which was also confirmed by other groups (29).
17. Depending on the acquisition speed and sensitivity of different MS instruments, optimal LC gradient and MS setting should be determined empirically. In our laboratory, more than 1,000 phosphopeptides can be usually identified from 100 μg of starting HeLa cytoplasmic lysate using a LC gradient of 120 min coupled with a LTQ-Orbitrap mass spectrometer (Thermo Fisher Scientific) (52, 53).
18. Only normal b, y-ions are considered. Dehydroxylated ions and neutral loss peaks cannot be used since they are indistinguishable from each other.

References

1. Hunter T (2000) Signaling–2000 and beyond. Cell 100:113–127
2. Mann M, Ong S-E, Grønborg M et al (2002) Analysis of protein phosphorylation using mass spectrometry: deciphering the phosphoproteome. Trends Biotechnol 20:261–268
3. Manning G, Whyte DB, Martinez R et al (2002) The protein kinase complement of the human genome. Science 298:1912–1934
4. Kersey P, Bower L, Morris L et al (2005) Integr8 and Genome Reviews: integrated views of complete genomes and proteomes. Nucleic Acids Res 33:D297–D302
5. Blume-Jensen P, Hunter T (2001) Oncogenic kinase signalling. Nature 411:355–365
6. Ashman K, Villar EL (2009) Phosphoproteomics and cancer research. Clin Transl Oncol 11:356–362
7. Dass C (2009) Recent developments in mass spectrometry analysis of phosphoproteomes. Curr Proteomics 6:32–42
8. Macek B, Mann M, Olsen JV (2009) Global and site-specific quantitative phosphoproteomics: principles and applications. Annu Rev Pharmacol Toxicol 49:199–221
9. Thingholm TE, Jensen ON, Larsen MR (2009) Analytical strategies for phosphoproteomics. Proteomics 9:1451–1468
10. Mayya V, Han DK (2009) Phosphoproteomics by mass spectrometry: insights, implications, applications and limitations. Expert Rev Proteomics 6:605–618
11. Lemeer S, Heck AJ (2009) The phosphoproteomics data explosion. Curr Opin Chem Biol 13:414–420
12. Matsuoka S, Ballif BA, Smogorzewska A et al (2007) ATM and ATR substrate analysis reveals extensive protein networks responsive to DNA damage. Science 316:1160–1166
13. Smolka MB, Albuquerque CP, Chen SH et al (2007) Proteome-wide identification of in vivo targets of DNA damage checkpoint kinases. Proc Natl Acad Sci USA 104:10364–10369
14. Stokes MP, Rush J, Macneill J et al (2007) Profiling of UV-induced ATM/ATR signaling pathways. Proc Natl Acad Sci USA 104:19855–19860
15. Andersen JN, Sathyanarayanan S, Di Bacco A et al (2010) Pathway-based identification of biomarkers for targeted therapeutics: personalized oncology with PI3K pathway inhibitors. Sci Transl Med 2:43ra55
16. Moritz A, Li Y, Guo A et al (2010) Akt-RSK-S6 kinase signaling networks activated by oncogenic receptor tyrosine kinases. Sci Signal 3:ra64
17. Rix U, Hantschel O, Durnberger G et al (2007) Chemical proteomic profiles of the BCR-ABL inhibitors imatinib, nilotinib, and dasatinib reveal novel kinase and nonkinase targets. Blood 110:4055–4063
18. Oppermann FS, Gnad F, Olsen JV et al (2009) Large-scale proteomics analysis of the human kinome. Mol Cell Proteomics 8:1751–1764

19. Ikeguchi Y, Nakamura H (1997) Determination of organic phosphates by column-switching high performance anion-exchange chromatography using on-line preconcentration on titania. Anal Sci 13:479–483
20. Pinkse MW, Uitto PM, Hilhorst MJ et al (2004) Selective isolation at the femtomole level of phosphopeptides from proteolytic digests using 2D-NanoLC-ESI-MS/MS and titanium oxide precolumns. Anal Chem 76:3935–3943
21. Sano A, Nakamura H (2004) Chemo-affinity of titania for the column-switching HPLC analysis of phosphopeptides. Anal Sci 20:565–566
22. Bodenmiller B, Mueller LN, Mueller M et al (2007) Reproducible isolation of distinct, overlapping segments of the phosphoproteome. Nat Methods 4:231–237
23. Sugiyama N, Masuda T, Shinoda K et al (2007) Phosphopeptide enrichment by aliphatic hydroxy acid-modified metal oxide chromatography for nano-LC-MS/MS in proteomics applications. Mol Cell Proteomics 6: 1103–1109
24. Tunesi S, Anderson M (1991) Influence of chemisorption on the photodecomposition of salicylic-acid and related-compounds using suspended TiO_2 ceramic membranes. J Phys Chem 95:3399–3405
25. Connor PA, McQuillan AJ (1999) Phosphate adsorption onto TiO2 from aqueous solutions: an in situ internal reflection infrared spectroscopic study. Langmuir 15:2916–2921
26. Tani K, Ozawa M (1999) Investigation of chromatographic properties of titania. I. On retention behavior of hydroxyl and other substituent aliphatic carboxylic acids: Comparison with zirconia. J Liq Chromatogr Relat Technol 22:843–856
27. Nuhse TS, Stensballe A, Jensen ON et al (2003) Large-scale analysis of in vivo phosphorylated membrane proteins by immobilized metal ion affinity chromatography and mass spectrometry. Mol Cell Proteomics 2:1234–1243
28. Larsen MR, Thingholm TE, Jensen ON et al (2005) Highly selective enrichment of phosphorylated peptides from peptide mixtures using titanium dioxide microcolumns. Mol Cell Proteomics 4:873–886
29. Jensen SS, Larsen MR (2007) Evaluation of the impact of some experimental procedures on different phosphopeptide enrichment techniques. Rapid Commun Mass Spectrom 21:3635–3645
30. Kyono Y, Sugiyama N, Imami K et al (2008) Successive and selective release of phosphorylated peptides captured by hydroxy acid-modified metal oxide chromatography. J Proteome Res 7:4585–4593
31. Kyono Y, Sugiyama N, Tomita M et al (2010) Chemical dephosphorylation for identification of multiply phosphorylated peptides and phosphorylation site determination. Rapid Commun Mass Spectrom 24:2277–2282
32. Sugiyama N, Nakagami H, Mochida K et al (2008) Large-scale phosphorylation mapping reveals the extent of tyrosine phosphorylation in *Arabidopsis*. Mol Syst Biol 4:193
33. Nakagami H, Sugiyama N, Mochida K et al (2010) Large-scale comparative phosphoproteomics identifies conserved phosphorylation sites in plants. Plant Physiol 153:1161–1174
34. Imami K, Sugiyama N, Tomita M et al (2010) Quantitative proteome and phosphoproteome analyses of cultured cells based on SILAC labeling without requirement of serum dialysis. Mol Biosyst 6:594–602
35. Perkins DN, Pappin DJ, Creasy DM et al (1999) Probability-based protein identification by searching sequence databases using mass spectrometry data. Electrophoresis 20:3551–3567
36. Shou W, Verma R, Annan RS et al (2002) Mapping phosphorylation sites in proteins by mass spectrometry. Methods Enzymol 351:279–296
37. Beausoleil SA, Villen J, Gerber SA et al (2006) A probability-based approach for high-throughput protein phosphorylation analysis and site localization. Nat Biotechnol 24:1285–1292
38. Olsen JV, Mann M (2004) Improved peptide identification in proteomics by two consecutive stages of mass spectrometric fragmentation. Proc Natl Acad Sci USA 101:13417–13422
39. Olsen JV, Blagoev B, Gnad F et al (2006) Global, in vivo, and site-specific phosphorylation dynamics in signaling networks. Cell 127:635–648
40. Cox J, Matic I, Hilger M et al (2009) A practical guide to the MaxQuant computational platform for SILAC-based quantitative proteomics. Nat Protoc 4:698–705
41. Cox J, Mann M (2008) MaxQuant enables high peptide identification rates, individualized p.p.b.-range mass accuracies and proteome-wide protein quantification. Nat Biotechnol 26:1367–1372

42. MacLean D, Burrell MA, Studholme DJ et al (2008) PhosCalc: a tool for evaluating the sites of peptide phosphorylation from mass spectrometer data. BMC Res Notes 1:30
43. Ruttenberg BE, Pisitkun T, Knepper MA et al (2008) PhosphoScore: an open-source phosphorylation site assignment tool for MS^n data. J Proteome Res 7:3054–3059
44. Savitski MM, Lemeer S, Boesche M et al (2011) Confident phosphorylation site localization using the mascot delta score. Mol Cell Proteomics 10:M110 003830
45. Tanner S, Shu H, Frank A et al (2005) InsPecT: identification of posttranslationally modified peptides from tandem mass spectra. Anal Chem 77:4626–4639
46. Wan Y, Cripps D, Thomas S et al (2008) PhosphoScan: a probability-based method for phosphorylation site prediction using MS2/MS3 pair information. J Proteome Res 7:2803–2811
47. Rappsilber J, Ishihama Y, Mann M (2003) Stop and go extraction tips for matrix-assisted laser desorption/ionization, nanoelectrospray, and LC/MS sample pretreatment in proteomics. Anal Chem 75:663–670
48. Rappsilber J, Mann M, Ishihama Y (2007) Protocol for micro-purification, enrichment, pre-fractionation and storage of peptides for proteomics using StageTips. Nat Protoc 2:1896–1906
49. Eng JK, Mccormack AL, Yates JR (1994) An approach to correlate tandem mass-spectral data of peptides with amino-acid-sequences in a protein database. J Am Soc Mass Spectrom 5:976–989
50. Imami K, Sugiyama N, Kyono Y et al (2008) Automated phosphoproteome analysis for cultured cancer cells by two-dimensional nanoLC-MS using a calcined titania/C18 biphasic column. Anal Sci 24:161–166
51. Kyono Y, Sugiyama N, Imami K et al (2010) Development of titania particles used for phosphopeptide enrichment in mass spectrometry-based phosphoproteomics. J Mass Spectrom Soc Jpn 58:129–138
52. Makarov A, Denisov E, Kholomeev A et al (2006) Performance evaluation of a hybrid linear ion trap/orbitrap mass spectrometer. Anal Chem 78:2113–2120
53. Scigelova M, Makarov A (2006) Orbitrap mass analyzer–overview and applications in proteomics. Proteomics 6(Suppl 2):16–21

Part II

Individual Protein Kinases

Chapter 4

Ca^{2+}/Calmodulin-Dependent Protein Kinase II (CaMKII)

Steven J. Coultrap and K. Ulrich Bayer

Abstract

The Ca^{2+}/calmodulin (CaM)-dependent protein kinase II (CaMKII) is a major regulator of synaptic plasticity. CaMKII function depends on complex regulation of its activity and localization by Ca^{2+}/CaM and several auto-phosphorylation reactions. Auto-phosphorylation at T286 makes the kinase "autonomous" (partially active even without Ca^{2+}/CaM), while auto-phosphorylation at T305/306 prevents subsequent Ca^{2+}/CaM-binding. These processes also regulate synaptic localization and binding to the NMDA-receptor subunit GluN2B (formerly known as NR2B). Here we discuss studying CaMKII by inhibition (including by the novel class of CN inhibitors) and by mutagenesis. We describe purification of CaMKII, activity assays, directing and probing auto-phosphorylation, and investigating CaMKII protein–protein binding in vitro and within cells (with the GluN2B interaction as example).

Key words: Autonomy, Auto-phosphorylation, Binding assay, CaMKII, Calcium/Calmodulin-dependent protein kinase II, Calcium, Calmodulin, Fluorescent microscopy, GluN2B, Kinase activity assay, Knock-down/re-expression, Mutational analysis, NMDA-receptor, NR2B, Phosphosite antibody, Phosphorylation, Protein purification, Scintillation, T286, T305/T306, Translocation, Western blot

1. Introduction

The Ca^{2+}/calmodulin (CaM)-dependent protein kinase II (CaMKII) (1) is well established as a major mediator of synaptic plasticity of glutamatergic synapses (2–5). More recently, CaMKII has also been implicated in pathological glutamate signaling related to stroke (6). CaMKII has a broad spectrum of substrates that it can phosphorylate at Ser/Thr residues. While CaMKII prefers an Arg in the −3 position of the phosphorylation site, even this minimal consensus is absent in some of its most prominent substrate sites, such as Ser831 on GluA1 (7). Four different CaMKII isoforms

Hideyuki Mukai (ed.), *Protein Kinase Technologies*, Neuromethods, vol. 68,
DOI 10.1007/978-1-61779-824-5_4, © Springer Science+Business Media, LLC 2012

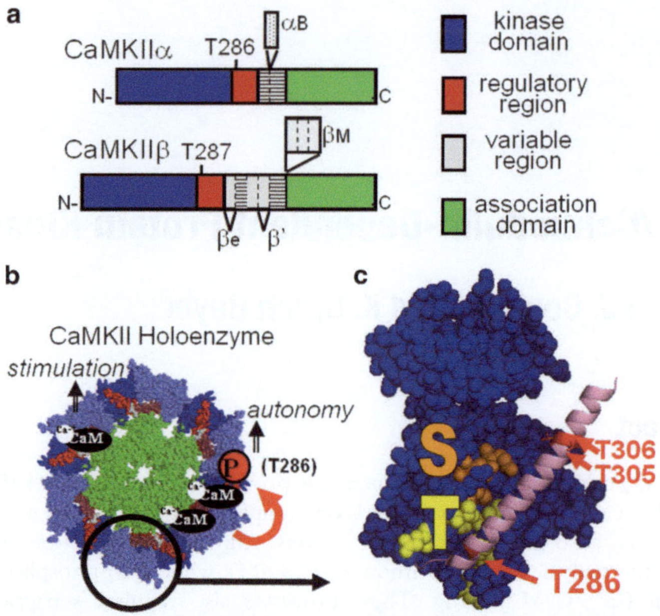

Fig. 1. Models of CaMKII structure (8, 9) and regulation. (**a**) Schematic representation of the CaMKIIα and β primary structure, with the N-terminal kinase domain (*blue*) followed by the regulatory region (*red*), the variable linker region (*gray*), and the C-terminal association domain (*green*). T286 in CaMKIIα is homologous to T287 in CaMKIIβ. Splice variants differ in the variable linker region, as indicated. (**b**) Model of the CaMKII holoenzyme structure (kindly provided by Drs. Chao and Kuriyan, University of Berkely). Stimulation by Ca^{2+}/CaM (of each subunit) and generation of autonomous, Ca^{2+}/CaM-independent activity by T286 auto-phosphorylation (inter-subunit) is indicated. (**c**) Structure of an individual CaMKII kinase domain (space fill) and its regulatory region (ribbon) (8). In the basal state, the regulatory α-helix binds to the T-site (*yellow*) and prevents access to the substrate binding S-site (*orange*). Auto-phosphorylation sites (*red*) generate autonomous activity (T286) or prevent CaM binding (T305/306); both can affect targeting.

(α, β, γ, and δ) are encoded by separate genes, and alternative splicing gives rise to additional diversity (Fig. 1). While CaMKIIγ and δ are ubiquitously expressed, CaMKIIα and β are largely restricted to the brain (10, 11), although β variants are additionally found at least also in skeletal muscle and pancreatic β cells (12–14). The CaMKIIα isoform is extremely highly expressed in brain (almost 1% of total protein) (15), and the CaMKIIα knockout mice were the first genetically manipulated animals reported to have a behavioral deficit in learning (16). The focus of this chapter is on the CaMKIIα isoform, but the methods herein are applicable to all isoforms.

1.1. CaMKII Structure and Regulation of Activity and Localization

CaMKII isoforms assemble into large, disc-shaped, 12meric holenzymes (~20 nM in diameter; 600–750 kDa), with their C-terminal association domains forming a central hub, and their N-terminal kinase domains radiating outward (8, 9) (Fig. 1). Situated between the kinase and association domains is the CaM-binding regulatory region, followed by a variable linker region that is subject to extensive alternative splicing in the β, γ, and δ isoforms (3). For CaMKIIα, only two splice variants have been described: α, the dominant variant, and α_B, a minor variant that contains a nuclear localization signal (17). However, alternative promoter usage creates an additional product from the CaMKIIα gene: αKAP, a CaMKII anchoring protein in skeletal muscle that contains the CaMKIIα linker region and association domain and (instead of the kinase domain and the regulatory region) an N-terminal hydrophobic membrane targeting region (12).

During a Ca^{2+}-stimuls, each kinase subunit of CaMKII holoenzyme is activated separately by Ca^{2+}/CaM binding. However, simultaneous binding of two Ca^{2+}/CaM molecules to two neighboring subunits can trigger an inter-subunit auto-phosphorylation at T286 (one Ca^{2+}/CaM is required for activation, the other to make T286 accessible as a substrate for the activated neighbor) (18) (Fig. 1). Interestingly, this dual requirement for Ca^{2+}/CaM makes the efficiency of T286 auto-phosphorylation dependent on the frequency of Ca^{2+} oscillations, enabling CaMKII to act as a molecular frequency decoder (19). Functionally, T286 auto-phosphorylation makes CaMKII one of the highest-affinity Ca^{2+}/CaM-binders within the cell (<0.02 nM), while un-phosphorylated CaMKII has one of the lowest Ca^{2+}/CaM affinities (15–100 nM) (20). Maybe even more importantly, T286 auto-phosphorylation generates "autonomous" (Ca^{2+}/CaM-independent) CaMKII activity (21–23). However, this "autonomy" is only partial (20–40%), and activity can still be significantly further stimulated by Ca^{2+}/CaM (at least for phosphorylation of "regular" substrates) (24). This mechanism can still provide a molecular memory of previous Ca^{2+} signals, but may prevent complete uncoupling from subsequent stimuli. Many protein kinases require phophorylation within their activation loop for full activity. Notably, however, T286 is not located within the activation loop of the core kinase domain, but within the regulatory region just N-terminal of the CaM-binding site. Additionally, T286 phosphorylation is not required for CaMKII activity. However, CaMKII T286 phosphorylation is required for induction of long term potentiation (LTP) of synaptic strength and for learning (25, 26).

Another regulatory but inhibitory auto-phosphorylation occurs at T305/306, within the CaM-binding site (27, 28). T305/306 phosphorylation inhibits Ca^{2+}/CaM-binding, and, vice versa, Ca^{2+}/CaM-binding inhibits T305/306 phosphorylation. Thus, efficient T305/306 phosphorylation only occurs after generation of autonomous activity by T286 phosphorylation followed by dissociation of Ca^{2+}/CaM. The resulting triple-phosphorylated kinase would be partially active, but could not be further stimulated. Interestingly, mice with a CaMKII T305/306A mutation (which cannot be phosphorylated at these residues) are impaired in un-learning of no longer advantageous behavior (29).

Neuronal stimulation also regulates CaMKII localization to the postsynaptic sites. LTP- or LTD-like stimuli can induce CaMKII translocation to excitatory or inhibitory synapses, respectively (30, 31). While there are many CaMKII-binding partners at excitatory synapses (32, 33), the NMDA-type glutamate receptor subunit GluN2B appears to be of special importance. Like synaptic translocation, GluN2B binding requires an initial Ca^{2+}/CaM-stimulation and is further enhanced by T286 auto-phosphorylation (which can also induce binding on its own in the absence of Ca^{2+}/CaM) (34, 35); it is decreased by T305/306 auto-phosphorylation and by GluN2B S1303 phosphorylation (36, 37). Addition of ATP overall enhances binding to GluN2B (38), in part through positive regulation by direct nucleotide binding (38, 39), as the effect of ADP or AMP-PNP is even greater. Once CaMKII is bound to GluN2B, Ca^{2+}/CaM, or T286 auto-phosphorylation are no longer required for maintenance of the binding (35, 40). Such a switch-like regulation would be required for a molecular memory mechanism. Functionally, the regulated interaction of CaMKII with the NMDA-receptor complex is indeed important in synaptic plasticity (41, 42).

1.2. Studying CaMKII Functions by Inhibition and Mutagenesis

The traditional CaMKII inhibitors KN62 and KN93 (along with the inactive control compound KN92) provided useful tools for studying cellular functions of CaMKII (43, 44). However, they cannot distinguish between CaMKII and CaMKIV (45), and also inhibit voltage-gated Ca^{2+} and K^{+} channels (46, 47). Maybe more importantly, they only inhibit stimulated but not autonomous CaMKII activity (43, 44, 48), a hallmark feature of CaMKII regulation. As KN inhibitors are competitive with CaM, their IC50 depends on the CaM concentration (43, 44). Peptides derived from the CaMKII auto-regulatory region, such as AIP or AC3-I, additionally inhibit PKC, MLCK, and PKD (49–51), and some studies found a relatively low potency (3 μM IC50) (52, 53). Peptides derived from the natural CaMKII inhibitory protein CaM-KIIN (CN peptides) provide an alternative, as they selectively and potently (50–100 nM IC50) inhibit both stimulated and autonomous CaMKII activity (54). Fusion with the tat sequence allows peptides to penetrate cells (55) as well as the blood brain barrier (56, 57). Indeed, the fusion peptide tatCN21 (54) effectively inhibited CaMKII functions in the brain even after systemic injection (26, 48). 5 μM tatCN21 was an effective working concentration in cell culture and hippocampal slices (26, 48, 54).

Introducing specific mutations into CaMKII provides a powerful tool to study the function of specific aspects of CaMKII regulation, both in transgenic mice and after overexpression in neurons and/or heterologous cells. Some mutational analysis may provide more mechanistic information when done at the single cell level. For instance, the effect of constitutively autonomous T286D mutants is at least partial uncoupling of CaMKII activity from the input stimulation. Indeed, experiments in mice with temporally controlled overexpression of T286D showed that such uncoupling from upstream regulation is detrimental for the system (58). On the single cell level, however, the same uncoupling from input stimulation can provide mechanistic insight into the downstream regulation of synaptic properties caused directly by autonomous CaMKII (59). In order to provide the best possible interpretation of functional experiments with mutants, a careful biochemical characterization of such mutants is desirable. For instance, careful characterization of the T305/306A mutant mice showed that prevention of the inhibitory phosphorylation also enhanced synaptic localization of this mutant (29). Is the resulting phenotype due to the effect on CaMKII activity or localization or both? Further mutational dissection of the effects may provide an answer, for instance by comparing T305/306 phospho-mimick mutants with other specific mutants that either only fail to bind CaM or only fail to localize. Again, this would likely require single cell analysis as well as careful biochemical characterization of the specific mutants used.

2. Material

For assembling the required material, also carefully read Sect. 3. Not detailed are general laboratory equipment, materials and methods needed for mammalian cell culture, transfection, bacterial culture, protein expression and plasmid purification, as well as for Western detection, immuno-cytochemistry, and radiation work.

2.1. CaMKII Purification

Centrifuge, ultracentrifuge, vacuum setup, PAGE set up, glass homogenizers, Buchner funnel with fritted disk (glass filter) (Pyrex), Columns (10 × 1.5 cm glass), P11 cellulose phosphate action exchange resin (Whatman), Protease inhibitor cocktail, (such as complete™ tablets, Roche) Calmodulin sepharose 4B (GE Healthcare).

For CaMKII activity assays of the fractions of the purification, see Sect. 2.3+3.2.

Solutions. The following solutions should be made in advance and kept refrigerated, but β-mercaptoethanol (BME) should be added freshly the day of use.

– Brickey buffer (10 mM Tris pH 7.5, 2.5% betaine, 1 mM EDTA, 1 mM EGTA, 1 mM BME)

All P11 Column buffers contain 50 mM PIPES pH 7.0, 1 mM EGTA, 1 mM BME.

- P11 column wash buffer 1 (+100 mM NaCl)
- P11 column wash buffer 2 (+180 mM NaCl)
- P11 column elution buffer (+500 mM NaCl)
- 0.5 M PIPES, pH 7.0

The following buffers are for the CaM-sepharose purification step.

- Column packing buffer (25 mM PIPES pH 7.0, 100 mM NaCl, 1 mM $CaCl_2$)
- Equilibration buffer (25 mM PIPES pH 7.0, 100 mM NaCl, 1 mM $CaCl_2$, 10% glycerol)
- Wash buffer (25 mM PIPES pH 7.0, 500 mM NaCl, 1 mM $CaCl_2$, 10% glycerol)
- Elution buffer (25 mM PIPES pH 7.0, 400 mM NaCl, 1 mM EGTA, 10% glycerol)

2.2 + 2.3. CaMKII Activity Assays and Auto-phosphorylation

Fully equipped radiation work station (for ^{32}P) including shielding, Geiger counter, waste collection; heat block, scintillation counter and vials, P81 cation exchange chromatography paper (Whatman), PIPES, BSA, $MgCl_2$, $CaCl_2$, EDTA, EGTA, ATP, [γ-^{32}P]ATP, CaM, CaMKII, substrate peptide.

The most commonly used CaMKII substrate peptides are AC2 and Syntide 2 (derived from the CaMKII T286 auto-phosphorylation site and from glycogen syntase, respectively). AC2 is more selective for CaMKII, and thus may be preferable when assessing CaMKII activity in tissue extracts. However, AC2 is an unusual T-site binding substrate that will report a higher degree autonomous activity compared to regular substrates such as Syntide 2 (24). An alternative substrate may be Optimide 2, which is a Syntide 2 mutant designed to enhance selectivity for CaMKII over CaMKI. However, this substrate still awaits better characterization.

AC2: `KKALRRQETVDAL`

Syntide 2: `PLARTLSVAGLPGKK`

Optimide 2: `PLARQLSVDGKPGKK` (Q, D and K underlined)

2.3. Western-Analysis of CaMKII

Standard equipment and materials for SDS-PAGE, Western-transfer, and immuno-chemiluminescence detection (such as Western Lightning; Perkin Elmer).

A variety of monoclonal and polyclonal antibodies are available for detection of CaMKII, specific isoforms, and specific auto-phosphorylation at T286 or T305/306 by Western-blot (or immuno-cytochemistry). We have mainly used antibodies from

BD Bioscience and PhosphoSolutions, but good CaMKII antibodies can be obtained from most major suppliers; Abcam has currently an especially large selection of isoform-selective antibodies. For detection of CaMKIIα and β isoforms, we have generally used the monoclonal antibodies CBα2 and CBβ1, (available from Invitrogen) respectively, but other good alternatives are available.

Solutions:

- 3×–5× loading buffer stock; DTT added freshly (final concentration of 1× loading buffer when added to sample: 67.5 mM Tris pH 6.8, 2% SDS, 10% glycerol, 0.25 mM bromophenol blue, 50 mM DTT).
- 10× TBS (200 mM Tris pH 7.55, 1.5 M NaCl). Dilute tenfold for 1×TBS.
- TBS-T (1× TBS, 0.1% Tween-20).
- TBS, 1% Tween-20.
- Blocking solution: 5% non-fat dry milk (or BSA) in TBS-T.
- Antibody incubation solution: 2.5% non-fat dry milk (or BSA) in TBS-T.

2.4. CaMKII Phosphorylation of Protein

Materials and solutions as in Sects. 2.2–2.3. Substrate protein (or tissue extract) of choice, corresponding phospho-selective antibodies (if available).

MAP2 is an excellent CaMKII substrate protein with multiple phosphorylation sites. However, we are not aware of any phospho-selective antibodies for MAP2. Such antibodies are available for GluR1 Ser831, GluN2B Ser1303, and tyrosine hydroxylase Ser19 (the latter is also well suitable for immuno-cytochemistry).

2.5. CaMKII Binding to GluN2B

Equipment and materials for Western detection (as in Sect. 2.3) and/or CaMKII activity assays (as in Sect. 2.2), as desired.

Anti-GST coated clear strip plates (Thermo Scientific), GST-N2Bc or other desired GST fusion protein (expressed in bacteria).

Solutions:

- TBS-T (20 mM TRIS pH 7.55, 150 mM NaCl, 0.05–0.1% Tween-20)
- PS-T (50 mM PIPES pH 7.1, 150 mM NaCl, 0.05–0.1% Tween-20)
- Kinase binding buffer (PS-T with 0.1% BSA, 1 mM $CaCl_2$, 1 μM Calmodulin). Note: modify as desired in order to test for regulation of the binding
- Kinase wash buffer (PS-T with 1 mM EGTA)
- Final wash buffer (50 mM PIPES pH 7.1, 1 mM EGTA)

2.6. Imaging CaMKII Translocation to GluN2B in Cells

Microscope system capable of GFP imaging (preferably including in live cells, i.e., inverted fluorescence microscope). Glass-bottom culture dishes (3 cm) for live imaging. Standard mammalian tissue culture and transfection methods; appropriate expression vectors. Standard methods for immuno-cytochemistry (of GluN2B). Additionally for neurons: glutamate, glycine, APV.

Solutions:

- Hanks balanced salt solution (HBSS)
- Imaging buffer (0.87× HBSS, 25 mM Hepes pH 7.4, 2 mM glucose, 2 mM $CaCl_2$, 1 mM $MgCl_2$). Note: For imaging neurons, modify $CaCl_2$ and $MgCl_2$ as indicated
- 100 μM ionomycin in imaging buffer (freshly diluted from 3 mM stock in DMSO, stored in aliquots at −20°C)

3. Methods

3.1. Purification of CaMKII

Purified CaMKII and CaM is available from various commercial sources. Thus CaM purification is not described here, but can be done by differential ammonium sulfate precipitation followed by phenyl-sepharose columns (60). However, in house purification is required in order to study specific CaMKII mutants. This is most conveniently done after overexpression in mammalian cells, as this allows the use of the same vectors also for functional studies. The method described here is for purification from transfected HEK 293t cells, but the method can be adjusted for purification from other sources, such as brain tissue or baculovirus/Sf9 cell expression systems.

For each CaMKII mutant, transfect fifteen 10 cm plates of HEK cells with CaMKII expression vector (for instance by the Ca_2PO_4 method). After 72–96 h, rinse cells in PBS and harvest on ice by scraping in PBS, then pellet cells (5 min at ~1,000×*g*) and store at −80°C.

3.1.1. Activation of P11 Phosphocellulose for Column (Day Before Purification)

1. Add 2 g of P11 phosphocellulose cation exchange resin (Whatman) to 50 ml of 0.5 M NaOH.
2. Add phosphocellulose/NaOH mix to glass filtration device attached to a vacuum (Note: vacuum should not be on yet! For best results in all steps, allow phosphocellulose slurry to settle before turning the vacuum on. Otherwise, the finer particles will clog the glass filter).
3. Incubate for 3 min.
4. Remove fluid by vacuum. Stop vacuum when fluid is just above phosphocellulose level (do not allow the resin to dry at any point).

5. Add 50 ml of ddH_2O. Allow slurry to settle. Remove fluid by vacuum as in step 4.
6. Repeat step 5 five times or until pH of filtrate is below 10.
7. Add 50 ml of 0.5 M HCl.
8. Incubate 3 min.
9. Remove fluid by vacuum as in step 4.
10. Add 50 ml of ddH_2O. Allow slurry to settle. Remove fluid by vacuum as in step 4.
11. Repeat step 10 five times.
12. Add 50 ml of 0.5M PIPES at pH 7.0. Transfer phosphocellulose slurry to a beaker and store overnight at 4°C.

3.1.2. Preparation and Equilibration of P11 Phosphocellulose Column (Day of Purification)

Note: all column work is to be done in the cold room.

1. Pack ¼ of the P11 resin (prepared the prior day) in each column (10 × 1.5 cm).
2. Remove excess PIPES, allowing it to flow through the column. Stop flow when fluid is just above phosphocellulose level (do not allow the resin to dry at any point).
3. Equilibrate column with 12.5 ml P11 column wash buffer. Allow buffer to flow out at ~0.5 ml/min, stop the flow when fluid is just above the resin.

3.1.3. First Stage of Purification (P11 Phosphocellulose Column)

For each CaMKII mutant, have cell pellets from fifteen 10 cm plates of transfected HEK cells prepared previously (stored at −80°C).

1. Add protease inhibitors to 50 ml Brickey buffer (dissolve one tablet of protease inhibitors in 0.5 ml Brickey buffer, then add to 49.5 ml Brickey buffer).
2. Add 1 mM BME to Brickey Buffer and all P11 column buffers.
3. Re-suspend the cell pellet in 15 ml ice-cold Brickey buffer with protease inhibitors.
4. Lyse cells with a glass homogenizer and a Teflon pestel attached to a Stedfast stirrer (Model 1200SL; Fisher Scientific) in the cold room (five passes at 60% power). (Note: for CaMKIIβ, add 150 mM sodium perchlorate after lysis and incubate for 20 min at 4°C, in order to break up binding to the cytoskeleton).
5. Transfer the homogenate to an ultracentrifuge tube, and if necessary adjust the volume with Brickey buffer to at least the minimum acceptable volume for the tubes (ideally 15–16 ml). Balance pairs (with caps) to 0.01 g.

6. Clarify material by centrifugation at 100,000×*g* at 4°C for 30 min. Collect supernatant to a clean tube, discard pellet.
7. Save 100 μl of supernatant prior to loading on the column.
8. Load supernatant onto phosphocellulose column with a flow rate of ~1 ml/min. Collect and save the flow through.
9. Wash each column in 20 ml P11 column wash buffer 1. Collect and save the wash.
10. Wash column in 4 ml P11 column wash buffer 2. Collect and save the wash.
11. Elute the kinase from the column with 12.5 ml P11 column elution buffer. Collect the elution in 1–1.5 ml fractions.
12. Determine which fractions are enriched for kinase either by performing a kinase activity assay using AC2 as substrate (see Sect. 3.2) or by running a sample of each fraction on a gel and coomassie staining. Typically only the starting material, flow through and the elution fractions are assayed. The washes are kept in case the kinase is not found in the elution.

3.1.4. Preparation of Calmodulin Sepharose 4B for Batch Purification

For larger scale purifications the CaM-sepharose should be used in a column; however, for the scale described here batch purification often works better. The following prepares enough CaM-sepharose for purification of three kinase mutants in parallel. As with the previous P11 column, do not let the CaM-sepharose dry out. Throughout this procedure, when removing supernatants draw up liquid leaving just enough to cover the sepharose.

1. Add 0.625 ml of pre-swollen CaM-sepharose 4B to a 15 ml tube. Calmodulin-sepharose 4B is supplied pre-swollen in 20% ethanol. Allow it to settle and remove the ethanol.
2. Wash CaM-sepharose with 5 ml column packing buffer.
3. Centrifuge 500×*g* for 5 min at 4°C and discard wash.
4. Repeat wash (steps 2 and 3).
5. Prepare a 50% CaM-sepharose slurry in equilibration buffer (0.5 ml settled Calmodulin-sepharose 4B + 0.5 ml equilibration buffer).

3.1.5. Purification of CaMKII with Calmodulin Sepharose

1. Combine the CaMKII containing fractions from the P11 columns in a 15 ml tube (typically two fractions; 2–3 ml).
2. Dilute the combined fractions twofold with equilibration buffer and add $CaCl_2$ to a final concentration of 1 mM and glycerol to a final concentration of 10% (save a small sample of this starting material to run on a gel).
3. Add 300 μl of the 50% CaM-sepharose to each tube (150 μl bed volume) then incubate on rocker in cold room for 1 h.

4. Spin at 500×*g* for 5 min at 4°C. Remove and save the supernatant (flow-through).
5. Add 1.5 ml wash buffer (10 bed volumes) to the CaM-sepharose pellet, wash by rocking in cold room for 5 min.
6. Spin at 500×*g* for 5 min. Remove the supernatant.
7. Resuspend the pellet in 1 ml wash buffer, transfer to a 1.5 ml tube, and wash for 5 min.
8. Spin at 500×*g* for 5 min (now in microcentrifuge). Remove the supernatant.
9. Add 1 ml wash buffer, and wash for 5 min.
10. Spin at 500×*g* for 5 min. Remove the supernatant.
11. Add 200 μl elution buffer then incubate on cold room rocker for 15 min.
12. Spin at 500×*g* for 5 min. Remove and save Elution 1.
13. Repeat steps 11 and 12 three more times, collecting Elutions 2–4.
14. Take a small sample from each fraction for analysis by gel.
15. Run a PAGE gel of the starting material, flow-through, elution fractions and if desired washes. The gel can be stained to determine the purity of the kinase; alternatively a western blot can be used to confirm the presence of CaMKII.
16. After determining CaMKII concentration, aliquot and store the kinase at −80°C. For subsequent applications, note that the kinase is now in a buffer containing 1 mM EGTA and 400 mM NaCl.

3.1.6. Determining the Concentration of Purified CaMKII

The concentration of purified CaMKII is best determined spectrophotometrically (by measuring the absorbance at a wavelength of 280 nM). This is done using undiluted CaMKII. Thus, use a well cleaned cuvette, and save the kinase after the measurement.

$$\text{Abs}_{(280\ \text{nm})} = \varepsilon l c$$

(ε = extinction coefficient of protein, 66,350 M^{-1} cm^{-1} for untagged CaMKIIα; l = path length of cuvette in cm; c = concentration of protein in M).

Expect CaMKII concentrations of 0.2–20 μM (depending on the starting material).

Alternatively, concentration can be determined by quantitative Western-analysis (see Sect. 3.4), using a CaMKII preparation of known concentration as standard.

3.2. CaMKII Activity Assays (Using Peptide Substrates)

Perform standard CaMKII assays at 30°C (compromise between stability and physiological temperature) for 1 min. This should result in linear reaction conditions for assays as described below.

However, linearity should be verified by stopping the reactions at various time points. While longer reaction times (10–30 min) can be useful for determining if a substrate is phosphorylated, the saturation of activity may complicate the interpretation of some results. For consideration of the peptide substrates that can be used, see Sect. 2.2.

Non-radioactive phosphorylation assays have been improved, and can be used in adaptations of the method described below. However, we use radioactive ^{32}P incorporation, which provides excellent linearity of detection. If using ^{32}P, be sure to be trained and familiar with all safety measures and regulations, including (but not limited to) adequate plexiglass shielding, personal protection, waste collection and disposal, and the detection of contaminated areas, materials, and liquids.

Composition of the reaction (50 μl):

- 2–20 nM CaMKII subunits (5–50 ng for CaMKIIα)
- 50 mM PIPES, pH 7.2 (or pH 7.0–7.4; alternatively 50 mM Hepes + 100 mM KCl)
- 0.1% BSA
- 1–2 μM CaM (this is ~10- to 20-fold Ka for CaMKIIα)
- 0.5–1 mM $CaCl_2$
- (For assessing "autonomous" activity: 0.5 mM EGTA instead of $CaCl_2$ and CaM)
- 10 mM $MgCl_2$ (5–15 mM)
- 100 μM ATP (this is ~tenfold Km)
- 0.125 mCi/ml [γ-^{32}P]ATP (for 1.25 mCi/μmol total ATP)
- 75 μM substrate peptide (see Sect. 2.2)

Consider any amounts of chelators or Ca^{2+} carried into the reaction when adjusting the $CaCl_2$ or EGTA concentration (i.e., make sure that the reactions contain the desired concentration of free Ca^{2+} and Mg^{2+}; see Sect. 3.3 for more detail).

Typically, we combine three components for each of such 50 μl reaction:

- 10 μl 5× kinase mix (purified CaMKII diluted in 5 mM buffer, 0.1% BSA; or CaMKII from cell or tissue extracts)
- 10 μl 5× "variable mix" (with inhibitor, substrate, or whichever components are to be varied in the particular experiment)
- 30 μl 1.67× reaction mix with either Ca^{2+}/CaM or EGTA (containing [γ-^{32}P]ATP and any other component not included in the "variable mix")

Reaction start. Typically, by adding 10 μl kinase (diluted in 5 mM buffer, 0.1% BSA; on ice) to 40 μl reaction mix (in 1.5 ml tubes in a heatblock at 30°C; pre-equilibrated for ~5 min). Mix by pipetting 2–4 times. For determining the background of the assay, also include some reactions without kinase added.

Reaction stop. Typically after 1 min. Directly pipette 40 μl of the reaction onto Whatman P81 cation exchange chromatography paper squares (~2.5 × 2.5 cm), and immediately drop the paper into a 600 ml beaker with ~400 ml wash solution (tap water or 0.5% phosphoric acid). P81 squares should be previously numbered (for identification after the wash) with pencil and laid out in order on a non-adsorbent surface (such as aluminum foil). Bending one of their corners upwards facilitates picking them up with forceps for transfer into the beaker.

Alternatively, stop reaction with 10 μl ice-cold 15% TCA, set on ice for 20 min, spin for 2 min in a microcentrifuge at ~14,000×*g* at 4°C, and pipette 40 μl of the supernatant onto P81 squares as above. This procedure precipitates protein, and should be considered if significant background from protein phosphorylation is expected, for instance when using crude cell extracts instead of purified CaMKII (see also Sect. 2.2 for choice of substrate peptide). In this case, a negative control reaction containing CaMKII inhibitors should be considered, in order to estimate possible contribution of other kinases within such crude extracts.

In another alternative, reactions can be stopped by adding sufficient EDTA to chelate Mg^{2+} (see Sect. 3.3) then placing the reaction on ice until further processing as desired. This should be done if protein substrate is used instead of peptide (see Sect. 3.5).

Wash and detection. Rinse P81 squares ~6 times with ~300 ml tap water (collect all wash solutions for radioactive waste): agitate gently (do not use a stir bar, as it may damage the P81 rectangles), then let sit in water for 5–15 min after each rinse (except omit the incubation after first and last rinse). Then pick P81 squares out with forceps, blot onto paper towels, and place in scintillation vials. In order to determine specific activity, pipette the equivalent of 1 μl of the kinase reaction (for instance 0.6 μl reaction mix if using 30 μl reaction mix for 50 μl reactions) on dry P81 squares (six to eight), and place in scintillation vials (without washing). Measure in a scintillation counter either without scintillation fluid using the tritium window (Cherenkov method), or add scintillation fluid and use the P32 window.

Analysis. Average CPM of the background measurements and subtract this from each sample. The specific activity is the average CPM in 1 μl kinase reaction (measured as described above) divided by the concentration of ATP in μM [pmol/μl] present in the kinase reaction (typically 100 μM). The kinase activity [in pmol phospho-substrate/pmol CaMKII subunits/minute] is: (CPM sample/specific activity [CPM/pmol ATP])/(concentration of CaMKII subunits in μM [pmol/μl] × μl of reaction spotted on paper)/reaction time in minutes.

This measure of activity essentially provides the number of phosphorylation reactions per kinase subunit per minute. To report the activity instead in nmol of phosphorylated substrate/minute/mg of CaMKII, use the same calculation with the concentration of CaMKII in mg/ml. Alternatively, activity can be reported as percent of a reference condition (usually the maximal Ca^{2+}/CaM stimulated activity).

Tips and further considerations. Do 4–6 replicates of each condition to enable meaningful analysis of the results. Four reactions can be done in parallel relatively easily (started then stopped in 15 s intervals); processing six reactions in parallel is possible (with 10 s intervals), but requires more practice and invites mistakes. Use of cap-less tubes is most convenient when the reactions are directly spotted onto P81 squares (use capped tubes if longer reactions are done, or if storage in the reaction tube is required, such as for subsequent PAGE analysis of protein substrates). Adding small amounts of water into the heat block wells accelerates heat exchange (but be sure to avoid splashing when inserting the tubes).

3.3. CaMKII Auto-phosphorylation

The same principle reaction buffer as for activity assays can be used (leaving out radioactivity as desired). Use at least 0.1 pmol of CaMKII in each reaction (for Western detection, Sect. 3.4), or a sufficient concentration to allow for dilution into subsequent activity assays. The auto-phosphorylation reactions are stopped by chelating Mg^{2+} (with EDTA) and/or Ca^{2+} (with EGTA). Note that chelating only Ca^{2+} will allow continued autonomous activity (which will be generated by T286 auto-phosphorylation during Ca^{2+}/CaM stimulation); chelating Mg^{2+} will prevent ATP binding and stop both stimulated and autonomous activity. For subsequent activity assays, the amounts of chelators carried over need to be considered, and the Ca^{2+} and/or Mg^{2+} concentrations may need to be adjusted accordingly. The chelator effects on concentration of free Ca^{2+} and Mg^{2+} should be calculated (using programs such as MaxChelator, currently freely available online).

For proceeding to subsequent kinase activity assays (with 2 nM CaMKII), we typically start with 50 nM CaMKII (and 10 mM Mg^{2+}) in the auto-phosphorylation reaction. The reactions are stopped by fivefold dilution (to 10 nM CaMKII and 2 mM Mg^{2+}) in ice-cold EDTA/0.1%BSA (to 5 mM EDTA endconcentration); this solution is then used as the 5× kinase mix (thus carrying 2 nM CaMKII and 1 mM EDTA into the kinase reaction).

For Western-analysis of phosphorylation (see Sect. 3.4), the final EDTA stop buffer can additionally be the loading buffer. Western-analysis after different reaction times is recommended, in order to verify maximal phosphorylation.

T286 auto-phosphorylation:

- Perform reaction in the presence of CaM (at least 1 μM more than CaMKII concentration) and Ca^{2+} (0.2–1 mM) on ice for 1–5 min.
- Stop reaction with sufficient EDTA to chelate Mg^{2+} (or directly proceed to activity assay).

Additional T305/306 auto-phosphorylation:

- Stop the T286 auto-phosphorylation instead with sufficient EGTA to chelate Ca^{2+}.
- Transfer to room temperature or 30°C for 5–10 min (note that this secondary burst auto-phosphorylation will also target sites other than T305/306).
- Stop with EDTA.

T306 auto-phosphorylation:

- Perform reaction in the absence of Ca^{2+}/CaM at 30°C for 30 min.
- Stop with EDTA (or directly proceed to activity assays).

3.4. Western-Analysis of CaMKII

Standard procedures can be used for Western detection of CaMKII and its T286 or T305/306 phosphorylated states. Based on the molecular weight of CaMKII (50–70 kDa) polyacrylamide concentrations of ~10% should be used in SDS-PAGE (gradient gels or 7.5–15% polyacrylamide can be used alternatively). Before loading, samples (extracts of purified CaMKII, after desired manipulation) should be boiled for ~10 min in loading buffer (does not need to be repeated for previously boiled frozen samples). Protein separated by SDS-PAGE can be electro-blotted onto nitrocellulose or PVDF membranes, but the latter have higher protein binding capacities. Follow manufacturer's instructions, especially for PVDF membranes, which require activation. Our following standard immuno-detection method on membranes may need to be optimized for specific antibodies, and can be substituted with similar established protocols.

3.4.1. Immuno-Detection on Membranes

1. Block membrane overnight at 4°C or for 0.5–1 h at room temperature in 5% dry milk in TBS-T (TBS with 0.1% Tween-20) (some antibodies may require 5% BSA as blocking agent instead; in this case substitute dry milk with BSA until the second antibody incubation).
2. Rinse briefly in TBS-T.

3. Incubate for 1 h at room temperature with desired antibody concentration (1:500 to 1:10,000) in 2.5% dry milk in TBS-T; gently agitate on vertical shaker.
4. Rinse briefly in TBS-T.
5. Wash three times 5 min in TBS-T; gently agitate on horizontal (rotation) shaker for all wash steps.
6. Incubate for 1 h at room temperature with appropriate secondary antibody (~1:4,000 anti-mouse, -rabbit, or -goat horseradish peroxidase-conjugate; GE Healthcare); gently agitate on vertical shaker.
7. Rinse briefly in TBS-T.
8. Wash 5 min in TBS-T.
9. Wash 15–20 min TBS with 1% Tween-20.
10. Wash 4 times 10 min in TBS-T.
11. Proceed with chemiluminescence detection (for instance by Western Lightning; Perkin Elmer).
12. Detect chemiluminescence by CCD camera system (such as Alpha Innotech) or film (such as Hyperfilm ECL; GE Healthcare).

3.4.2. Analysis and Quantification

For quantitative (or semi-quantitative) analysis, chemiluminescence detection by a CCD camera system is preferable (due to good linear detection range and objective detection of saturated exposure), but densitometry of exposed film is also possible. Each condition should be run at least in triplicate in order to allow statistical analysis. The two principle alternative procedures for band intensity quantification can be performed with various imaging programs; the second procedure is described here specifically for the freely available software Image J. In either case, rotate the image file so that lanes as exactly vertical as possible (for use with Image J, save as non-compressed TIFF).

(a) Draw a box that barely fits the largest band to be quantified. Copy and paste the box in order to mark each individual band to be quantified. Additionally, mark a background area for each lane, above each band. Then quantify the pixel intensity for each box (it is important that all are the same size), and subtract the background for each band. The disadvantage is that background selection is arbitrary in this method, and Western-blot background is often unequally distributed in the lanes.

(b) Using the box tool, draw vertical rectangle that ideally covers a whole lane (without including any of the neighboring lanes). Move this box to the first lane and mark (Ctrl 1). Move box to the second lane and mark (Ctrl 2), and repeat as necessary (mark the last lane with "Ctrl 3" instead of "Ctrl 2"). A window with intensity profiles for each lane will appear (the horizontal pixel intensities are summed). Using the line tool, draw a base line that follows the background for each profile. Make sure that the peaks corresponding to the bands to be quantified form an area enclosed by this background base line (if necessary, enclose the area with additional lines). Using the wand tool, select each peak area to be quantified in order. After the first selection, a window with quantitative measures of the peak area will appear. Select the numbers, copy, and paste into a spreadsheet program (or store and analyze numbers as desired). This method provides a more accurate background subtraction, even though it involves manual tracing of the base line.

For blots with low background and a good signal to noise ratio, both methods will provide equally good measurements. For more marginal blots, the second method may be the only way to generate any meaningful measurement. Measurements with either method will provide arbitrary "immuno-detection values" (IDV). These IDVs are not truly quantitative, as a twofold difference in IDV does not mean that there is a twofold difference in protein amount (the true difference may be greater or smaller). Even though detection of the output signal (by CCD camera) may be linear over a wide range, the antibody binding in previous steps of the immuno-detection may not be linear. True quantification requires running a standard curve (different known quantities of purified protein) on each blot, in order to assign an actually quantitative value to each IDV (provided that they fall within the range of the standard curve). However, the "raw" IDVs can be used for statistical analysis (when measured at least in triplicate) in order to determine if there is a significant difference between conditions. Many questions can be sufficiently addressed in a meaningful way by this latter "semi-quantitative" approach.

Tips and other considerations. Note that T286 auto-phosphorylation will cause a slight, usually only barely detectable bandshift. Additional T305/306 auto-phosphorylation will cause an obvious bandshift compared to unphosphorylated CaMKII. The full bandshift (corresponding to an apparent molecular weight gain of up to 10 kDa) after complete "burst" auto-phosphorylation is caused by auto-phosphorylation including additional residues. Thus, several bands may appear after auto-phosphorylation even for purified CaMKIIα. Efficient protein transfer onto membranes can be estimated by using pre-stained molecular weight markers. When loading cell extracts, it may be desirable to verify efficient transfer by staining both the gel (with coomassie) and the blot (PonceauS or Coomassie) (staining the blot will also allow to compare the band pattern from Western-analysis with the total protein pattern).

3.5. CaMKII Phosphorylation of Protein

Protein phosphorylation reactions can be performed as in Sect. 3.2, with substituting the peptide substrate for purified protein (or cell extracts), and subsequent analysis after SDS-PAGE. Saturating substrate concentrations can usually not be achieve when using protein, but concentration should be at least above ~0.5 μM. Perform control reactions without added CaMKII (and with CaMKII but without substrate), especially when using cell extracts instead of purified protein as substrate. In this case, also consider increasing the CaMKII concentration (for enhancing the signal over the "background" phosphorylation by other kinases present in the extract).

If a phospho-specific antibody for the substrate site in question is available, reactions can be performed without radioactivity, followed by Western-analysis (essentially as described in Sect. 3.4). Otherwise, autoradiography after SDS-PAGE is the preferred analysis. In this case, it may be beneficial to reduce the reaction volume and increase the relative amount of labeled ATP. Additionally, the molecular weight of the substrate protein needs to be considered, as the signal from auto-phosphorlated CaMKII may interfere with the detection. This potential problem can be reduced by pre-phosphorylating CaMKII (in the absence of radioactive ATP). Autoradiography can be done by film or, preferably, by more linear detection methods, using a PhosphorImager or Storm System (Molecular Dynamics). Quantification of the bands can be done as described in Sect. 3.4.

Tips and other considerations. In most SDS-PAGE systems, ATP runs approximately with the bromophenol blue dye front. In order to keep unincorporated ^{32}P as contained as possible, stop the gel before the dye front runs out, and cut the bottom of the gel away before washing or blotting the rest of the gel. This will also help decrease background from unincorporated ^{32}P. When boiling radioactive samples in loading buffer, make sure to secure tube caps with appropriate clamps (alternatively, puncture the caps to avoid pressure build up).

3.6. CaMKII Binding to GluN2B In Vitro

The following method was designed to determine CaMKII binding to a bacterially expressed GST fusion protein with the cytoplasmic GluN2B C terminus (GST-N2Bc, which includes the GluN2B amino acids 1,120–1,482), but can be adapted for other GST fusion proteins. This method immobilizes the GST fusion protein in anti-GST coated microtiter well (instead of more common immobilization on glutathione-coated beads). In addition to testing for binding, this approach allows convenient access to the bound CaMKII for activity assays or other manipulations. It also allows convenient testing of multiple conditions and replicates in parallel. However, the approach may be less sensitive, and may not be as suitable for detecting low affinity binding interactions.

CaMKII binding to GluN2B requires initial stimulation by Ca^{2+}/CaM (or by T286 auto-phosphorylation), but Ca^{2+}/CaM is not required for maintaining the binding once it is established (35). Thus, our kinase binding buffer contains Ca^{2+}/CaM, while our wash buffers do not. For testing the regulation of CaMKII binding to other proteins, it may be useful to compare wash buffers with and without Ca^{2+}/CaM (as binding in the presence of Ca^{2+}/CaM might be reversed during the washes in the absence of Ca^{2+}/CaM), i.e., in initial experiments, the binding and wash buffers should contain the same inducing agents to be tested.

1. Remove the desired number of wells from the anti-GST coated clear strip plates, then wash each well three times with 200 μl TBS-T.
2. Add the GST-fused protein (purified or in bacterial extract), diluted in TBS-T to each well and allow to bind for 1 h at room temperature. Use an amount of GST-protein pre-determined to saturate binding to the plate to reduce well to well variability; use GST only (or GST fusion with a non-CaMKII-binding protein) as negative control, at least for initial experiments.
3. Wash each well twice with 200 μl TBS-T.
4. Wash each well an additional three times with 200 μl PS-T.
5. Block the wells for at least 30 min with 5% BSA in PS-T, in order to prevent non-specific kinase binding.
6. Remove the blocking solution then add CaMKII (or kinase containing extract) diluted in binding buffer (PS-T, 0.1% BSA, 1 mM $CaCl_2$ and 1 μM CaM) and incubate for 15 min at room temperature. (Note: modify binding buffer to test regulation of the binding as desired; for GluN2B, including a binding reaction in the presence of EGTA instead of Ca^{2+}/CaM is a good negative control.)
7. Wash each well five times with 200 μl kinase wash buffer.
8. Elute the proteins from the well using SDS-PAGE loading buffer and heating to 95°C for 10 min then performing a western blot of the samples.
9. Determine amount of bound CaMKII by Western-analysis of the eluted protein (as described in Sect. 3.4).

Alternatively to step 8, perform CaMKII activity assays or auto-phosphorylation reactions after step 7, essentially as described in Sect. 3.2 or Sect. 3.3 directly in the wells. In this case, rinse twice with 50 mM PIPES, pH 7.2, before the reaction (or with other buffer system chosen for the CaMKII reaction).

Tips and other considerations. For testing new CaMKII interactions, it may be useful to optimize the binding and wash buffer, i.e., lowering their ionic strength (to no less than 50 mM) and/or substituting NaCl with KCl. Generally, higher ionic strength should provide more stringent binding conditions. When using KCl, it should be omitted from the last wash (in order to prevent precipitation of the SDS in the loading buffer as PDS). For the washes, removal of the solution from the wells is achieved most efficiently by a flicking motion followed by tapping the strips onto a paper towel. In order to avoid drying, immediately apply the next solution, ideally with a multi-pipette device.

3.7. Imaging CaMKII Translocation to GluN2B in Cells

Co-transfect non-neuronal adherent mammalian cells (such as HEK 293 or Cos7) with expression vectors for GluN2B (or other protein of interest) and mGFP-CaMKII (wild type or mutants of interest) by the method of choice (we use Ca_2PO_4 or lipofectamine plus). mGFP (monomeric GFP) is an A207K mutant of EGFP that eliminates GFP-dimerization, which is especially important when studying multimeric proteins such as CaMKII (in order to prevent formation of large aggregates). GluN2B will largely target to the ER membrane in HEK 293 cells; when testing interaction with a cytoplasmic protein instead, a targeting tag should be fused in order to generate a localization pattern distinct from individually expressed CaMKII. For transfection, cells should be plated on glass-bottom dishes (or on cover slips) to allow microscopy. If using cover slips, make sure that the focal length of the objective (40–100×) allows live imaging of the cells (i.e., when cover slips are placed into glass-bottom dishes, increasing total thickness of the glass). Optimize the ratio of co-transfected plasmids, using 1:10 mGFP-CaMKII to GluN2B as approximate start point, and include a negative control without GluN2B. Image 16–32 h after transfection on suitable microscope setup. Medium should be replaced with ~1 ml imaging buffer (for a 3 cm glass-bottom dish). To induce CaMKII translocation, create a Ca^{2+} stimulus by adding 111 µl of 100 µM ionomycin in imaging buffer (freshly prepared from ionomycin stock solution and well mixed) for an endconcentration of 10 µM ionomycin. Alternatively, a perfusion system can be used. Take images before and after 1–15 min stimulation. Translocation should become obvious after 1–5 min in >60% of the cells. Imaging (confocal) z-stacks at each time point is advantageous, but not required. If possible, conduct imaging at ~33°C, but room temperature should work. Quantification can be done by determining the percentage of monitored cells that show translocation at the different time points. For additional (or alternative) quantification, correlation of GFP-CaMKII and GluN2B co-localization can be determined in fixed cells after immunocytochemical staining of GluN2B. Fix and stain cells after 3–5 min of ionomycin treatment (or at optimal time as determined by live imaging as described above).

Synaptic translocation of GFP-CaMKII in primary hippocampal neurons in dissociated culture can be determined essentially as described above for non-neuronal cells, but without co-transfection of GluN2B. Ideally, use neurons with well established spine synapse (after ~14 days in culture). Use lipofectamine 2000 (Invitrogen) for transfection (or another method successfully used with these neurons). After transfection, add 100 μM APV (NMDA-receptor inhibitor) to the culture medium, in order to reduce synaptic CaMKII localization triggered by basal neuronal activity. Remove APV prior imaging, for instance by replacing medium with imaging buffer. Ideally, the imaging buffer should be adjusted to the osmolarity of the culture medium (likely ~260 mOsm), as neurons are more sensitive to sudden changes in osmolarity. Stimulate neurons with glutamate or NMDA instead of ionomycin (for instance with 100/10 μM glutamate/glycine; glycine is required as co-agonist for NMDA-type glutamate receptors). The relatively low Mg^{2+} concentration in the imaging buffer allows efficient stimulation of NMDA-receptors, which provides the Ca^{2+} signal. CaMKII translocation should become obvious within 1–2 min. If a perfusion system is used for removing the stimulus, the "wash" buffer should contain low Ca^{2+}/high Mg^{2+} (0.6/5 mM, instead of 1.2/1 mM in the neuronal imaging buffer).

Tips and other considerations. Plasmid preparations should generally be free of bacterial endotoxins; this is even more important for transfection of neurons. Upon ionomycin treatment, CaMKII can form aggregates that lead to formation of clusters even in the absence of GluN2B. Formation of such aggregates before ~12–15 min is an indication that the Hepes buffer has gone bad (as aggregation requires a drop in pH while GluN2B binding does not). Aggregation and GluN2B binding in HEK cells is morphologically distinct: Aggregation leads to sharply defined and mostly round CaMKII clusters, while CaMKII localization by GluN2B binding covers larger areas and typically includes a "string of pearls" pattern around the nucleus (due to ER localization of GluN2B in these cells). If no GFP-CaMKII translocation is observed after ionomycin treatment, consider transfecting a larger amount of GluN2B expression vector. Alternatively, there may be a problem with the ionomycin (gone bad or not in solution). This is likely the case if the ionomycin treatment does not induce obvious morphological changes in the shape of the cells within 5–10 min.

References

1. Schulman H, Greengard P (1978) Stimulation of brain membrane protein phosphorylation by calcium and an endogenous heat-stable protein. Nature 271:478–479
2. Lisman J, Schulman H, Cline H (2002) The molecular basis of CaMKII function in synaptic and behavioural memory. Nat Rev Neurosci 3:175–190
3. Hudmon A, Schulman H (2002) Neuronal CA2+/calmodulin-dependent protein kinase II: the role of structure and autoregulation in cellular function. Annu Rev Biochem 71:473–510
4. Sheng M, Kim MJ (2002) Postsynaptic signaling and plasticity mechanisms. Science 298:776–780
5. Colbran RJ, Brown AM (2004) Calcium/calmodulin-dependent protein kinase II and synaptic plasticity. Curr Opin Neurobiol 14:318–327
6. Coultrap SJ, Vest RS, Ashpole NM, Hudmon A, Bayer KU (2011) CaMKII in cerebral ischemia. Acta Pharmacol Sin 32(7):861–872
7. Derkach V, Barria A, Soderling TR (1999) Ca^{2+}/calmodulin-kinase II enhances channel conductance of alpha-amino-3-hydroxy-5-methyl-4-isoxazolepropionate type glutamate receptors. Proc Natl Acad Sci USA 96:3269–3274
8. Rosenberg OS, Deindl S, Sung RJ, Nairn AC, Kuriyan J (2005) Structure of the autoinhibited kinase domain of CaMKII and SAXS analysis of the holoenzyme. Cell 123:849–860
9. Rosenberg OS, Deindl S, Comolli LR, Hoelz A, Downing KH, Nairn AC, Kuriyan J (2006) Oligomerization states of the association domain and the holoenyzme of Ca/CaM kinase II. FEBS J 273:682–694
10. Tobimatsu T, Fujisawa H (1989) Tissue-specific expression of four types of rat calmodulin-dependent protein kinase II mRNAs. J Biol Chem 264:17907–17912
11. Bayer KU, Lohler J, Schulman H, Harbers K (1999) Developmental expression of the CaM kinase II isoforms: ubiquitous gamma- and delta-CaM kinase II are the early isoforms and most abundant in the developing nervous system. Brain Res Mol Brain Res 70:147–154
12. Bayer KU, Harbers K, Schulman H (1998) alphaKAP is an anchoring protein for a novel CaM kinase II isoform in skeletal muscle. EMBO J 17:5598–5605
13. Urquidi V, Ashcroft SJH (1995) A novel pancreatic [beta]-cell isoform of calcium/calmodulin-dependent protein kinase II ([beta]3 isoform) contains a proline-rich tandem repeat in the association domain. FEBS Lett 358:23–26
14. O'Leary H, Lasda E, Bayer KU (2006) CaMKIIbeta association with the actin cytoskeleton is regulated by alternative splicing. Mol Biol Cell 17:4656–4665
15. Erondu NE, Kennedy MB (1985) Regional distribution of type II Ca^{2+}/calmodulin-dependent protein kinase in rat brain. J Neurosci 5:3270–3277
16. Silva AJ, Paylor R, Wehner JM, Tonegawa S (1992) Impaired spatial learning in alpha-calcium-calmodulin kinase II mutant mice. Science 257:206–211
17. Brocke L, Srinivasan M, Schulman H (1995) Developmental and regional expression of multifunctional Ca^{2+}/calmodulin-dependent protein kinase isoforms in rat brain. J Neurosci 15:6797–6808
18. Hanson PI, Meyer T, Stryer L, Schulman H (1994) Dual role of calmodulin in autophosphorylation of multifunctional CaM kinase may underlie decoding of calcium signals. Neuron 12:943–956
19. De Koninck P, Schulman H (1998) Sensitivity of CaM kinase II to the frequency of Ca^{2+} oscillations. Science 279:227–230
20. Meyer T, Hanson PI, Stryer L, Schulman H (1992) Calmodulin trapping by calcium-calmodulin-dependent protein kinase. Science 256:1199–1202
21. Miller SG, Kennedy MB (1986) Regulation of brain type II Ca^{2+}/calmodulin-dependent protein kinase by autophosphorylation: a Ca^{2+}-triggered molecular switch. Cell 44:861–870
22. Lou LL, Lloyd SJ, Schulman H (1986) Activation of the multifunctional Ca^{2+}/calmodulin-dependent protein kinase by autophosphorylation: ATP modulates production of an autonomous enzyme. Proc Natl Acad Sci USA 83:9497–9501
23. Schworer CM, Colbran RJ, Soderling TR (1986) Reversible generation of a Ca^{2+}-independent form of Ca^{2+}(calmodulin)-dependent protein kinase II by an autophosphorylation mechanism. J Biol Chem 261:8581–8584
24. Coultrap SJ, Buard I, Kulbe JR, Dell'Acqua ML, Bayer KU (2010) CaMKII autonomy is substrate-dependent and further stimulated by Ca^{2+}/calmodulin. J Biol Chem 285: 17930–17937
25. Giese KP, Fedorov NB, Filipkowski RK, Silva AJ (1998) Autophosphorylation at Thr286 of

the alpha calcium-calmodulin kinase II in LTP and learning. Science 279:870–873

26. Buard I, Coultrap SJ, Freund RK, Lee YS, Dell'Acqua ML, Silva AJ, Bayer KU (2010) CaMKII "autonomy" is required for initiating but not for maintaining neuronal long-term information storage. J Neurosci 30:8214–8220
27. Hanson PI, Schulman H (1992) Inhibitory autophosphorylation of multifunctional Ca^{2+}/calmodulin-dependent protein kinase analyzed by site-directed mutagenesis. J Biol Chem 267:17216–17224
28. Colbran RJ (1993) Inactivation of Ca^{2+}/calmodulin-dependent protein kinase II by basal autophosphorylation. J Biol Chem 268: 7163–7170
29. Elgersma Y, Fedorov NB, Ikonen S, Choi ES, Elgersma M, Carvalho OM, Giese KP, Silva AJ (2002) Inhibitory autophosphorylation of CaMKII controls PSD association, plasticity, and learning. Neuron 36:493–505
30. Shen K, Meyer T (1999) Dynamic control of CaMKII translocation and localization in hippocampal neurons by NMDA receptor stimulation. Science 284:162–166
31. Marsden KC, Shemesh A, Bayer KU, Carroll RC (2010) Selective translocation of Ca^{2+}/calmodulin protein kinase IIalpha (CaMKIIalpha) to inhibitory synapses. Proc Natl Acad Sci USA 107:20559–20564
32. Colbran RJ (2004) Targeting of calcium/calmodulin-dependent protein kinase II. Biochem J 378:1–16
33. Merrill MA, Chen Y, Strack S, Hell JW (2005) Activity-driven postsynaptic translocation of CaMKII. Trends Pharmacol Sci 26:645–653
34. Strack S, Colbran RJ (1998) Autophosphorylation-dependent targeting of calcium/calmodulin-dependent protein kinase II by the NR2B subunit of the N-methyl-D-aspartate receptor. J Biol Chem 273: 20689–20692
35. Bayer KU, De Koninck P, Leonard AS, Hell JW, Schulman H (2001) Interaction with the NMDA receptor locks CaMKII in an active conformation. Nature 411:801–805
36. Strack S, McNeill RB, Colbran RJ (2000) Mechanism and regulation of calcium/calmodulin-dependent protein kinase II targeting to the NR2B subunit of the N-methyl-D-aspartate receptor. J Biol Chem 275: 23798–23806
37. Leonard AS, Bayer KU, Merrill MA, Lim IA, Shea MA, Schulman H, Hell JW (2002) Regulation of calcium/calmodulin-dependent protein kinase II docking to N-methyl-D-aspartate receptors by calcium/calmodulin and alpha-actinin. J Biol Chem 277: 48441–48448
38. O'Leary H, Liu WH, Rorabaugh JM, Coultrap SJ, Bayer KU (2011) Nucleotides and phosphorylation bi-directionally modulate Ca^{2+}/calmodulin-dependent protein kinase II (CaMKII) binding to the N-Methyl-D-aspartate (NMDA) receptor subunit GluN2B. J Biol Chem 286:31272–31281
39. Robison AJ, Bartlett RK, Bass MA, Colbran RJ (2005) Differential modulation of Ca^{2+}/calmodulin-dependent protein kinase II activity by regulated interactions with N-methyl-D-aspartate receptor NR2B subunits and alpha-actinin. J Biol Chem 280:39316–39323
40. Bayer KU, LeBel E, McDonald GL, O'Leary H, Schulman H, De Koninck P (2006) Transition from reversible to persistent binding of CaMKII to postsynaptic sites and NR2B. J Neurosci 26:1164–1174
41. Barria A, Malinow R (2005) NMDA receptor subunit composition controls synaptic plasticity by regulating binding to CaMKII. Neuron 48:289–301
42. Zhou Y et al (2007) Interactions between the NR2B receptor and CaMKII modulate synaptic plasticity and spatial learning. J Neurosci 27:13843–13853
43. Tokumitsu H, Chijiwa T, Hagiwara M, Mizutani A, Terasawa M, Hidaka H (1990) KN-62, 1-[N, O-bis(5-isoquinolinesulfonyl)-N-methyl-L-tyrosyl]-4-phenylpiperazi ne, a specific inhibitor of Ca^{2+}/calmodulin-dependent protein kinase II. J Biol Chem 265:4315–4320
44. Sumi M, Kiuchi K, Ishikawa T, Ishii A, Hagiwara M, Nagatsu T, Hidaka H (1991) The newly synthesized selective Ca^{2+}/calmodulin dependent protein kinase II inhibitor KN-93 reduces dopamine contents in PC12h cells. Biochem Biophys Res Commun 181:968–975
45. Enslen H, Sun P, Brickey D, Soderling SH, Klamo E, Soderling TR (1994) Characterization of Ca^{2+}/calmodulin-dependent protein kinase IV. Role in transcriptional regulation. J Biol Chem 269:15520–15527
46. Li G, Hidaka H, Wollheim CB (1992) Inhibition of voltage-gated Ca^{2+} channels and insulin secretion in HIT cells by the Ca^{2+}/calmodulin-dependent protein kinase II inhibitor KN-62: comparison with antagonists of calmodulin and L-type Ca^{2+} channels. Mol Pharmacol 42:489–498

47. Ledoux J, Chartier D, Leblanc N (1999) Inhibitors of calmodulin-dependent protein kinase are nonspecific blockers of voltage-dependent K+ channels in vascular myocytes. J Pharmacol Exp Ther 290:1165–1174
48. Vest RS, O'Leary H, Coultrap SJ, Kindy MS, Bayer KU (2010) Effective post-insult neuroprotection by a novel Ca(2+)/calmodulin-dependent protein kinase II (CaMKII) inhibitor. J Biol Chem 285:20675–20682
49. Hvalby O et al (1994) Specificity of protein kinase inhibitor peptides and induction of long-term potentiation. Proc Natl Acad Sci USA 91:4761–4765
50. Smith MK, Colbran RJ, Soderling TR (1990) Specificities of autoinhibitory domain peptides for four protein kinases. Implications for intact cell studies of protein kinase function. J Biol Chem 265:1837–1840
51. Backs J et al (2009) The delta isoform of CaM kinase II is required for pathological cardiac hypertrophy and remodeling after pressure overload. Proc Natl Acad Sci USA 106:2342–2347
52. Patel R, Holt M, Philipova R, Moss S, Schulman H, Hidaka H, Whitaker M (1999) Calcium/calmodulin-dependent phosphorylation and activation of human Cdc25-C at the G2/M phase transition in HeLa cells. J Biol Chem 274:7958–7968
53. Chen HX, Otmakhov N, Strack S, Colbran RJ, Lisman JE (2001) Is persistent activity of calcium/calmodulin-dependent kinase required for the maintenance of LTP? J Neurophysiol 85:1368–1376
54. Vest RS, Davies KD, O'Leary H, Port JD, Bayer KU (2007) Dual mechanism of a natural CaMKII inhibitor. Mol Biol Cell 18:5024–5033
55. Schwartz JJ, Zhang S (2000) Peptide-mediated cellular delivery. Curr Opin Mol Ther 2:162–167
56. Schwarze SR, Ho A, Vocero-Akbani A, Dowdy SF (1999) In vivo protein transduction: delivery of a biologically active protein into the mouse. Science 285:1569–1572
57. Aarts M et al (2002) Treatment of ischemic brain damage by perturbing NMDA receptor-PSD-95 protein interactions. Science 298:846–850
58. Mayford M, Bach ME, Huang Y-Y, Wang L, Hawkins RD, Kandel ER (1996) Control of memory formation through regulated expression of a CaMKII transgene. Science 274:1678–1683
59. Pi HJ, Otmakhov N, Lemelin D, De Koninck P, Lisman J (2010) Autonomous CaMKII can promote either long-term potentiation or long-term depression, depending on the state of T305/T306 phosphorylation. J Neurosci 30:8704–8709
60. Singla SI, Hudmon A, Goldberg JM, Smith JL, Schulman H (2001) Molecular characterization of calmodulin trapping by calcium/calmodulin-dependent protein kinase II. J Biol Chem 276:29353–29360

Chapter 5

CASK: A Specialized Neuronal Kinase

Konark Mukherjee

Abstract

CASK, a scaffolding protein present in neuronal synapses and other cell junctions, contains a CaM-kinase domain at its N terminus. Due to lack of adequate biochemical data and based on bioinformatics, this domain was classified as a pseudokinase. But new evidence suggests that CASK is a unique kinase which is independent of divalent cofactors. Surprisingly, it is inhibited by many divalent ions which include the essential kinase cofactor magnesium. The inability of CASK to use a cofactor makes it a very slow enzyme; however, this disadvantage of CASK is partially compensated by substrate docking to its scaffolding domains. Indeed, the only characterized substrate to date is Neurexin, which is recruited to CASK via its PDZ domain. Synaptic activity inhibits Neurexin phosphorylation by CASK due to acute influx of divalent ions indicating the divalent ion sensitivity might be a regulatory mechanism. The biological role of this kinase activity remains unclear. It is quite possible that similar to CASK other classified pseudokinases might also turn out to be specialized kinases working in particular physiological niche. Since the discovery of CASK as an active kinase, at least another pseudokinase has turned out to be an active enzyme.

Key words: Pseudokinase, CASK, MAGUK, Neurexin, Phosphorylation, Magnesium, Presynaptic scaffold, Protein kinase

1. Overview

CASK belongs to a family of scaffolding molecules called *M*embrane *A*ssociated *Gu*anylate *K*inase (MAGUK) proteins (1). MAGUK scaffolding proteins evolved in premetazoans by the merging of guanylate kinase (GuK) domain, the Src homology 3 (SH3) domain, and the PSD, Dlg, ZO1 (PDZ) domain (2). Through their PDZ domain, the MAGUKs interact with adhesion molecules and orchestrate the formation of cytosolic signaling complexes in cell–cell junctions. Neuronal MAGUKs like CASK are fundamental to components of synapse development and physiology, including synaptogenesis, synaptic fate determination, and synaptic plasticity (3–5). In addition to the canonical MAGUK domains, CASK harbors a unique CaM-kinase domain and two L27 domains at its

Hideyuki Mukai (ed.), *Protein Kinase Technologies*, Neuromethods, vol. 68,
DOI 10.1007/978-1-61779-824-5_5, © Springer Science+Business Media, LLC 2012

N terminus. Mutations in the CASK gene have been linked to developmental disorders like mental retardation, microcephaly, hypoplasia of the brain stem and cerebellum, and cleft palate (6–8). Mice lacking CASK (CASK knockout) die within 2 h of birth and exhibit cleft palate indicating the indispensability of this molecule in development and survival (9).

CASK is expressed in almost all tissues tested; however, it is highly concentrated in the central nervous system (1). It forms an evolutionary conserved tripartite complex with two other synaptic proteins, MINT and Veli (10, 11). In addition, the complex domain arrangement in CASK ensures a plethora of protein–protein interaction. Some of the described protein interactors include, Calmodulin, Neurexin, Calcium channel, Parkin, Tbr-1, Id-1, Caskin, Dlg, and Liprin (1, 12–16). A direct consequence of this intricate interaction profile of CASK is that it might be involved in multiple cellular processes. At the molecular level, various roles of CASK have been suggested since its 1996 discovery in three experimental animal models, the worm, the fly, and the mouse. Such functions range from synaptogenesis and ion channel trafficking to transcription activation and localization of LET-23 kinase (12, 15, 17–19). In flies, CASK has been also shown to play a role in synaptic plasticity by gating CaM-kinaseII (20, 21) and regulating the phosphorylation status of *ether a go-go potassium* channel (22). Interestingly, CASK was discovered as an active CaM-kinase in the *Drosophila* model (23); experimental data from studies in *Caenorhabditis elegans* also strongly suggest that mutations in CaM-kinase domain of CASK affecting its kinase function render the protein biologically less potent (19). However, based on bioinformatics and due to lack of credible biochemical data, it was dubbed as a pseudokinase. Intriguingly, phosphorylation of CASK, among other posttranslational modifications like Sumoylation (24) and ubiquitination (25), is thought to affect its biological function. Although CDK5 has been demonstrated as a possible kinase for CASK (15), an autophosphorylation event remains a distinct possibility.

Recently, enzymatic activity of CASK has been characterized and quantified. Unlike any known protein kinase, CASK performs phosphotransfer only from free ATP. Divalent ions that chelate ATP, including magnesium, counterintuitively inhibit this kinase (26). Here, I will describe the available evolutionary, structural, biochemical, and biological data that shed light on this enigmatic kinase.

2. Structure and Evolution of CASK

CASK is a multidomain protein that includes a CaM-kinase domain, two L27 domains, a PDZ domain, a SH3 domain and a GuK domain in that order (Fig. 3b and (1, 29)). CASK evolved either simultaneously or just after the evolution of MAGUKs in basal metazoans like *Trichoplax adhaerens* (28). The only MAGUK coding for a CaM-kinase domain, CASK is represented by a single gene in all animals studied. The CaM-kinase domain of CASK falls close to the CaM-kinase II cluster in the CaM-kinase family of the human kinome.

The CaM-kinase domain of CASK folds into a bilobal structure typical of eukaryotic protein kinases. The N-terminal lobe exhibits predominantly β-sheets and the C-terminal lobe mostly exhibits α-helices (Fig. 1a, b and (26)). The relative position of these two lobes in eukaryotic protein kinases results in two extreme conformations—an active "closed conformation" and an inactive "open conformation." Most kinases adopt an inactive "open conformation" in absence of the correct activator; however, the CASK CaM-kinase domain adopts a "closed conformation" constitutively (Fig. 1e–g). Conventional CaM-kinases typically have autoregulatory domains following their catalytic cores. The autoregulatory domain works by either interfering with the nucleotide binding pocket or by acting as a pseudosubstrate (27, 31, 32). Following the canonical kinase domain, CASK also exhibits an autoregulatory domain homologous to that of its closest paralog CaM-kinaseII. The autoregulatory domain of CASK, however, does not hinder the nucleotide binding pocket (26), furthermore, this domain displays a critical substitution (arginine to leucine) compared to CaM-kinaseII which makes it a weak pseudosubstrate (28, 33). Unlike the more distant CaM-kinase I, the structure of CaM-kinase II reveals an intrinsically active kinase domain which is inhibited by a coiled-coil dimeric strut formed by the autoregulatory domain. Therefore, the formation of such a dimer seems to be a structural requirement for autoinhibition in CaM-kinase II family (32). None of the structures solved for CASK CaM-kinase exhibited dimerization of autoregulatory domain (26, 28). The autoregualtory domain seems to be readily available Ca^{2+}/calmodulin binding and it has been reported that CASK can bind to calmodulin, but the consequence of this interaction is not known.

Overall, the CaM-kinase domain of CASK adopts a structure similar to activated protein kinases, indicating that it is constitutively active. Both the substrate binding site and the nucleotide pockets are well formed and readily accessible to their respective ligands (Fig. 1 (26)). These features allow a steady phosphorylation by CASK in absence of any specific signal, thereby mitigating the need of high burst-phase enzymatic velocity. Moreover, CASK is anchored to its substrate via its scaffolding domains like PDZ domain. Such interaction not only raises the reaction rate by increasing local concentration of substrate, but also helps by sustaining enzyme-substrate stoichiometry. Despite the overall homology with CaM-kinase II, CASK and its ortholog exhibit critical changes in the nucleotide binding pocket (Fig. 2). Since the nucleotide

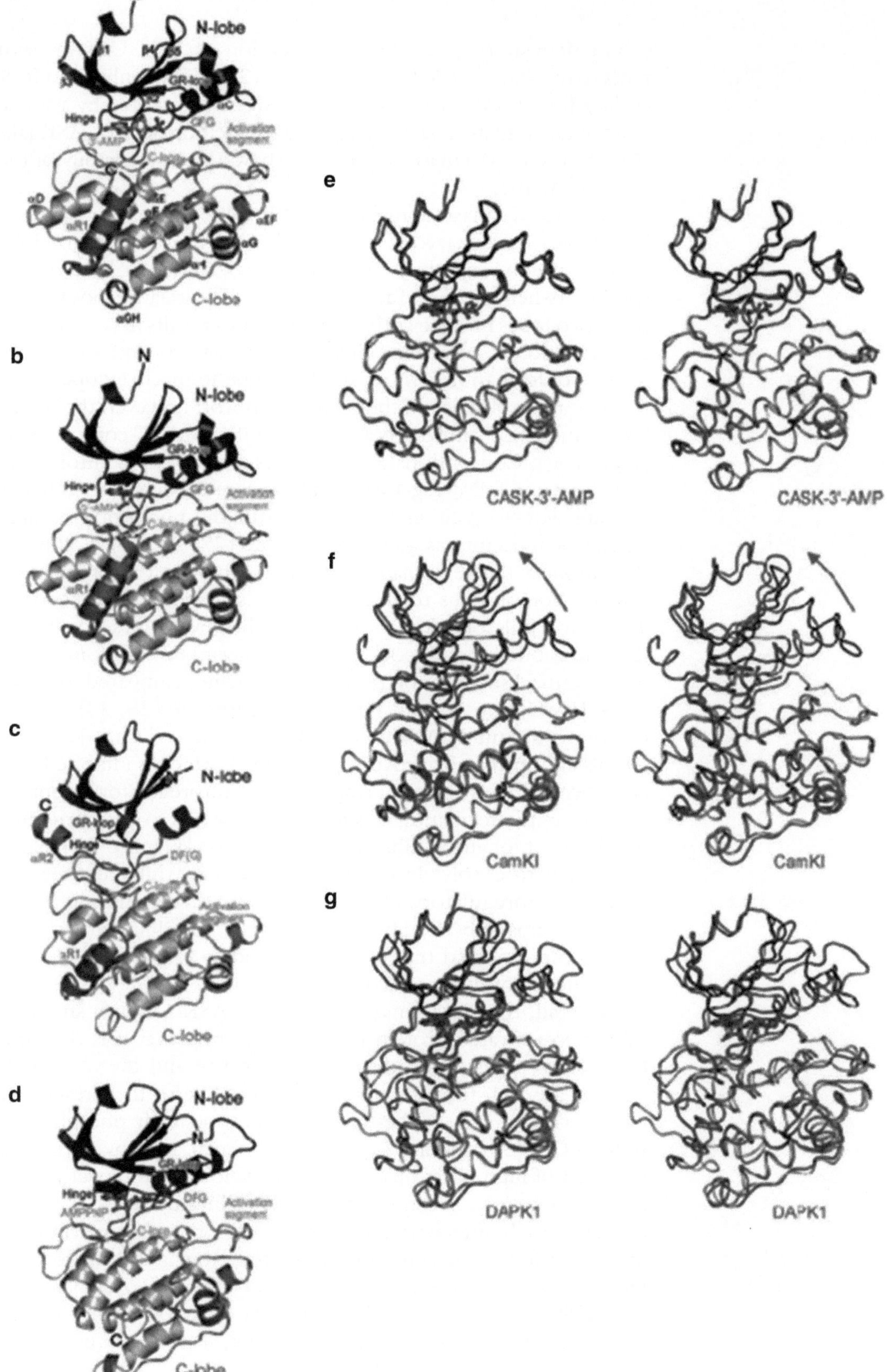
a
N
N-lobe
Hinge
GFG
Activation segment
αD
αR1
αEF
αG
C-lobe
αGH
b
N
N-lobe
Hinge
GFG
Activation segment
αR1
C-lobe
c
N-lobe
C
GR-loop
Hinge
αR2
DF(G)
Activation segment
αR1
C-lobe
d
N-lobe
N
Hinge
DFG
Activation segment
AMPPNP
C
C-lobe
e
CASK-3'-AMP
CASK-3'-AMP
f
CamKI
CamKI
g
DAPK1
DAPK1

binding pocket of the most primitive CASK ortholog in Trichoplax bears a strong similarity with conventional CaM-kinases, one can speculate that ancient CASK was a synthesis of a conventional, magnesium-coordinating fast CaM-kinase, fused with domains of a MAGUK protein (28). Presumably, this fusion event severely limited the movement of this enzyme by recruiting it to the membrane via its interaction through the PDZ domain to cell adhesion molecules. Under these circumstances the substrate profile of CASK would be limited to its immediate interactors. It is tempting to speculate that in such a scenario, the evolutionary pressure to maintain fast kinetics by effectively recruiting an essential divalent cofactor is removed.

The mechanism of the notable negative effect of divalent ions on CASK can be narrowed down to only four residues lining its nucleotide binding pocket (Fig. 2). The residue involved in this negative regulation includes the two known invariant magnesium-coordinating amino acids, the aspartate of the magnesium binding loop and the asparagine of the catalytic loop, which are substituted in CASK by a glycine and cysteine, respectively. Two additional changes are an alanine to proline substitution in the glycine-rich loop, and a critical glutamate to histidine substitution in the catalytic loop (Fig. 2). The histidine in the catalytic loop of CASK serves as the divalent ion sensor and inhibits CASK-kinase activity in the presence of these ions (28). This histidine protrudes into the nucleotide binding pocket close to the putative coordination sphere of magnesium. Mutating this histidine to the canonical glutamate creates an alternate glutamate–arginine bond with the regulatory domain similar to that of CaM-kinase II. These substitutions begun very early on in animal evolution, since CASK orthologs of relatively simple animals like cnidarians (sea anemone) also harbor

Fig. 1. *Overall structure.* (**a–d**) Ribbon plots depicting the overall folds of the CASK CaM-kinase domain in complex with 3′-AMP (orthorhombic form), the CASK CaM-kinase domain in complex with 5′-AMP (triclinic form), rat CaM-kinase I (PDB ID 1A06; (27)), and DAPK1 in complex with AMPPNP (PDB ID 1IG1; (30)), respectively. The structures are shown in the same orientation with the N-terminal lobes on top and the C-terminal lobes at the bottom. The N-terminal lobes are in *dark gray*, the C-terminal lobes in *light gray*. N- and C-termini are labeled. Specific elements are *color coded* and labeled: portion of the glycine-rich loop (GR-loop)—*brown*; catalytic loop (C-loop)—*yellow*; D/GFG of the magnesium binding loop—*orange* (the third residue is disordered in the CaM-kinase I structure); activation segment—*green*; and C-terminal Ca^{2+}/calmodulin binding segment (Cam segment)—*red*. The region between α R1 and α R2 folds into the ATP binding cleft between N- and C-terminal lobe in CaM-kinase I and interacts with the tip of the GR-loop. The same element is partially disordered and does not interact with the ATP binding pocket in the CASK CaM-kinase domain. Bound nucleotides in A (3′-AMP), B (5′-AMP), and C (AMPPNP) are shown in ball-and-sticks. (**e–g**) Overlays of the CASK CaM-kinase domain in complex with 3′-AMP (orthorhombic form), of CaM-kinase I and of DAPK1, respectively, on the structure of the CASK CaM-kinase domain in complex with 5-AMP (stereo plots). The CaM-kinase domain in complex with 5′-AMP is *color-coded* in the same fashion as in *panel B*. The overlaid molecules are in *purple*. *Arrows* in the CaM-kinase I comparison (*panel F*) indicate a relative movement of the CaM-kinase I N-terminal lobe with respect to the C-terminal lobe as a consequence of the positioning of the autoinhibitory Ca^{2+}/calmodulin binding segment (26).

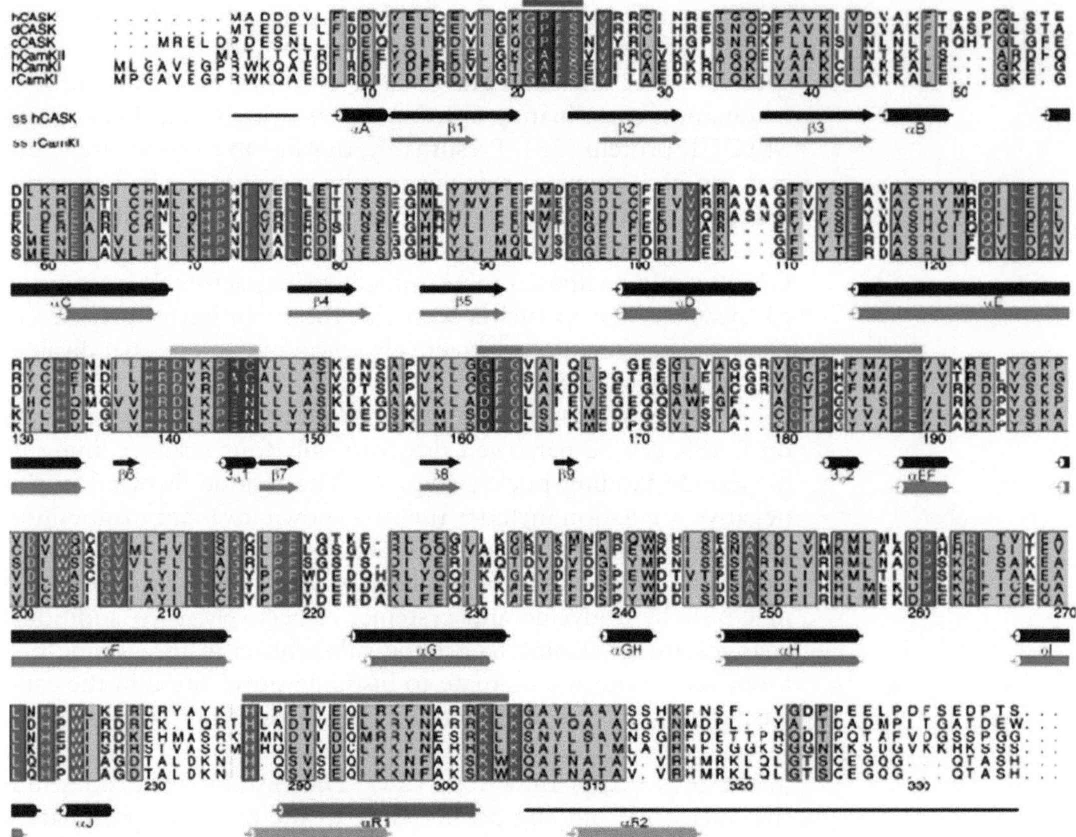

Fig. 2. *Sequence alignment*. Sequence alignment of CASK CaM-kinase domains from human (hCASK), *Drosophila* (dCASK) and *C. elegans* (cCASK) with human CaM-kinase II (hCamKII), human CaM-kinase I (hCmaKI) and rat CaM-kinase I (rCamKI). Sequence numbering below the alignment refers to hCASK. Identical residues are *colored red*, conserved positions are *yellow*. The three residues in *blue* (P22 of the glycine-rich loop, H145 and C146 of the catalytic loop, and G162 of the magnesium binding loop) line the ATP binding pocket and are notably different from corresponding residues in other Cam-kinases. *Black symbols* below the alignment designate secondary structure elements found in hCASK ("ss (hCASK)"), *gray symbols* indicate secondary structure elements found in rCaM-kinase I (PDB ID 1A06; (27); "ss (rCamKI)"). The secondary structure elements are labeled according to kinase nomenclature. α-helices in the C-terminal Ca^{2+}/calmodulin binding tail of hCASK and rCaM-kinase I (α R1 and α R2) are depicted in *red* and *pink*, respectively. Only R1 is seen in the CASK CaM-kinase domain, the region encompassing α R2 is disordered in the structures. *Straight lines* below the alignment denote regions that are disordered in the CaM-kinase domain structures. *Colored bars* above the alignment denote portions of the glycine-rich loop (*brown*), the catalytic loop (*yellow*), the magnesium binding loop (*orange*), and the activation segment (*green*) (26).

the critical glutamate to histidine substitution suggesting an inhibitory effect of divalent ion on enzymatic activity of these CASK orthologs (Fig. 3a). Excluding ecdysozoans (molting animals), which curiously show differences in the four residues involved in the divalent ion sensing, these four residues are substituted in all complex multicellular animals (Fig. 3a and (28)). In particular, the insect CASK sequence suggests that it should not be inhibited by divalent ions. The functional relevance for this evolutionary differences has not been tested. It is quite possible that CASK might have a different function in such organisms.

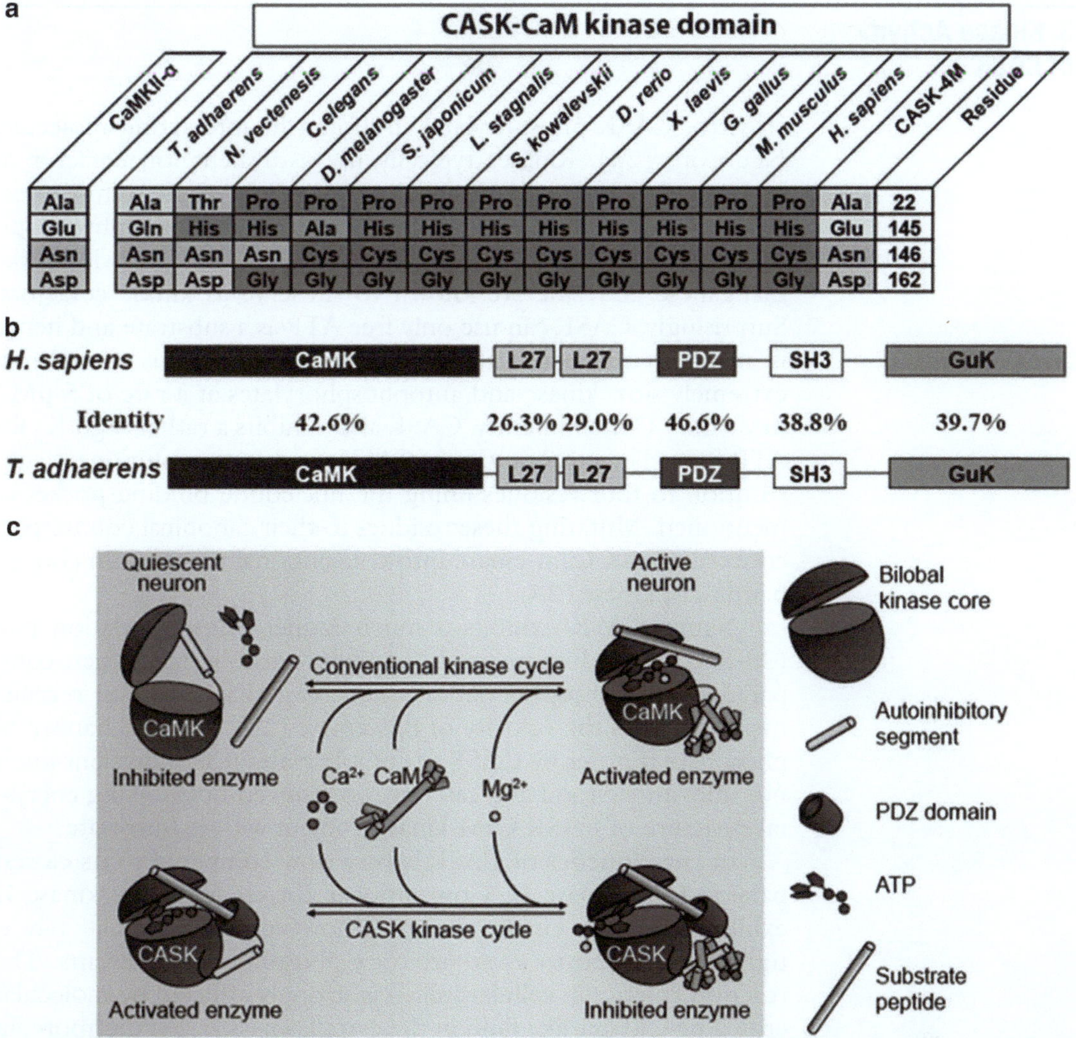

CaMKII-α	T. adhaerens	N. vectenesis	C.elegans	D. melanogaster	S. japonicum	L. stagnalis	S. kowalevskii	D. rerio	X. laevis	G. gallus	M. musculus	H. sapiens	CASK-4M	Residue
Ala	Ala	Thr	Pro	Pro	Pro	Pro	Pro	Pro	Pro	Pro	Pro	Pro	Ala	22
Glu	Gln	His	His	Ala	His	His	His	His	His	His	His	His	Glu	145
Asn	Asn	Asn	Asn	Cys	Cys	Cys	Cys	Cys	Cys	Cys	Cys	Cys	Asn	146
Asp	Asp	Asp	Gly	Gly	Gly	Gly	Gly	Gly	Gly	Gly	Gly	Gly	Asp	162

Fig. 3. *CASK evolution*. (**a**) Evolutionary changes in the nucleotide-binding pocket of CASK CaM-kinase domain. CASK CaM-kinase domain sequences from various animal species were aligned, and the residues corresponding to those CaM-kinases were identified (Fig. 2) and are shown. Corresponding human CaM-kinase IIα residues are shown on the *left* for comparison. (**b**) Sequence conservation (identities) of CASK domains between human and placozoan CASK (from *Trichoplax adhaerens*). (**c**) Model comparing CASK and CaM-kinase II catalytic cycles. Typically, CaM-kinases are held in an autoinhibited conformation by the autoregulatory domain (*yellow*) with an open, inactive nucleotide binding cleft. Upon binding of Ca^{2+} (*purple*)-CaM (*green*), this autoinhibition is relieved and the enzyme attains an active closed conformation amenable to Mg^{2+} (*yellow*)-ATP (*blue*) binding and substrate binding CASK CaM-kinase domain, on other hand, constitutively binds ATP, and is regulated by the recruitment of its substrates through the MAGUK scaffolding domains, especially the PDZ-domain (28).

Indeed, the major phenotype in CASK null *C. elegans* is in the formation of the vulva, and is not neuronal in nature.

High-resolution atomic structures for all domains of CASK are available (26, 34, 35). Future work aiming at acquiring the structure of the full molecule displaying the structure of the entire protein and the domain arrangement in three dimensions will be critical to understand the function of this protein.

3. Kinase Activity of CASK

In vitro, CASK phosphrylates itself and its interacting molecule, Neurexins (26). Kinases typically use a divalent ion cofactor to potentiate their enzymatic activity (36). Only magnesium is present in cytosol in high-enough concentration to act as a physiological cofactor (37). However, in vitro other divalent ions like manganese and zinc are known to subserve as kinase cofactors. Surprisingly, CASK can use only free ATP as a substrate and hence is unable to utilize a divalent ion as cofactor. CASK is naturally an extremely slow kinase and autophosphorylates at a rate of 5 μM/mM/min. Concomitantly, CASK also exhibits a rather high K_m for ATP (26). These properties of CASK are due to evolutionary substitution to four residues lining the nucleotide binding pocket as mentioned. Mutating these residues to their canonical counterpart converts CASK Cam-kinase into conventional magnesium coordinating fast kinase (28).

Native CASK exhibits a much higher phosphorylation rate (30-fold) toward complexed Neurexin cytosolic tails when compared to autophosphorylation, indicating that substrate recruitment increases the velocity of this kinase (26). The mechanism of phosphate transfer by CASK in complete absence of divalent ion is not fully understood and can only be resolved by a solving cocrystal structure of CASK CaM-kinase domain with its substrate.

In vitro kinetics of CASK is very slow compared to its closest paralog CaM-kinase II. Conventional kinases like CaM-kinase II exhibit very high kinetics in vitro; however, their cellular rate is tightly controlled to avoid arbitrary phosphorylation events. The reaction kinetics in cellular milieu is strongly affected by molecular crowding and gel-like fluid in the cytoplasm (38). Furthermore, in cellular milieu, enzymes are spatially restricted by their intracellular localization and temporally restricted by availability of substrate and activators. Most of the known eukaryotic protein kinases have evolved their own mechanism of controlling their activity (39). Since MAGUKs, like CASK, are tightly anchored to the membrane, their movement is severely restricted, and thus any enzymatic activity associated with such molecules will be slowed down. Lack of evolutionary pressure to maintain fast kinetics can then allow mutations in otherwise invariant residues in these enzymes, thereby either slowing their catalysis or completely abolishing it. Indeed, most of the GuK domains in MAGUK proteins harbor such changes, making them either slow or inactive (40, 41). CASK evolved as a fusion between a conventional magnesium binding constitutively active kinase and a MAGUK protein. The regulation in such an enzyme would be spatial confinement, reducing the overall kinetics of the enzyme. Subsequently, CASK would have lost the need for fast catalysis and magnesium coordination. Despite its slow rate in vitro, CASK cotransfection significantly increases steady-state phosphorylation of Neurexin, both in hetrologous cells as well as in cultured cortical neurons (26). Importantly, the steady-state phosphorylation of Neurexin is not enhanced any further despite including four mutations to convert CASK into a magnesium-coordinating kinase (28). This indicates that CASK has been evolutionarily optimized to serve as a Neurexin kinase.

Interestingly, CASK activity is inhibited by divalent ions which reduce the availability of free ATP in the milieu (8, 26). This could act as a second mechanism by which the enzymatic activity of CASK can be controlled in excitable tissues like neurons. In neurons, divalent ion fluxes like calcium and magnesium can limit the supply of free ATP to CASK, thereby inhibiting its function (Fig. 3c). In neuronal cultures, Neurexin phosphorylation by CASK is indeed negatively affected by network activity (Fig. 3c and (26)). However, other factors and regulators might also be involved in such a negative regulation, and this will require further investigation. Besides regulation, further work is required to uncover additional substrates of CASK and to study the downstream signaling of CASK-mediated phosphorylation events.

4. Biological Function of CASK

Though CASK is an essential protein in vertebrates, its molecular function remains elusive. Investigation into its biological function has suggested a variety of different possibilities, indicating that it might have multiple functions which are supported by its ubiquitous distribution and multidomain structure. CASK was discovered in mammals as an interactor of the synaptic adhesion molecule Neurexin (1); moreover, it can interact with the active zone protein liprin, indicating that it might play a role in synaptogenesis (15). CASK can also interact with calcium channels and might be important for recruiting calcium channels at presynapses (17). However, studies in CASK-knockout mice showed neither a defect in synapse formation nor defects in the calcium currents (9). In mammals, CASK has also been shown to participate in intracellular trafficking and polarization of cells (42). However, studies in knockouts did not support such interpretations either (43). One of the reasons for this discrepancy could be functional redundancy with other MAGUKs. Mice lacking CASK die within 2 h of birth. At birth, the gross structure of the brains of CASK knockout mice appeared to be comparable to that of wild-type mice. However, 80% of the knockout mice exhibited cleft palate, indicating that CASK is necessary for proper development.

Furthermore, this data indirectly indicates that the cleft-palate per se is not the cause of neonatal lethality in mice. Similar to the Neurexin 1, 2, and 3 triple knockout mice, CASK-null mice exhibited a deficit in ventilation. It is not clear if the reduced ventilation is the immediate reason of the lethality. Neuronal cultures from CASK-null mice form normal number of synapses, all of which are structurally normal. Evoked inhibitory and excitatory synaptic responses in these cultures also appear to be normal. The only obvious synaptic phenotype is an alteration in the frequency of inhibitory and excitatory miniature currents (caused by spontaneous vesicle fusion events) frequency (9). The mechanism of this phenotype is not understood. Furthermore, thalamus from the CASK null brain exhibited a higher number of apoptotic cells. Thus, CASK might play a critical role in maintenance and survival of neurons (9).

CASK was identified in *Drosophila* as a possible CaM-kinase II isoform due to a frame shift mutation during cloning leading to loss of PDZ, SH3, and GuK domains. The authors found that, similar to other CaM-kinases, expression of CASK increased phosphorylation of transcription factors and, consequently, gene expression. However, the rate of this phosphorylation was much lower than that which was mediated by CaM-kinase II. Deleting the CASK gene from flies affects their locomotor behavior and activity (23, 44). In part, this function of CASK seems to be mediated by its interaction with Neurexin (45). In most animals, the CASK gene makes several alternate transcripts, some of these transcripts lack the CaM-kinase domain. The CASK gene in invertebrates, like *Drosophila* and *C. elegans*, is also capable of producing a protein with alternate N-terminal domain instead of the standard CaM-kinase domain. It is pertinent to mention that the product of the CASK gene encoding the CaM-kinase domain specifically is required for maintenance of the appropriate level of locomotor activity in flies (44). Together, the available data indicate that the CaM-kinase domain of *Drosophila* CASK might exhibit a slow kinase activity, and might be important for its biological activity.

CASK mutant *C. elegans* exhibits a vulval phenotype due to improper localization of the Let23 tyrosine kinase. Rescue of this phenotype was near complete (92%) with either the wild-type CASK transgene or the CASK transgene bearing point mutations in GuK domain. Empirical mutations aimed at lowering the kinase activity of CASK, however, reduced its ability to rescue significantly (77 and 49%) (19). Despite the evidence of signaling pathways downstream of CASK CaM-kinase, its catalytic function has not been worked out in the invertebrate models. Rescue experiments with the mutant form of CASK that is unable to coordinate ATP would be valuable.

5. Conclusion

CASK evolved early in animal evolution by fusion of a CaM-kinase domain with a MAGUK protein. It is unique among kinases, because it does not utilize a divalent cofactor. The biological effect of the kinase activity has yet to be established. Like most other MAGUKs, CASK has continually evolved at a molecular level. The CASK gene seems to have undergone a duplication event, followed by a loss of the CaM-kinase domain, giving rise to the MPP family in vertebrates (46). Therefore, unlike other MAGUKs, CASK itself has a single isoform in the mammalian genome. This might be one of the reasons that CASK is absolutely essential for survival. CASK seems to have acquired a more prominent function in higher animals, since deleting this gene in invertebrates does not compromise their survival. Mutations in CASK in humans are either lethal or may cause severe mental retardation. Despite detailed structural and biochemical understanding of individual domains of CASK, its biological function remains an enigma. One of the compounding factors is the very diverse phenotype obtained in different animal models. A major challenge ahead in the field would be to connect these various findings, in order to gain a comprehensive understanding of CASK function. That CASK encodes a protein kinase, regulated by synaptic activity together with a transcription factor, may put it in the center of activity-dependent neuronal development and survival. Further experiments in animal models, together with careful biochemical analysis, would shed light on the molecular function of CASK in health and disease.

References

1. Hata Y, Butz S, Sudhof TC (1996) CASK: a novel dlg/PSD95 homolog with an N-terminal calmodulin-dependent protein kinase domain identified by interaction with neurexins. J Neurosci 16:2488–2494
2. te Velthuis AJ, Admiraal JF, Bagowski CP (2007) Molecular evolution of the MAGUK family in metazoan genomes. BMC Evol Biol 7:129
3. Hsueh YP (2006) The role of the MAGUK protein CASK in neural development and synaptic function. Curr Med Chem 13: 1915–1927
4. Cuthbert PC, Stanford LE, Coba MP, Ainge JA, Fink AE, Opazo P, Delgado JY, Komiyama NH, O'Dell TJ, Grant SG (2007) Synapse-associated protein 102/dlgh3 couples the NMDA receptor to specific plasticity pathways and learning strategies. J Neurosci 27: 2673–2682
5. Mendoza-Topaz C, Urra F, Barria R, Albornoz V, Ugalde D, Thomas U, Gundelfinger ED, Delgado R, Kukuljan M, Sanxaridis PD, Tsunoda S, Ceriani MF, Budnik V, Sierralta J (2008) DLGS97/SAP97 is developmentally upregulated and is required for complex adult behaviors and synapse morphology and function. J Neurosci 28:304–314
6. Najm J, Horn D, Wimplinger I, Golden JA, Chizhikov VV, Sudi J, Christian SL, Ullmann R, Kuechler A, Haas CA, Flubacher A, Charnas LR, Uyanik G, Frank U, Klopocki E, Dobyns WB, Kutsche K (2008) Mutations of CASK cause an X-linked brain malformation phenotype with microcephaly and hypoplasia of the brainstem and cerebellum. Nat Genet 40(90):1065–1067
7. Froyen G, Van Esch H, Bauters M, Hollanders K, Frints SG, Vermeesch JR, Devriendt K, Fryns JP, Marynen P (2007) Detection of

genomic copy number changes in patients with idiopathic mental retardation by high-resolution X-array-CGH: important role for increased gene dosage of XLMR genes. Hum Mutat 28:1034–1042

8. Piluso G, D'Amico F, Saccone V, Bismuto E, Rotundo IL, Di Domenico M, Aurino S, Schwartz CE, Neri G, Nigro V (2009) A missense mutation in CASK causes FG syndrome in an Italian family. Am J Hum Genet 84: 162–177
9. Atasoy D, Schoch S, Ho A, Nadasy KA, Liu X, Zhang W, Mukherjee K, Nosyreva ED, Fernandez-Chacon R, Missler M, Kavalali ET, Sudhof TC (2007) Deletion of CASK in mice is lethal and impairs synaptic function. Proc Natl Acad Sci USA 104:2525–2530
10. Butz S, Okamoto M, Sudhof TC (1998) A tripartite protein complex with the potential to couple synaptic vesicle exocytosis to cell adhesion in brain. Cell 94:773–782
11. Kaech SM, Whitfield CW, Kim SK (1998) The LIN-2/LIN-7/LIN-10 complex mediates basolateral membrane localization of the *C. elegans* EGF receptor LET-23 in vulval epithelial cells. Cell 94:761–771
12. Hsueh YP, Wang TF, Yang FC, Sheng M (2000) Nuclear translocation and transcription regulation by the membrane-associated guanylate kinase CASK/LIN-2. Nature 404: 298–302
13. Qi J, Su Y, Sun R, Zhang F, Luo X, Yang Z (2005) CASK inhibits ECV304 cell growth and interacts with Id1. Biochem Biophys Res Commun 328:517–521
14. Sanford JL, Mays TA, Rafael-Fortney JA (2004) CASK and Dlg form a PDZ protein complex at the mammalian neuromuscular junction. Muscle Nerve 30:164–171
15. Samuels BA, Hsueh YP, Shu T, Liang H, Tseng HC, Hong CJ, Su SC, Volker J, Neve RL, Yue DT, Tsai LH (2007) Cdk5 promotes synaptogenesis by regulating the subcellular distribution of the MAGUK family member CASK. Neuron 56:823–837
16. Tabuchi K, Biederer T, Butz S, Sudhof TC (2002) CASK participates in alternative tripartite complexes in which Mint 1 competes for binding with caskin 1, a novel CASK-binding protein. J Neurosci 22:4264–4273
17. Maximov A, Sudhof TC, Bezprozvanny I (1999) Association of neuronal calcium channels with modular adaptor proteins. J Biol Chem 274:24453–24456
18. Maximov A, Bezprozvanny I (2002) Synaptic targeting of N-type calcium channels in hippocampal neurons. J Neurosci 22:6939–6952
19. Hoskins R, Hajnal AF, Harp SA, Kim SK (1996) The *C. elegans* vulval induction gene lin-2 encodes a member of the MAGUK family of cell junction proteins. Development 122:97–111
20. Lu CS, Hodge JJ, Mehren J, Sun XX, Griffith LC (2003) Regulation of the Ca^{2+}/CaM-responsive pool of CaMKII by scaffold-dependent autophosphorylation. Neuron 40: 1185–1197
21. Hodge JJ, Mullasseril P, Griffith LC (2006) Activity-dependent gating of CaMKII autonomous activity by Drosophila CASK. Neuron 51:327–337
22. Marble DD, Hegle AP, Snyder ED II, Dimitratos S, Bryant PJ, Wilson GF (2005) Camguk/CASK enhances Ether-a-go-go potassium current by a phosphorylation-dependent mechanism. J Neurosci 25: 4898–4907
23. Martin JR, Ollo R (1996) A new Drosophila Ca^{2+}/calmodulin-dependent protein kinase (Caki) is localized in the central nervous system and implicated in walking speed. EMBO J 15:1865–1876
24. Chao HW, Hong CJ, Huang TN, Lin YL, Hsueh YP (2008) SUMOylation of the MAGUK protein CASK regulates dendritic spinogenesis. J Cell Biol 182:141–155
25. Sun Q, Kelly GM (2010) Post-translational modification of CASK leads to its proteasome-dependent degradation. Int J Biochem Cell Biol 42:90–97
26. Mukherjee K, Sharma M, Urlaub H, Bourenkov GP, Jahn R, Sudhof TC, Wahl MC (2008) CASK Functions as a Mg^{2+}-independent neurexin kinase. Cell 133:328–339
27. Goldberg J, Nairn AC, Kuriyan J (1996) Structural basis for the autoinhibition of calcium/calmodulin-dependent protein kinase I. Cell 84:875–887
28. Mukherjee K, Sharma M, Jahn R, Wahl MC, Sudhof TC (2010) Evolution of CASK into a Mg^{2+}-sensitive kinase. Sci Signal 3:ra33
29. Dimitratos SD, Woods DF, Bryant PJ (1997) Camguk, Lin-2, and CASK: novel membrane-associated guanylate kinase homologs that also contain CaM-kinase domains. Mech Dev 63:127–130
30. Tereshko V, Teplova M, Brunzelle J, Watterson DM, Egli M (2001) Crystal structures of the catalytic domain of human protein kinase associated with apoptosis and tumor suppression. Nat Struct Biol 8:899–907
31. Colbran RJ, Smith MK, Schworer CM, Fong YL, Soderling TR (1989) Regulatory domain of calcium/calmodulin-dependent protein

kinase II. Mechanism of inhibition and regulation by phosphorylation. J Biol Chem 264: 4800–4804

32. Rosenberg OS, Deindl S, Sung RJ, Nairn AC, Kuriyan J (2005) Structure of the autoinhibited kinase domain of CaMKII and SAXS analysis of the holoenzyme. Cell 123: 849–860
33. Fong YL, Soderling TR (1990) Studies on the regulatory domain of Ca^{2+}/calmodulin-dependent protein kinase II. Functional analyses of arginine 283 using synthetic inhibitory peptides and site-directed mutagenesis of the alpha subunit. J Biol Chem 265:11091–11097
34. Daniels DL, Cohen AR, Anderson JM, Brunger AT (1998) Crystal structure of the hCASK PDZ domain reveals the structural basis of class II PDZ domain target recognition. Nat Struct Biol 5:317–325
35. Li Y, Spangenberg O, Paarmann I, Konrad M, Lavie A (2002) Structural basis for nucleotide-dependent regulation of membrane-associated guanylate kinase-like domains. J Biol Chem 277:4159–4165
36. Adams JA (2001) Kinetic and catalytic mechanisms of protein kinases. Chem Rev 101: 2271–2290
37. Waas WF, Rainey MA, Szafranska AE, Cox K, Dalby KN (2004) A kinetic approach towards understanding substrate interactions and the catalytic mechanism of the serine/threonine protein kinase ERK2: identifying a potential regulatory role for divalent magnesium. Biochim Biophys Acta 1697:81–87
38. Schnell S, Turner TE (2004) Reaction kinetics in intracellular environments with macromolecular crowding: simulations and rate laws. Prog Biophys Mol Biol 85:235–260
39. Huse M, Kuriyan J (2002) The conformational plasticity of protein kinases. Cell 109:275–282
40. Dimitratos SD, Woods DF, Stathakis DG, Bryant PJ (1999) Signaling pathways are focused at specialized regions of the plasma membrane by scaffolding proteins of the MAGUK family. Bioessays 21:912–921
41. Olsen O, Bredt DS (2003) Functional analysis of the nucleotide binding domain of membrane-associated guanylate kinases. J Biol Chem 278:6873–6878
42. Borg JP, Straight SW, Kaech SM, de Taddeo-Borg M, Kroon DE, Karnak D, Turner RS, Kim SK, Margolis B (1998) Identification of an evolutionarily conserved heterotrimeric protein complex involved in protein targeting. J Biol Chem 273:31633–31636
43. Lozovatsky L, Abayasekara N, Piawah S, Walther Z (2009) CASK deletion in intestinal epithelia causes mislocalization of LIN7C and the DLG1/Scrib polarity complex without affecting cell polarity. Mol Biol Cell 20: 4489–4499
44. Slawson JB, Kuklin EA, Ejima A, Mukherjee K, Ostrovsky L, Griffith LC (2011) Central regulation of locomotor behavior of *Drosophila melanogaster* depends on a CASK isoform containing CaMK-like and L27 domains. Genetics 187:171–184
45. Sun M, Liu L, Zeng X, Xu M, Fang M, Xie W (2009) Genetic interaction between Neurexin and CAKI/CMG is important for synaptic function in Drosophila neuromuscular junction. Neurosci Res 64:362–371
46. de Mendoza A, Suga H, Ruiz-Trillo I (2010) Evolution of the MAGUK protein gene family in premetazoan lineages. BMC Evol Biol 10:93

Chapter 6

Cyclin-Dependent Kinase 5 (Cdk5): Preparation and Measurement of Kinase Activity

Seiji Minegishi, Taro Saito, and Shin-ichi Hisanaga

Abstract

Cyclin-dependent kinase 5 (Cdk5) is a versatile protein kinase that plays a role in a variety of neuronal activities including neuronal migration during brain development, synaptic activities in mature neurons, and neuronal cell death in neurodegenerative diseases. However, the functions of Cdk5 in brains have not been elucidated fully and its extraneuronal functions are being explored. It is important to identify the functional substrates of Cdk5 and to elucidate the mechanism regulating its activity. Cdk5 is a catalytic subunit of an active kinase complex with the regulatory subunit, p35 or p39. Because the kinase activity of Cdk5 is regulated mainly by synthesis and degradation of the regulatory subunits, care is needed when detecting and measuring its kinase activity. We describe here how to prepare kinase-active Cdk5, to measure its kinase activity, and to distinguish Cdk5 phosphorylation from other types of kinase-dependent phosphorylation.

Key words: Cdk5, p35, p25, p39, Protein kinase, Neuronal migration, Neurite outgrowth, Synaptic activity, Alzheimer disease

1. Introduction

Cyclin-dependent kinase 5 (Cdk5) is a member of the cyclin-dependent kinase family. Cdk5 is a catalytic subunit of the kinase-active heterodimeric complex of Cdk5 and its regulatory subunit. As in other Cdks, Cdk5 alone does not display kinase activity and requires a regulatory subunit for kinase activity. p35 and p39 are the regulatory subunits of Cdk5. In contrast to other Cdks, which are activated in proliferating cells, and although Cdk5 is expressed in many cells and tissues, Cdk5 activity is detected predominantly in postmitotic neurons because p35 and p39 are expressed mainly in neurons (1–3). Cdk5 plays a role in a variety of neuronal functions,

Hideyuki Mukai (ed.), *Protein Kinase Technologies*, Neuromethods, vol. 68,
DOI 10.1007/978-1-61779-824-5_6, © Springer Science+Business Media, LLC 2012

including neuronal migration during brain development, synaptic activities in mature neurons, and neuronal cell death in neurodegenerative diseases (2–6). Cyclin I was recently reported to bind to and activate Cdk5 (7), but the activation and function of Cdk5 have not been investigated thoroughly.

The presence of Cdc2-like kinase in mammalian brains was suggested by the observed phosphorylation of neuron-specific neurofilament-H (NF-H) and -M (NF-M) subunits by Cdc2-cyclin B kinase (8). This protein kinase was purified from bovine brain as tau protein kinase II (TPKII) (9) or neuron-cdc2-like kinase (nclk) (10), proline-directed protein kinase (PDPK) from rat brain (11), and NF-H kinase from porcine brain (12). The kinase was named Cdk5 according to its sequence similarity to Cdc2 (Cdkl) and Cdk2 (13, 14). Cdk5 was purified as a complex with p25, the C-terminal fragment of p35 (15–17). During purification, p35 is cleaved to p25 by calpain, a cytoplasmic Ca^{2+}-activated cysteine protease (18, 19). The complex of Cdk5–p35 has not been purified from mammalian brain. The Cdk5–p35 complex is a membrane-bound protein. Cdk5 binds to the membrane via myristoylation of p35, whereas monomeric Cdk5 is a soluble cytoplasmic protein (Fig. 1). A part of Cdk5–p35 and Cdk5–p39 also associates with the cytoskeleton, actin filaments, microtubules, and neurofilaments (9, 12, 21, 22), although it is not clear how the Cdk5 complex associates with these different cytoskeletal components. When bound to the membrane, Cdk5–p35 displays moderate

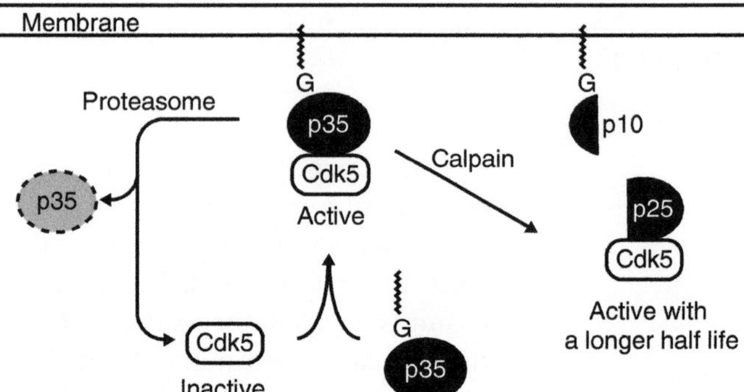

Fig. 1. The activation mechanism of Cdk5. Cdk5 alone does not exhibit kinase activity and requires the p35 regulatory subunit for activation. p35 is a myristoylated protein, and Cdk5–p35 binds to membranes via myristoylation of p35. p35 is a protein with a short half-life that is degraded by proteasomes. Degradation of p35 inactivates Cdk5, whereas cleavage of p35 to p25 by calpain dissociates Cdk5–p25 from membrane and stabilizes p25, producing the mislocalized and hyperactive kinase complex. Another regulatory subunit p39 is not described in this figure, but p39 is synthesized, myristoylated, and degraded similarly to p35. p39 is also cleaved to p29 by calpain, but the extent of cleavage is less than that of p35 (20).

kinase activity. The kinase activity of Cdk5–p35 is stimulated by treatment with nonionic detergent (23), whereas Cdk5–p39 is inactivated by detergent treatment (24). Neither the function nor kinase properties of Cdk5–p39 have been characterized fully.

Cdk5 is a proline-directed serine/threonine kinase that phosphorylates Ser/Thr residues followed by Pro with (S/T)PX(K/R) sequences as the preferred consensus (11, 25). Cytoskeletal proteins, membrane proteins, and signaling proteins are major categories of substrates for Cdk5. Proline-directed phosphorylation is abundant in brain, as detected by comprehensive mass spectroscopic analysis (26). The number of substrate proteins for Cdk5 continues to increase (2, 4, 27) as do the number of identified neuronal functions of Cdk5. Because there are several PDPKs, it is important to distinguish which kinase phosphorylates which site. In some cases, multiple PDPKs phosphorylate the same sites depending on the cellular context. To understand the specific functions of Cdk5, it is important to know how to identify Cdk5-specific substrates and to handle the active Cdk5 complex.

2. Materials

Many Cdk5 and p35 antibodies are commercially available. The anti-p35 antibody C19, and anti-Cdk5 antibodies C8 and DC17, which have been used in many previous studies, were obtained from Santa Cruz Biotechnology (Sant Cruz, CA). We generated anti-p39 antibody by immunizing rabbits with the 14 C-terminal amino acids (24). We have not examined commercially available anti-p39 antibodies. The Cdk5 inhibitor roscovitine was obtained from Calbiochem (San Diego, CA). Histone H1, a substrate for the in vitro Cdk5 assay, was obtained from Roche Diagnostics (Mannheim, Germany). The tail domain of NF-H was cloned into pCMV2–FLAG and used for in situ phosphorylation by Cdk5 (Fig. 3) (28). Cdk5, p35, and p39 cDNA tagged with FLAG, HA, or myc were inserted into pcDNA3 or adenovirus expression vectors (20, 24, 29, 30). Phos-tag was purchased from Wako Pure Chemical (Osaka, Japan).

HEK293 or COS-7 cells were used for overexpression and immunoprecipitation of the Cdk5 complexes and were cultured in Dulbecco's modified Eagle's medium (DMEM) supplemented with 10% fetal bovine serum. Baculovirus encoding Cdk5, kinase dead (KD)-Cdk5, p35his, p25his, or p39his were prepared using BakPAK Baculovirus Expression System (Clontech, Palo Alto, CA) according to the manufacturer's protocol, and their overexpression in Sf9 cells and purification were performed using Ni-NTA agarose beads (Qiagen, Hilden, Germany) (24, 29).

Porcine brains, which are used for biochemical purification of Cdk5–p25, were obtained from a local slaughterhouse. Wistar rats or Charles River mice were from Sankyo Lab Service (Tokyo, Japan). Primary cortical neurons were prepared from embryonic day 15–16 (E15–16) fetal mouse brains or E17–18 fetal rat brains (20).

2.1. Buffers

PEM buffer: 0.1 M Pipes, pH 6.8, 1 mM EGTA, 1 mM $MgCl_2$, and protease inhibitors (10 μg/mL leupeptin, 0.4 mM Pefablock SC (Roche Diagnostics), and 1 mM E64)

Phosphate buffer: 20 mM Na-phosphate, pH 6.8, 0.5 M KCl, 1 mM $MgCl_2$, 0.1 mM EDTA, 0.1 mM EGTA, and protease inhibitors

MOPS buffer: 20 mM MOPS, pH 7.2, 1 mM $MgCl_2$, 0.1 mM EGTA, 0.1 mM EDTA, 5% glycerol, 0.5% Nonidet P-40, protease inhibitors, and 0.1 mM dithiothreitol (DTT)

Homogenization buffer I: 20 mM MOPS, pH 7.2, 2 mM $MgCl_2$, 0.1 mM EDTA, 5 mM EGTA, 1 mM DTT, and protease inhibitors

Homogenization buffer II: 10 mM Tris-HCl, pH 8.0, 10 mM imidazole, 1 mM $MgCl_2$, 10 mM β-mercaptoethanol, and protease inhibitors

Washing buffer A: 10 mM Tris-HCl, pH 7.5, 25 mM imidazole, 1 mM $MgCl_2$, and 0.75 M NaCl

Washing buffer B: 10 mM Tris-HCl, pH 7.5, 25 mM imidazole, 1 mM $MgCl_2$, and 1% Triton X-100

Washing buffer C: 10 mM Tris-HCl, pH 7.5, 5 mM imidazole, 1 mM $MgCl_2$, and 1% Triton X-100

Washing buffer D: 10 mM MOPS, pH 6.8, and 1 mM $MgCl_2$

Washing buffer E: 50 mM Tris-HCl, pH 7.5, 150 mM NaCl, and 0.05% Tween-20

Elution buffer: 10 mM Tris-HCl, pH 7.5, 1 M NaCl, and 350 mM imidazole

Kinase assay buffer: 20 mM MOPS, pH 7.2, 1 mM $MgCl_2$, 0.1 mM EDTA, and 0.1 mM EGTA

RIPA buffer: 20 mM Tris-HCl, pH 7.5, 1 mM EGTA, 1 mM EDTA, 0.15 M NaCl, 1% Nonidet P-40, 0.1% SDS, 0.1% sodium deoxycholate, 10 mM β-glycerophosphate, 5 mM NaF, 1 mM *p*-nitrophenylphosphate, 1 mM DTT, and protease inhibitors

3. Methods

3.1. Preparation of the Kinase-Active Cdk5 Complex

3.1.1. Purification of Cdk5–p25 from Mammalian Brains by Column Chromatography

Cdk5 and its regulatory subunits are highly expressed in mammalian brain, and brain is the only tissue where the active endogenous Cdk5 has been purified by column chromatography. Cdk5–p25 has been purified from bovine, porcine, and rat brains. It is important to use fresh brain preparations, but after the outbreak of bovine spongiform encephalopathy, it has become difficult to obtain fresh bovine brain. Porcine brain is recommended for large-scale purification of Cdk5 kinase. In our laboratory, a slaughterhouse obtains the brains from the slaughtered pigs and keeps them in a refrigerator. We transport them on ice or in cold water or buffer to our laboratory. Thereafter, all procedures are performed at 4°C, unless otherwise indicated.

Porcine brain is homogenized in an equal volume of PEM buffer and centrifuged at 32,000×*g* for 45 min (12). The supernatant is adjusted to 25% glycerol and 1 mM ATP, and warmed at 35°C for 45 min. After centrifugation at 100,000×*g* for 45 min, the pelleted microtubular cytoskeletal components are suspended in PEM buffer containing 0.75 M KCl at 35°C and cleared by centrifugation at 200,000×*g* for 1 h. After twofold dilution with cold PEM buffer, the supernatant is applied to an anion exchange column such as Q-Sepharose, SP-Sepharose, or DEAE Sepharose fast flow (GE Healthcare, Buckinghamshire, UK). The flow-through fractions are applied to a hydroxyapatite column (Bio-Gel HT or Bio-Gel HTP, BioRad, Hercules, CA) and, after washing with phosphate buffer, the active Cdk5 is eluted with a linear gradient of Na-phosphate from 20 to 200 mM. The kinase fractions are pooled and brought to 50% saturation with ammonium sulfate. The precipitate is suspended in a small volume of MOPS buffer containing 0.5 M NaCl, dialyzed against the same buffer, and fractionated by a gel filtration column such as Superdex 75 (GE Healthcare). The kinase-active fractions are pooled and dialyzed against MOPS buffer for several hours. The dialyzed kinase fraction is applied to a Mono Q (GE Healthcare) column equilibrated with MOPS buffer. The flow-through fractions are applied to a Protein-Pak SP (Waters, Milford, MA) or Mono Q column equilibrated with the same buffer and eluted with a linear gradient of NaCl (0–0.2 M).

In published column purification procedures, Cdk5–p25 was purified from the high-speed supernatant of brain homogenate (9–12). However, it was later found that Cdk5–p35 is a membrane-bound kinase (31, 32). Although it might be possible to purify Cdk5–p35 from the membrane fraction of brain, there has been no report of the column purification of Cdk5–p35. We tried but could not protect p35 from cleavage. It was effective to start with the microtubule pellet as the first step of purification. Although the reason is not known exactly, the microtubule pellet contains a large amount of membranes, with which the Cdk5 complex associates.

3.1.2. Preparation of the Kinase-Active Cdk5 Complex from Sf9 Cells

Presently, the kinase-active complex of recombinant Cdk5 and its regulatory subunit are purified from Sf9 cells infected by baculovirus encoding Cdk5 and its regulatory subunits. The commercially available kinase-active Cdk5 (Cosmo Bio, Tokyo, Japan; Millipore, Bedford, MA) must be prepared from Sf9 cells. In establishing the system to prepare active Cdk5 in Sf9 cells, the most time-consuming process is constructing the baculovirus encoding Cdk5 or its regulatory subunit. We use the BacPAK Baculovirus Expression system (Clontech) (24, 29). Once constructed, the baculovirus can be used for the preparation of Cdk5 complexes with high kinase activity.

Sf9 cells are infected with baculovirus encoding Cdk5 and its regulatory subunit, p35, p25, or p39 depending on the experimental objectives. Sf9 cells expressing Cdk5 and p35, p25, or p39 are collected 48 h after infection and disrupted in homogenization buffer II. The extract is obtained from the supernatant after centrifugation at 15,000×*g* for 15 min and is incubated with Ni-NTA agarose for 2 h with rotation. After incubation with Ni-NTA agarose, the beads are washed sequentially twice with washing buffer A, twice with washing buffer B, twice with washing buffer C, and twice with washing buffer D by the method recommended by the manufacturer (Clontech) with some modifications. Finally, the Cdk5 complex is eluted with elution buffer and used for the phosphorylation experiments after dialysis against kinase assay buffer containing 10% glycerol, 0.1% Triton X-100, and protease inhibitors.

3.1.3. Preparation of Cdk5 from Escherichia coli

Expression and purification of recombinant Cdk5 and p25 or p35 from *Escherichia coli* have been reported (33–35). However, even if GST-fusion protein is used, we have found that it can be difficult to obtain enough of the kinase-active Cdk5 complex. In particular, the preparation of p35 or p25 is difficult because these can form inclusion bodies.

3.1.4. Immunoprecipitation of the Kinase-Active Cdk5 Complex

Immunoprecipitation is another method used frequently to prepare active Cdk5 complexes recently. The endogenous kinase-active Cdk5–p35 can be prepared from mouse or rat brain, or from cultured neurons derived from these brains. Rat and mouse brains contain three different pools of Cdk5: free monomeric Cdk5, Cdk5–p35, and Cdk5–p39. Monomeric Cdk5 is a soluble protein, whereas Cdk5–p35 and Cdk5–p39 are membrane bound. Cdk5–p35 is a stable complex, whereas Cdk5–p39 is unstable. Therefore, the choice of the extraction methods and buffers is important.

Immunoprecipitation of Endogenous Cdk5–p35 from Brain or Cultured Neurons

Rat or mouse brain is homogenized in 10 vol (v/w) of homogenization buffer I with a Teflon pestle glass homogenizer at 4°C and centrifuged at 15,000×*g* for 10 min at 4°C. The supernatant is precleared with protein A/G PLUS–Agarose beads (Santa Cruz Biotechnology) or Protein A-Sepharose (GE Healthcare), and incubated with anti-Cdk5 C8 or anti-p35 C19 antibody for 1~2 h at 4°C. Protein A/G PLUS–Agarose or Protein A-Sepharose beads are added to the mixture and incubated further for 1~2 h at 4°C. Alternatively, after preclearance, the extract is incubated with anti-Cdk5 C8 or anti-p35 C19 bound to Protein A/G PLUS–Agarose or Protein A-Sepharose for 1–2 h at 4°C. After washing three times with MOPS buffer containing 0.75 M NaCl and 1% Triton X-100 and twice with MOPS buffer, the agarose beads are suspended in kinase assay buffer.

Cultured neurons are lysed in 10 vol (v/w) of homogenization buffer I containing 0.75 M NaCl and 1% Triton X-100 and centrifuged at 10,000×*g* for 15 min to collect the extract as the supernatant. Immunoprecipitation is carried out as described above. The beads are removed from the extract and the extract is washed three times with kinase assay buffer.

Immunoprecipitation of the Cdk5 Complex from Cultured Cells Transiently Transfected with Cdk5 and p35 (or p25)

HEK293 or COS-7 cells are often used for this purpose. Other cell lines could be used if the transfection efficiency and protein expression are high. These cells are maintained in a recommended medium and transfected with mammalian cell expression vectors encoding Cdk5 and p35 (or p25) by a lipofection method. Stable transfectants of p35 (or p25) have not been obtained. In the case of recombinant Cdk5 and p35 (or p25), a purification tag can be added. When p35 is tagged, the position of the tag must be considered (see Sect. 5.2). Cdk5–p35 (or –p25) is prepared from cells expressing this complex by immunoprecipitation with anti-Cdk5, anti-p35, or anti-tag antibodies.

For example, HEK293 cells are transfected with pcDNA3 encoding Cdk5 and p35 (or p25) using PolyFect Transfection Reagent (Qiagen). Cells are harvested 24 h after transfection and suspended in homogenization buffer I at 4°C in the presence of 0.75 M NaCl and 1% Triton X-100. The extract is isolated from the supernatant after centrifugation at 15,000×*g* for 20 min at 4°C. The supernatant is precleared with Protein A Sepharose or Protein A/G PLUS–Agarose and then incubated with anti-Cdk5, -p35, or -tag antibody for 1 h at 4°C. Protein A-Sepharose or Protein A/G PLUS–Agarose is added to the mixture and incubated further for 1 h at 4°C. The Sepharose or agarose beads are collected by brief centrifugation and used for the kinase assay after washing five times with kinase assay buffer containing 0.75 M NaCl and 1% Triton X-100, and then three times with kinase assay buffer at 4°C.

Immunoprecipitation of Cdk5–p39

Endogenous Cdk5–p39 can be prepared from mouse or rat brain by immunoprecipitation. However, because of the unstable property of the complex, the Cdk5–p39 complex dissociates during immunoprecipitation in the presence of salt and detergent. Therefore, the concentrations of NaCl and Triton X-100 in the buffer solutions used for extraction and immunoprecipitation should be reduced to less than 0.15 M and 0.01%, respectively. If possible, to avoid contamination of Cdk5–p35 in the mild washing conditions (see Sect. 4.5), recombinant Cdk5 and p39 are recommended. Recombinant Cdk5–p39 expressed in cultured cells can be prepared by immunoprecipitation as described in section "Immunoprecipitation of Endogenous Cdk5–p35 from Brain or Cultured Neurons," except for using low salt and low detergent buffer.

3.2. In Vitro Kinase Assay of the Cdk5–Regulatory Subunit Complexes

The kinase activity of the Cdk5 complexes can be measured in physiological buffer conditions using 0.1–0.5 mg/mL histone H1 and 0.1 mM [γ-^{32}P]ATP as substrates at 30–35°C. Cdk5 has many substrate proteins, but histone H1 is the best reported to date and is used frequently in in vitro Cdk5 kinase assays. Within the pH range of 6.5–8.0, Cdk5–p25 has a higher activity at lower pH, and the activity decreases gradually with increasing pH. Therefore, we use MOPS buffer at pH 6.8 or 7.2 more than we use HEPES at pH 7.4 or Tris buffer at pH 7.5 or 8.0. Mg^{2+}, which is used as Mg-ATP for the substrate, should be included in the reaction mixture at a higher concentration than that of ATP. Mg^{2+} concentrations of several millimolars appear to be enough to produce saturated activity. Because the active Cdk5 is a sticky protein, to prevent adhesion to the tube wall, we usually add 0.1 or 0.5% nonionic detergent (Nonidet P-40 or Triton X-100) and 50 or 100 mM NaCl in the reaction mixture. A higher concentration of NaCl > 0.15 M tends to inhibit the kinase activity. The incubation time should be determined in preliminary experiments, but ~60 min reaction time is sufficient if Cdk5 is prepared properly.

The reaction is stopped by adding Laemmli's sample buffer and boiling for 3–5 min. The samples are separated on SDS-PAGE using a 10 or 12.5% polyacrylamide gel, and the radioactivity incorporated into histone H1 is detected with a phosphorimager (FLA7000 image analyzer; Fuji-Film, Tokyo, Japan). Quantification can be obtained using photostimulated luminescence (PSL) on the phosphorimager. Radioactivity is also counted for the excised histone H1 bands as Cerenkov radiation using a liquid scintillation counter. Radioactivities measured by PSL and Cerenkov counting are proportional. Anti-phospho-histone H1 antibody (Millipore; Sigma-Aldrich Laboratories, St. Louis, MO) is used to detect histone H1 phosphorylation, although less frequently (36, 37). Relative, but not absolute, kinase activity can be measured by this method without using radioisotopes.

Several peptide substrates, including a (S/T)P sequence synthesized after the amino acid sequence of histone H1, bradykinin, pp60c-src, tau, or the NF-H subunit have been used to measure the kinase activity (9, 10, 25, 38–40). When the peptide substrate is used in the kinase assay, Whatman P81 phosphocellulose filter paper (GE Healthcare) is used for measuring phosphorylation. The phosphorylation reaction is terminated by spotting onto a P81 filter paper, and the filters are dipped in a solution of 1% phosphoric acid. After washing several times with 1% phosphoric acid, the filter paper is dried and the radioactivity is counted in scintillation vials with water for Cerenkov counting or in ACSII scintillation fluid (GE Healthcare).

3.3. In Vitro Phosphorylation of Substrate Proteins by Cdk5

Many proteins including (S/T)P sequence are phosphorylated by Cdk5 kinase. The phosphorylation conditions for histone H1 (see Sect. 3.2) can be applied to other substrate proteins. In these cases, however, a sufficient amount of proteins is not always prepared. The objective here is to detect phosphorylation of the protein of interest and not to measure kinase activity of Cdk5. Therefore, an excess of substrate proteins is not required except for measuring the K_m value of the substrate. The following detection methods are used: (1) upward electrophoretic mobility shift, (2) immunoblotting with anti-phospho antibodies, and (3) incorporation of ^{32}P-phosphate into the substrate protein using [γ-^{32}P]ATP.

(1) Phosphorylation by Cdk5 is often accompanied by the upward electrophoretic mobility shift of substrate proteins. Immunoblotting is performed with phosphorylation-independent antibody after SDS-PAGE. An upward shift is a sign of phosphorylation, but this depends on the phosphorylation site and may not always be observed. (2) Cdk5 phosphorylation can sometimes be detected with anti-phospho SP or TP antibodies, which are commercially available. After the phosphorylation site(s) are determined, the phosphorylation site-specific antibody can be produced. (3) If the two methods described above do not work, the substrate proteins must be labeled using [γ-^{32}P]ATP.

Phosphorylation of new substrate proteins should be checked by several control experiments. Cdk5-dependent phosphorylation is tested with Cdk5 inhibitors. Roscovitine is used most often, but olomoucine and butyrolactone I can also be used as inhibitors. However, these chemicals show similar inhibition of the cell cycle Cdks, Cdk1, and Cdk2. If contamination of cell cycle Cdks is suspected, another control is needed; for example, without p35 expression or coexpression in a dominant-negative Cdk5 (K33T or D144N) (24, 41).

3.4. Cellular Phosphorylation of Substrate Proteins by Cdk5

When studying a new substrate protein, it is necessary to show cellular phosphorylation in addition to in vitro phosphorylation. Cultured cell lines are the first choice. Although cultured proliferating cells express endogenous Cdk5, they do not exhibit Cdk5 activity because they do not express its regulatory subunits. Therefore, to show Cdk5-dependent phosphorylation, the regulatory subunit must be transfected exogenously. Endogenous Cdk5 can be activated by transfection of the regulatory subunit alone, but cotransfection of Cdk5 is recommended to increase the Cdk5 activity. The phosphorylation can be detected by one of the methods described above (Sect. 3.2), i.e., (1) upward shift of electrophoretic mobility, (2) anti-phospho-antibodies, or (3) isotope labeling.

For isotope labeling, substrate proteins are metabolically labeled with [^{32}P]phosphate. Cells expressing Cdk5–p35 and its substrate are cultured in the presence of [^{32}P]orthophosphate in phosphate-free medium for several hours and harvested in RIPA buffer. The extract is obtained as a supernatant of centrifugation at 10,000×*g* for 10 min. The substrate protein is isolated from the supernatant by immunoprecipitation, and ^{32}P incorporation is detected by autoradiography with a phosphorimager after SDS-PAGE. Negative control experiments must also be performed to exclude phosphorylation by other kinases (see Sect. 3.3).

4. Experimental Variables

4.1. Extraction of Cdk5–p35

Cdk5–p35 is a membrane-bound kinase in neurons and the brain. The extraction efficiency depends on the extraction buffers and starting materials. Cdk5–p35 is recovered in the particulate fractions of centrifugation but can be extracted in the soluble fraction using a buffer containing nonionic detergent and high salt. However, their use as the homogenization buffer increases the amount of protein extracted particularly in brain samples; otherwise many of them are not extracted. The membrane-enriched fraction may be used as the starting material, but the question which membrane fraction is best for Cdk5–p35 preparation has not been examined systematically. We usually use the crude supernatant obtained by centrifugation at 20,000×*g* for 20 min; microsomes in this fraction may provide a source of Cdk5–p35.

4.2. Expression Levels of p35 in Cultured Neurons

p35 is a labile protein. Degradation of p35 is easily induced in cultured neurons by subtle stimuli such as a change of medium. If the consistent expression of p35 is not observed during culture, the neurons should be handled carefully, gently, and quickly. To prevent cleavage to p25 during preparation, buffers used for preparation should contain EGTA to chelate Ca^{2+} and protease inhibitors including leupeptin as a cysteine protease inhibitor.

4.3. Expression Levels of the Active Cdk5 in Cell Line Cultures

Because most types of cells express Cdk5, active Cdk5 complexes can be obtained only by transfecting the regulatory subunit exogenously into cells. To obtain higher Cdk5 activity, however, Cdk5 should be cotransfected. The high expression level may be important if contamination of cycling Cdks would be considered because cell cycle Cdks have similar substrate specificity to Cdk5 and are inhibited similarly by Cdk5 inhibitors (12). The best way to avoid this problem is to increase the expression levels of Cdk5 and p35 (see Sect. 3.3).

4.4. Levels of p25

p25 is a product of cleavage of p35 by calpain. Cdk5–p25 dissociates from membranes and has a longer half-life. The cleavage is called abnormal activation of Cdk5. However, there is only a marginal amount of p25 in healthy neurons and brain. A detectable amount of p25 observed in brain lysates or cell lysates is a sign of calpain activation, suggesting the postmortem delay in the case of brain and cell death in the case of cultured neurons. Brain should be dissected and put into ice-cold buffer immediately after killing of the animal, or the culture conditions for primary neurons should be improved.

4.5. Studies on Cdk5–p39

Cdk5–p39 is an unstable complex. p39 dissociates from Cdk5 in a solution containing high concentrations of nonionic detergent and NaCl, resulting in loss of Cdk5 activity. These properties make the study of Cdk5–p39 difficult. Care must be taken when handling Cdk5–p39. When Cdk5–p39 is prepared by immunoprecipitation, the concentration of NaCl (or KCl) and Triton X-100 (Nonidet P-40 or other nonionic detergent) in all solutions must be reduced to <0.15 M and 0.01%, respectively. Under such buffer conditions, Cdk5–p39 can be prepared from brain tissue (Fig. 2). Alternatively,

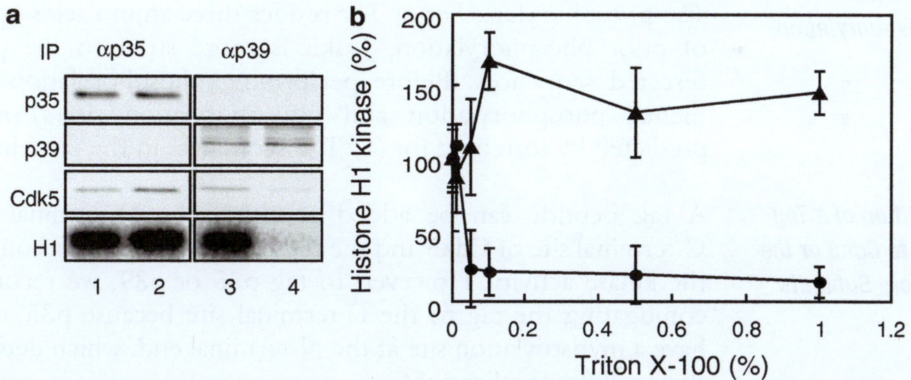

Fig. 2. Immunoprecipitation and kinase activity of Cdk5–p35 and Cdk5–p39. (**a**) Cdk5–p35 (*lanes 1 and 2*) and Cdk5–p39 (*lanes 3 and 4*) were immunoprecipitated with anti-p35 and anti-p39 antibodies, respectively, from rat brain extract. The immunoprecipitates were washed with MOPS buffer (*lanes 1 and 3*) or MOPS buffer containing 0.75 M NaCl and 1% Triton X-100 (*lanes 2 and 4*). p35 (*top panel*), p39 (*second panel*), or Cdk5 (*third panel*) were examined by immunoblotting. The kinase activity is shown by histone H1 phosphorylation (*bottom panel*). (**b**) The effects of Triton X-100 on the kinase activities of Cdk5–p35 and Cdk5–p39. Histone H1 phosphorylation by Cdk5–p35 (*triangles*) and Cdk5–p39 (*circles*) is shown as a function of Triton X-100 concentration. The kinase activity is expressed as a percentage of that in the absence of Triton X-100. Reproduced with permission (24).

cultured cells expressing only Cdk5–p39, but not p35, are recommended as the preparation materials. A phosphorylation assay for Cdk5–p39 should also be performed in the absence of detergent. Because Cdk5–p39 shows similar substrate specificity to Cdk5–p35, Cdk5–p35 (or –p25) can be used in place of Cdk5–p39 in in vitro experiments even if Cdk5–p39 is the complex of interest.

4.6. Choice of the Reaction Mixture for the In Vitro Kinase Assay

To obtain a strong signal of phosphorylation in the in vitro phosphorylation assay, we change the concentration of ATP to suit the detection method. When we detect phosphorylation by electrophoretic mobility shift, we put 1 mM ATP in the reaction mixture. When phosphorylation is assessed with [γ-^{32}P]ATP, we reduce the concentration of ATP to 0.1 mM to increase the specific activity of ^{32}P, usually by including 1,000–1,700 Bq/nmol [γ-^{32}P]ATP. We have found some papers in which phosphorylation is performed with radiolabeled ATP alone without the addition of cold ATP. In these cases, the concentration of ATP must be very low. We do not recommend decreasing ATP concentration to <0.1 mM. As described in Sect. 3.2, because ATP is used as Mg^{2+}-ATP by kinases, the concentration of Mg^{2+} should be higher than that of ATP in the reaction mixture, but a concentration of several millimolars is enough.

5. Typical/ Anticipated Results

5.1. Prediction of Substrates and Phosphorylation Sites

Cdk5 is a PDPK. The phosphorylation sites are (Ser/Thr)-Pro sequences with (Ser/Thr)-Pro-X-(Lys/Arg) as a preferred consensus. Compared with GSK3β, another proline-directed kinase that also phosphorylates Ser or Thr resides three amino acids upstream of prior phosphorylation, Cdk5 is more strict to the proline-directed sequences. Before performing phosphorylation experiments, phosphorylation and phosphorylation site(s) may be predicted by searching for (S/T)P sequences in the proteins.

5.2. Addition of a Tag Peptide to Cdk5 or the Regulatory Subunits

A tag peptide can be added at either the N-terminal or the C-terminal site of Cdk5 and the regulatory subunits without losing the kinase activity. However, to tag p35 or p39, we recommend conjugating the tag to the C-terminal site because p35 and p39 have a myristoylation site at the N-terminal end, which determines their cellular localization.

5.3. In Vitro Kinase Assay with Immunoprecipitated Cdk5

Figure 2a shows the results of immunoprecipitation of Cdk5–p35 and Cdk5–p39 using two different buffer solutions with or without high salt and nonionic detergent. Cdk5–p35 and Cdk5–p39 were prepared without cross-contamination (Fig. 2a). However, Cdk5–p39 immunoprecipitated in the buffer containing 0.75 M NaCl and 0.5% Triton X-100 does not exhibit kinase activity. Figure 2b shows the concentration dependency of Cdk5–p35 and

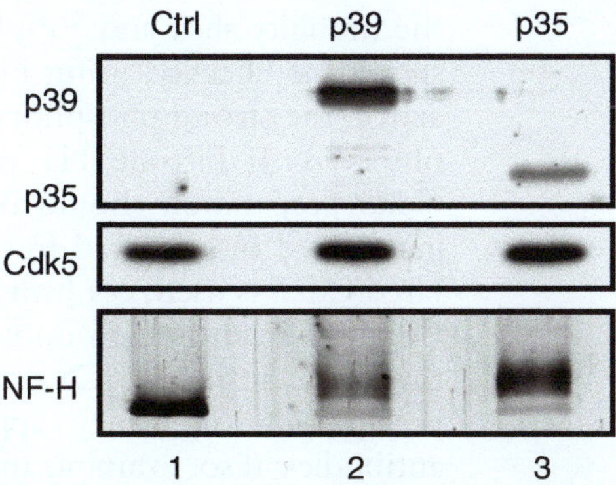

Fig. 3. Detection of the NF-H tail domain phosphorylation by Cdk5–p35 and Cdk5–p39 in cultured cells. The NF-H tail domain, which contains many SPXK sequences, was cotransfected into HEK293 cells with p35 (*lane 3*) or p39 (*lane 2*). The NF-H tail domain was detected by immunoblotting with phosphorylation-independent antibody (*bottom panel*). The NF-H tail domain shifted the electrophoretic mobility upward after coexpression with p35 or p39 compared with that in the absence of the Cdk5 regulatory subunit (*lane 1*).

Cdk5–p39 on Triton X-100 in the kinase assay buffer. Triton X-100 stimulates Cdk5–p35 but inactivates Cdk5–p39. We have seen the use of RIPA for immunoprecipitation of Cdk5–p35 in the literature. Cdk5–p35 loses its kinase activity in the presence of SDS. Therefore, RIPA buffer containing 0.1% SDS should not be used in the preparation of kinase-active Cdk5–p35.

5.4. Phosphorylation of Substrate Proteins in Cultured Cells

Phosphorylation of the NF-H tail domain by Cdk5 in cells is shown in Fig. 3 as an example of cellular phosphorylation. HEK293 cells were cotransfected with the NF-H tail domain and p35 or p39 expression plasmids. In this experiment, endogenous Cdk5 expressed in HEK293 cells was used. Phosphorylation is detected by the upward shift of NF-H in immunoblotting with phosphorylation-independent anti-NF-H antibody.

6. Troubleshooting

6.1. No Mobility Shift of Substrate Proteins on SDS-PAGE After In Vitro Phosphorylation

Proline-directed phosphorylation tends to shift the electrophoretic mobility more than that observed for other kinases (8). Therefore, phosphorylation is detected easily by the upward shift in the electrophoretic mobility of substrate proteins on SDS-PAGE. When the upward shift is not seen distinctly in Laemmli's SDS-PAGE, we recommend using Phos-tag SDS-PAGE (42, 43), which expands the phosphorylation-dependent shift markedly. Most types of phosphorylation can be detected by this method. Even though the phosphorylation-dependent shift is observed, confirmation using [γ-^{32}P]ATP may be need. When phosphorylation is not detected by the mobility shift and ^{32}P-phosphate incorporation, Cdk5 activity should be checked using histone H1 and [γ-^{32}P)ATP. If Cdk5 is active, the strong phosphorylation signal of histone H1 should be observed. If histone H1 phosphorylation is not observed, the Cdk5 preparation should be changed. The Cdk5 complexes are inactivated by repeated freezing–thawing. If the immunoprecipitated Cdk5 is used, confirm the efficiency of immunoprecipitation. Cdk5–p35 can be immunoprecipitated from brain or neuron cultures with anti-Cdk5 C8 or anti-p35 C19, although the immunoprecipitation efficiency appears to be dependent on lots of antibodies; if so, examine another lot or antibody. The preparation of brain or cell extract should also be checked. Sections 3.1.4, 3.2, 4.5, and 5.3 discuss the buffer and extraction conditions. If the Cdk5 complex is prepared from transfected cultured cells, Cdk5–p25 may be a better choice as the kinase source.

6.2. No Mobility Shift of Substrate Proteins in Cellular Phosphorylation

When no mobility shift of substrate proteins is observed in cellular phosphorylation, check the expression of the regulatory subunit. Cdk5 needs the regulatory subunit for activation, and there is no constitutively active form of Cdk5. Even though Cdk5 and the regulatory subunit may be expressed sufficiently, if the mobility shift is not observed, the phosphorylation may not affect the mobility shift in Laemmli's SDS-PAGE. Phos-tag SDS-PAGE is an alternative method (see Sect. 6.1). Anti-phospho-SP or TP antibody may also be used. We have used anti-phospho-TP antibody and others have also reported its use.

6.3. Checking the Immunoprecipitation of Cdk5 and p35

Coimmunoprecipitation of p35 with anti-Cdk5 or Cdk5 with anti-p35 can be confirmed by the immunoblot detection of p35 in the Cdk5 immunoprecipitates or of Cdk5 in the p35 immunoprecipitates. Cdk5 immunoprecipitated with rabbit anti-Cdk5 C8 or anti-p35 C19 can be detected with mouse monoclonal anti-Cdk5 DC17. However, because there are no effective mouse monoclonal antibodies for p35, the same rabbit polyclonal antibody used for immunoprecipitation must be used again for immunoblotting. The large amount of IgG light chain hampers the detection of p35 or Cdk5. In such case, the crosslinking of anti-p35 or anti-Cdk5 with Protein A or G beads by disuccinimidyl suberate (Thermo Scientific/Pierce, Rockford, IL) would greatly reduce the amount of IgG eluted from beads, which would reduce the reaction of rabbit IgG light and heavy chains.

6.4. Use of Cdk5–p25

When you have trouble obtaining sufficient amounts of Cdk5-p35, you may use Cdk5-p25. Cdk5-p25 is a stable complex and is easier to handle than Cdk5-p35 and considerably easier than Cdk5-p39. Cdk5-p25 has similar substrate specificity to Cdk5-p35 and Cdk5-p39, at least, in in vitro phosphorylation.

7. Conclusion

Cdk5 is a multifunctional kinase in mammalian brain whose diverse functions cover the range of normal brain development to neurodegenerative diseases. Our understanding of the actions and regulation of Cdk5 is far from complete. New substrates and functions are being reported increasingly and extraneuronal functions have also been reported recently. Cdk5 has overlapping substrate specificity with other PDPKs, showing the importance of identifying the Cdk5 substrates and functions in a context-dependent manner. In contrast to other signaling kinases, Cdk5 needs a regulatory subunit for activation, and this presents some difficulties in the study of Cdk5. By describing our experience with Cdk5 studies, we have discussed the best methods for handling and assessing Cdk5. We hope this chapter provides helpful information for your research.

References

1. Tang D, Wang JH (1996) Cyclin-dependent kinase 5 (Cdk5) and neuron-specific Cdk5 activators. Prog Cell Cycle Res 2:205–216
2. Dhavan R, Tsai LH (2001) A decade of CDK5. Nat Rev Mol Cell Biol 2:749–759
3. Hisanaga S, Endo R (2010) Regulation and role of cyclin-dependent kinase activity in neuronal survival and death. J Neurochem 115: 1309–1321
4. Shelton SB, Johnson GVW (2004) Cyclin-dependent kinase-5 in neurodegeneration. J Neurochem 88:1313–1326
5. Angelo M, Plattner F, Giese KP (2006) Cyclin-dependent kinase 5 in synaptic plasticity, learning and memory. J Neurochem 99:353–370
6. Cheung ZH, Fu AK, Ip NY (2006) Synaptic roles of Cdk5: implications in higher cognitive functions and neurodegenerative diseases. Neuron 50:13–18
7. Brinkkoetter PT, Olivier P, Wu JS, Henderson S, Krofft RD, Pippin JW, Hockenbery D, Roberts JM, Shankland SJ (2009) Cyclin I activates Cdk5 and regulates expression of Bcl-2 and Bcl-XL in postmitotic mouse cells. J Clin Invest 119:3089–3101
8. Hisanaga S, Kusubata M, Okumura E, Kishimoto T (1991) Phosphorylation of neurofilament H subunit at the tail domain by CDC2 kinase dissociates the association to microtubules. J Biol Chem 266:21798–21803
9. Ishiguro K, Takamatsu M, Tomizawa K, Omori A, Takahashi M, Arioka M, Uchida T, Imahori K (1992) Tau protein kinase I converts normal tau protein into A68-like component of paired helical filaments. J Biol Chem 267:10897–10901
10. Lew J, Beaudette K, Litwin CM, Wang JH (1992) Purification and characterization of a novel proline-directed protein kinase from bovine brain. J Biol Chem 267:13383–13390
11. Shetty KT, Link WT, Pant HC (1993) cdc2-like kinase from rat spinal cord specifically phosphorylates KSPXK motifs in neurofilament proteins: isolation and characterization. Proc Natl Acad Sci USA 90:6844–6848
12. Hisanaga S, Uchiyama M, Hosoi T, Yamada K, Honma N, Ishiguro K, Uchida T, Dahl D, Ohsumi K, Kishimoto T (1995) Porcine brain neurofilament-H tail domain kinase: its identification as cdk5/p26 complex and comparison with cdc2/cyclin B kinase. Cell Motil Cytoskeleton 31:283–297
13. Hellmich MR, Pant HC, Wada E, Battey JF (1992) Neuronal cdc2-like kinase: a cdc2-related protein kinase with predominantly neuronal expression. Proc Natl Acad Sci USA 89:10867–10871
14. Meyerson M, Enders GH, Wu CL, Su LK, Gorka C, Nelson C, Harlow E, Tsai LH (1992) A family of human cdc2-related protein kinases. EMBO J 11:2909–2917
15. Lew J, Huang QQ, Qi Z, Winkfein RJ, Aebersold R, Hunt T, Wang JH (1994) A brain-specific activator of cyclin-dependent kinase 5. Nature 371:423–426

16. Tsai LH, Delalle I, Caviness VS Jr, Chae T, Harlow E (1994) p35 is a neuralspecific regulatory subunit of cyclin-dependent kinase 5. Nature 371:419–423
17. Uchida T, Ishiguro K, Ohnuma J, Takamatsu M, Yonekura S, Imahori K (1994) Precursor of cdk5 activator, the 23 kDa subunit of tau protein kinase II: its sequence and developmental change in brain. FEBS Lett 355:35–40
18. Kusakawa G, Saito T, Onuki R, Ishiguro K, Kishimoto T, Hisanaga S (2000) Calpain-dependent proteolytic cleavage of the p35 cyclin-dependent kinase 5 activator to p25. J Biol Chem 275:17166–17172
19. Lee MS, Kwon YT, Li M, Peng J, Friedlander RM, Tsai LH (2000) Neurotoxicity induces cleavage of p35 to p25 by calpain. Nature 405:360–364
20. Minegishi S, Asada A, Miyauchi S, Fuchigami T, Saito T, Hisanaga S (2010) Membrane association facilitates degradation and cleavage of the cyclin-dependent kinase 5 activators p35 and p39. Biochemistry 49:5482–5493
21. Humbert S, Dhavan R, Tsai LH (2000) p39 activates cdk5 in neurons, and is associated with the actin cytoskeleton. J Cell Sci 113:975–983
22. He L, Zhang Z, Yu Y, Ahmed S, Cheung NS, Qi RZ (2011) The neuronal p35 activator of Cdk5 is a novel F-actin binding and bundling protein. Cell Mol Life Sci 68:1633–1643
23. Zhu YS, Saito T, Asada A, Maekawa S, Hisanaga S (2005) Activation of latent cyclin-dependent kinase 5 (Cdk5)-p35 complexes by membrane dissociation. J Neurochem 94:1535–1545
24. Yamada M, Saito T, Sato Y, Kawai Y, Sekigawa A, Hamazumi Y, Asada A, Wada M, Doi H, Hisanaga S (2007) Cdk5-p39 is a labile complex with the similar substrate specificity to Cdk5-p35. J Neurochem 102:1477–1487
25. Beaudette KN, Lew J, Wang JH (1993) Substrate specificity characterization of a cdc2-like protein kinase purified from bovine brain. J Biol Chem 268:20825–20830
26. Ballif BA, Villén J, Beausoleil SA, Schwartz D, Gygi SP (2004) Phosphoproteomic analysis of the developing mouse brain. Mol Cell Proteomics 3:1093–1101
27. Dhariwala FA, Rajadhyaksha MS (2008) An unusual member of the Cdk family: Cdk5. Cell Mol Neurobiol 28:351–369
28. Sasaki T, Taoka M, Ishiguro K, Uchida A, Saito T, Isobe T, Hisanaga S (2002) In vivo and in vitro phosphorylation at Ser-493 in the glutamate (E)-segment of neurofilament-H subunit by glycogen synthase kinase 3beta. J Biol Chem 277:36032–36039
29. Saito T, Onuki R, Fujita Y, Kusakawa G, Ishiguro K, Bibb JA, Kishimoto T, Hisanaga S (2003) Developmental regulation of the proteolysis of the p35 cyclin-dependent kinase 5 activator by phosphorylation. J Neurosci 23:1189–1197
30. Endo R, Saito T, Asada A, Kawahara H, Ohshima T, Hisanaga S (2009) Commitment of 1-methyl-4-phenylpyrinidinium ion-induced neuronal cell death by proteasome-mediated degradation of p35 cyclin-dependent kinase 5 activator. J Biol Chem 284:26029–26039
31. Patrick GN, Zukerberg L, Nikolic M, de la Monte S, Dikkes P, Tsai LH (1999) Conversion of p35 to p25 deregulates Cdk5 activity and promotes neurodegeneration. Nature 402: 615–622
32. Asada A, Yamamoto N, Gohda M, Saito T, Hayashi N, Hisanaga S (2008) Myristoylation of p39 and p35 is a determinant of cytoplasmic or nuclear localization of active cyclin-dependent kinase 5 complexes. J Neurochem 106:1325–1336
33. Qi Z, Huang QQ, Lee KY, Lew J, Wang JH (1995) Reconstitution of neuronal Cdc2-like kinase from bacteria-expressed Cdk5 and an active fragment of the brain-specific activator. J Biol Chem 270:10847–10854
34. Tang D, Chun AC, Zhang M, Wang JH (1997) Cyclin-dependent kinase 5 (Cdk5) activation domain of neuronal Cdk5 activator. J Biol Chem 272:2318–2327
35. Amin ND, Albers W, Pant HC (2002) Cyclin-dependent kinase 5 (cdk5) activation requires interaction with three domains of p35. J Neurosci Res 67:354–362
36. Lin H, Chen MC, Chiu CY, Song YM, Lin SY (2007) Cdk5 regulates STAT3 activation and cell proliferation in medullary thyroid carcinoma cells. J Biol Chem 282:2776–2784
37. Sanchez AM, Flamini MI, Fu XD, Mannella P, Giretti MS, Goglia L, Genazzani AR, Simoncini T (2009) Rapid signaling of estrogen to WAVE1 and moesin controls neuronal spine formation via the actin cytoskeleton. Mol Endocrinol 23:1193–1202
38. Hisanaga S, Ishiguro K, Uchida T, Okumura E, Okano T, Kishimoto T (1993) Tau protein kinase II has a similar characteristic to cdc2 kinase for phosphorylating neurofilament proteins. J Biol Chem 268:15056–15060
39. Lee KY, Rosales JL, Tang D, Wang JH (1996) Interaction of cyclin-dependent kinase 5 (Cdk5) and neuronal Cdk5 activator in bovine brain. J Biol Chem 271:1538–1543
40. Ohshima T, Ward JM, Huh CG, Longenecker G, Veeranna PHC, Brady RO, Martin LJ,

Kulkarni AB (1996) Targeted disruption of the cyclin-dependent kinase 5 gene results in abnormal corticogenesis, neuronal pathology and perinatal death. Proc Natl Acad Sci USA 93:11173–11178

41. Nikolic M, Dudek H, Kwon YT, Ramos YF, Tsai LH (1996) The cdk5/p35 kinase is essential for neurite outgrowth during neuronal differentiation. Genes Dev 10:816–825

42. Kinoshita E, Kinoshita-Kikuta E, Takiyama K, Koike T (2006) Phosphate-binding tag, a new tool to visualize phosphorylated proteins. Mol Cell Proteomics 5:749–757

43. Hosokawa T, Saito T, Asada A, Fukunaga K, Hisanaga S (2010) Quantitative measurement of in vivo phosphorylation states of Cdk5 activator p35 by Phos-tag SDS-PAGE. Mol Cell Proteomics 9:1133–1143

Chapter 7

ErbB Membrane Tyrosine Kinase Receptors: Analyzing Migration in a Highly Complex Signaling System

Nicole M. Brossier, Stephanie J. Byer, Lafe T. Peavler, and Steven L. Carroll

Abstract

The erbB membrane tyrosine kinases (EGF receptor, erbB2, erbB3, and erbB4) are a family of structurally similar transmembrane proteins that act as receptors for the EGF and neuregulin (NRG) families of growth factors. ErbB receptors and their ligands are widely expressed by neurons and glia throughout the peripheral and central nervous system, where they promote proliferation, survival, migration, differentiation, and other effects. Precisely which effects are induced by erbB activation depends on a variety of factors. These factors include which erbB receptors are recruited to the signaling complex, whether key non-erbB intramembranous proteins (e.g., mucin 1, CD44, integrins) are present and interact with the erbB kinases and the lineage and developmental state of the cell. Different erbB ligands acting through the same receptors often also elicit distinct or even diametrically opposed effects. In this chapter, we review key aspects of the complexity intrinsic to signaling by erbB kinases and their ligands in the nervous system. We then relate this knowledge to the rational design of experiments examining erbB signaling in migration, a common response induced by erbB receptor activation. The proper performance of Boyden chamber assays is considered, together with the procedures used to identify appropriate migration substrates and to optimize key experimental parameters (cell number, migration time, comparisons of the effects of different erbB ligands, and establishing optimal concentrations of these growth factors). The use of pharmacologic inhibitors and RNA interference to establish whether specific erbB kinases are required for the migration of nervous system cells is also described.

Key words: Epidermal growth factor, Neuregulin, Membrane tyrosine kinase receptors, Boyden chamber, RNA interference, Pharmacologic inhibitors

1. Introduction

1.1. ErbB Actions in the Central and Peripheral Nervous System

The erbB family of membrane tyrosine kinases is widely expressed by neurons and glia throughout the central (CNS) and peripheral (PNS) nervous system (1). This family of kinases consists of four structurally conserved proteins—the EGF receptor (EGFR; also

Hideyuki Mukai (ed.), *Protein Kinase Technologies*, Neuromethods, vol. 68,
DOI 10.1007/978-1-61779-824-5_7, © Springer Science+Business Media, LLC 2012

known as erbB1 or HER1), erbB2 (also referred to as HER2 in humans and c-neu in rats), erbB3 (HER3), and erbB4 (HER4)—which serve as receptors for multiple membrane-bound and secreted ligands. Despite their structural similarity, specific erbB receptors mediate distinct and highly variable responses in neurons at different developmental stages. For instance, in the developing nervous system erbB receptors regulate neural crest migration (2), mediate interactions between radial glia and migrating neuronal precursors (3–6), promote the chemotaxis of tangentially migrating interneurons (7, 8) and regulate neural progenitor migration in the rostral migratory stream (9). In contrast, erbB receptor activation in mature neurons triggers distinctly different responses which vary between different adult neuron populations. ErbB-mediated effects observed in adult neurons include enhancement of neurite outgrowth and branching (10–13), promotion of the expression of neurotransmitter receptors [e.g., *N*-methyl-D-aspartate (NMDA) receptor subunit 2C (14, 15), γ-amino butyric acid (GABA) receptor subunit 2B (12), cholinergic receptor subunits (α5, α7, and β4) (16)] modulation of calcium-activated potassium channel function (17, 18), and suppression of long-term potentiation (19).

The effects of erbB receptor activation in glia also vary between different glial cell types and within each glial lineage at different developmental stages. ErbB signaling has been most extensively studied in Schwann cells, where considerable evidence indicates that the erbB ligand neuregulin-1 (NRG1) promotes the proliferation, survival, and migration of axon-associated Schwann cell precursors and immature Schwann cells (20–27). As these cells mature, they no longer require NRG1 for survival (20) and at least some NRG1 isoforms instead promote myelination (28, 29). However, other NRG1 isoforms have been shown to induce demyelination in Schwann cell-neuron cocultures (30) and in mice overexpressing this NRG1 isoform in peripheral nerve (31), suggesting that different NRG1 isoforms may have distinct effects in these cells. Activation of erbB receptors similarly promotes the proliferation, survival, and differentiation of oligodendrocyte progenitors (32–37). Curiously, however, NRG1-induced erbB signaling promotes (38–40) but is not required for CNS myelination (41). Although less thoroughly explored, there is evidence that erbB signaling also influences astrocyte physiology, with astrocytes from different regions of the brain often demonstrating distinct responses to erbB ligands (42–45). Even microglia are responsive to these kinases, as shown by the recent finding that microglial proliferation and migration is enhanced by erbB signaling (46).

The erbB receptor-mediated actions noted above represent only a partial list of the known effects these molecules exert on neuronal and glial physiology. As a result of these and other studies, it is increasingly apparent that the consequences of erbB signaling are highly cell type-dependent and must be empirically determined in different populations of neurons and glia. The variability of the effects elicited by erbB receptors in different nervous system cell types in turn raises the question of precisely how this family of kinases elicits such diverse responses. The answer lies in part in the fact that the erbB receptors are part of a richly complex intercellular signaling network. In this network, cellular responses can be fine-tuned based on which erbB receptors are expressed and interact within the cell, which of the numerous erbB ligands activate these receptors, and whether key non-erbB membranous proteins that interact with erbB receptors are present in the targeted cell. Below, we will first consider how signaling complexity is generated by erbB signaling cascades. We will then turn to how this knowledge is applied in investigations of erbB effects on migration, a response that is observed in a wide variety of cells intrinsic to the peripheral and central nervous system.

1.2. Structure and Functional Characteristics of ErbB Membrane Tyrosine Kinases

The four erbB receptors share a common structure. These molecules are embedded in and pass through the cell membrane, resulting in the protrusion of a large region into the extracellular space. The extracellular domains of the erbB kinases contain two subdomains which mediate ligand binding (subdomains I and III; Fig. 1a) and two cysteine-rich subdomains (subdomains II and IV) which facilitate receptor dimerization (47–50). The extracellular domain is linked via a single membrane-spanning segment to a large intracellular domain. This intracellular domain contains both a highly conserved segment with tyrosine kinase activity and a carboxy-terminal autophosphorylation domain which, when phosphorylated on specific tyrosine, serine and/or threonine residues, provides docking sites for cytoplasmic signaling molecules. Precisely which residues within the autophosphorylation domain are phosphorylated (and consequently what cytoplasmic signaling molecules are recruited to the signaling complex) depends on a number of factors including the sequence of the receptor, which erbB receptors are present in the signaling complex, and the identity of the ligand activating the receptor.

Ligand binding to erbB receptors promotes receptor dimerization and the subsequent transphosphorylation of tyrosine residues within the autophosphorylation domains of the dimerized receptors. All four erbB receptors can form heterodimers with other erbB receptors; the EGF receptor and erbB4 can also homodimerize (51–55). Although virtually any combination of erbB kinases can hypothetically occur, in practice erbB2 is the preferred heterodimerization partner for the other three erbB kinases (56, 57). As cells in the nervous system often simultaneously express multiple erbB receptors, EGFR-erbB2, erbB2-erbB3 and erbB2-erbB4 heterodimers can all potentially form, with the specific erbB heterodimers occurring in a given cell type being dependent on the

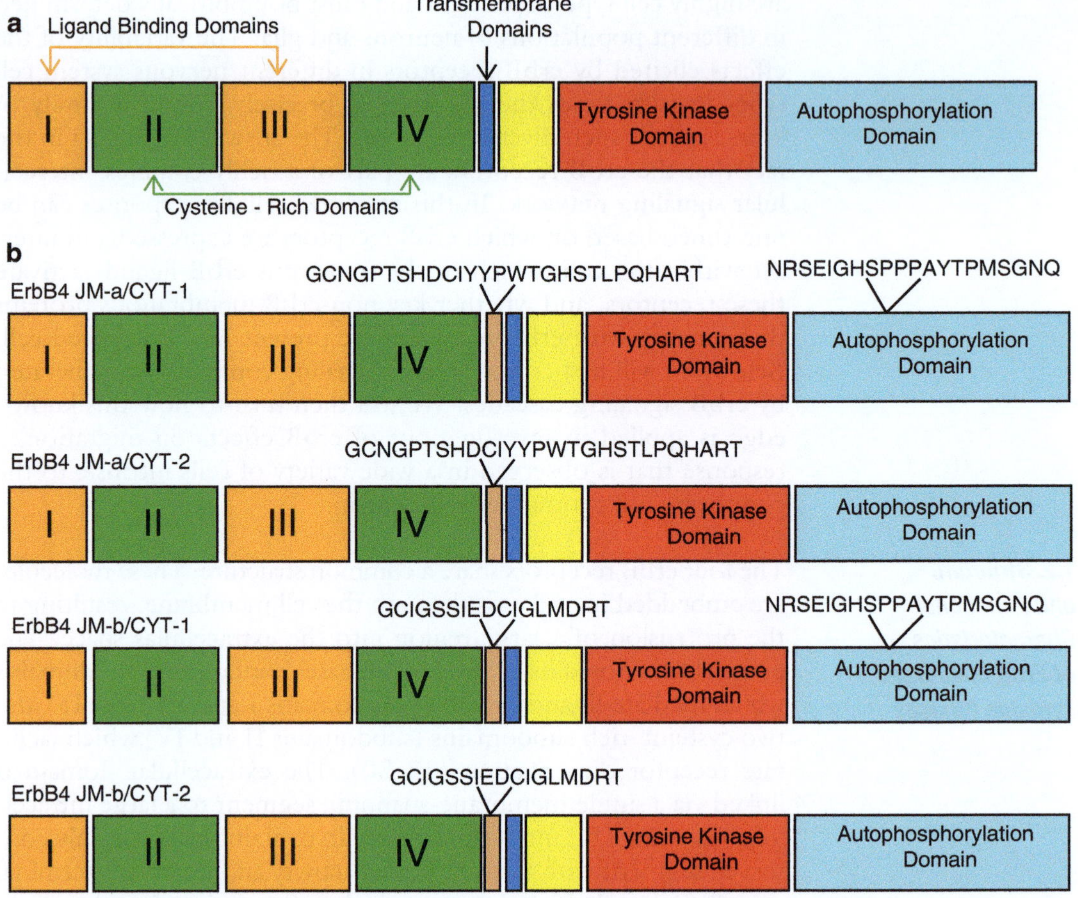

Fig. 1. The structure of erbB membrane receptor tyrosine kinases. (**a**) This diagram illustrates the shared structure of all four erbB kinases. The extracellular domain of these receptors (domains I–IV) is linked by a single transmembrane domain to a large cytoplasmic region which contains the tyrosine kinase activity (the most highly conserved domain in the erbB kinases) and the autophosphorylation domain. This latter domain shows the least degree of sequence conservation among the erbB kinases and contains multiple residues which, when phosphorylated, serve as docking sites for cytoplasmic signaling molecules. (**b**) ErbB4 kinases contain additional regions of structural variability that are not present in the EGF receptor, erbB2, or erbB3. These include an alternatively spliced exon in the autophosphorylation domain which encodes a PI-3 kinase docking site (CYT-1 variants) and a region of sequence variability in the "stalk" domain which is N-terminal of the transmembrane domain (JM-a and JM-b variants).

complement of erbB receptors expressed in the cell. Multiple lines of evidence indicate that each of the different erbB dimers elicits distinct signals in neurons and glia. First, each erbB kinase has distinct functional characteristics. For instance, erbB2 is a critically important component of erbB signaling complexes, but does not directly bind NRG proteins or EGF receptor ligands (58); direct binding of ligands is instead mediated by the EGF receptor, erbB3 or erbB4 (59). Further, erbB3, unlike the other three erbB kinases, has little endogenous tyrosine kinase activity (59). Second, specific erbB receptors differ in their ability to recruit cytoplasmic signaling molecules to the active signaling complex. For example, erbB3 contains multiple binding sites for the p85 subunit of phosphatidylinositol-3 kinase (PI3K), while EGFR and erbB2 are incapable of direct PI3K activation (59); some, but not all, erbB4 splice variants (see below) contain a single PI3K docking site (60). In contrast, erbB2 and EGFR contain docking sites for phospholipase Cγ (PLCγ), the ubiquitin ligase c-Cbl, Ras GAP, and the adaptor protein Grb2 (61, 62), whereas erbB3 does not directly interact with any of these molecules (62, 63). Third, the cytoplasmic residues which are phosphorylated in activated erbB receptors vary depending on their dimerization partner (64). Different erbB heterodimers thus elicit signals unique to that heterodimer rather than simply "summing up" standard sets of signals generated by each erbB kinase. Fourth, the level of erbB expression can be a factor in determining whether cells proliferate or exhibit other responses (e.g., apoptosis) when stimulated with ligand (65). Finally, although it does not bind NRG-1, the EGFR can dimerize with other erbB receptors, allowing the EGFR and NRG-1 receptor systems to "cross-talk" and thus modify cellular responses to their ligands (66).

Signaling by erbB4 is further diversified by the occurrence of splice variants of erbB4 mRNA which differ in two regions (51, 67, 68). First, some erbB4 mRNAs contain an alternatively spliced exon (exon 26) which encodes a novel 16 amino acid segment within the autophosphorylation domain that provides a binding site for the p85 subunit of PI3K (68); erbB4 variants containing this domain are referred to as CYT-1 isoforms, while those lacking it are designated CYT-2 isoforms (Fig. 1b). In addition, alternative splicing to include either exon 15 or 16, which encodes the "stalk" region (a small domain N-terminal to the transmembrane domain), gives rise to erbB4 variants referred to as JM-a and JM-b isoforms. The JM-a stalk region contains a proteolytic cleavage site, which allows the erbB4 ectodomain to be shed from the cell surface as a 120 kDa fragment (Fig. 2). ErbB4 cleavage, which is performed by tumor necrosis factor-α converting enzyme (TACE, also known as ADAM17) (69, 70), can be stimulated via distinct pathways involving either NRG1 (71) or protein kinase C activation (72). The remaining membrane-bound 80 kDa erbB4 fragment is then cleaved within the transmembrane domain by γ-secretase (73, 74), releasing the carboxy terminus and allowing it to translocate to the nucleus where it forms a complex with TAB2 and N-CoR that modulates gene expression (75). Alternatively, the erbB4 intracellular domain can accumulate within mitochondria, where it triggers apoptosis (76, 77).

Other non-erbB transmembrane proteins can also associate with erbB receptors and modify their signaling responses. Molecules demonstrated to have such effects include mucin 1 (Muc1) (78), CD44 (79), and integrins (80); this undoubtedly represents an

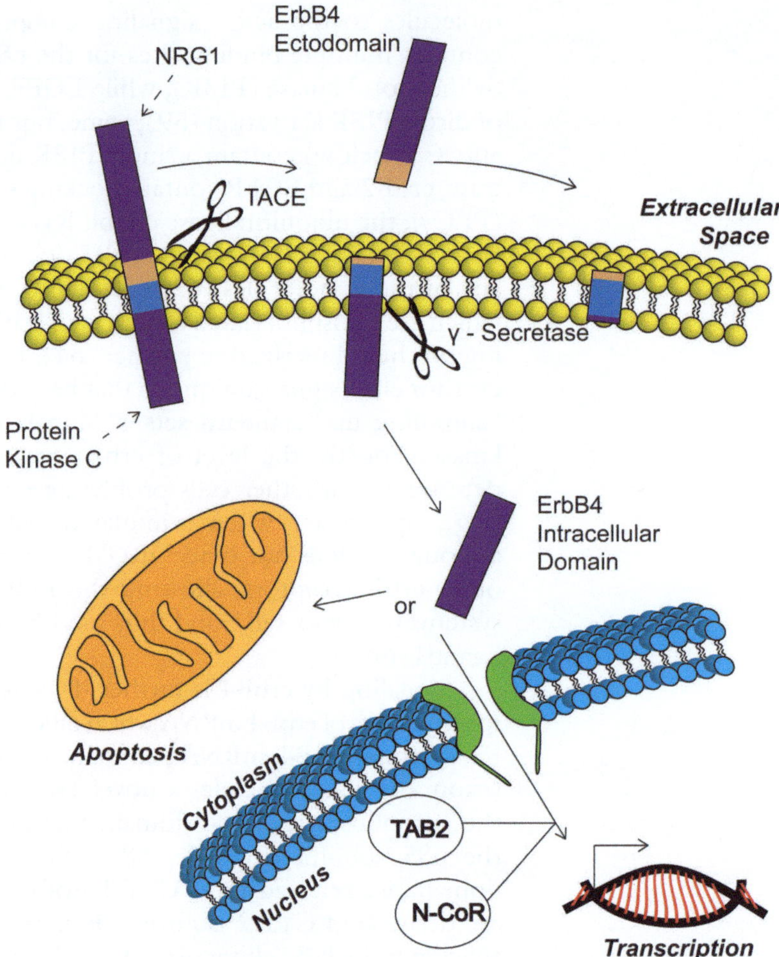

Fig. 2. Backsignaling by erbB4 isoforms containing a JM-a stalk region. These receptors are first cleaved by TACE within the stem region, releasing the erbB4 ectodomain into the extracellular space. The remaining fragment of the erbB receptor subsequently undergoes intramembranous cleavage by γ-secretase, releasing the erbB4 intracellular domain into the cytoplasm. The erbB4 intracellular domain can then either translocate into the nucleus where it forms a transcriptional regulatory complex with TAB2 and N-CoR or it can accumulate within mitochondria and trigger apoptosis.

incomplete list of such erbB associated transmembrane proteins and it is likely that others remain to be discovered. These associations play a particularly important role in responses such as erbB-mediated migration. Consequently, interactions between these non-erbB transmembrane proteins and substrate molecules must be factored in when investigating signaling mechanisms involved in erbB-stimulated migration.

1.3. ErbB Receptor Ligands

With the exception of erbB2, the erbB receptors serve as receptors for multiple EGF family ligands. These ligands can be divided into three groups, based on their relative affinity for specific erbB receptors.

Table 1
Specificity of ErbB receptor ligands

	EGFR	ErbB2	ErbB3	ErbB4
Ligands binding only EGFR				
Epidermal growth factor (EGF)	Yes	No	No	No
Transforming growth factor α (TGFα)	Yes	No	No	No
Amphiregulin	Yes	No	No	No
Epigen	Yes	No	No	No
Ligands binding both EGFR and ErbB4				
Epiregulin	Yes	No	No	Yes
Betacellulin	Yes	No	No	Yes
Heparin-binding epidermal growth factor (HB-EGF)	Yes	No	No	Yes
Ligands binding ErbB3 and/or ErbB4				
Neuregulin-1 (NRG1)	No	No	Yes	Yes
Neuregulin-2 (NRG2)	No	No	No	Yes
Neuregulin-3 (NRG3)	No	No	No	Yes
Neuregulin-4 (NRG4)	No	No	No	Yes
Ligands binding ErbB2				
Mucin 4 (Muc4)	No	Yes	No	No

The first group of erbB ligands, which consists of EGF, transforming growth factor α (TGFα), amphiregulin, and epigen, binds only the EGF receptor (Table 1). The second group of factors (heparin binding-EGF, epiregulin, and betacellulin) can bind directly to either the EGF receptor or erbB4. The third group, which interacts directly with erbB3 and/or erbB4, is represented by multiple forms of NRG1, NRG2, NRG3, and NRG4. In addition to these classic erbB ligands, mucin 4 (Muc4), a transmembrane protein containing two EGF-like domains, can interact with and activate erbB2 when these two molecules are coexpressed in the same cell (81).

A complete description of the structural variants that have been identified for every member of the large family of erbB ligands and how each of these variations impacts on inter- and intracellular signaling pathways is beyond the scope of this chapter. However, the mechanisms by which they act can be illustrated by considering our current understanding of the structure and function of NRG1, one of the most extensively studied erbB ligands. Via the use of alternative promoters and extensive alternative mRNA splicing, the NRG1 gene produces at least 31 different proteins (82). All of these NRG1 isoforms contain a single EGF-like domain in which an invariant N-terminal region is linked to one of two possible variable sequences (α or β domains; Fig. 3). This EGF-like domain is both necessary and sufficient for binding to erbB3 or erbB4;

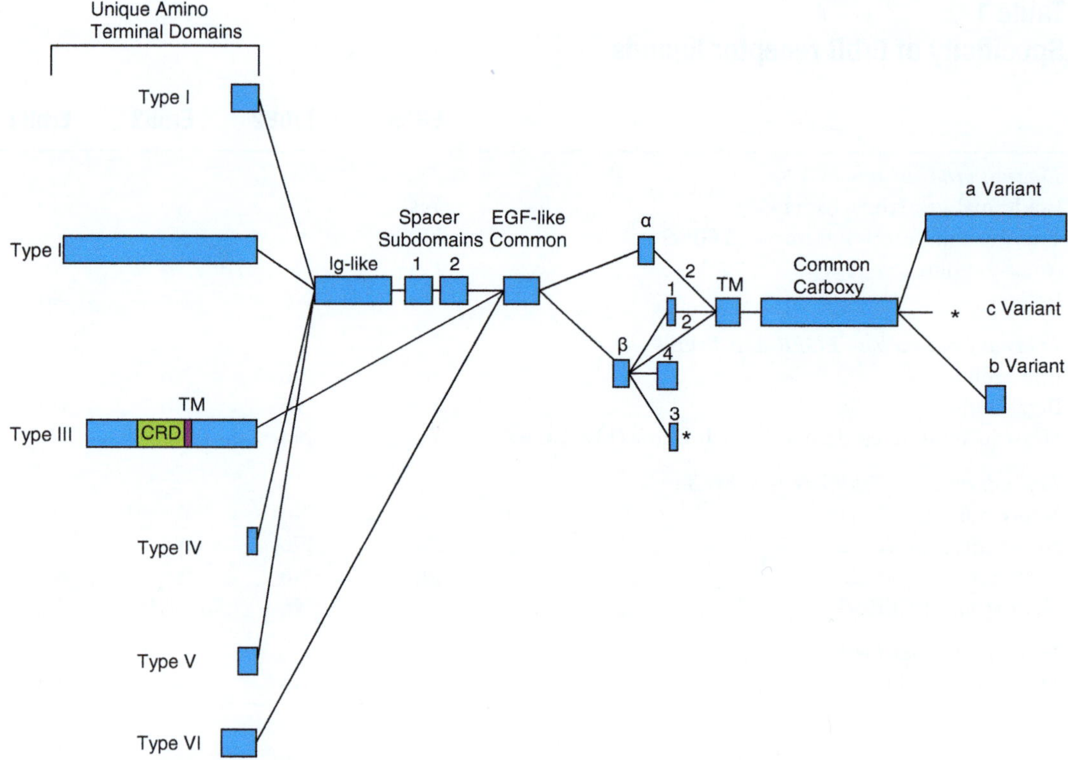

Fig. 3. Schematic illustrating the structural variability of NRG1 isoforms and the different domains present in the six major classes of NRG1 protein. All NRG1 proteins contain an EGF-like domain (the EGF-like common domain linked to either α or β domains) that mediates erbB binding. Six unique amino termini, which are generated by alternative promoter usage, have been identified and are used to classify these proteins as types I–VI. Types I, II, IV, and V contain Ig-like domains and 1 or 2 spacer subdomains. Type III NRG1 isoforms contain a unique amino terminus with a membrane-spanning segment and a cysteine-rich domain (CRD). In type VI isoforms, the amino terminal segment is linked directly to the EGF-like domain. These proteins may be synthesized as transmembrane precursors with one of three variant carboxy terminal domains (a, b, or c variants) or as directly secreted proteins (variants with a "3" juxtamembrane domain). *Asterisk*, sites of termination.

indeed, recombinant protein containing only the NRG1 EGF-like domain binds erbB receptors with an affinity similar to that of the holofactor and induces the same responses (83). Other regions within NRG1 proteins are much more variable. Six distinct NRG1 amino-terminal domains have been identified and are used to classify NRG1 proteins as types I through VI. The amino termini of type I [also known as neu differentiation factor, acetylcholine receptor inducing activity (ARIA) or heregulin], type II (glial growth factor; GGF), type IV, and type V NRG1 isoforms are followed by a heparin-binding immunoglobulin-like (Ig-like) domain that facilitates the concentration of these proteins in extracellular matrix. In many type I, II, IV, and V NRG1 isoforms, the Ig-like domain is linked to 1 or 2 heavily glycosylated "spacer" subdomains located N-terminal to the EGF-like domain. In contrast, the amino terminal domain of type VI NRG1 isoforms is connected directly to the EGF-like domain without intervening Ig-like or spacer domains. The large amino terminal domain of type III NRG1 isoforms differs even more radically, lacking both Ig-like and spacer domains and instead containing a centrally located cysteine-rich domain (CRD) flanked by a hydrophobic membrane-spanning segment.

NRG1 isoforms can be synthesized as membrane-spanning proproteins with the EGF-like domain oriented extracellularly. In NRG1 transmembrane precursors, variably spliced juxtamembrane domains ("1," "2," or "4" variants) link the extracellular domain to a hydrophobic transmembrane domain and a large structurally variable intracellular domain ("a," "b," or "c" carboxy terminal variants). Proteolytic cleavage of NRG1 transmembrane precursors occurs within the juxtamembrane domain and is performed by β-amyloid precursor protein cleaving enzyme (BACE) (84, 85), tumor necrosis factor-α converting enzyme (TACE, ADAM17) (86, 87), or meltrin-β (ADAM19) (88). In the case of type I, II, IV, V, and VI NRG1 transmembrane precursors, proteolytic cleavage releases soluble NRG1 into the extracellular space, where it diffuses to act on other cell types (Fig. 4). In contrast, cleavage within the juxtamembrane domain of type III NRG1 isoforms does not release these proteins from the cell membrane as they contain a second membrane-spanning domain within their amino terminus; juxtamembrane domain cleavage instead frees the type III NRG1 EGF-like domain to be displayed on the cell surface (Fig. 4), where it can act as a juxtacrine signaling molecule. NRG1 isoforms can also be synthesized as directly secreted factors (variants with a "3" juxtamembrane domain sequence) (89).

The large size and highly conserved nature of the NRG1 intracellular domain led to the suggestion that pro-NRG1 could also "back-signal" (Fig. 4) when stimulated by erbB receptors on adjacent cells or when binding erbB4 extracellular domain released by proteolytic cleavage (see Sect. 1.2) (90). Indeed, following pro-NRG1 cleavage within the juxtamembrane domain, the transmembrane domain of this molecule can be cleaved by γ-secretase thereby releasing the intracellular domain into the cytoplasm. Neuronal depolarization or challenge with the erbB4 extracellular domain both induce translocation of the NRG1 intracellular domain to the nucleus (90), where it interacts with the transcription factor Eos to regulate the transcription of PSD95 (91), a major postsynaptic density protein. There is also evidence that NRG1 backsignaling promotes redistribution of preexisting α7 nicotinic acetylcholine receptors via a PI3-kinase-dependent mechanism (92). Thus, under at least some circumstances, erbB receptors and their ligands can signal bidirectionally.

One question that is often asked is why so many ligands exist for each of the erbB receptors. Although the answer to this question is not completely clear, it may lie in the fact that different erbB ligands can induce very different responses even when activating

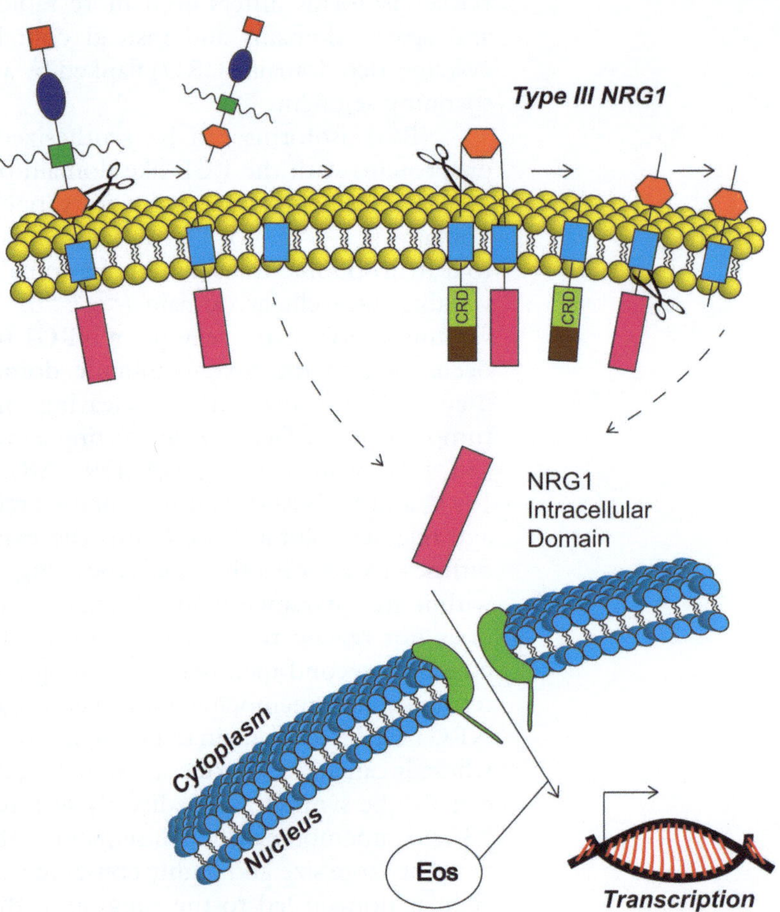

Fig. 4. Backsignaling by NRG1 transmembrane precursors. NRG1 transmembrane precursors are cleaved within the juxtamembrane domain by TACE, BACE or meltrin-β. For types I, II, IV, V, and VI NRG1 isoforms, this releases the NRG1 ectodomain into the extracellular space, where it acts as a diffusible growth factor; isoforms with a heparin-binding Ig-like domain (types I, II, IV, and V) may accumulate within basement membranes. In contrast, juxtamembrane domain cleavage of type III NRG1 isoforms allows the NRG1 EGF-like domain to be displayed on the cellular surface where it can participate in juxtacrine signaling. For all NRG1 transmembrane precursors, subsequent intramembranous cleavage by γ-secretase releases the NRG1 intracellular domain into the cytoplasm. This domain translocates into the nucleus where it associates with the transcription factor Eos to form a transcription regulatory complex. Although not indicated in this figure, the NRG1 cytoplasmic domain can also interact with and backsignal via modulation of LIMK1 activity. Type I NRG1 is illustrated as a representative of this group of isoforms. The following shapes in the type I NRG1 schematic indicate specific NRG1 extracellular domains: hexagon, EGF-like domain; square with wavy lines, glycosylated spacer domain; oval, Ig-like domain; square at N terminus, N-terminal domain.

the same receptors. For example, we have found that NRG-1β potently induces the migration of neoplastic Schwann cells, whereas NRG-1α actually inhibits their migration. In this setting, the differences in the responses elicited by NRG-1β and NRG-1α reflect differences in their ability to activate key cytoplasmic signaling molecules and the time course with which they do so (80). Similar differences have been noted in comparisons of NRG1 and NRG2 action (93). Unfortunately, studies systematically comparing the responses induced by multiple members of the erbB ligand family in specific cell types are virtually nonexistent and so little guidance can be offered in predicting how specific erbB ligands will behave in a given setting.

1.4. Designing Assays to Assess ErbB Action During Migration

Transwell migration assays are a modification of the Boyden chamber assay, a classic means of assessing cellular migration. Transwell assays are performed using disposable sterile tissue culture inserts whose undersurfaces are sealed by filters perforated by numerous pores of defined size. The undersurface of the filters is coated with a suitable migration substrate and the tissue culture plate well beneath the coated insert is filled with growth factor- or vehicle-containing medium (Fig. 5). Cells suspended in medium lacking growth factor are plated inside the insert and allowed to migrate for a defined period. At the end of this period, the number of cells that have migrated to the filter undersurface are counted to assess the effectiveness of migration in response to the growth factor.

Although conceptually simple, analyses of migratory responses stimulated by erbB ligands are actually quite complex due to the

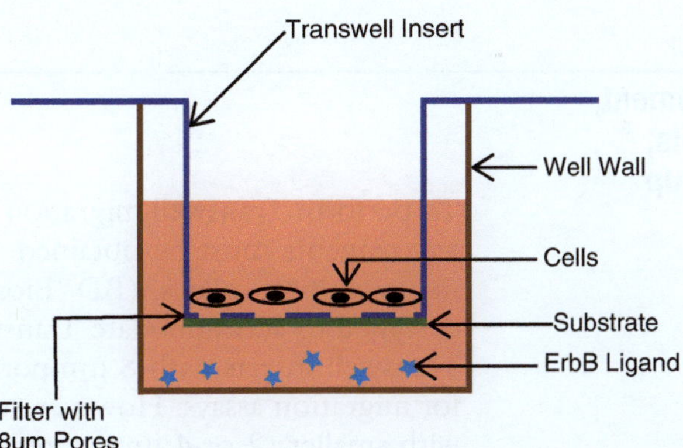

Fig. 5. The components of a Transwell migration assay and their locations. In this assay, a Transwell insert is placed in a tissue culture well. The *bottom* of this insert is covered by a filter pierced by numerous 8 μm pores that has been coated with migration substrate. The well beneath the insert is filled with tissue culture medium containing erbB ligand. Cells are plated inside the filter insert in medium lacking the growth factor and, if the erbB ligand has a promigratory effect, migrate through the filter pores and substrate onto the filter undersurface.

number of variables involved. For instance, as cell lineage, the state of growth of the cells and their health all can influence responsiveness to erbB ligands, these variables must be rigidly controlled. Non-erbB molecules that physically interact with both erbB receptors and substrate (e.g., integrins) can also influence erbB action. In settings where such interactions occur, erbB ligands will only induce migration on the substrate recognized by associated non-erbB molecule. Consequently, it is not sufficient to establish that a candidate cell type will migrate on a particular substrate—it must also be determined whether the growth factor of interest enhances migration on the substrate. Effects elicited by erbB ligands are often strongly concentration dependent. Consequently, a range of concentrations must be tested to establish whether a factor has promigratory activity. Further, given the distinct actions elicited by different erbB ligands, these parameters must be determined empirically for each cell type and growth factor.

The protocol outlined below is a modified Transwell migration assay protocol that we have developed to address these issues. Although originally developed to examine the role of erbB signaling in the migration of Schwann cells, astrocytes, and their neoplastic counterparts [malignant peripheral nerve sheath tumors (MPNST) and glioblastoma cells, respectively], it is widely applicable to a variety of nervous system cell types. It includes experimental variations designed to compare the relative effectiveness of different erbB ligands, establish the role individual erbB receptor subunits play in this process, assess chemotactic versus chemokinetic effects, and identify contributions made by other non-erbB membrane-associated molecules recruited to the mature erbB signaling complex.

2. Equipment, Materials, and Setup

To perform Transwell migration assays, a number of materials and key reagents must be obtained. For these assays, we use 24-well tissue culture plates (BD Biosciences; catalog #BD 353504) designed to accommodate Transwell filter inserts. We typically use Transwell inserts with 8 μm pores (catalog number BD 353097) for migration assays. However, investigators may wish to try filters with smaller (2 or 4 μm) pores, particularly if working with neoplastic cells, as some feel that migration through smaller pores is a more rigorous test of migratory responses. Inserts must be coated with an appropriate substrate (see Sect. 3.2). Substrates we have successfully used in migration assays include poly-L-lysine/laminin [Sigma (St. Louis, MO) #P5899 and Invitrogen #23017-015,

Table 2
Composition of complete and defined media used to prepare Schwann and MPNST cells for migration assays

Complete medium
Dulbecco's minimal essential medium (DMEM)
10% fetal bovine serum
10 IU/ml penicillin
10 μg/ml streptomycin
Serum-free Schwann cell-defined medium
1:1 mixture of DMEM and Ham's F-12 media
5 μg/ml insulin
100 μg/ml transferrin
60 μg/ml progesterone
16 μg/ml putrescine
5.2 ng/ml selenite
50 ng/ml thyroxine
50 ng/ml triiodothyronine
38 ng/ml dexamethasone

respectively], fibronectin (Sigma #F2518), and collagen type I (Sigma, #C7624).

If primary cultures are to be used, the method of isolation and complete culture medium used to maintain these cultures must be empirically established by the investigator; when maintaining permanent cell lines, we use the complete culture medium recommended by the investigators from whom we obtained the line. Prior to performing migration assays, cells must be serum-starved by incubation for 6–18 h in defined medium lacking the growth factor of interest as well as other growth factors that might stimulate migration. As an example, the composition of the complete medium and defined medium (94, 95) we use in Schwann cell and MPNST cell migration assays is presented in Table 2. During the migration assay itself, cells are maintained in Migration Assay Buffer (MAB), which is composed of a 1:1 mix of serum-free DMEM and Ham's F-12 medium supplemented with 0.1% fatty acid-free bovine serum albumin (BSA; Sigma #A6003). It is critically important that fatty acid-free BSA be used when preparing MAB as standard BSA preparations often contain large quantities of bound lysophosphatidic acid, which promotes the migration of many cell types.

Cells are stained with crystal violet to quantify the number of cells migrating through the Transwell filter. To make the staining reagent, a stock solution is first prepared by dissolving 1 g of crystal violet (Sigma #C3886) in 99 mL of 20% ethanol. The working solution for staining the cells is then prepared by mixing 20 mL of the crystal violet stock solution with 40 mL 95% ethanol and 150 mL of deionized distilled water.

EGF receptor ligands can be obtained from a number of commercial sources. We have used EGF (#E4127), betacellulin (#B3670), TGFα (#T7924), HB-EGF (#E4643), amphiregulin (#A7080), and epiregulin (#E8780) purchased from Sigma-Aldrich (St. Louis, MO) and found that these factors quite effectively stimulated migration under our assay conditions. Epigen (#1127-EP) purchased from R&D Systems (Minneapolis, MN) also works well under these same conditions. A number of laboratories have previously used the NRG1 β1 EGF-like domain from R&D Systems (#396-HB) in migration assays. However, we have found that it is also quite simple to produce large quantities of biologically active NRG1 in our own laboratory (95). We do not have experience with NRG2, NRG3, or NRG4 available from commercial sources.

Pan-erbB inhibitors, used to confirm that erbB activity is required for migration [PD168393 (# 513033) or PD158780 (# 513035)], can be purchased from Calbiochem (La Jolla, CA). If it is desirable to use a stably integrated, doxycycline-inducible shRNA system to ablate the expression of individual erbB kinases, we have found that the pSLIK lentiviral system (96), which is available from the American Type Culture Collection (see http://www.signaling-gateway.org/data/plasmid/ for details), works quite well for this purpose. For transient transfections, we have successfully used SureSilencing shRNA plasmids (SABiosciences; Valencia, CA) targeting human EGFR (#KH00137G), erbB2 (#KH00209G), and erbB3 (#KH00436G); we have transfected shRNA plasmids into cells using either lipofection or electroporation prior to performing migration assays.

3. Procedures

3.1. Basic Migration Assay Protocol

In brief, Transwell migration is accomplished by plating serum-starved cells onto Transwell migration filters coated with an appropriate substrate. Growth factors or inhibitors are then added and cells allowed to migrate for a defined period of time. At the end of this period, cells are stained with crystal violet and the cells that have migrated to the filter undersurface are visualized with light microscopy. Proper coating of filters with substrate is the first step in this process and it is one that is often plagued with technical difficulties.

The sequence in which Transwell filter inserts are coated with substrate depends upon the particular substrate that is to be used. For instance, when preparing poly-L-lysine/laminin-coated inserts, the filters are first coated with poly-L-lysine and then with laminin. When using other substrates (e.g., fibronectin or collagen type I), filters are coated directly with substrate without poly-L-lysine precoating. To coat Transwell filter inserts, the inserts are placed into tissue culture wells under sterile conditions. Using a pipetter equipped with a sterile tip, the pipet tip is inserted between the filter insert and the side of the well and 400 μL of substrate is introduced into the bottom of the well. Plates containing the inserts and substrate solution are then placed in a tissue culture incubator at 37°C for 1 h. At the end of this time, the substrate solution is aspirated off and discarded. Sterile phosphate buffered saline (PBS; 400 μL) is added to the well in the same manner as the substrate solution and then immediately aspirated; this is repeated twice to wash off excess substrate. For dual substrates such as poly-L-lysine/laminin, this procedure is repeated to apply the second substrate. Filters can be left in substrate overnight if desired. After application of substrate and PBS washes, nonspecific binding sites are blocked by incubating filter inserts with 1% fatty acid-free BSA for 1 h at 37°C. We use the filters immediately after completing BSA blocking.

Each substrate has its own optimal concentration for Transwell migration filter coating, as well as its own quirks in storage and handling. Poly-L-lysine, unlike other substrates, is dissolved in sterile deionized water rather than PBS at a 10 mg/mL concentration. As poly-L-lysine can be difficult to get into solution, it may be necessary to heat this solution to 37°C and rigorously pipet it. Poly-L-lysine should be dissolved and stored in plastic rather than glass containers, as poly-L-lysine deposition onto the walls of glass containers may alter the stock's concentration. Sterile filtration of poly-L-lysine stocks is also not recommended, as the concentration of the stock can also be altered by binding to the filtration apparatus. Diluted poly-L-lysine stocks are aliquoted and frozen at −20°C; repeated freeze–thaw cycles should be avoided. It is also important to realize that the molecular weight of the poly-L-lysine polymer may affect migration assays. We typically use 300,000 MW poly-L-lysine as a substrate. However, many manufacturers of poly-L-lysine accept a molecular weight range of 300,000–500,000 as acceptable for "300,000 MW" lysine polymers. We have found that lots of poly-L-lysine with molecular weights toward the upper end of this range (450,000–500,000) produce stainable films on filter inserts that interfere with cell counts (Fig. 6). As poly-L-lysine can be metabolized by mammalian cells, higher molecular weight polymers may also be lethal for some cell types.

Other substrates are typically easier to handle than poly-L-lysine. Undiluted (1 mg/mL) laminin purchased from the

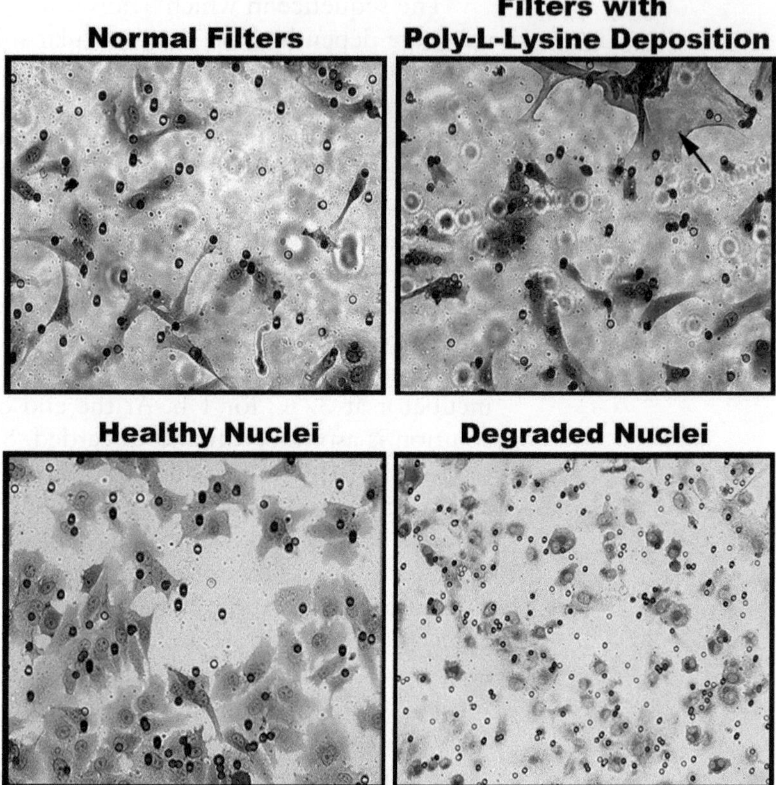

Fig. 6. Examples of technical artifacts commonly encountered in Transwell migration assays. The *top two panels* compare a typical result obtained with normally coated filters (*left*) and filters covered with a densely staining film (*right, arrow*) resulting from coating the filter with higher than normal molecular weight poly-L-lysine. The *bottom two panels* compare the cells in a recently completed migration assay (*left*) and cells on the undersurface of filters that have been stored for more than 2 weeks. Note the nuclear degradation in the latter cells.

manufacturer is aliquoted and stored at –80°C. To coat filters, laminin is dissolved in PBS at 10 μg/mL and used immediately. We never store aliquots of diluted laminin and we routinely discard any material remaining from an undiluted laminin aliquot. Fibronectin and collagen type I are both used as 40 μg/mL solutions and are handled similarly to laminin.

To perform migration assays, 400 μL of MAB is added to the well beneath the coated Transwell filter insert. Confluent cells (70%) that have been maintained overnight in serum-free defined medium are removed from their flask using prewarmed Cell Stripper (Mediatech, Inc.; Manassas, VA), centrifuged to form a pellet, washed with PBS, recentrifuged, and then resuspended in MAB. Cells (see Sect. 3.3 for a procedure to optimize the number of cells used) are plated directly onto the upper filter surface and allowed to adhere for 30 min at 37°C. Growth factor is then added to the well beneath the filter insert, after which plates are returned to the tissue culture incubator and cells allowed to migrate for an empirically determined time (see Sect. 3.3). Filters are washed twice with PBS and then stained overnight at 4°C with crystal violet. Plates must be tightly parafilmed during staining to avoid evaporation. The next morning, filters are rinsed four times with PBS and the cells on the top surface of the filter gently removed using a cotton-tipped applicator dipped in PBS.

Cells on the lower surface of the filter are visualized by bright field microscopy and the number of cells in distinct 20× fields counted (we typically count nine randomly chosen fields per filter). Only cells with visible nuclei are counted. Of note, filters must be photographed within 2 weeks of the experiment. We have found that storing filters for longer than this results in nuclear degradation (Fig. 6), which makes counts quite difficult.

3.2. Choosing an Optimal Substrate

The choice of what substrate to use in migration assays is critically important as choosing the wrong substrate may lead the investigator to incorrectly conclude that erbB receptor activation does not induce migration. It is not sufficient to determine whether a candidate substrate supports the migration of unstimulated cells; it must also be established that the growth factor of interest enhances migration on the substrate. As an example, in our studies of MPNST migration we tested fibronectin, collagen type I and poly-L-lysine/laminin as substrates and found that all three of these substrates supported MPNST migration. However, NRG-1β only enhanced MPNST migration on a poly-L-lysine/laminin substrate, and we subsequently found that the integrins that act as laminin receptors are physically associated with activated erbB kinases in these cells (80). Also, it cannot be assumed that the same substrate will be optimal for all erbB ligands in a specific cell type. As evidence of this, we have found that induction of MPNST cell migration by EGF receptor ligands occurs on substrates that do not support NRG-1β-induced migration (unpublished observations).

To determine what substrates should be tested, we look to the in vivo environment of the cell type of interest and ask what substrates are normally available to the cells. We then perform preliminary experiments in which 20,000 cells per filter are plated on the candidate substrates, challenged for 6 h (a time which works well for a variety of cell types) with a range of erbB ligand concentrations (we typically use 1, 5, and 10 nM concentrations of ligand for this purpose) and then processed and counted as described in Sect. 3.1. Using a range of erbB ligand concentrations is essential as migration assays often show bell-shaped curves in which concentrations higher than the optimal concentration produce decreased levels of migration. Suitable substrates will both support migration of the cells and support enhanced migration in the presence of erbB ligand relative to that seen in cells challenged with vehicle.

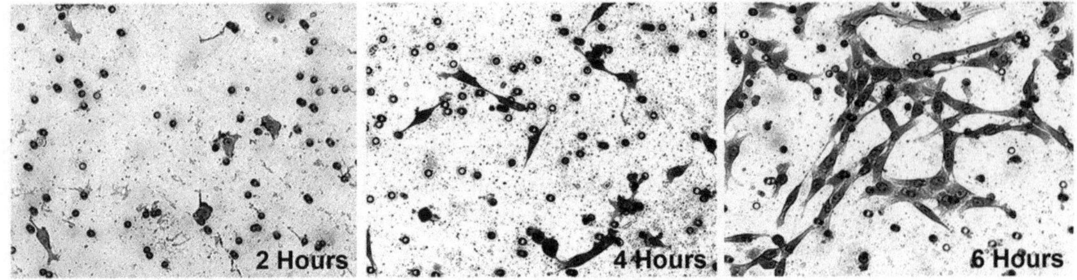

Fig. 7. Optimization of migration time in Transwell migration assays. Shown are representative fields of filter undersurfaces following cell migration for 2, 4, or 6 h. Note that cellular processes are just beginning to protrude through the filter pores at 2 h and that relatively few cells have migrated through the pores by 4 h. For this line, 6 h is the optimal migration time.

3.3. Optimization of Migration Time and Cell Numbers

After selecting a suitable substrate, the optimal period allowed for migration is determined. Using the optimal substrate determined as described above, migration assays are performed in which 20,000 cells are plated in the filter insert, with MAB alone in the chamber beneath the insert. Cells are allowed to migrate for 2, 4, or 6 h and then processed as described in Sect. 3.1. If there are no cells on the filter, they may have already passed through the membrane and into the well bottom; alternatively, not enough time has passed for migration to occur. To distinguish between these possibilities, wash the bottom of well and check the wash to see if cells are present. If cells are present, migration has gone too long. Before staining with crystal violet solution, the number of cells remaining on the upper surface of the filter should also be examined; if numerous cells are present on top of the filter, migration has not gone long enough. Examples of suboptimal and optimal results are presented in Fig. 7.

After establishing an optimal migration time, the optimal number of cells to be plated must be determined. We do this by plating 5,000, 10,000, 20,000, and 40,000 cells per filter and allowing the cells to migrate in the absence of erbB ligand for the optimal migration period. The optimal number of plated cells will result in 10–20 cells per 20× field migrating to the filter undersurface. Representative images illustrating the desired result are presented in Fig. 8.

3.4. Assessment of ErbB Ligand Effects on Migration

We have found that erbB ligand effects on migration are highly concentration dependent. Graphs relating the number of cells per microscopic field to ligand concentration typically produce a bell-shaped curve in which migration is decreased with higher concentrations of erbB ligand. Given this, it is evident that testing only a single concentration of erbB ligand is hazardous, as a randomly chosen concentration higher or lower than the optimal concentration will lead to the erroneous conclusion that the factor has no promigratory activity. Consequently, we typically test a range of erbB ligand concentrations between 0.1 and 20 nM.

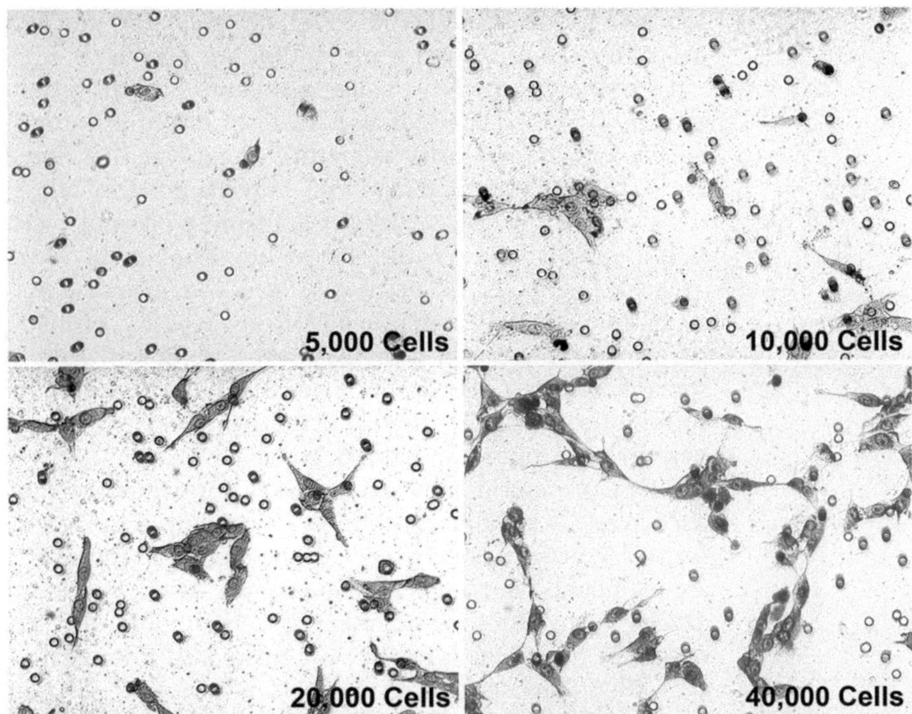

Fig. 8. Optimization of cell numbers in Transwell migration assays. Shown are representative fields of filter undersurfaces 6 h after 5,000, 10,000, 20,000, and 40,000 cells were plated on the upper surface of the filter with MAB alone present in the well. Note that relatively few cells are present in each field when either 5,000 or 10,000 cells are plated. In contrast, larger numbers of cells are present in each field when either 20,000 or 40,000 cells were plated on the upper surface of the filter; these cells can be individually distinguished in each of these conditions, which facilitates counts of the number of cells per field. In this case, we selected 20,000 cells per field as optimal. This number of cells allows us to readily assess basal levels of migration in the absence of erbB ligand, while still insuring that accurate quantification of cell numbers can be performed if stimulation with erbB ligand enhances migration.

After determining the concentration of erbB ligand that optimally induces migration, it is often desirable to establish whether the factor has chemotactic activity (directed movement toward a higher concentration of factor) or chemokinetic activity (a non-directed increase in movement). To distinguish between these possibilities, migration is compared in three conditions: (1) erbB ligand added only to the well beneath the Transwell insert; (2) erbB ligand added only to media within the Transwell insert; and (3) erbB ligand added to both chambers. If the factor has chemotactic activity, the first condition will produce the greatest increase in migration. If the factor has predominantly chemokinetic activity, all three conditions will produce similar results.

3.5. Assessment of the Role of Specific ErbB Kinases in Migratory Responses

To verify that erbB kinase activity is required for migration induced by erbB ligands, pan-erbB inhibitors are placed in both the top and bottom chambers of the Transwell apparatus 30 min prior to the addition of growth factor and assays performed as described above.

We have found that PD168393 (1 nM to 1 μm) and PD158780 (10 nM to 10 μm) both work well for this purpose.

Combining shRNA-mediated ablation of erbB receptors with migration assays allows an assessment of the contribution individual kinases make to the migratory phenotype. As noted above, erbB shRNAs can be introduced into mammalian cells by either transient transfection of shRNA-containing plasmid vectors or by stable transduction with lentiviruses containing doxycycline-inducible shRNAs. In either case, it is essential that the optimal timeframe for erbB receptor knockdown be established prior to performing migration assays. As erbB receptors have relatively short half-lives (<10 h), 72 h of shRNA expression is usually sufficient to produce a good knock-down. However, this should be determined empirically for each cell line.

Ease of experimental set-up makes transient transfection ideal, particularly for primary cultures. Transient transfection is performed in plasticware coated with the migration substrate and cells are maintained in complete tissue culture media post-transfection. Twelve to twenty-four hours prior to the empirically determined optimum timepoint for shRNA-mediated erbB knockdown (typically 48–60 h post-transfection), serum starvation is begun. At this time, cell density should be no more than 70%. Seventy-two hours post-transfection, the serum-starved cells are plated in substrate-coated Transwell inserts and migration assays conducted as described above.

If the variability often encountered in transient transfections is an issue, a doxycycline-inducible stable transduction system may be a better approach. This technique also has the advantage of a built-in negative control (namely, the cell line in the uninduced state). However, doxycycline-inducible systems require a large amount of "up-front" work and doxycycline itself has phenotypic effects in many mammalian cell types. Consequently, we recommend that investigators test the effects of doxycycline on their cells prior to investing considerable effort in this approach. We have utilized the pSLIK lentiviral system (96) for this purpose. In this system, shRNAs are designed to target the erbB receptors using the RNAi Codex algorithm (http://katahdin.cshl.org:9331/portal/scripts/main2.pl); the design of these oligonucleotides is modified to incorporate overhangs complementary to those created in the pen_TTGmiRc2 entry vector by *Bfu*A1 digestion. Annealed shRNA oligonucleotides are cloned into the digested entry vector, and the product of this ligation recombined with the pSLIK vector using LR clonase (Invitrogen). We prefer the pSLIK-hygromycin vector, as hygromycin is a relatively inexpensive antibiotic and works well for selection in mammalian cells. However, neomycin- and puromycin-resistant forms of this vector are also available.

To make packaged virus, the recombined lentiviral vector is co-transfected into 239FT cells with PLP1, PLP2, and VSVg helper plasmids using Polyfect reagent (Qiagen). PLP1 plasmid (10 μg), 2 μg PLP2 plasmid, 1 μg VSVg plasmid, and 6 μg of erbB-shRNA pSLIK-hygro plasmid are diluted in serum-free, antibiotic-free DMEM (40 μL per μg DNA) and then incubated with Polyfect reagent (10 μL per μg DNA) for 10 min. The reaction mixture is then carefully pipetted over 50–60% confluent 239FT cells plated in a T75 flask containing fresh DMEM-10. Twenty-four hours later, the medium is replaced with fresh DMEM-10 containing 1 μM sodium butyrate, which increases viral titers. Seventy-two post-transfection, lentivirus-containing supernatant is collected. Lentiviral supernatant can be stored at –80°C in cryovials for future use. To avoid freeze–thaw cycles (which reduce viral titer), 1.5–2.0 mL lentiviral supernatant should be aliquoted per cryovial. BSL2 guidelines (http://www.cdc.gov/od/ohs/biosfty/bmbl4/bmbl4toc.htm) must be followed when handling the supernatant.

To make stably transduced cells, cells are plated in tetracycline-free DMEM-10 in a six-well plate at a density that allows them to reach confluency within 72–96 h. After allowing the cells to adhere overnight, the media is removed and replaced with 1.5 mL of lentiviral supernatant plus 0.5 mL of fresh tetracycline-free DMEM-10 and 6 μg/ml Polybrene. Twenty-four hours later, the lentivirus-containing media is replaced with fresh tetracycline-free DMEM-10. After an additional 24–48 h, the cells are split into tetracycline-free DMEM-10 supplemented with the selection antibiotic. Selection antibiotic concentrations must be determined empirically for each cell line; with hygromycin, we find that concentrations in the range of 50 μg–100 μg/ml are usually appropriate. Good selection depends on the cells being singly dispersed in the selection medium. Seeding density is also critical; we have found that splitting each well of a six-well plate into three 100 mm tissue culture dishes is optimal. Cell death should begin within 6 h of exposure to antibiotic, with maximal cell death achieved by 3 days. Subsequently, single cells should be visibly dispersed throughout the plates. Fresh tetracycline-free DMEM-10 is added every 2–3 days for 1–2 weeks, until small colonies begin to form from the singly dispersed cells. Colonies are removed from plates using cloning disks soaked in Cell Stripper, with each clone being transferred into a single well of a 24-well plate. Most clones will take 1–2 weeks to expand in the 24-well plate, after which they can be expanded into six-well plates.

Expanded clones must be screened for appropriate reduction in the target mRNA and/or protein after shRNA induction. We initially screen with real-time PCR, as large numbers of samples can be screened at one time. Clones demonstrating a reduction in the target mRNA following shRNA induction are then rescreened by immunoblotting. As we have observed clonal differences in erbB expression, both uninduced and doxycycline-induced samples should be screened for each clone rather than screening against the parent line or a single uninduced clone. The pSLIK vector also expresses GFP upon doxycycline induction, but we have found that neither the presence nor the relative level of green fluorescence correlates well to protein knockdown. Doxycycline (2 μg/ml) generally produces optimal knockdown in cells transduced with the pSLIK vector, but this should be verified by performing a doxycycline concentration curve in newly generated stable cell lines.

Migration assays using stably transduced cell lines are performed identically to those described above except for a period of doxycycline induction prior to the assay. As with all shRNA studies, key controls should be performed. Two shRNAs directed against different sequences within the targeted erbB mRNA should be tested to identify potential off-target effects. A nonsense shRNA sequence should also be tested to potential nonspecific activation of RISC. Also, as erbB receptors share significant sequence homology, the expression of all erbB receptors should be examined in doxycycline-free and doxycycline-induced samples to verify that knockdown is specific for the targeted erbB kinase.

4. Conclusion

ErbB membrane tyrosine kinases and their multitude of ligands utilize a surprisingly complex variety of signaling mechanisms to exert their effect. The protocols described in this chapter were developed with these varied mechanisms in mind and allow the rational design of experiments testing the role erbB receptors play in migratory responses. However, our understanding of erbB signaling is undoubtedly still incomplete. Given this, it is hoped that the material included in this chapter will provide a useful basis for examining new novel erbB signaling mechanisms as they are identified.

Acknowledgments

The development of the experimental protocols described in this chapter was supported by grants from the National Institute of Neurological Diseases and Stroke (R01 NS048353 to S.L.C.; F30 NS063626 to N.M.B.), the National Cancer Institute (R01 CA122804 to S.L.C.; R01 CA134773 to Kevin A. Roth and S.L.C.), and the Department of Defense (X81XWH-09-1-0086 to S.L.C.). L.T.P. was supported in part by a grant to the University of Alabama at Birmingham from the Howard Hughes Medical

Institute through the Med into Grad Initiative. We thank the Alabama Neuroscience Blueprint Core Center (P30 NS57098) and the UAB Neuroscience Core Center (P30 NS47466) for technical assistance with studies from our laboratory that are described in this chapter. The content is solely the responsibility of the authors and does not necessarily represent the official views of the National Institutes of Health or the Department of Defense.

References

1. Gerecke KM, Wyss JM, Karavanova I, Buonanno A, Carroll SL (2001) ErbB transmembrane tyrosine kinase receptors are differentially expressed throughout the adult rat central nervous system. J Comp Neurol 433:86–100
2. Golding JP, Trainor P, Krumlauf R, Gassmann M (2000) Defects in pathfinding by cranial neural crest cells in mice lacking the neuregulin receptor ErbB4. Nat Cell Biol 2:103–109
3. Rio C, Rieff HI, Qi P, Khurana TS, Corfas G (1997) Neuregulin and erbB receptors play a critical role in neuronal migration. Neuron 19:39–50
4. Anton ES, Marchionni MA, Lee KF, Rakic P (1997) Role of GGF/neuregulin signaling in interactions between migrating neurons and radial glia in the developing cerebral cortex. Development 124:3501–3510
5. Gierdalski M, Sardi SP, Corfas G, Juliano SL (2005) Endogenous neuregulin restores radial glia in a (ferret) model of cortical dysplasia. J Neurosci 25:8498–8504
6. Schmid RS, McGrath B, Berechid BE, Boyles B, Marchionni M, Sestan N, Anton ES (2003) Neuregulin 1-erbB2 signaling is required for the establishment of radial glia and their transformation into astrocytes in cerebral cortex. Proc Natl Acad Sci U S A 100:4251–4256
7. Yau HJ, Wang HF, Lai C, Liu FC (2003) Neural development of the neuregulin receptor ErbB4 in the cerebral cortex and the hippocampus: preferential expression by interneurons tangentially migrating from the ganglionic eminences. Cereb Cortex 13:252–264
8. Flames N, Long JE, Garratt AN, Fischer TM, Gassmann M, Birchmeier C, Lai C, Rubenstein JL, Marin O (2004) Short- and long-range attraction of cortical GABAergic interneurons by neuregulin-1. Neuron 44:251–261
9. Anton ES, Ghashghaei HT, Weber JL, McCann C, Fischer TM, Cheung ID, Gassmann M, Messing A, Klein R, Schwab MH, Lloyd KC, Lai C (2004) Receptor tyrosine kinase ErbB4 modulates neuroblast migration and placement in the adult forebrain. Nat Neurosci 7: 1319–1328
10. Gerecke KM, Wyss JM, Carroll SL (2004) Neuregulin-1beta induces neurite extension and arborization in cultured hippocampal neurons. Mol Cell Neurosci 27:379–393
11. Bermingham-McDonogh O, McCabe KL, Reh TA (1996) Effects of GGF/neuregulins on neuronal survival and neurite outgrowth correlate with erbB2/neu expression in developing rat retina. Development 122:1427–1438
12. Rieff HI, Raetzman LT, Sapp DW, Yeh HH, Siegel RE, Corfas G (1999) Neuregulin induces GABA(A) receptor subunit expression and neurite outgrowth in cerebellar granule cells. J Neurosci 19:10757–10766
13. Lopez-Bendito G, Cautinat A, Sanchez JA, Bielle F, Flames N, Garratt AN, Talmage DA, Role LW, Charnay P, Marin O, Garel S (2006) Tangential neuronal migration controls axon guidance: a role for neuregulin-1 in thalamocortical axon navigation. Cell 125:127–142
14. Ozaki M, Sasner M, Yano R, Lu HS, Buonanno A (1997) Neuregulin-beta induces expression of an NMDA-receptor subunit. Nature 390:691–694
15. Ozaki M, Tohyama K, Kishida H, Buonanno A, Yano R, Hashikawa T (2000) Roles of neuregulin in synaptogenesis between mossy fibers and cerebellar granule cells. J Neurosci Res 59:612–623
16. Yang X, Kuo Y, Devay P, Yu C, Role L (1998) A cysteine-rich isoform of neuregulin controls the level of expression of neuronal nicotinic receptor channels during synaptogenesis. Neuron 20:255–270
17. Subramony P, Dryer SE (1997) Neuregulins stimulate the functional expression of Ca^{2+}-activated K^{+} channels in developing chicken parasympathetic neurons. Proc Natl Acad Sci U S A 94:5934–5938
18. Cameron JS, Lhuillier L, Subramony P, Dryer SE (1998) Developmental regulation of neuronal

K+ channels by target-derived TGF beta in vivo and in vitro. Neuron 21:1045–1053

19. Kwon OB, Paredes D, Gonzalez CM, Neddens J, Hernandez L, Vullhorst D, Buonanno A (2008) Neuregulin-1 regulates LTP at CA1 hippocampal synapses through activation of dopamine D4 receptors. Proc Natl Acad Sci U S A 105:15587–15592
20. Dong Z, Brennan A, Liu N, Yarden Y, Lefkowitz G, Mirsky R, Jessen KR (1995) Neu differentiation factor is a neuron-glia signal and regulates survival, proliferation, and maturation of rat Schwann cell precursors. Neuron 15:585–596
21. Grinspan JB, Marchionni MA, Reeves M, Coulaloglou M, Scherer SS (1996) Axonal interactions regulate Schwann cell apoptosis in developing peripheral nerve: neuregulin receptors and the role of neuregulins. J Neurosci 16:6107–6118
22. Syroid DE, Maycox PR, Burrola PG, Liu N, Wen D, Lee KF, Lemke G, Kilpatrick TJ (1996) Cell death in the Schwann cell lineage and its regulation by neuregulin. Proc Natl Acad Sci U S A 93:9229–9234
23. Li Y, Tennekoon GI, Birnbaum M, Marchionni MA, Rutkowski JL (2001) Neuregulin signaling through a PI3K/Akt/Bad pathway in Schwann cell survival. Mol Cell Neurosci 17:761–767
24. Maurel P, Salzer JL (2000) Axonal regulation of Schwann cell proliferation and survival and the initial events of myelination requires PI 3-kinase activity. J Neurosci 20:4635–4645
25. Meintanis S, Thomaidou D, Jessen KR, Mirsky R, Matsas R (2001) The neuron-glia signal beta-neuregulin promotes Schwann cell motility via the MAPK pathway. Glia 34:39–51
26. Ogata T, Iijima S, Hoshikawa S, Miura T, Yamamoto S, Oda H, Nakamura K, Tanaka S (2004) Opposing extracellular signal-regulated kinase and Akt pathways control Schwann cell myelination. J Neurosci 24:6724–6732
27. Vartanian T, Goodearl A, Lefebvre S, Park SK, Fischbach G (2000) Neuregulin induces the rapid association of focal adhesion kinase with the erbB2-erbB3 receptor complex in schwann cells. Biochem Biophys Res Commun 271:414–417
28. Michailov GV, Sereda MW, Brinkmann BG, Fischer TM, Haug B, Birchmeier C, Role L, Lai C, Schwab MH, Nave KA (2004) Axonal neuregulin-1 regulates myelin sheath thickness. Science 304:700–703
29. Taveggia C, Zanazzi G, Petrylak A, Yano H, Rosenbluth J, Einheber S, Xu X, Esper RM, Loeb JA, Shrager P, Chao MV, Falls DL, Role L, Salzer JL (2005) Neuregulin-1 type III determines the ensheathment fate of axons. Neuron 47:681–694
30. Zanazzi G, Einheber S, Westreich R, Hannocks MJ, Bedell-Hogan D, Marchionni MA, Salzer JL (2001) Glial growth factor/neuregulin inhibits Schwann cell myelination and induces demyelination. J Cell Biol 152:1289–1299
31. Huijbregts RP, Roth KA, Schmidt RE, Carroll SL (2003) Hypertrophic neuropathies and malignant peripheral nerve sheath tumors in transgenic mice overexpressing glial growth factor beta3 in myelinating Schwann cells. J Neurosci 23:7269–7280
32. Canoll PD, Musacchio JM, Hardy R, Reynolds R, Marchionni MA, Salzer JL (1996) GGF/neuregulin is a neuronal signal that promotes the proliferation and survival and inhibits the differentiation of oligodendrocyte progenitors. Neuron 17:229–243
33. Milner R, Anderson HJ, Rippon RF, McKay JS, Franklin RJ, Marchionni MA, Reynolds R, Ffrench-Constant C (1997) Contrasting effects of mitogenic growth factors on oligodendrocyte precursor cell migration. Glia 19:85–90
34. Raabe TD, Suy S, Welcher A, DeVries GH (1997) Effect of neu differentiation factor isoforms on neonatal oligodendrocyte function. J Neurosci Res 50:755–768
35. Shi J, Marinovich A, Barres BA (1998) Purification and characterization of adult oligodendrocyte precursor cells from the rat optic nerve. J Neurosci 18:4627–4636
36. Fernandez PA, Tang DG, Cheng L, Prochiantz A, Mudge AW, Raff MC (2000) Evidence that axon-derived neuregulin promotes oligodendrocyte survival in the developing rat optic nerve. Neuron 28:81–90
37. Calaora V, Rogister B, Bismuth K, Murray K, Brandt H, Leprince P, Marchionni M, Dubois-Dalcq M (2001) Neuregulin signaling regulates neural precursor growth and the generation of oligodendrocytes in vitro. J Neurosci 21:4740–4751
38. Wang Z, Colognato H, Ffrench-Constant C (2007) Contrasting effects of mitogenic growth factors on myelination in neuron-oligodendrocyte co-cultures. Glia 55:537–545
39. Roy K, Murtie JC, El-Khodor BF, Edgar N, Sardi SP, Hooks BM, Benoit-Marand M, Chen C, Moore H, O'Donnell P, Brunner D, Corfas G (2007) Loss of erbB signaling in oligodendrocytes alters myelin and dopaminergic function, a potential mechanism for neuropsychiatric disorders. Proc Natl Acad Sci U S A 104:8131–8136
40. Taveggia C, Thaker P, Petrylak A, Caporaso GL, Toews A, Falls DL, Einheber S, Salzer JL (2008)

Type III neuregulin-1 promotes oligodendrocyte myelination. Glia 56:284–293

41. Brinkmann BG, Agarwal A, Sereda MW, Garratt AN, Muller T, Wende H, Stassart RM, Nawaz S, Humml C, Velanac V, Radyushkin K, Goebbels S, Fischer TM, Franklin RJ, Lai C, Ehrenreich H, Birchmeier C, Schwab MH, Nave KA (2008) Neuregulin-1/ErbB signaling serves distinct functions in myelination of the peripheral and central nervous system. Neuron 59:581–595
42. Prevot V, Cornea A, Mungenast A, Smiley G, Ojeda SR (2003) Activation of erbB-1 signaling in tanycytes of the median eminence stimulates transforming growth factor beta1 release via prostaglandin E2 production and induces cell plasticity. J Neurosci 23:10622–10632
43. Prevot V, Lomniczi A, Corfas G, Ojeda SR (2005) erbB-1 and erbB-4 receptors act in concert to facilitate female sexual development and mature reproductive function. Endocrinology 146:1465–1472
44. Prevot V, Rio C, Cho GJ, Lomniczi A, Heger S, Neville CM, Rosenthal NA, Ojeda SR, Corfas G (2003) Normal female sexual development requires neuregulin-erbB receptor signaling in hypothalamic astrocytes. J Neurosci 23:230–239
45. Sharif A, Duhem-Tonnelle V, Allet C, Baroncini M, Loyens A, Kerr-Conte J, Collier F, Blond S, Ojeda SR, Junier MP, Prevot V (2009) Differential erbB signaling in astrocytes from the cerebral cortex and the hypothalamus of the human brain. Glia 57:362–379
46. Calvo M, Zhu N, Tsantoulas C, Ma Z, Grist J, Loeb JA, Bennett DL (2010) Neuregulin-ErbB signaling promotes microglial proliferation and chemotaxis contributing to microgliosis and pain after peripheral nerve injury. J Neurosci 30:5437–5450
47. Lax I, Bellot F, Howk R, Ullrich A, Givol D, Schlessinger J (1989) Functional analysis of the ligand binding site of EGF-receptor utilizing chimeric chicken/human receptor molecules. EMBO J 8:421–427
48. Summerfield AE, Hudnall AK, Lukas TJ, Guyer CA, Staros JV (1996) Identification of residues of the epidermal growth factor receptor proximal to residue 45 of bound epidermal growth factor. J Biol Chem 271:19656–19659
49. Garrett TP, McKern NM, Lou M, Elleman TC, Adams TE, Lovrecz GO, Zhu HJ, Walker F, Frenkel MJ, Hoyne PA, Jorissen RN, Nice EC, Burgess AW, Ward CW (2002) Crystal structure of a truncated epidermal growth factor receptor extracellular domain bound to transforming growth factor alpha. Cell 110:763–773
50. Ogiso H, Ishitani R, Nureki O, Fukai S, Yamanaka M, Kim JH, Saito K, Sakamoto A, Inoue M, Shirouzu M, Yokoyama S (2002) Crystal structure of the complex of human epidermal growth factor and receptor extracellular domains. Cell 110:775–787
51. Plowman GD, Culouscou JM, Whitney GS, Green JM, Carlton GW, Foy L, Neubauer MG, Shoyab M (1993) Ligand-specific activation of HER4/p180erbB4, a fourth member of the epidermal growth factor receptor family. Proc Natl Acad Sci U S A 90:1746–1750
52. Culouscou JM, Plowman GD, Carlton GW, Green JM, Shoyab M (1993) Characterization of a breast cancer cell differentiation factor that specifically activates the HER4/p180erbB4 receptor. J Biol Chem 268:18407–18410
53. Plowman GD, Green JM, Culouscou JM, Carlton GW, Rothwell VM, Buckley S (1993) Heregulin induces tyrosine phosphorylation of HER4/p180erbB4. Nature 366:473–475
54. Ferguson KM, Darling PJ, Mohan MJ, Macatee TL, Lemmon MA (2000) Extracellular domains drive homo- but not hetero-dimerization of erbB receptors. EMBO J 19:4632–4643
55. Mendrola JM, Berger MB, King MC, Lemmon MA (2002) The single transmembrane domains of ErbB receptors self-associate in cell membranes. J Biol Chem 277:4704–4712
56. Tzahar E, Waterman H, Chen X, Levkowitz G, Karunagaran D, Lavi S, Ratzkin BJ, Yarden Y (1996) A hierarchical network of interreceptor interactions determines signal transduction by Neu differentiation factor/neuregulin and epidermal growth factor. Mol Cell Biol 16:5276–5287
57. Graus-Porta D, Beerli RR, Daly JM, Hynes NE (1997) ErbB-2, the preferred heterodimerization partner of all ErbB receptors, is a mediator of lateral signaling. EMBO J 16:1647–1655
58. Sliwkowski MX, Schaefer G, Akita RW, Lofgren JA, Fitzpatrick VD, Nuijens A, Fendly BM, Cerione RA, Vandlen RL, Carraway KL III (1994) Coexpression of erbB2 and erbB3 proteins reconstitutes a high affinity receptor for heregulin. J Biol Chem 269:14661–14665
59. Carraway KL III, Cantley LC (1994) A neu acquaintance for erbB3 and erbB4: a role for receptor heterodimerization in growth signaling. Cell 78:5–8
60. Sawyer C, Hiles I, Page M, Crompton M, Dean C (1998) Two erbB-4 transcripts are expressed in normal breast and in most breast cancers. Oncogene 17:919–924
61. Kassis J, Moellinger J, Lo H, Greenberg NM, Kim HG, Wells A (1999) A role for phospholipase C-gamma-mediated signaling in tumor cell invasion. Clin Cancer Res 5:2251–2260

62. Fedi P, Pierce JH, di Fiore PP, Kraus MH (1994) Efficient coupling with phosphatidylinositol 3-kinase, but not phospholipase C gamma or GTPase-activating protein, distinguishes ErbB-3 signaling from that of other ErbB/EGFR family members. Mol Cell Biol 14:492–500
63. Yarden Y, Sliwkowski MX (2001) Untangling the ErbB signalling network. Nat Rev Mol Cell Biol 2:127–137
64. Olayioye MA, Neve RM, Lane HA, Hynes NE (2000) The ErbB signaling network: receptor heterodimerization in development and cancer. EMBO J 19:3159–3167
65. Tikhomirov O, Carpenter G (2004) Ligand-induced, p38-dependent apoptosis in cells expressing high levels of epidermal growth factor receptor and ErbB-2. J Biol Chem 279:12988–12996
66. Tzahar E, Yarden Y (1998) The ErbB-2/HER2 oncogenic receptor of adenocarcinomas: from orphanhood to multiple stromal ligands. Biochim Biophys Acta 1377:M25–M37
67. Elenius K, Corfas G, Paul S, Choi CJ, Rio C, Plowman GD, Klagsbrun M (1997) A novel juxtamembrane domain isoform of HER4/ErbB4. Isoform-specific tissue distribution and differential processing in response to phorbol ester. J Biol Chem 272:26761–26768
68. Elenius K, Choi CJ, Paul S, Santiestevan E, Nishi E, Klagsbrun M (1999) Characterization of a naturally occurring ErbB4 isoform that does not bind or activate phosphatidyl inositol 3-kinase. Oncogene 18:2607–2615
69. Vecchi M, Carpenter G (1997) Constitutive proteolysis of the ErbB-4 receptor tyrosine kinase by a unique, sequential mechanism. J Cell Biol 139:995–1003
70. Rio C, Buxbaum JD, Peschon JJ, Corfas G (2000) Tumor necrosis factor-alpha-converting enzyme is required for cleavage of erbB4/HER4. J Biol Chem 275:10379–10387
71. Zhou W, Carpenter G (2000) Heregulin-dependent trafficking and cleavage of ErbB-4. J Biol Chem 275:34737–34743
72. Vecchi M, Baulida J, Carpenter G (1996) Selective cleavage of the heregulin receptor ErbB-4 by protein kinase C activation. J Biol Chem 271:18989–18995
73. Ni CY, Murphy MP, Golde TE, Carpenter G (2001) gamma-Secretase cleavage and nuclear localization of ErbB-4 receptor tyrosine kinase. Science 294:2179–2181
74. Lee HJ, Jung KM, Huang YZ, Bennett LB, Lee JS, Mei L, Kim TW (2002) Presenilin-dependent gamma-secretase-like intramembrane cleavage of ErbB4. J Biol Chem 277:6318–6323
75. Sardi SP, Murtie J, Koirala S, Patten BA, Corfas G (2006) Presenilin-dependent ErbB4 nuclear signaling regulates the timing of astrogenesis in the developing brain. Cell 127:185–197
76. Naresh A, Long W, Vidal GA, Wimley WC, Marrero L, Sartor CI, Tovey S, Cooke TG, Bartlett JM, Jones FE (2006) The ERBB4/HER4 intracellular domain 4ICD is a BH3-only protein promoting apoptosis of breast cancer cells. Cancer Res 66:6412–6420
77. Naresh A, Thor AD, Edgerton SM, Torkko KC, Kumar R, Jones FE (2008) The HER4/4ICD estrogen receptor coactivator and BH3-only protein is an effector of tamoxifen-induced apoptosis. Cancer Res 68:6387–6395
78. Schroeder JA, Thompson MC, Gardner MM, Gendler SJ (2001) Transgenic MUC1 interacts with epidermal growth factor receptor and correlates with mitogen-activated protein kinase activation in the mouse mammary gland. J Biol Chem 276:13057–13064
79. Yu WH, Woessner JF Jr, McNeish JD, Stamenkovic I (2002) CD44 anchors the assembly of matrilysin/MMP-7 with heparin-binding epidermal growth factor precursor and ErbB4 and regulates female reproductive organ remodeling. Genes Dev 16:307–323
80. Eckert JM, Byer SJ, Clodfelder-Miller BJ, Carroll SL (2009) Neuregulin-1 beta and neuregulin-1 alpha differentially affect the migration and invasion of malignant peripheral nerve sheath tumor cells. Glia 57:1501–1520
81. Carraway KL, Theodoropoulos G, Kozloski GA, Carothers Carraway CA (2009) Muc4/MUC4 functions and regulation in cancer. Future Oncol 5:1631–1640
82. Mei L, Xiong WC (2008) Neuregulin 1 in neural development, synaptic plasticity and schizophrenia. Nat Rev Neurosci 9:437–452
83. Holmes WE, Sliwkowski MX, Akita RW, Henzel WJ, Lee J, Park JW, Yansura D, Abadi N, Raab H, Lewis GD et al (1992) Identification of heregulin, a specific activator of p185erbB2. Science 256:1205–1210
84. Hu X, Hicks CW, He W, Wong P, Macklin WB, Trapp BD, Yan R (2006) Bace1 modulates myelination in the central and peripheral nervous system. Nat Neurosci 9:1520–1525
85. Willem M, Garratt AN, Novak B, Citron M, Kaufmann S, Rittger A, DeStrooper B, Saftig P, Birchmeier C, Haass C (2006) Control of peripheral nerve myelination by the beta-secretase BACE1. Science 314:664–666

86. Loeb JA, Susanto ET, Fischbach GD (1998) The neuregulin precursor proARIA is processed to ARIA after expression on the cell surface by a protein kinase C-enhanced mechanism. Mol Cell Neurosci 11:77–91
87. Montero JC, Rodriguez-Barrueco R, Yuste L, Juanes PP, Borges J, Esparis-Ogando A, Pandiella A (2007) The extracellular linker of pro-neuregulin-alpha2c is required for efficient sorting and juxtacrine function. Mol Biol Cell 18:380–393
88. Yokozeki T, Wakatsuki S, Hatsuzawa K, Black RA, Wada I, Sehara-Fujisawa A (2007) Meltrin beta (ADAM19) mediates ectodomain shedding of Neuregulin beta1 in the Golgi apparatus: fluorescence correlation spectroscopic observation of the dynamics of ectodomain shedding in living cells. Genes Cells 12: 329–343
89. Buonanno A, Fischbach GD (2001) Neuregulin and ErbB receptor signaling pathways in the nervous system. Curr Opin Neurobiol 11: 287–296
90. Bao J, Wolpowitz D, Role LW, Talmage DA (2003) Back signaling by the Nrg-1 intracellular domain. J Cell Biol 161:1133–1141
91. Bao J, Lin H, Ouyang Y, Lei D, Osman A, Kim TW, Mei L, Dai P, Ohlemiller KK, Ambron RT (2004) Activity-dependent transcription regulation of PSD-95 by neuregulin-1 and Eos. Nat Neurosci 7:1250–1258
92. Hancock ML, Canetta SE, Role LW, Talmage DA (2008) Presynaptic type III neuregulin1-ErbB signaling targets {alpha}7 nicotinic acetylcholine receptors to axons. J Cell Biol 181:511–521
93. Crovello CS, Lai C, Cantley LC, Carraway KL III (1998) Differential signaling by the epidermal growth factor-like growth factors neuregulin-1 and neuregulin-2. J Biol Chem 273: 26954–26961
94. Frohnert PW, Stonecypher MS, Carroll SL (2003) Lysophosphatidic acid promotes the proliferation of adult Schwann cells isolated from axotomized sciatic nerve. J Neuropathol Exp Neurol 62:520–529
95. Frohnert PW, Stonecypher MS, Carroll SL (2003) Constitutive activation of the neuregulin-1/ErbB receptor signaling pathway is essential for the proliferation of a neoplastic Schwann cell line. Glia 43:104–118
96. Shin KJ, Wall EA, Zavzavadjian JR, Santat LA, Liu J, Hwang JI, Rebres R, Roach T, Seaman W, Simon MI, Fraser ID (2006) A single lentiviral vector platform for microRNA-based conditional RNA interference and coordinated transgene expression. Proc Natl Acad Sci U S A 103:13759–13764

Chapter 8

The Extracellular Signal-Regulated Kinase (ERK) Cascade in Neuronal Cell Signaling

Daniel Orellana, Ilaria Morella, Marzia Indrigo, Alessandro Papale, and Riccardo Brambilla

Abstract

The Ras-controlled extracellular signal-regulated kinase (ERK) pathway mediates a large number of cellular events, from proliferation to survival, from synaptic plasticity to memory formation. In order to study the role of the two major ERK isoforms in the brain, ERK1 and ERK2, we have generated GFP fusion proteins of both protein kinases. In addition, we have produced two swapped constructs in which the N-term tail of ERK1, the domain responsible for its unique properties, has either been removed from ERK1 or attached to ERK2. We demonstrated that all four GFP proteins are properly expressed in vitro in mouse embryo fibroblasts. However, only ERK1 and ERK2 > 1 overexpression resulted in a significant growth retardation. In addition, we have expressed all four GFP fusion constructs in vivo, in the adult brain, using lentiviral vector-assisted transgenesis and found that they are predominantly neuronal. Finally, we have devised an ex-vivo system, in which brain slices prepared from adult mice can be stimulated with glutamate and the activation of both cytoplasmic and nuclear substrates of ERK can be detected. Since phosphorylation of both the ribosomal protein S6 and of the histone H3 is completely prevented by a chemical inhibition of the ERK pathway, this ex-vivo system can be exploited in the future to investigate the regulation of the ERK cascade using slices from LV-injected or conventional transgenic animals.

Key words: Ras-ERK signaling, Cell proliferation, Lentiviral vectors, Glutamate, MEK inhibitors, Brain slices, Immunofluorescence

1. Introduction

Cell communication is a major process controlling life of both unicellular and multicellular organisms. Within cells, a broad range of different signaling pathways is involved in these communications. Among them, special attention has been pointed to the mitogen-activated protein (MAP) kinase system, a major signal transduction pathway mediating the transfer of extracellular stimuli to the

Hideyuki Mukai (ed.), *Protein Kinase Technologies*, Neuromethods, vol. 68,
DOI 10.1007/978-1-61779-824-5_8, © Springer Science+Business Media, LLC 2012

nucleus. The MAPK system is highly conserved throughout eukaryotic evolution, from yeast to humans. The core component of this pathway is made up of a three-tier protein kinase system which is upstream regulated, at the cell membrane, by members of the Ras superfamily of small GTPases. The MAPK family of cellular pathways in mammalian cells is divided into four distinct subfamilies named following the last protein kinase of the signal transduction cascade: ERK (extracellular signal-regulated kinase), JNK (c-Jun *N-terminal* kinase), p38 SAPK (stress-activated protein kinase), and ERK5/BMK1 (big MAP kinase1) subfamily (1–4).

Once activated, ERK signaling may be involved, depending on the cell type and on the physiological context, in a wide range of processes, ranging from embryogenesis, cell differentiation, cell survival, cell proliferation, and cancer. In the brain, this cascade provides a link between ionotropic, metabotropic, and neurotrophin receptors to cytosolic (regulation of ion channels and of protein translation) and nuclear events, leading to gene transcription, de novo protein synthesis and changes either in synaptic remodeling and plasticity, memory formation, or neuronal survival, depending on the context. Once activated by neurotransmitter receptors through GTP/GDP exchange factors, the small GTPases belonging to the Ras class (p21 H-, K- and N-Ras gene products) stimulate sequentially the cascade of protein kinases consisting of serine/threonine kinases of the Raf subfamily (mainly c-Raf and B-Raf, the MAPK Kinase Kinase tier), the threonine/tyrosine dual-specificity kinases MEK1/2 (MAPK Kinase), and finally ERK1/2 proteins (the MAPK component) (5). More specifically, activation of MEK1/2 leads to a selective interaction with ERK proteins through specific docking domains which result in phosphorylation of the conserved recognition motif of threonine and tyrosine (TEY domain), within the activation loop of ERK1/2 (6). ERK1 and ERK2, also known as p44/p42MAPK, respectively, are homologous isoforms produced by two genes, erk1 (MAPK3) and erk2 (MAPK1). Both of them have nearly 85% amino acid identity with much greater identity in the core regions and are expressed in essentially all cells, even if ERK2 is the predominant isoform in brain and hematopoietic cells (7). Once activated, ERK1 and ERK2, the two major MAPK in the brain, are able to translocate into the nucleus. There, they can activate, either directly or indirectly (via the kinases of the RSK families), transcription factors, including the CREB-like class of transcriptional regulators, or to regulate chromatin remodeling (via the kinases of the MSK family) by for instance phosphorylating histone H3 (8–10). The ability of ERK in regulating gene expression and chromatin organization is believed to be a crucial step not only in the processes of neural adaptations underlying normal cognitive processes but also in the onset of several neuropsychiatric disorders (11–15). In addition, this signaling pathway also controls protein translation in the cytosol and in the dendrites by activating the mTor pathway that regulates a number of factors including the S6 kinase which directly phosphorylate the S6 ribosomal protein (16).

ERK1 and ERK2 are activated by the same stimuli and are believed to bear similar substrate recognition properties and subcellular localization (17). Because of their great similarity, the two isoforms have been considered for a long time as largely interchangeable, with mostly overlapping functions, and are traditionally designated as a single ERK1/2 entity; this concept is also derived from the use, in several studies of the ERK pathway, of MEK inhibitors (such as U0126 or SL327) that do not discriminate between MEK1 and MEK2 (18–20). Although some studies still claim that ERK isoforms contribute to ERK signaling according to the ratio of their expression levels and not to their specific properties, recent evidence has highlighted critical functional differences between these two isoforms. A strong support in this direction derives from the results obtained using ERK1 or ERK2 knockout (KO) mice. ERK2 KO mice die early in development, showing that ERK1 cannot compensate for ERK2 in the embryo (21–26). On the contrary, *erk1* gene ablation is fully compatible with the adult life (27–29). Experiments at the molecular level in a variety of cell types showed that in ERK1-deficient cells (taken from ERK1 KO mice) the stimulus-dependent activation of ERK2 was enhanced, without any compensatory increase in ERK2 protein levels (27, 29). Other studies have used RNA interference techniques to silence the activity of ERK1, obtaining its knockdown in cells: as a result, cell growth was facilitated (7, 30). Interestingly, the phenotype caused by ERK1 knockdown, i.e., an enhancement of ERK2 activity, strongly parallel that observed when a constitutive active form of MEK1 is overexpressed via lentiviral vectors into mouse embryo fibroblasts (MEF) cells. On the contrary, the ERK2 knockdown phenotype is similar to what observed with a dominant negative form of MEK1 (30). Importantly, ERK2 activity enhancement in the nervous system of ERK1 mutant mice has been linked to improvement of certain forms of learning and memory (29, 31). The model put forward to explain these functional differences between ERK1 and ERK2 is that while ERK2 appears to be the main MAPK isoform transducing the signal to the nucleus, ERK1 seems to act like a partial agonist, antagonizing the binding of ERK2 with interacting partners. Importantly, the structural basis of these differences has recently been located to the unique N-terminal moiety of ERK1, which causes a significant delay in the nuclear/cystoplamic shuttling. Indeed, a final confirmation that the N-terminal portion of ERK1 is responsible for the observed phenotype comes from the swap of this domain between the two ERK isoforms: ERK1 without its N-term behaves in proliferating cells like ERK2 while ERK2 fused to the ERK1 N-term causes a ERK1-like phenotype (17).

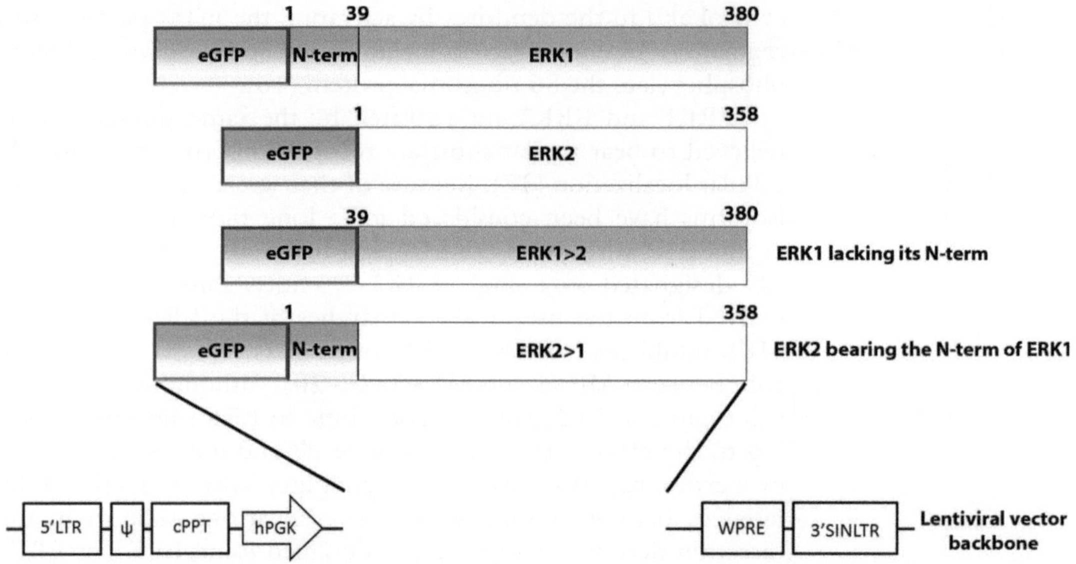

Fig. 1. Lentiviral vectors for ERK1, ERK2, and their swapped GFP fusion proteins. Schematic representation of the different ERK constructs in which the N-term moiety (1-39aa) of ERK1, conferring unique properties, is either present or absent. The ERK constructs were subcloned in a LV vector (pCCLsin.PPT.hPGK.eGFP.PRE) for in vivo expression studies.

In order to study in vivo in the brain the functional differences between ERK1 and ERK2, we generated GFP fusion proteins not only of ERK1 and ERK2 but also of their swapped counterparts: ERK1 > 2 (ERK1 deprived of its N-term) and ERK2 > 1 (ERK2 with the ERK1 N-term attached) (Fig. 1). These constructs have been introduced in a lentiviral vector (LV) and high-titer viral preparations have been obtained. LV is a powerful and efficient tool to modify gene expression in primary cells without the need to use a selection marker, which is a major complication for in vitro study of cell proliferation. Supporting the competition model described above, here we show that the overexpression of ERK1 and of ERK2 > 1 fusion proteins into MEF causes a reduction in the cell growth (Fig. 2). On the contrary, overexpression of ERK2 and of ERK1 > 2 fusion proteins results in a growth advantage. Importantly, once validated their activity in vitro, these LV constructs can now be used to study the function of ERK1 and ERK2 kinases in the brain. Stereotactic injections at the dorsal striatum of the mouse brain of such LV lead to a strong visualization of all four GFP fusion proteins. These proteins can be observed within 4 weeks from the surgical procedure. Superimposition of the fluorescence signals of GFP expressing cells (in green) with either a neuron-specific marker (NeuN, in red, Fig. 3a) or with a glial-specific staining (GFAP, in red, Fig. 3b) demonstrates a preferential neuronal localization of all four LV-expressed ERK–GFP fusion constructs. In order to study biochemically the function of these

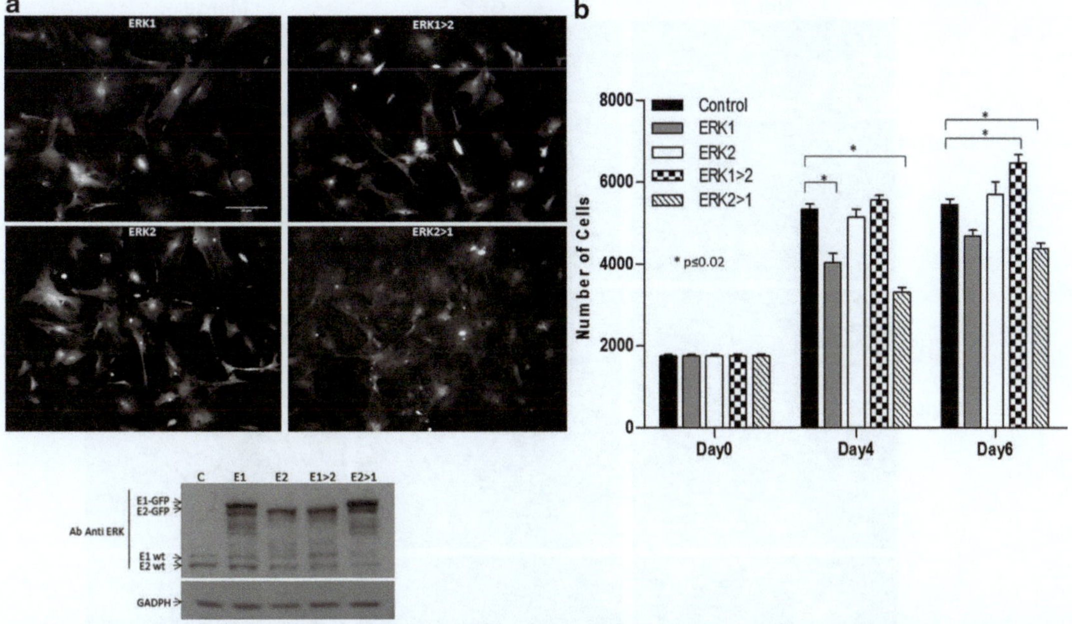

Fig. 2. Effects of ERK constructs in MEF proliferation. (**a**) Images of MEF cells infected with the LV expressing different GFP fusion proteins (MOI = 5). MEF cells were infected with LV vectors expressing ERK1, ERK2, and their swapped counterparts ERK1 > 2 and ERK2 > 1. Four days after infection, the GFP protein was found expressed ubiquitously in the cell. Bar = 50 μm. In the low left insert a representative Western blot is shown. A whole-cell extract was immunoblotted for ERK and GADPH (loading control). The GFP fusion proteins of ERK1, ERK2, and their swapped counterparts were easily detected by immunoblotting. (**b**) Proliferation assay on MEF transduced with LV for ERK1, ERK2, and their swapped versions. Proliferation was assessed by counting every day the cells using a Coulter Counter. 1.25×10^5 cells/well were seeded in six-well plate and maintained in culture medium. The following day (day 0), cells were transduced at MOI = 5 with lentiviral vectors expressing ERK1, ERK2, and their swapped counterparts or GFP as a control. At the end of the experiment (day 6), a significant growth retardation was observed in cells infected with LV for ERK1 and ERK2 > 1. Results are the mean of $n = 6$.

GFP fusion proteins, we have also set up an ex-vivo system in which brain slices can be freshly prepared from adult mice, incubated in a perfusing chamber and stimulated with appropriate agonist and antagonists. As shown in Figs. 4 and 5, striatal slices can be challenged with glutamate and induction of ERK signaling can be monitored at the single cell level using phospho-specific antibodies either against ribosomal protein S6 (pS6) or histone H3 (pH3). Quantification of double pS6 or pH3 (in red) and anti-NeuN staining (in green) allow us to monitor changes in ERK activity. Importantly, pretreating the slices with U0126, a specific MEK1/2 inhibitor, completely blocks S6 and H3 phosphorylation. Having validated this ex-vivo system, brain slices can in the future be prepared from LV-injected mice and/or from other transgenic animals and the activity of the ERK pathway can be easily assessed with a tight spatio-temporal resolution.

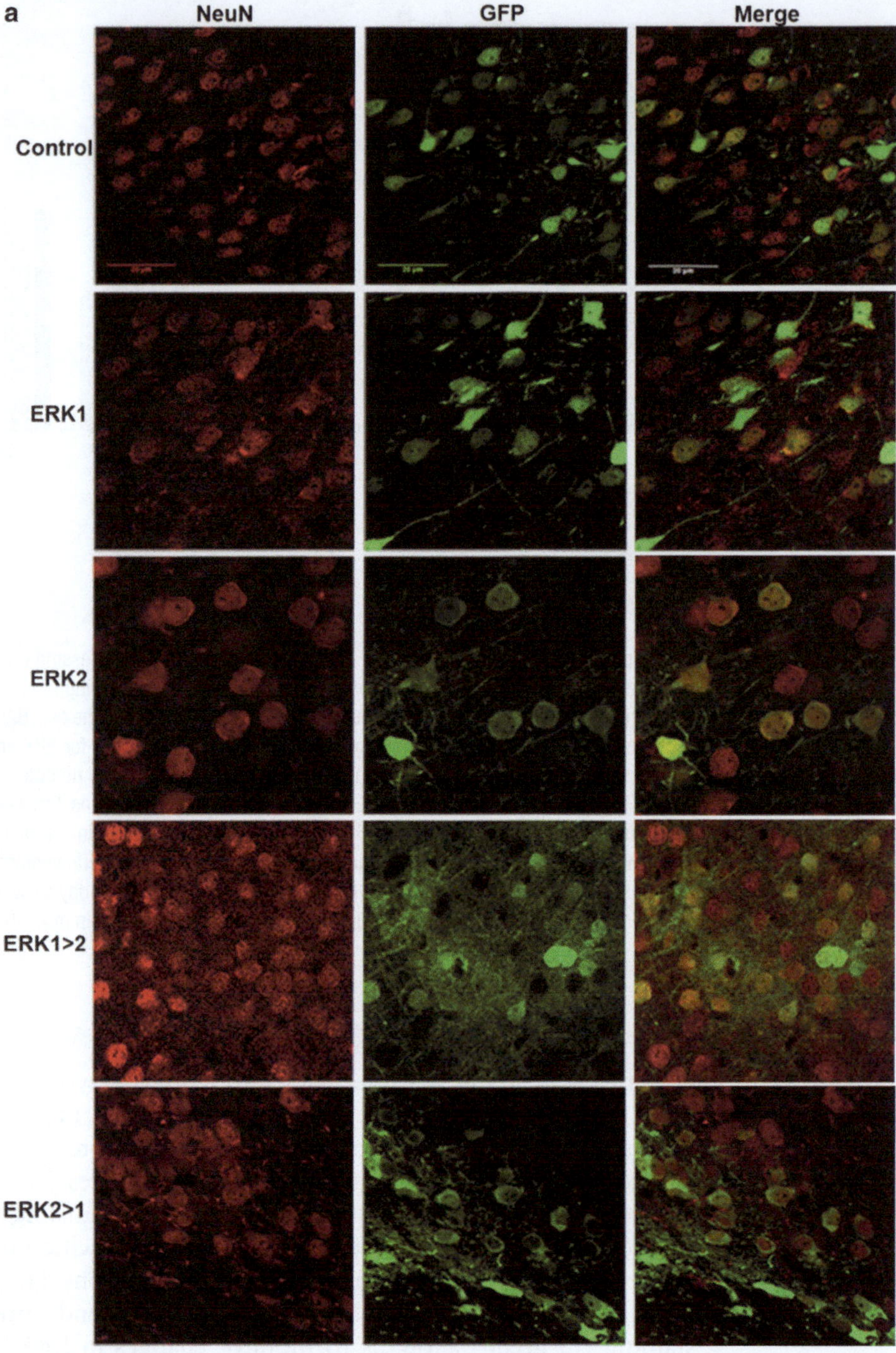

Fig. 3. In vivo expression of the LV constructs in the mouse brain. Images from the mouse striatum infected with LV for the different ERK constructs. Sections of the mouse brain were studied by immunofluorescence at 4 weeks following dorsal striatum injections. Immunofluorescence for either (**a**) NeuN (a neuron-specific marker, in *red*) or, (**b**) GFAP (a glial-specific marker, in *red*) demonstrated a preferential neuronal expression mediated by the lentiviral vectors. Smaller neurons show obvious colocalization of GFP and the primarily nuclear NeuN expression. Some neurons do not include the nucleus in the confocal optical plane of section and thus do not appear NeuN-positive in these images, although they would show co-localization in subsequent focal planes. Bar = 20 μm.

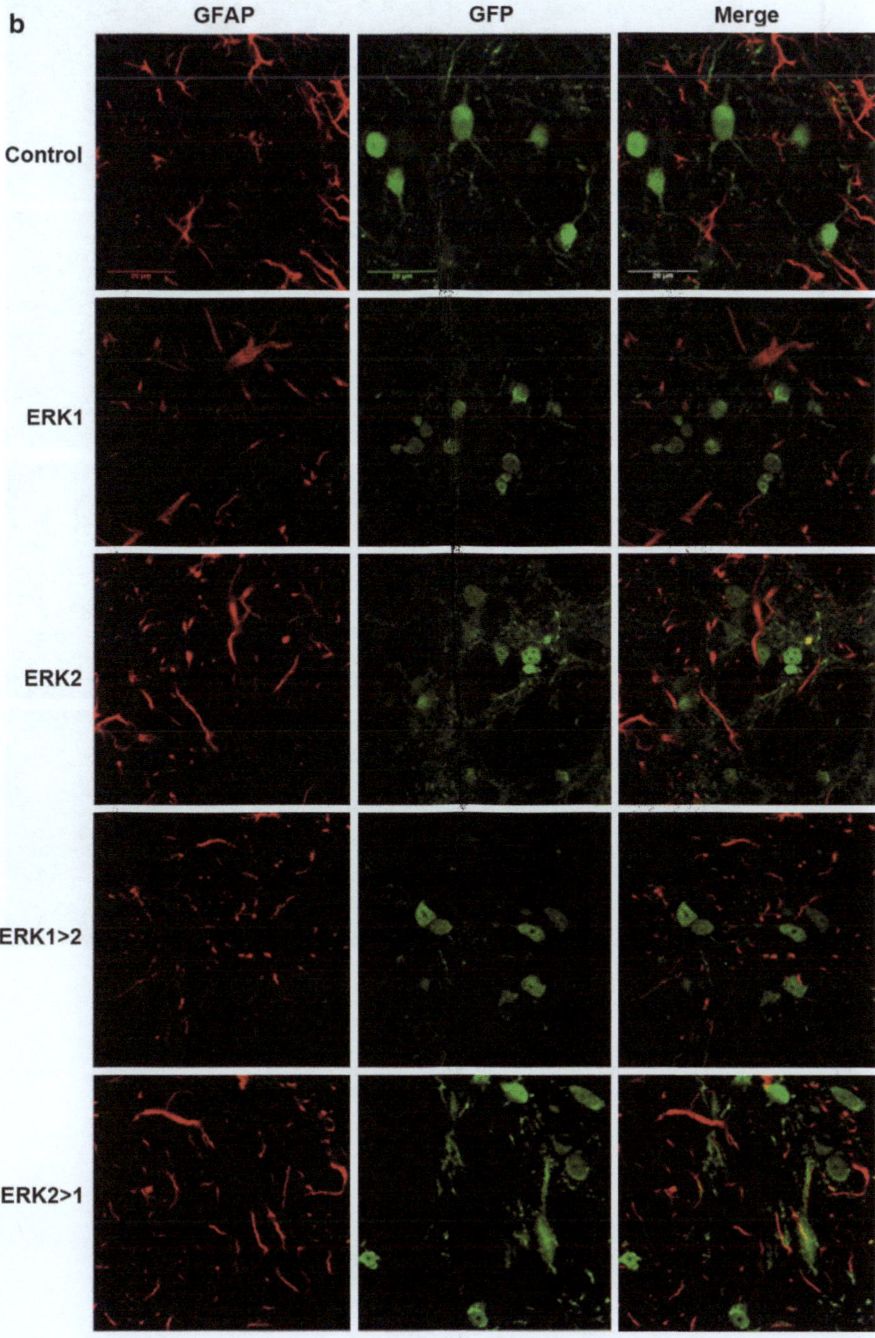

Fig. 3. (continued)

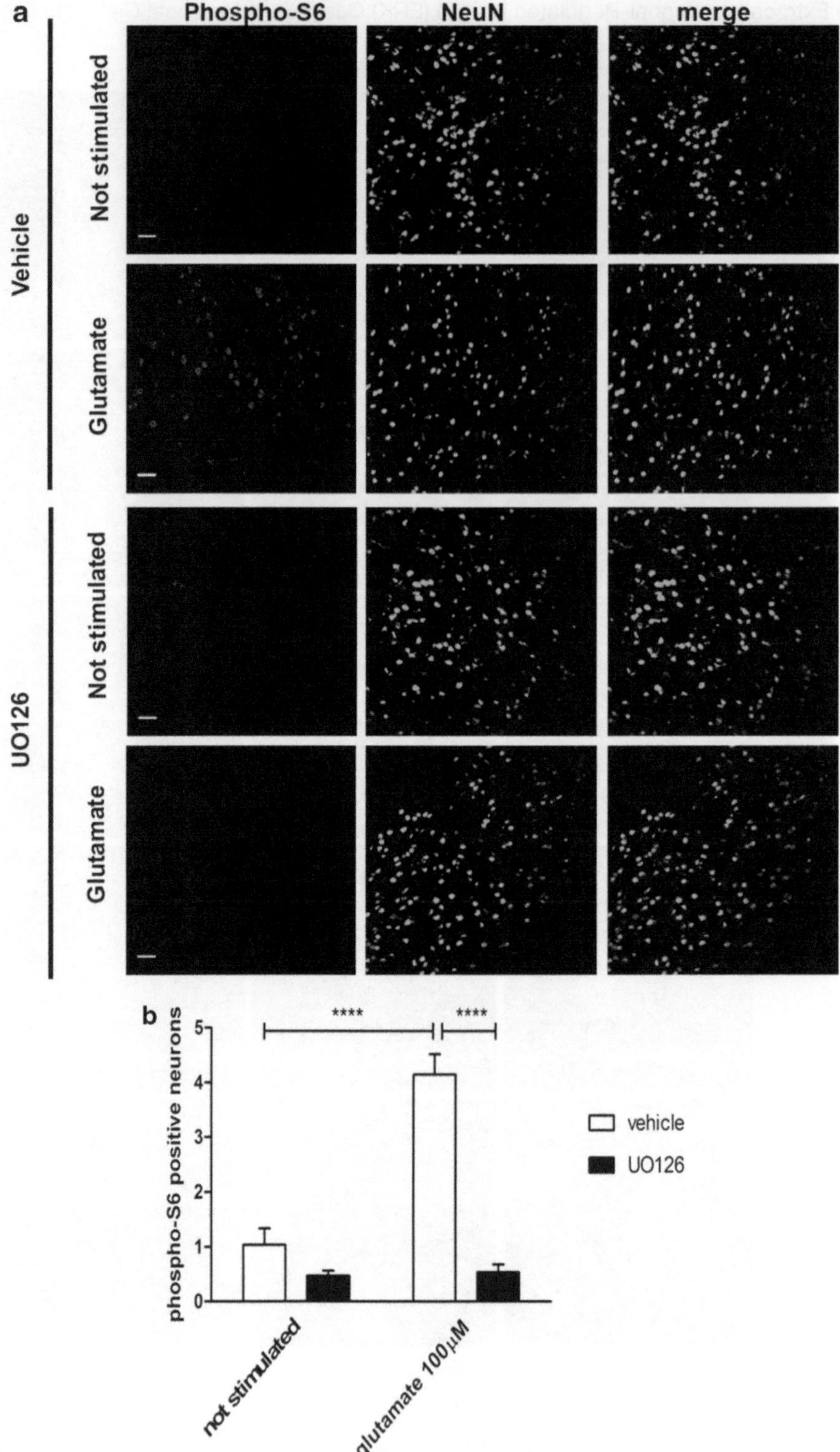

Fig. 4. S6 ribosomal protein phosphorylation is prevented by MEK inhibitor U0126 in a model of mature striatal slices. (**a**) Double immunolabeling of phospho S6 ribosomal protein (Ser235/236) (*red*) and NeuN (*green*) in adult striatal slices stimulated or not with glutamate 100 μM for 10 min in the presence of the vehicle or 20 μM U0126. (**b**) The data from the quantification are represented in the graph as mean ± SEM ($n=7$ and $n=8$ for the groups treated with vehicle and respectively stimulated or not with glutamate, $n=9$ for the groups treated with U0126). Statistical analysis was performed using two-way ANOVA and post-hoc comparisons between groups using Bonferroni test (****$p<0.0001$). Scale bars: 30 μm.

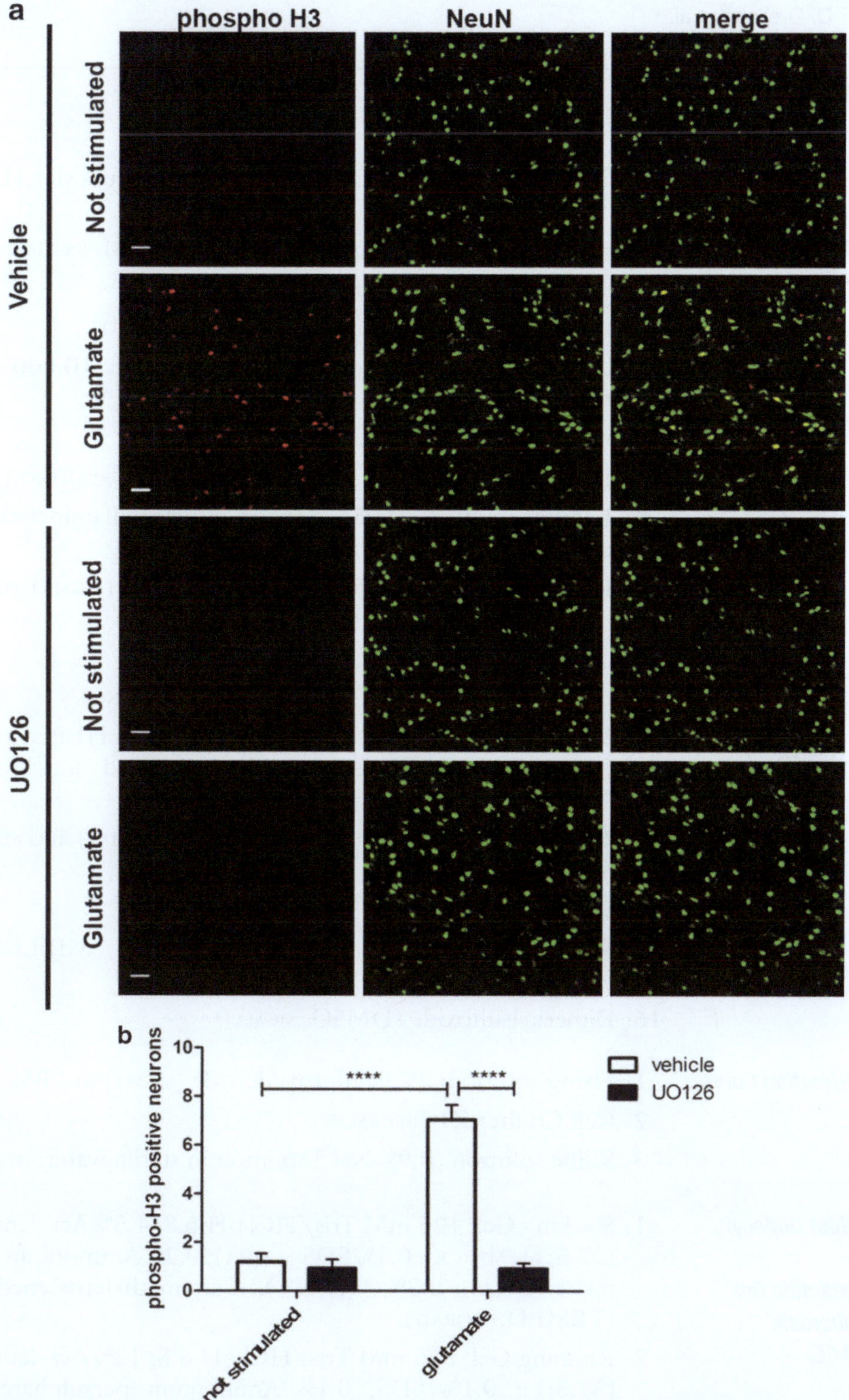

Fig. 5. Histone H3 phosphorylation is prevented by MEK inhibitor U0126 in a model of mature striatal slices. (**a**) Double immunolabeling of phospho-histone H3 (*red*) and NeuN (*green*) in adult striatal slices stimulated or not with glutamate 100 μM for 10 min in the presence of the vehicle or 20 μM U0126. (**b**) The data from the quantification are represented in the graph as mean ± SEM ($n=4$ and $n=3$ for the groups treated with vehicle and respectively stimulated or not with glutamate, $n=6$ and $n=5$ for the groups treated with U0126 vehicle and respectively stimulated or not with glutamate). Statistical analysis was performed using two-way ANOVA and post-hoc comparisons between groups using Bonferroni test (****$p<0.0001$). Scale bars: 30 μm.

2. Materials

2.1. Cell Culture, Transfection, and Viral Vector Production

1. Dulbecco's Modified Eagle Medium with GlutaMax (DMEM GlutaMax, GIBCO).
2. Iscove's Modified Dulbecco Medium (IMDM, SIGMA).
3. Fetal Bovine Serum (FBS, EUROCLONE).
4. 100× Glutamine (200 mM Glutamine) (GIBCO).
5. 100× Pen/Strept (10,000 U/ml Penicillin, 10,000 μg/ml Streptomycin, GIBCO).
6. Sodium butyrate (SIGMA).
7. Dulbecco's Phosphate Saline Buffer (PBS, EUROCLONE).
8. Trypsin solution (0.05% Trypsin/ethylenediaminetetraacetic acid) (GIBCO).
9. Tris–EDTA (TE) buffer 1× [10 mM Tris (SIGMA) pH 8.0; 1 mM EDTA, SIGMA].
10. Collagenase (SIGMA).
11. 100 mm, 150 mm and six-well plates (CORNING).
12. 2× HBS pH 7.09: 281 mM NaCl, 100 mM HEPES (SIGMA), 1.5 mM Na_2HPO_4 (SIGMA), filter sterilized and stored at -20°C.
13. 2.5 M $CaCl_2$ solution (SIGMA) filtered (0.22 μm) and stored at -20°C.
14. Sterile water (DIACO).
15. Syringe-driven filter units (0.22 μm) MILLEX GP (MILLIPORE).
16. Dimethylsulfoxide (DMSO, SIGMA).

2.2. Proliferation Curve

1. Fixing solution: 2% Paraformaldehyde (SIGMA) in PBS.
2. Cell Coulter Z1 (BECKMAN).
3. Saline solution (0.9% NaCl solution in sterile water, DIACO).

2.3. Sodium Dodecyl Sulfate–PolyAcrylamide Gel Electrophoresis (SDS–PAGE)

1. Stacking Gel: 125 mM Tris/HCl pH 6.8; 4.5% Acrylamide-bis (37.5:1) (MERCK); 0.1% SDS (SIGMA); 0.2% Ammonium persulphate (SIGMA); 0.2% *N,N,N′,N′*-Tetramethylethylenediamine (TEMED, BIORAD).
2. Running Gel: 375 mM Tris/HCl pH 8.8; 12% Acrylamide-bis (37.5:1); 0.1% SDS; 0.1% Ammonium persulphate; 0.1% TEMED.
3. Running Buffer pH 8.8: 25 mM Tris; 190 mM Glycine (SIGMA); 0.1% SDS.

4. Sample Buffer 5×: 250 mM Tris/HCl pH 6.8; 10% SDS; 50% Glycerol; 0.02% Bromophenol Blue (SIGMA); 110 mM Dithiothreitol (DTT, SIGMA).
5. Lysis Buffer: 125 mM Tris/HCl pH 6.8, 2.5% SDS.

2.4. Western Blot

1. Transfer Buffer: 25 mM Tris; 190 mM Glycine, 20% Methanol (MERCK).
2. Protran Nitrocellulose membrane (WHATMAN).
3. Tris Buffered Saline buffer (TBS): 20 mM Tris/HCl pH 7.4; 150 mM NaCl, (SIGMA)) with 0.1% Tween (SIGMA) (TBS-TW).
4. Blocking buffer: 5% Bovine Serum Albumin, (SIGMA) in TBS-TW.
5. Antibody buffer: 3% BSA in TBS-TW.
6. Antibodies: p44-42 MAPK #SC-153 (SANTA CRUZ BIOTECHNOLOGY), GAPDH #SC-25778 (SANTA CRUZ BIOTECHNOLOGY).

2.5. In Vivo Injection

1. C57/BL6 male mice 26g (Charles River). All experimental procedures must be performed in accordance with National and Institutional Guidelines for the use of laboratory animals.
2. Anesthetic solution: 100 mg/kg Ketamine (KETAVET, INTERVET), 10 mg/kg Xilazina (Rompun, BAYER) in physiologic solution.
3. Microscope (L-0940SD, INAMI).
4. Stereotaxic Frame (Kopf Instruments).

2.6. Perfusion and Preparation of Free-Floating Slices

1. Phosphate Buffer (PB): pH 7.4 (Na_2HPO_4 81 mM, NaH_2PO_4 18.84 mM).
2. Perfusion solution: Paraformaldehyde (PFA) 4% in PB.
3. Sucrose 30% in PB.
4. Anti-Freeze: $Na_2HPO_4 \cdot 2H_2O$ 30.62 mM, $NaH_2PO_4 \cdot H_2O$ 11.38 mM, 33.33% ethylene glycol, 33.33% glycerol.
5. Microtome (MICROM HM 430).

2.7. Preparation of Acute Striatal Slices

1. C57/BL6 mice 2–3 months old (Charles River) (see Note 1). All experimental procedures must be performed in accordance with National and Institutional Guidelines for the use of laboratory animals.
2. Artificial cerebrospinal fluid (ACSF): 124 mM NaCl, 5 mM KCl, 1.3 mM $MgSO_4 \cdot 7H_2O$, 1.2 mM $NaH_2PO_4 \cdot H_2O$, 25 mM $NaHCO_3$, 10 mM D-glucose 2.4 mM $CaCl_2$ (1 M stock) in dH_2O. This solution should be prepared fresh on the day of the experiment and saturated with carbogen (95% O_2, 5% CO_2) for 10 min before use.

3. Sucrose-based dissecting solution: 87 mM NaCl, 2.5 mM KCl, 7 mM $MgCl_2$ 1 mM NaH_2PO_4, 75 mM sucrose, 25 mM $NaHCO_3$, 10 mM D-glucose, 0.5 mM $CaCl_2$ (1 M stock), and 2 mM kynurenic acid in dH_2O. This solution should be prepared fresh on the day of the experiment and saturated with carbogen for 10 min before use.
4. L-glutamic acid 100 μM (>99%, Sigma-Aldrich).
5. UO126 20 μM (Tocris Bioscience) diluted in DMSO.
6. Fixative solution: 4% paraformaldehyde in 0.1 M sodium phosphate buffer pH 7.4. This solution should be prepared fresh on the day of the experiment. This solution is harmful, and it must be used in ventilated hood wearing gloves.
7. Cryoprotectant solution: 30% sucrose in 0.1 M sodium phosphate buffer pH 7.4. This solution should be prepared fresh on the day of the experiment.
8. Dissection tools for removing the brains (scissors, forceps, and spatula).
9. Cyanoacrylate glue.
10. Vibratome (VT1000S, Leica Microsystems).
11. Brain slice chamber-BSC1 (Scientific System design Inc., Mississauga, ON, Canada).
12. Proportional temperature controller (Scientific System design Inc., Mississauga, ON, Canada).
13. Two peristaltic pumps (Minipuls 3, Gilson) with R8 medium flow pump heads (see Note 2).
14. Silicone peristaltic tubing and connectors (Gilson).

2.8. Slices Sectioning

1. Cryostat.
2. Embedding medium for frozen sectioning.
3. Superfrost slides.

2.9. Immunofluorescence

1. Dulbecco's Phosphate Saline Buffer (PBS, EUROCLONE).
2. Blocking solution:
   (a) NeuN antibody: 5% NGS (normal goat serum), 0.1% Triton100 in PBS.
   (b) GFAP antibody: 10% NGS, 0.1% Triton100, BSA 1 mg/ml in PBS.
3. Primary Antibodies: anti-phospho-S6 ribosomal protein (Thr235/236) (Cell signaling Technology) 1:100, anti-phospho (Ser10)-acetyl (Lys14)-histone H3 (1:1000) (Upstate), Alexa Fluor 488 conjugate antiGFP (1:500) (A21311, Invitrogen), NeuN (1:500) (MAB-377, CHEMICON), rabbit anti-GFAP (1:500) (DAKO).

4. Secondary Antibodies: Alexa Fluor 546 goat anti-mouse (1:200) (A11003, Invitrogen), Alexa Fluor 555 goat anti-mouse (1:1000) (A21422, Molecular Probes), Alexa Fluor 546 goat anti-rabbit (1:200) (A-11035, Molecular Probes), Alexa Fluor 488 goat anti-mouse (1:200) (A11001, Invitrogen).
5. PapPen (Sigma).
6. Fluorescence Mounting Medium (S3023, Dako).
7. Microscope Slide Superfrost Plus (Thermo Scientific) and coverslips.
8. Leica confocal microscope.

3. Methods

3.1. Mouse Embryonic Fibroblasts Preparation

MEF cultures are prepared from wild-type E13.5 embryos obtained from commercially available C57 B/6 mice (Charles River) (see Note 3).

1. Dissection is carried out in PBS.
2. Remove the internal organs and the head, reduce the remaining tissues in small pieces and transfer in a 50 ml tube.
3. Centrifuge at 3,000 rcf for 5 min.
4. Wash in PBS and incubate for 1 h at 37°C in a solution of 0.25% collagenase and 20% FBS in PBS.
5. Dissociate tissues using a syringe.
6. Centrifuge at 3,000 rcf for 5 min at room temperature.
7. Resuspend the cells in DMEM GlutaMax containing 10% FBS, 1× Pen/Strept.
8. Count the cells and seed 5×10^6 cells on 15 mm plates.
9. Freeze the cells at confluence in 95% serum and 5% DMSO.

3.2. Lentiviral Vectors Production

We use all third-generation LV, modifications of the originally described backbone (see Note 4).

1. Seed and incubate 9×10^6 HEK 293T cells [ATCC, CRL-11268] in 150 mm dishes, approximately 24 h before transfection. The medium used is DMEM GlutaMax containing 10% FBS, 1× Pen/Strept. Use low passage cells (not more than P12–15) and do not ever let cells grow to confluence.
2. Change medium 2 h before transfection: IMDM supplemented with 10% FBS, 1× Pen/Strept, 1× glutamine (22 ml final volume).
3. Prepare the plasmid DNA mix by adding together: 9 μg ENV plasmid (VSV-G), 12.5 μg packaging plasmid (pMDLg/pRRE

or CMV R8.74), 6.25 μg of pRSV-REV, and 32 μg of gene transfer plasmid. The plasmid mix solution is made up to a final volume of 1,125 μl with 0.1× TE Buffer (1: 10 mM Tris pH 8.0; 1 mM EDTA pH 8.0 in water). Finally, 125 μl of 2.5 M $CaCl_2$ is added.

4. Leave the mix 15 min at room temperature.
5. The precipitate is formed by dropwise addition of 1,250 μl of 2× HBS (281 mM NaCl, 100 mM HEPES, 1.5 mM Na_2PO_4, pH 7.06–7.12) solution to the 1,250 μl DNA–TE–$CaCl_2$ mixture from step 3 while vortexing at full speed. The precipitate should be added to HEK 293T-cell immediately following the addition of the 2 HBS. High magnification microscopy of the cells should reveal a very small granular precipitate of $CaPO_4$-precipitate plasmid DNA, initially above the cell monolayer and after incubation in 37°C incubator overnight, on the bottom of the plate in the large spaces between the cells.
6. The $CaPO_4$-precipitate plasmid DNA should be allowed to stay on the cells for 14–16 h, after which the media should be replaced with fresh medium (IMDM with 10% FBS, 1× Pen/Strept, 1× glutamine and 1 M sodium butyrate).
7. Collect the cell supernatants at 36 h after changing the medium, filter (0.22 μm) and centrifuge at 20,000 rpm, at 20°C for 2 h (Beckman Ultracentrifuge, SW32Ti rotor).
8. Discard the supernatant and resuspend the pellet in sterile PBS 1×.
9. Aliquot and store at −80°C.

3.3. Titration of the Lentiviral Vectors

1. Seed and incubate 5×10^4 HEK 293T cells in 35 mm dishes, 12–14 h before the infection.
2. Make serial dilution of the LV in the growing medium (DMEM GlutaMAX, 10% FBS, 1× Pen/Strept) and transduce the cells with the desired dilutions in a final volume of 1 ml.
3. After 5 days, collect the cells, wash in PBS with 1% FBS, and resuspend in 2% paraformaldehyde in PBS 1×.
4. The titer (transforming units, TU) is determined by using FACS calibur (BD BIOSCIENCE) and counting the percentage of GFP-positive cells in each dilution. When the percentage is between 2.5 and 25%, the titer can be determined using the following formula:

$$\text{Titer(T.U./ml)} = \frac{\text{Percentage of GFP positive cells} \times \text{Cells plated the first day}(5 \times 10^4)}{\text{Dilution}}.$$

3.4. Proliferation Assays

Cells are initially plated in the exponential phase and followed for 5 days, till they reached a subconfluent stage (see Note 5).

1. The day before the infection seed 1.25×10^5 MEF cells/well in six-well plates. It is better to use low passage cells, in this case P4–5 cells were used. Prepare one plate for each day of counting for each condition.
2. The day after (day 0) count one plate for each condition. The count is performed following the next protocol:
   (a) Detach the cells with the trypsin solution for 5 min at 37°C.
   (b) Collect the cells and wash carefully the wells with PBS 1× with 1% serum to be sure to take all the cells.
   (c) Centrifuge at 1,500 rcf for 5 min at RT.
   (d) Resuspend the cells in 1 ml of fixing solution.
   (e) Dilute 200 μl of cell suspension in 10 ml of saline and determine the cell number with the coulter counter.
3. Tranduce (MOI=5 in 1 ml of fresh medium) cells in the remaining plates.
4. Every day count one plate for each condition and change medium in the others.
5. At day 2 after infection, cells are controlled for GFP expression in order to check the efficiency of the process.

3.5. In Vivo Lentiviral Injection

1. Anesthetize the mouse.
2. Secured the animal in a steroetaxic frame equipped with a mouse adaptor.
3. Unilateral LV injection is performed into two different sites of the motor dorsal striatum:
   (a) 1° SITE: AP +1; L −2.1; DV −2.6.
   (b) 2° SITE: AP +0.3; L −2.3; DV −2.4.

3.6. Brain Perfusion and Preparation of Free-Floating Slices

1. Anesthetize the mouse.
2. Perfuse the mice with cold physiologic solution at 35 rpm for 2 min.
3. Perfuse the mice with cold perfusion solution at 35 rpm for 5 min.
4. Carefully, extract the brain and post-fix it in cold perfusion solution, overnight at 4°C.
5. Discard the perfusion solution, wash the brain with PB and incubate with sucrose 30% for 24 h at 4°C.
6. Cut the brain at 35 μm with a microtome and stock the slices in antifreeze at −20°C.

3.7. Free Floating Immunofluorescence

1. Rinse the slices three times for 10 min in PBS.
2. Preincubate the slices for 1 h at room temperature with blocking solution.
3. Incubated overnight at 4°C with primary antibodies.
4. Rinse the slices three times for 10 min in PBS.
5. Incubated 1 h at room temperature with secondary antibodies.
6. Rinse the slices three times for 10 min in PBS.
7. Mount the slices on microscope slides, leave them dry and cover with a cover slip. Close it with the mounting medium.

3.8. Preparation of Acute Striatal Slices

1. Prepare the ACSF solutions containing 20 μM UO126 or the vehicle. Carboxygenate these solutions for at least 10 min and let them flow into the brain slice chamber at a constant rate of 2 ml/min at 32°C. The height of the perfusion fluid should be adjusted so that it forms a thin film over the slices.
2. Anesthetize a mouse and decapitate it with scissors, rapidly remove the brain in ice-cold sucrose-based dissecting solution oxygenated with 95% O_2 and 5% CO_2 (see Note 6).
3. Mount the brain on the vibratome stage on a drop of cyanoacrylate glue.
4. Cut 200 μm-thick slices keeping the brain submerged in ice-cold carboxygenated sucrose-based dissecting solution. The vibration rate of the blade should be fast and the speed nearly minimal.
5. Transfer the slices into the brain slice chamber placing them upon a lens paper (up to four slices in every chamber) and let them recover for 1 h at 32°C, constantly perfusing carboxygenated ACSF in the presence of the inhibitor or the vehicle.
6. After the recovery period, infuse 100 μM glutamate in the chamber for 10 min.
7. Remove the slices from the chamber and rapidly fix them in 4% PFA for 15 min at room temperature (see Note 7).
8. Rinse extensively (three times for 20 min with gentle agitation) in 0.1 M sodium phosphate buffer pH 7.4 at room temperature.
9. Incubate the slices in the cryoprotectant overnight at 4°C. Slices should not be incubated longer than overnight.

3.9. Slices Sectioning

1. Set the temperature of the cryostat chamber to −22°C.
2. Mount a piece of embedding medium on a metal specimen holder and cut the surface to make it flat.
3. Wrap a glass slide with parafilm and place the slice on this surface on a drop of embedding medium.

4. Transfer the slice onto the embedding medium flat surface and let it freeze.
5. Cut 18 μm cryosections and collect them onto Superfrost slides. Store in the cryostat until the cutting is completed (see Note 8).

3.10. Acute Striatal Slices Immunofluorescence

1. Draw a hydrophobic ring around the slices using a PapPen.
2. Rinse the slides with PBS three times for 10 min.
3. Pipette 100–200 μl blocking solution on each slide and incubate for 1 h at room temperature in a humid chamber (see Note 9).
4. Discard the blocking solution and pipette 100–200 μl primary antibodies solution on each slide. Incubate in a humid chamber overnight at 4°C.
5. Rinse the slides with PBS three times for 10 min.
6. Pipette 100–200 μl secondary antibodies on each slide and incubate in a humid chamber for 1 h at room temperature.
7. Rinse the slides with PBS three times for 10 min.
8. Discard PBS without drying the sections and mount the coverslips using the fluorescent mounting medium.
9. Store the slides at 4°C overnight in the dark before the examination at the microscope.

3.11. Microscopy and Image Analysis

Single and double-labeled images were obtained using a laser scanning confocal microscopy (Leica SP2), equipped with the corresponding lasers and the appropriate filters sets to avoid the cross-talk between the fluorochromes. Images were obtained with a 40× and 63× objectives. Neuronal quantification was performed with ImageJ software by counting phospho-S6/phospho-H3 immunoreactive neurons among NeuN-positive neurons in each slice. Statistical analysis was performed using two-way ANOVA and post-hoc comparisons between groups were made using Bonferroni test (****$p<0.0001$).

4. Notes

1. Brain slices obtained from young animals are more viable.
2. The height of the perfusion fluid in the brain slice chamber can be better regulated when the inflow and the outflow in the chamber are controlled by two independent pumps.
3. Mouse embryonic fibroblasts, prepared following this protocol, are cells with a relatively low proliferative potential and they should exclusively be used at low passages (maximum two

times after thawing) since they rapidly undergo senescence. We never refreeze MEF after the initial freezing. Although MEF normally grow at 10% FBS and can be starved at 0.5–1% FBS for inducing entering into the G_0 phase of the cell cycle, they can also be grown at 5% FBS, as done in our proliferation assays. Since MEF are resistant to plasmid transfection, we use for the experiments in Figs. 2 and 3 LV-assisted infection with the constructs described in Note 4. Using our infection conditions (multiplicity of infection, MOI = 5), transduction efficiency of MEF is very high (<90% of GFP-positive cells).

4. All lentiviral vectors used for the MEF proliferation assays and in-vivo injections (Figs. 2 and 3) are self-inactivating, third-generation vectors (32). The plasmids used for packaging, kindly provided by Luigi Naldini, were pREV (expressing REV protein), pVSVG (expressing the envelope: G glycoprotein of Vesicular Stomatitis Virus), and pRRE (expressing capsid, polymerase, protease, and integrase proteins). The lentiviral vector used as control in the MEF proliferation and in-vivo injection assay is pCCLsin.PPT.hPGK.eGFP.PRE (33). This vector allows the expression of eGFP under the human phosphoglycerate kinase (hPGK) promoter. The four LV used for overexpressing ERK1, ERK2, ERK1 > 2, and ERK2 > 1 were derived from the previously described constructs (7). The expression of the ERK constructs in the LV is under the hPGK promoter.

5. At the proliferation assays, the following cell densities (mean ± SEM) were reached. For Day 4: control vector, 5,338.06 ± 65.95 cells/mm^2; ERK1, 4,035.72 ± 116.09 cells/mm^2; ERK2, 5,142.39 ± 99.53 cells/mm^2; ERK1 > 2, 5,570.61 ± 56.04 cells/mm^2; ERK2 > 1, 3,310.61 ± 58.43 cells/mm^2. For Day 6: control vector, 5,453.83 ± 71.30 cells/mm^2; ERK1, 4,681.61 ± 74.49 cells/mm^2; ERK2, 5,700.06 ± 156.69 cells/mm^2; ERK1 > 2, 6,475.61 ± 94.57 cells/mm^2; ERK2 > 1, 4,376.17 ± 68.44 cells/mm^2. Statistical analysis was performed using two-way ANOVA and post-hoc comparisons between groups was made using Bonferroni test ($*p \le 0.02$).

6. Speed is essential for good results: the brain should be removed within 1–2 min after the decapitation.

7. The duration of fixation can be changed depending on the primary antibody and the brain region.

8. The cryosections should not be stored too long in the freezer. Best results are achieved processing the sections for immunofluorescence immediately after cutting.

9. The sections must never dry during the entire staining procedure in order to avoid high levels of background fluorescence.

Acknowledgments

This work was supported by the Michael J Fox Foundation for Parkinson's Research and the Parkinson's UK as well as by the Italian Ministry of Health, the Fondazione CARIPLO and the Compagnia di San Paolo (to RB). Daniel Orellana conducted this study as partial fulfillment of his PhD in Molecular Medicine, Program in Neuroscience, Vita-Salute San Raffaele University, Milan, Italy. Part of this work was carried out in ALEMBIC, an advanced microscopy laboratory established by the San Raffaele Scientific Institute and the Vita-Salute San Raffele University.

References

1. Pearson G et al (2001) Mitogen-activated protein (MAP) kinase pathways: regulation and physiological functions. Endocr Rev 22(2):153–183
2. Rubinfeld H, Seger R (2005) The ERK cascade: a prototype of MAPK signaling. Mol Biotechnol 31(2):151–174
3. Shaul YD, Seger R (2006) The MEK/ERK cascade: from signaling specificity to diverse functions. Biochim Biophys Acta 1773(8):1213–1226
4. Ramos JW (2008) The regulation of extracellular signal-regulated kinase (ERK) in mammalian cells. Int J Biochem Cell Biol 40(12):2707–2719
5. Orban PC, Chapman PF, Brambilla R (1999) Is the Ras-MAPK signalling pathway necessary for long-term memory formation? Trends Neurosci 22(1):38–44
6. Chuderland D, Seger R (2005) Protein-protein interactions in the regulation of the extracellular signal-regulated kinase. Mol Biotechnol 29(1):57–74
7. Vantaggiato C et al (2006) ERK1 and ERK2 mitogen-activated protein kinases affect Ras-dependent cell signaling differentially. J Biol 5(5):14
8. Thomas GM, Huganir RL (2004) MAPK cascade signalling and synaptic plasticity. Nat Rev 5(3):173–183
9. Sweatt JD (2004) Mitogen-activated protein kinases in synaptic plasticity and memory. Curr Opin Neurobiol 14(3):311–317
10. Davis S, Laroche S (2006) Mitogen-activated protein kinase/extracellular regulated kinase signalling and memory stabilization: a review. Genes Brain Behav 5(Suppl 2):61–72
11. Samuels IS, Saitta SC, Landreth GE (2009) MAP'ing CNS development and cognition: an ERKsome process. Neuron 61(2):160–167
12. Santini E, Valjent E, Fisone G (2008) Parkinson's disease: levodopa-induced dyskinesia and signal transduction. FEBS J 275(7): 1392–1399
13. Girault JA, Valjent E, Caboche J, Herve D (2007) ERK2: a logical AND gate critical for drug-induced plasticity? Curr Opin Pharmacol 7(1):77–85
14. Kim EK, Choi EJ (2010) Pathological roles of MAPK signaling pathways in human diseases. Biochim Biophys Acta 1802(4):396–405
15. Tidyman WE, Rauen KA (2009) The RASopathies: developmental syndromes of Ras/MAPK pathway dysregulation. Curr Opin Genet Dev 19(3):230–236
16. Klann E, Dever TE (2004) Biochemical mechanisms for translational regulation in synaptic plasticity. Nat Rev 5(12):931–942
17. Marchi M et al (2008) The N-terminal domain of ERK1 accounts for the functional differences with ERK2. PLoS One 3(12):e3873
18. Dudley DT, Pang L, Decker SJ, Bridges AJ, Saltiel AR (1995) A synthetic inhibitor of the mitogen-activated protein kinase cascade. Proc Natl Acad Sci U S A 92(17):7686–7689
19. Favata MF et al (1998) Identification of a novel inhibitor of mitogen-activated protein kinase kinase. J Biol Chem 273(29):18623–18632
20. Sebolt-Leopold JS, Herrera R (2004) Targeting the mitogen-activated protein kinase cascade to treat cancer. Nat Rev Cancer 4(12):937–947
21. Saba-El-Leil MK et al (2003) An essential function of the mitogen-activated protein kinase

Erk2 in mouse trophoblast development. EMBO Rep 4(10):964–968

22. Yao Y et al (2003) Extracellular signal-regulated kinase 2 is necessary for mesoderm differentiation. Proc Natl Acad Sci U S A 100(22):12759–12764
23. Hatano N et al (2003) Essential role for ERK2 mitogen-activated protein kinase in placental development. Genes Cells 8(11): 847–856
24. Meloche S, Vella FD, Voisin L, Ang SL, Saba-El-Leil M (2004) Erk2 signaling and early embryo stem cell self-renewal. Cell Cycle 3(3):241–243
25. Fischer AM, Katayama CD, Pages G, Pouyssegur J, Hedrick SM (2005) The role of erk1 and erk2 in multiple stages of T cell development. Immunity 23(4):431–443
26. Satoh Y et al (2007) Extracellular signal-regulated kinase 2 (ERK2) knockdown mice show deficits in long-term memory; ERK2 has a specific function in learning and memory. J Neurosci 27(40):10765–10776
27. Pagès G et al (1999) Defective thymocyte maturation in p44 MAP kinase (Erk 1) knockout mice. Science 286(5443):1374–1377
28. Selcher JC, Nekrasova T, Paylor R, Landreth GE, Sweatt JD (2001) Mice lacking the ERK1 isoform of MAP kinase are unimpaired in emotional learning. Learn Mem 8(1):11–19
29. Mazzucchelli C et al (2002) Knockout of ERK1 MAP kinase enhances synaptic plasticity in the striatum and facilitates striatal-mediated learning and memory. Neuron 34:807–820
30. Indrigo M, Papale A, Orellana D, Brambilla R (2010) Lentiviral vectors to study the differential function of ERK1 and ERK2 MAP kinases. Methods Mol Biol 661:205–220
31. Tronson NC et al (2008) Regulatory mechanisms of fear extinction and depression-like behavior. Neuropsychopharmacology 33(7):1570–1583
32. Naldini L, Blomer U, Gage FH, Trono D, Verma IM (1996) Efficient transfer, integration, and sustained long-term expression of the transgene in adult rat brains injected with a lentiviral vector. Proc Natl Acad Sci U S A 93(21):11382–11388
33. Follenzi A, Ailles LE, Bakovic S, Geuna M, Naldini L (2000) Gene transfer by lentiviral vectors is limited by nuclear translocation and rescued by HIV-1 pol sequences. Nat Genet 25(2):217–222

Chapter 9

Glycogen Synthase Kinase-3 in Neurological Diseases

Oksana Kaidanovich-Beilin and James Robert Woodgett

Abstract

Glycogen synthase kinase-3 (GSK-3) occupies an unusual niche in cellular regulation via its negative regulation of a series of important cellular target proteins coupled with its own sensitivity to several major signaling pathways including the Wnt, Notch, Hedgehog, and insulin systems. This protein kinase thus has multiple physiological functions depending on the cell or tissue type. Mouse genetic models in which the two mammalian isoforms (known as GSK-3α and GSK-3β) have been globally or tissue-specifically inactivated or activated have revealed roles in behavior, neurodegenerative diseases, and cognition. The activity of the protein kinase is also influenced directly or indirectly by a series of neuroactive drugs including lithium and psychotropic agents. In this review, we describe the modes of regulation, the molecular targets lying downstream and the association of dysfunction of GSK-3 in various neurological disorders including bipolar disorder, schizophrenia, and neurodegeneration as well as discuss possible strategies that may target GSK-3 for therapeutic benefit.

Key words: Protein phosphorylation, Glycogen synthase kinase, Signal transduction, Lithium

1. Introduction

Protein kinases comprise a family of enzymes that assist in transducing both intrinsic and extrinsic cellular stimuli by promoting phosphorylation of target proteins (substrates), which leads to modulation of the functions of the substrates in a reversible manner. Two percent of the human genetic complement is devoted to encoding over 500 protein kinases. Their essential role in normal development and physiology is apparent from the consequences of their dysfunction which is linked to a diverse array of human diseases. Consequently, members of the protein kinase superfamily are major targets for drug discovery by the pharmaceutical industry. This chapter provides an overview of the features, functions, and regulation of a key

Hideyuki Mukai (ed.), *Protein Kinase Technologies*, Neuromethods, vol. 68,
DOI 10.1007/978-1-61779-824-5_9, © Springer Science+Business Media, LLC 2012

member of the CMCG family of proline-directed serine/threonine protein kinases (*C*yclin-dependent kinases (*C*DKs), *M*itogen-activated protein kinases (*M*APKs), *G*lycogen synthase kinases (*G*SKs), and CDK-like kinases (*C*LKs); reviewed in (1))—*Glycogen Synthase Kinase-3* (GSK-3). This enzyme derives its name from its initial discovery as an important regulator of glycogen metabolism. However, data accumulated since its first identification have revealed a far wider role for GSK-3, as a crucial negative regulator of a plethora of cellular processes including cytoskeletal regulation, transcriptional control, and cell fate determination.

2. From Humble Beginnings: A Brief History of GSK-3

GSK-3 is a monomeric, second messenger-independent protein kinase that was first discovered through its ability to activate the ATP-Mg-dependent form of type-1 protein phosphatase (Factor A activity) and to phosphorylate the key rate-limiting metabolic enzyme that catalyzes the last step of glycogen synthesis, glycogen synthase (2–4). An important effect of insulin on intracellular metabolism is its ability to stimulate the synthesis of glycogen in muscle and liver (5). As it was suggested in late 1960–1970s, several distinct GSKs may underlie the insulin-induced dephosphorylation and activation of glycogen synthase, such as glycogen synthase kinase-1 and -2 (cyclic AMP-dependent protein kinase/GSK-1 and phosphorylase kinase/GSK-2, respectively) (6–11). However, several years later, it was discovered that insulin-induced activation of glycogen synthase involves its dephosphorylation at the same serine residues that are targeted by GSK-3 (12), indicating that the insulin signaling cascade may promote dephosphorylation and activation of glycogen synthase by selectively inhibiting GSK-3. That prediction was proven correct in the mid-1990s, once protein kinase B (PKB, also termed Akt) was discovered and shown to be responsible for mediating insulin-induced phosphorylation and inhibition of GSK-3 (13). This finding was subsequently confirmed by several studies which showed that insulin leads to acute inhibition of GSK-3 (e.g., (14, 15)).

A decade following its initial discovery, GSK-3 was shown to be involved in the Wnt signaling pathway (Wnt) (16–18), as loss-of-function mutations in the fruit fly orthologues of GSK-3 zeste-white 3/shaggy (Zw3/Sgg), lead to the loss of denticles and wing patterning defects, whereas loss-of-function mutations in wingless, the fly orthologues of mammalian Wnts, showed the opposite phenotype (18). This genetic analysis also demonstrated that Wnt likely acted through suppression of GSK-3. In 1995, studies in *Xenopus laevis* showed that the expression of a catalytically inactive mutant of GSK-3 induces duplication of the dorsal axis (19, 20), a phenotype similar to earlier studies of lithium ion treatment of Xenopus embryos (21). A molecular explanation for this relationship was discovered in 1996 when GSK-3 was discovered to be inhibited by lithium (22, 23). These findings proved to be important milestones for awakening of interest in GSK-3 in the field of psychiatry.

3. Cloning of GSK-3 and Comparison Between Two Genes

GSK-3 is a highly conserved protein kinase and has orthologues in plants, fungi, worms, flies, sea squirts, and vertebrates; isoenzymes from species as distant as flies and humans display more than 90% sequence similarity within the protein kinase domain (reviewed in (24)).

For example, the zeste-white3/shaggy ($zw3^{sgg}$) gene of *Drosophila* is involved in embryonic development and also encodes a protein kinase with significant homology to mammalian GSK-3 (16, 17, 25). One of the mammalian isoforms of GSK-3 (β) has been shown to substitute functionally for mutant $zw3^{sgg}$, demonstrating that it is a true homolog of the insect kinase (18, 26). In the budding yeast *Saccharomyces cerevisiae*, the MCK1, RIM11 and MDS1[2] genes encode proteins with 44–58% identity to the kinase domain of GSK-3, and these gene products are capable of phosphorylating substrates characteristics of those recognized of mammalian GSK-3 (reviewed in (27)). Screening of a chick embryo cDNA library resulted in isolation of more than 10 GSK-3β cDNAs but no GSK-3α cDNAs (Nikolakaki, E. and Woodgett, J., unpublished data); indeed a recent study has revealed that the GSK-3α gene is missing in birds (28).

GSK-3/F_A has been purified by several groups (29–31), and was molecularly cloned based on partial peptide sequencing (32). Two rat brain cDNAs encoding GSK-3 were isolated that corresponded to isoenzymes designated GSK-3 alpha and GSK-3 beta, with apparent M_r of 51,000 and 47,000, respectively (32). Genes encoding GSK-3α and β are located on mouse chromosome 7/ human chromosome 19 and on mouse chromosome 16/human chromosome 3, respectively (based on http://genome.ucsc.edu) (Fig. 1).

Thus, mammalian GSK-3 is represented by two paralogous proteins. Overall, the GSK-3α and β proteins share 85% sequence homology, including 98% amino acid sequence identity within their kinase domains (32). Despite their similarity, GSK-3 isoenzymes diverge in their N- and C-termini. For example, the two gene products share only 36% identity in the last 76 C-terminal residues (32). GSK-3α has an extended glycine-rich N-terminal region that has been proposed to function as a pseudosubstrate (Fig. 2) (33).

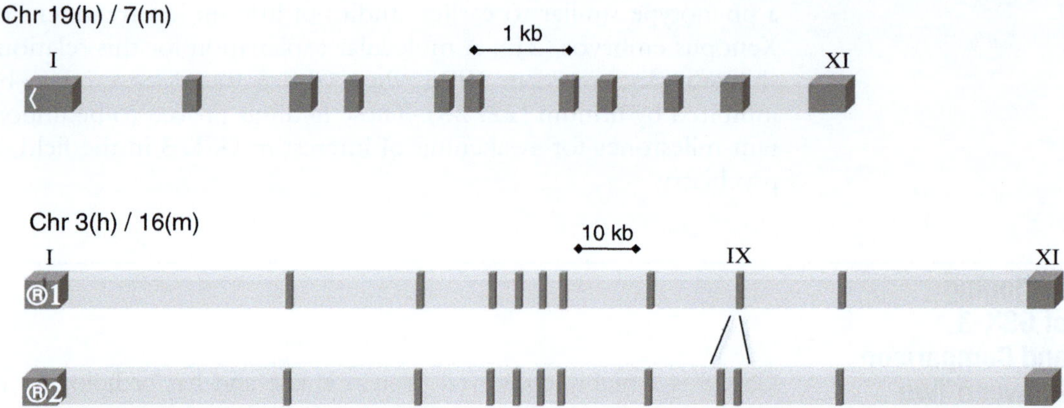

Fig. 1. Cartoon of the exon/intron structure of GSK-3. The chromosome localization and differential splice of GSK-3β are indicated.

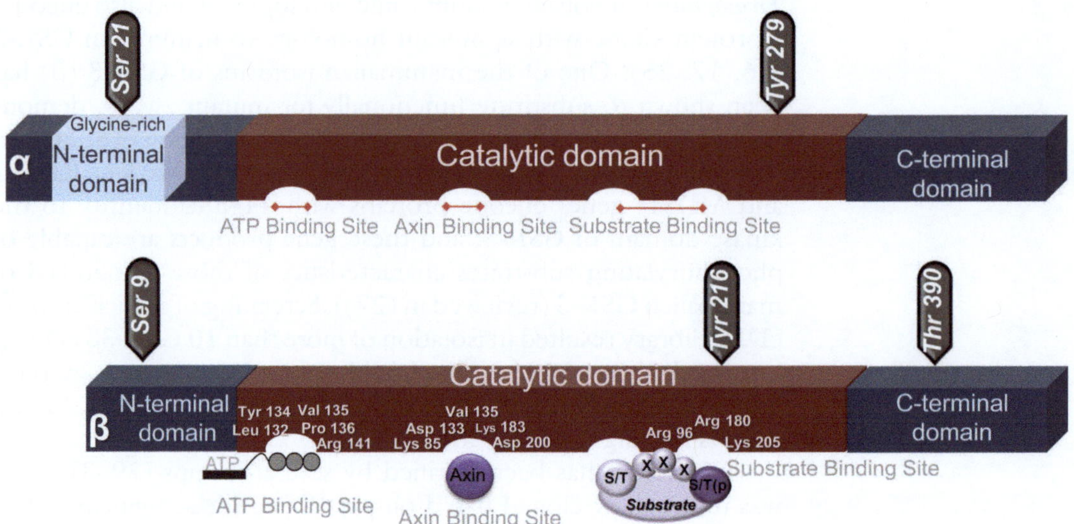

Fig. 2. Topographical organization of GSK-3 functional domains. The two isoforms are shown together with sites of phosphorylation, scaffold protein binding, ATP binding, and substrate binding. Key residues involved in these interaction sites are indicated for GSK-3β.

Whether the two isoenzymes have differential roles is still not fully understood, however studies in conventional GSK-3 animal knockout models have revealed functional divergence and absence of compensation between the isoforms, as GSK-3α KO mice are viable (34), contrasting with embryonic lethality observed upon deletion of GSK-3β (35).

GSK-3 appears to be expressed ubiquitously and both gene products are found in nearly all mammalian tissues. The kinase is highly expressed in the brain (32, 36, 37), both in neurons and glia (38). GSK-3α is especially abundant in the hippocampus, cerebral cortex, striatum, and cerebellum (based on Allen Brain Atlas).

GSK-3β is expressed in nearly all brain regions, although there are marked regional differences of GSK-3β mRNA levels in the human brain (39). As a caution, the glycine-rich (and hence purine-rich) region of GSK-3α may distort comparative analysis of RNA expression between it and GSK-3β.

In the brain, there is evidence for two alternatively spliced exons of GSK-3β which generate a short form (GSK-3β1) and a long form containing a 13 amino acid insert in the catalytic domain (GSK-3β2) (Fig. 1). This insert is between residues 303 and 304 of GSK-3β1, flanked by two proximal α-helices of kinase subdomains X and XI (40, 41). This alternative spliced isoform of GSK-3β in rodents (37, 41) and in human (42–44) has been implicated in neuronal-specific functions. GSK-3β1 is ubiquitously expressed including the developing and adult nervous system (45, 46). By contrast, GSK-3β2 is predominantly expressed in the nervous system, with highest levels in the developing brain with persistence into adulthood (41, 47).

A rabbit polyclonal antibody raised to the 13-aa insert of GSK-3β2 (anti-8A) that specifically recognizes the GSK-3β2 variant, labeled the cell body, including the nucleus, neurites, and growth cones of embryonic neurons in culture (47) as well as in the axon, growth cone, and the nucleus of differentiated neurons (48, 49). In undifferentiated cells undergoing cell division, only GSK-3β1 is expressed, whereas when cells cease mitosis and differentiate along a neuronal pathway following exposure to nerve growth factor, they maintain GSK-3β1 expression but increase expression of GSK-3β2 (50). These findings suggest that GSK-3β2 expression correlates with neuronal differentiation. In the adult central nervous system the highest expression of GSK-3β2 occurs in the regions with a high proportion of white matter, suggesting that GSK-3β2 is preferentially localized in axons (47).

4. GSK-3 Regulation

An unusual feature of GSK-3 is that the kinase is constitutively active in cells under resting conditions (51–53) and is primarily regulated through inhibition of its activity. It is one of the few protein kinases that can be inactivated by phosphorylation. However, the mechanisms of GSK-3 regulation are more varied and not fully understood; precise control appears to be achieved by a combination of phosphorylation, localization, and sequestration by a number of GSK-3-binding proteins (reviewed in (54–57)). Of note, GSK-3 is sensitive to several distinct signaling pathways. Hence, to maintain appropriate signaling specificity the kinase must link to each upstream signal through distinct means, providing effective insulation between the pathways. This is achieved through specific pools of GSK-3, which differ in intracellular localization, binding partner affinity, and amount. In this way, the same protein can actually be differentially deployed and employed for multiple signaling systems (reviewed in (58)).

The activity of GSK-3 at the posttranslational level is positively regulated by phosphorylation on a "T loop" tyrosine residue within subdomain VIII (Tyr279 for GSK-3α and Tyr216 for GSK-3β) (59, 60) and negatively regulated by N-terminal phosphorylation of serine residues of the enzyme (Ser 21 for GSK-3α and Ser 9 for GSK-3β) (Fig. 2) (52, 61).

From the crystal structure, it has been proposed that unphosphorylated Tyr276/Tyr 216 blocks the access of primed substrates (as discussed below). Indeed, the structure of phosphorylated GSK-3β (62) shows that phosphorylated Tyr216 undergoes a conformational change that allows substrates to bind the enzyme. Previous studies, however, led to conflicting conclusions as to whether the tyrosine phosphorylation of GSK-3 is catalyzed by GSK-3 itself (autophosphorylation) or by a distinct tyrosine kinase (59, 63–66). In *Dictyostelium*, the Zaphod Kinase (Zak1) activates GSK-3 by tyrosine phosphorylation (65, 67), while in mammals, Fyn tyrosine kinase (66) or a related tyrosine kinase may be involved in this process. Moreover, two tyrosine residues in the activation loop of GSK-3 appear to be modulated by cAMP in *Dictyostelium* (68). Both of these residues, Tyr214 and Tyr220, are conserved in mammals (Tyr216 and Tyr219 in GSK-3β, respectively), such that, in addition to Tyr216, Tyr219 may play a role in the regulation of GSK-3β (and GSK-3α) in mammals. In support of the autophosphorylation model, Lochhead et al. (60) showed that newly synthesized GSK-3β autophosphorylated itself on tyrosine and that this event could be prevented by exposure to GSK-3 inhibitors. This process is similar to tyrosine autophosphorylation of the DYRK family of kinases that, like GSK-3, strictly phosphorylate Ser/Thr residues on exogenous substrates (69).

In contrast to the tyrosine phosphorylation events, regulation of N-terminal serine phosphorylation is only conserved in GSK-3 homologues from mammals, *Xenopus* and *Drosophila*, but not in yeast, higher plants, *Dictyostelium* or *Caenorhabditis elegans*. The phosphorylation state of serine residues of both isoenzymes is dynamic, involving phosphorylation by several protein kinases and dephosphorylation by protein phosphatase-1 (PP-1) (70). N-terminal domain serine phosphorylation of GSK-3α and GSK-3β leads to inhibition of its activity (52, 71). Phosphorylation of GSK-3 within its N-terminal region creates a "pseudosubstrate" which intramolecularly binds to the active site of the kinase, suppressing activity (Fig. 3) (72).

This inhibitory mechanism is induced by agonists such as neurotrophins and growth factors that activate protein kinases that act on the N-terminal domain of GSK-3 such as PKB/Akt, p90rsk, cyclic

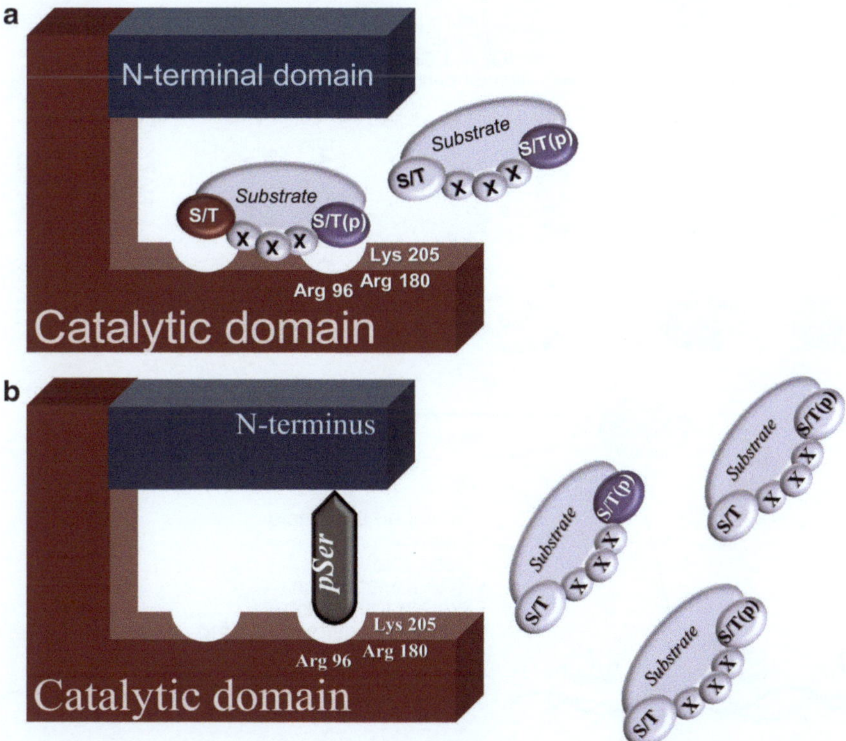

Fig. 3. Role of substrate priming and mode of N-terminal domain inhibition. In (**a**), prephosphorylated substrates are shown with the priming phosphate interacting with a pocket of basic amino acids to align the N-4 serine or threonine into the active site. In (**b**), the phosphorylated amino terminal domain (Ser21 or 9) acts as an intramolecular pseudosubstrate competitor for the priming site pocket, excluding access to exogenous substrates hence reducing activity.

AMP-dependent protein kinase, p70 S6 kinase, as well as regulators of phosphatase 1 (Fig. 4) (73–75). For example, insulin leads to inhibition of GSK-3 via insulin receptor substrate-1-dependent induction of phosphatidylinositol 3′ kinase (PI3K), which then stimulates PKB/Akt (13). GSK-3 has previously been shown to catalyze serine phosphorylation of IRS-1 and IRS-2, interfering with receptor-mediated tyrosine phosphorylation by the insulin receptor, effectively attenuating insulin receptor signaling via a negative feedback loop (76, 77).

Growth factors, such as EGF and PDGF, can also inhibit GSK-3 activity through the PI3K pathway (51, 78), as well as through induction of the MAPK cascade (79, 80). Serine 9/21 phosphorylation of GSK-3 can be modified by amino acids through mammalian target of rapamycin (mTOR) (81–83) or in response to agonists that elevate the intracellular levels of cAMP through cyclic-AMP-dependent protein kinase (PKA) (74, 84). The PKA-anchoring protein 220 binds both GSK-3 and PKA and hence facilitates GSK-3 phosphorylation by this protein kinase (85). PKC agonists can also regulate GSK-3 (86, 87), however certain PKCs preferentially regulate GSK-3β but not GSK-3α (88).

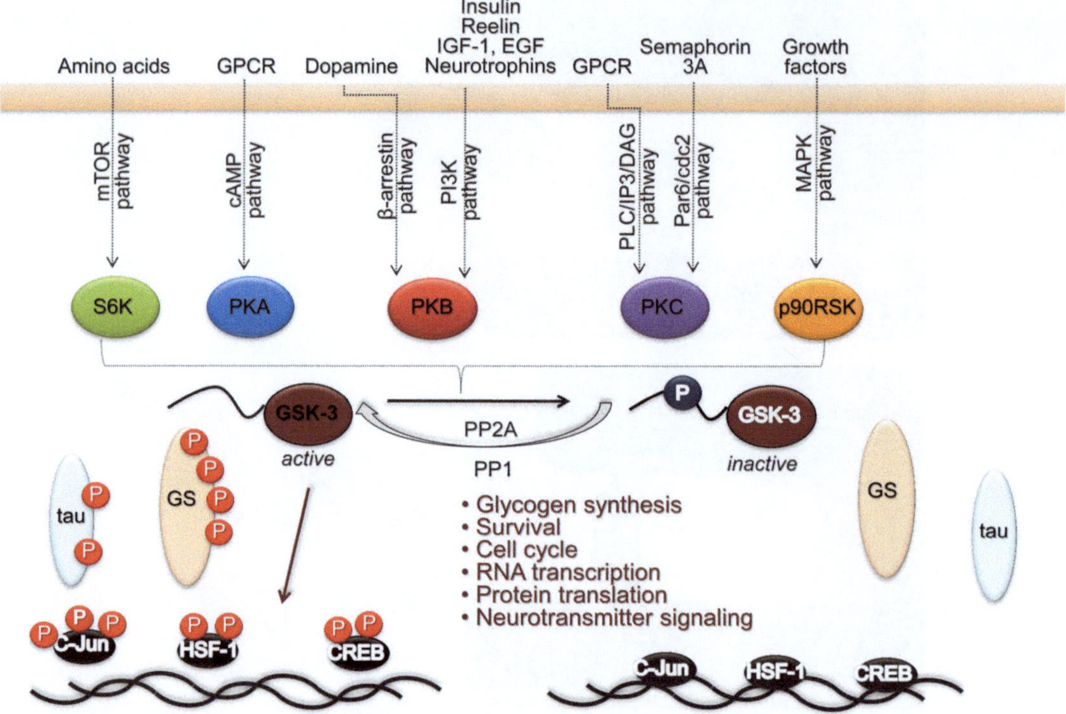

Fig. 4. Regulation of GSK-3 N-terminal phosphorylation by several signaling pathways. GSK-3 lies downstream of several pathways that independently induce Ser 21/9 phosphorylation and inactivation of GSK-3 resulting in dephosphorylation of GSK-3 substrates.

The Wnt/Wingless signaling pathway employs a distinct mechanism for regulating GSK-3 that is independent of N-terminal domain phosphorylation and, instead, relies on protein:protein interactions and intracellular sequestration (Fig. 5). In resting cells, in the absence of a Wnt signal, approximately 5–10% of cellular GSK-3 is associated with a scaffolding protein termed Axin (89, 90). These molecules are joined by others to create a "destruction complex" comprising Axin, adenomatous polyposis coli (APC), casein kinase I (CK1), and β-catenin (91–93). Within this machine, casein kinase I phosphorylates Ser45 of β-catenin, which generates a priming site for subsequent GSK-3 phosphorylation on Thr41 (94–99) and subsequently Ser37 and Ser33, the latter modifications signaling β-catenin for ubiquitin-mediated proteasomal degradation (94, 97, 100).

The Wnts comprise a family of secreted, cysteine-rich, glycosylated protein ligands that regulate cell growth, differentiation, migration, and fate (reviewed in (101–103)). Wnts bind to members of the seven transmembrane-pass family of frizzled receptors in cooperation with low-density lipoprotein receptor-related protein (LRP) 5 or 6 which promotes a conformational change that acts to recruit the destruction complex such that LRP5/6 become

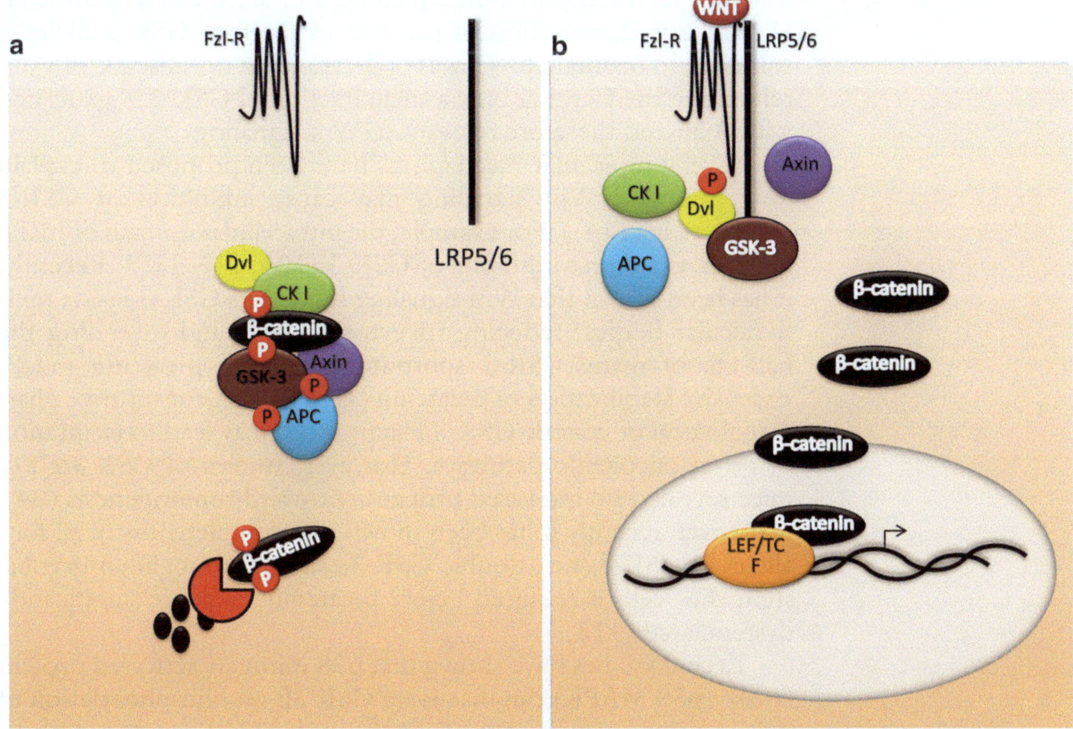

Fig. 5. Role of GSK-3 in Wnt signaling. In (**a**), GSK-3 bound to the Axin/APC scaffold phosphorylates cytoplasmic β-catenin targeting it for ubiquitinylation and degradation. This maintains very low levels of β-catenin. In (**b**), Wnt binding to Frizzled (Fzl-R) and LRP5/6 causes reorganization of the Axin/APC/GSK-3 destruction complex allowing β-catenin to escape destruction. The accumulated β-catenin binds to LEF/TCF DNA binding proteins and regulates transcription of target genes. Casein kinase I (CKI) acts as the priming kinase for β-catenin. Dishevelled (Dvl) acts to coordinate binding of the complexes.

phosphorylated by GSK-3 and casein kinase-1 (104–106). This creates a high-affinity Axin binding site on LRP5/6 which promotes dissolution of the destruction complex and sequestration of the Axin-associated GSK-3 within vesicles, resulting in escape of β-catenin from phosphorylation and, hence, destruction. An "LRP6-signalosome" (cytoplasmic destruction complex) that comprises aggregates of phospho-LRP6, Frizzled, Dvl, Axin and GSK-3, forms within 15 min after Wnt stimulation (107). Internalization of this complex of proteins is a requirement for canonical Wnt pathway signals (108, 109). As a consequence of the sequestration of GSK-3, β-catenin accumulates and translocates into the nucleus where interacts with the TCF/LEF (T-cell factor/lymphoid enhancer binding factor) family of transcription factors and promotes gene-specific transcription.

Both Axin and APC are phosphorylated by GSK-3. Phosphorylation of Axin by GSK-3 increases its stability and binding to β-catenin (92, 110, 111). Phosphorylation of APC increases its binding to β-catenin (112). Both mammalian GSK-3 isoforms function identically in Wnt signaling and are entirely redundant (113). Indeed, retention of just one of the four GSK-3 alleles is sufficient to maintain low levels of β-catenin in the absence of Wnt, reiterating the fact that only a small fraction of GSK-3 is associated with Axin and therefore relevant to Wnt signaling.

Several molecular mechanisms have been proposed to explain how canonical Wnt signaling may cause inhibition of GSK-3 (reviewed in (114)). For example, the intracellular domain of LRP6 may act as a direct inhibitor of GSK-3 (104, 115–117). Recently, it has been shown that Wnt signaling sequesters GSK-3 inside multivesicular bodies, reducing its cytosolic level and extending the half-life of many GSK-3 substrates including β-catenin (118). Extensive stabilization of β-catenin (as a result of constitutive pharmacological or genetic GSK-3 inactivation) may lead to facilitation of proto-oncogenic pathways. However, studies of GSK-3α KO mice revealed no increased propensity toward tumorigenesis (34). Moreover, recently it has been shown that complete loss of all four alleles of GSK-3 (i.e., double GSK-3α+β knockout) in embryonic stem (ES) cells is required before β-catenin levels are significantly dysregulated (113).

Recent studies have shown that p38 mitogen-activated protein kinase (p38 MAPK) can inactivate GSK-3β via phosphorylation of its C-terminal region (Fig. 2) (119). p38 MAPK-mediated phosphorylation of GSK-3β occurs primarily in the brain and thymocytes.

5. GSK-3 Substrates

Reversible phosphorylation of regulatory proteins is a widespread mechanism for the control of cellular processes (120). A critical aspect of this control is the ability of a given protein kinase to modify only selected target proteins as specific functionally relevant amino acid residues. Even though multiple phosphorylation of target proteins is common, there is a high degree of discrimination in the recognition of the residues modified by particular protein kinases.

The specificity of GSK-3 for its substrates is unusual in that is also often regulated by the phosphorylation state of these proteins. This was originally appreciated via studies of phosphorylation of native glycogen synthase by casein kinase II which, by itself, is without effect on enzyme activity. However, this modification potentiated the ability of GSK-3 to phosphorylate and inactivate glycogen synthase (121, 122). Moreover, it was shown that synergistic phosphorylation by casein kinase II and GSK-3 involved intramolecular interactions within this region of glycogen synthase (30). The determination of the crystal structure of GSK-3β provided further insight into the molecular nature of the regulation of GSK-3 and its predilection for primed, prephosphorylated, substrates (33, 62, 123).

The binding site for the substrate priming phosphate group within GSK-3 contains three crucial basic residues—lysine 205 (Lys205), arginine 96 and 180 (Arg96, Arg180) (Fig. 2) (33, 62, 123). These three residues are conserved in all GSK-3 homologues identified to date, suggesting conservation of the priming phosphate-binding site and hence the substrate specificity of GSK-3 in all organisms. Binding of the priming phosphate of the substrate to this pocket on GSK-3 induces a conformational change, aligning the substrate for subsequent phosphorylation.

The majority of GSK-3 substrates exhibit an absolute requirement for prior phosphorylation by another kinase at a "priming" residue located C-terminal to the site of subsequent phosphorylation by GSK-3 (124). GSK-3-catalyzed phosphorylation of these substrates occurs at the fourth (125) or fifth (126) serine or threonine residue N-terminal to the primed site ($pS/T_1XXXpS/T_2$), where the first pS/T_1 (Ser or Thr) is the target residue, X is any amino acid (but often Pro), and the last pS/T_2 is the site for priming phosphorylation. Thus, the primed Ser/Thr is recognized by the positively charged "binding pocket" on GSK-3 which facilitates the correct orientation of the substrates within the active site of the kinase (Figs. 2 and 3). Several protein kinases can act as priming enzymes for GSK-3 phosphorylation, including CDK-5 (127–129), PAR-1 (130), casein kinase I (94), casein kinase II (121, 122), and PKA (131). Priming phosphorylation increases the efficiency of substrate phosphorylation of most GSK-3 substrates by 100- to 1,000-fold (132).

Certain substrates avoid the requirement for prior phosphorylation including c-Jun (133) and c-Myc (134). In these cases, acidic residues may substitute for the priming phosphate. Typically, the stoichiometry of phosphorylation of the priming site is insensitive to cellular conditions. Thus, this requisite is likely more of specificity determinant than a regulatory mechanism.

To date, over 70 cytoplasmic and nuclear proteins have been identified as substrates of GSK-3 (reviewed in (54, 55, 57)) and can be classified into several groups such as transcription factors, structural/cytoskeletal proteins, and regulatory enzymes (see Table 1) (reviewed in (54–57)). In order to prove that an in vitro identified protein can be recognized as in vivo physiological substrate of GSK-3, the target has to meet several criteria described by Cohen and Frame (56). These include phosphorylation of the protein at the appropriate residues by the protein kinase in vitro and under conditions known to modulate that kinase in vivo and selective reduction in those phosphorylation sites upon treatment with a specific inhibitor of the protein kinase (or via gene knockout/RNAi).

Table 1
Listing of known GSK-3 substrates classified by general function

Metabolic proteins (6)
ATP-citrate lyase
Glycogen synthase
Pyruvate dehydrogenase
Acetyl CoA carboxylase
Protein phosphatase 1
Inhibitor-2
Signaling proteins (22)
Axin
APC (adenomatous polyposis coli)
Cubitus interruptus (Ci)/Gli
elF2B (eukaryotic initiation factor 2B)
Amyloid precursor protein
Presenilin-1
hnRNP D (heterogeneous nuclear ribonucleoprotein D)
p21 cdk inhibitor
Insulin receptor substrate 1
Insulin receptor substrate 2
NGF receptor (nerve growth factor)
Mcl-1
Cyclin-AMP-dependent protein kinase (PKA)—RII subunit
Cyclin D1
Cyclin E
Myelin basic protein (MBP)
Cry2
Per2
Nucleoporin p62
PTEN (phosphatase and tensin homologue)
LRP6 (lipoprotein receptor-related protein 6)
TSC2 (tuberous sclerosis 2)
Structural proteins (15)
DF3/MUC1 (high molecular weight mucin-like glycoprotein)
Mucin 1/DF3 antigen
Dynamin-like protein
Kinesin light chain
MAP1B (microtubule-associated protein 1B)
MAP2 (microtubule-associated protein 2)
Tau
Paxillin
CRMP-2 (collapsin response mediator protein 2)
NCAM (neural cell adhesion protein)
Neurofilament-heavy subunit
Ninein
Telokin (KRP) (kinase-related protein)
CLAST 1 (CLIP-associated protein 1)
CLAST 2 (CLIP-associated protein 2)

(continued)

Table 1 (continued)

Transcription factors (29)
β-Catenin
C/EBP α (CCAAT/enhancer-binding protein α)
C/EBP β (CCAAT/enhancer-binding protein β)
CREB (cyclic AMP response element-binding protein)
GATA4
HIF-1 (hypoxia-inducible factor-1)
HSF-1 (heat shock factor-1)
c-Myc
L-myc
c-Jun
c-Myb
NFATc (nuclear factor of activated T cells)
NF-κB (p65 and p105) (nuclear factor κB)
Notch
p53
Snail
AP-1 (activator protein-1)
Glucocorticoid receptor
MITF (microphthalmia-associated transcription factor)
NeuroD
BCL-3
Bmal1
Rev-erb α
Timeless
Clock
SMAD1
Neurogenin 2
BCLAF1 (Bcl-2 interacting transcriptional repressor)

As previously noted, GSK-3 phosphorylation of its substrates typically acts to negatively regulate their function. The GSK-3 sites on a number of substrates form phospho-degrons which promote ubiquitinylation and subsequent proteolytic degradation by the 26S proteasome. The classical example of this is β-catenin degradation but many other targets are similarly regulated including c-myc, cyclin D1, Smad1, CDC25, and Mcl1 (135–140).

Genetic inactivation of GSK-3β leads to embryonic lethality related to failure to activate NF-κB in response to inflammatory cytokines (35). In the absence of opportunistic infections, GSK-3β knockout animals die at birth from incorrect heart physiology (double outlet, right ventricle (141)). GSK-3α knockout mice are viable but show evidence of enhanced sensitivity to insulin and brain and behavioral changes (34, 142). These findings demonstrate that the two GSK-3 isoforms cannot fully compensate for each other. The nature of their differences may lie in differential subcellular localizations, overlapping but distinct substrate specificities, etc. There are data supporting distinct substrate preferences (143–148). As discussed above, within the protein kinase domain, GSK-3α and GSK-3β sequences are 98% identical (32). Therefore, it is likely that the NH_2- and/or COOH-terminal regions, which are more divergent, contribute to any differences in substrate preferences. For example, GSK-3β acts as a significantly better inhibitor 2 kinase compared with GSK-3α (145).

As mentioned previously, in brain GSK-3β is expressed as two splice variants. GSK-3β2 harbors a 13 amino acid insert within the highly conserved catalytic domain that likely forms a loop/hook, which may alter kinase activity toward certain substrates directly, or indirectly by interacting with other factors, such as scaffolding proteins that then expose the isoform to a distinct subset of target proteins (41). GSK-3β1 is more efficient at phosphorylating the microtubule-associated protein MAP1B than GSK-3β2 (47), consistent with previous findings with microtube-associated protein tau (41).

6. Methodological Approaches for Detection of GSK-3 Activity

Based on our current knowledge about structure of GSK-3 and the mechanisms pertaining to its regulation as discussed above, there are several methodological approaches for evaluating GSK-3 activity in vitro and in vivo.

(a) A classical in vitro approach is to perform an assay of catalytic activity of GSK-3, where GSK-3 activity toward specific substrates can be quantitated (113). It is important to maintain the phosphorylation state of GSK-3 during this procedure through the use of phosphatase inhibitors and EDTA to chelate cellular Mg^{2+}. The most essential step of this procedure is immunoprecipitation of endogenous GSK-3 from cell or tissue lysates using conditions that do not inactivate the protein but remove other protein kinases (e.g., avoidance of denaturants). Since most GSK-3 substrates require priming via prephosphorylation, it is important to use prephosphorylated substrates. These include proteins that have been purified from cells (e.g., glycogen synthase) or synthetic phosphopeptides. The most commonly used of the latter is GS-1 (e.g., GSM, Millipore, 12-533). Such substrates also increase the specificity of the kinase assay. However, this method has several limitations. For example, the cellular lysis required disrupts association of specific pools of GSK-3 and cannot be employed to evaluate the effects of Wnt on GSK-3. Moreover, only a small percentage of total GSK-3 in cells interacts with particular regulators (89, 90, 104).

(b) The most commonly used method to assess GSK-3 activity is measurement of its phosphorylation status in cells or tissues, using specific phospho-serine or phospho-tyrosine GSK-3 antibodies via immunoblot analysis (refer to Doble et al. (113)). Several companies provide excellent antibodies to Ser9, Ser21, and the phosphotyrosine in GSK-3 (e.g., Cell Signaling Technologies #9331 for pSer21/9 and Epitomics EPR933Y, respectively). It is also important to also blot the membranes with total GSK-3 antibodies to account for any loading and expression differences. Note, levels of phosphoserine in GSK-3 inversely correlate with activity, but this relationship is not linear and fully phosphorylated GSK-3 retains significant activity (~50%) compared with unphosphorylated GSK-3.

(c) It is becoming increasing apparent that changes in GSK-3 activity are not always accompanied by changes in its phosphorylation status, thus in addition to GSK-3 kinase assays, it is important to evaluate the status of targets downstream of GSK-3 using phospho-specific antibodies by immunoblot analysis. This type of approach becomes even more relevant when taking into consideration the fact that GSK-3 may not have equal access to all of its potential intracellular substrates in vivo, owing to subcellular localization and complex formation. Thus, subcellular fractionation techniques and appropriate lysate buffer conditions should be applied prior kinase assay and/or immunoblot analysis.

(d) If one of aforementioned methods detects potential changes in GSK-3 activity, it is strongly recommended to apply pharmacological and/or genetic approaches to inactivate GSK-3 to support the evidence for physiological relevance. There are numerous small molecule inhibitors of GSK-3, which have the typical caveat of off-target effects. However, using reagents of different chemical classes can mitigate the risk that observations are due to lack of specificity. Genetic models have the advantage of absolute specificity and isoform selectivity but the tissues and cells develop without GSK-3 such that adaptive mechanisms may emerge. Hence, combinatorial approaches are recommended for robust assessment of the role of this protein kinase in a given process (see Sect. 12).

7. Function of GSK-3 Substrates in the Brain

GSK-3 and several of its substrates (MAP1B, Tau, presenilin-1, CREB, β-catenin) have been implicated in fundamental brain functions such as neurogenesis, development of neuronal tissue, regulation of synaptogenesis and axonal growth cone collapse (149–152), cytoskeletal stabilization (153–156), cell adhesion

(157), energy metabolism (158), synaptic plasticity and memory formation (159–162), as well as neurotransmitter signaling (163–165) and circadian rhythms (166, 167).

8. GSK-3, Neurodevelopment, and Neurite Growth

GSK-3 has been shown to play critical roles during neurodevelopment, neurogenesis, neuronal migration, neuronal polarization, axonal growth, and guidance (reviewed in (168)). GSK-3 lies downstream of Wnt and Shh signaling pathways, which are each essential for normal brain formation. Moreover, several studies have shown that GSK-3 may directly regulate Notch signaling via phosphorylation of the intracellular domains Notch1 or Notch2, or regulate the actions of the γ-secretase regulator, presenilin-1 (169–171). In a recent study using a combination of floxed alleles of GSK-3β and a global knockout of GSK-3α and Nestin-Cre, all four alleles of GSK-3 were inactivated in neural progenitors which shifted the balance toward self-renewal and away from neuronal differentiation, resulting in dramatic morphological neuronal malformations (172).

GSK-3 negatively regulates several transcription factors that are important for critical steps of neurodevelopment such as β-catenin (97), c-Jun (173), c-Myc (174), SMAD1 (140), cyclic AMP response element binding protein (CREB) (71), nuclear factor of activated T cells (NFATc) (175), and neurogenin 2 (176). Many of these transcription factors undergo proteasomal degradation after GSK-3 phosphorylation. In addition to the transcriptional regulators, the most well-characterized substrates of GSK-3 involved in neural development are microtubule-associated proteins; GSK-3 may therefore control mitotic spindle reorganization during cell division, movement of the leading processes and soma during neuronal migration, and directed growth cone advancement during axon growth and guidance, all of which require coordinated control of microtubule dynamics. These substrates include adenomatosis polyposis coli (APC) (177), CLIP-associated protein 1 (CLASP1) and CLASP2 (178), collapsin response mediator protein 2 (CRMP2) (179), microtubule-associated protein 1B (MAP1B) (180), and Tau (181). Moreover, signaling molecules, such as PTEN and the Wnt co-receptor LRP6, are also phosphorylated by GSK-3 (182). Other GSK-3 substrates include kinesin light chain (KLC) (183), which regulates selective transport, and essential components of the translational machinery, such as eukaryotic initiation factor 2B (EIF2B) (15) and tuberous sclerosis 2 (TSC2) (184). Recent studies have identified GSK-3 as a novel Disc1 interactor, a molecule known to have an important role in neurodevelopment through regulation of neural progenitor proliferation (185). Transgenic mice expressing constitutively active forms of GSK-3α and β (GSK-3 S9A, S21A knockin mice) (186) have been shown to have deficiencies in neurogenesis due to reduced neurotrophin signaling and impairment of neural precursor cell proliferation (187).

Impaired neurogenesis has been implicated in the pathogenesis of depression (reviewed in (188–190)), bipolar disorder (reviewed in (191)), and schizophrenia (reviewed in (192)). Moreover, therapeutic interventions, including antidepressants (193–197), electroconvulsive shock (198–201), mood stabilizers (202–204), and antipsychotics (205–208) have been shown to influence neurogenesis and are also known to modulate the activity of GSK-3 (75, 163, 209–211). Thus, GSK-3 is an important intracellular mediator involved in neurogenesis and neuronal functions; dysfunction of GSK-3-dependent signaling pathways may contribute to molecular mechanisms of neurogenesis-related psychiatric disorders.

9. GSK-3 and Neuronal Apoptosis

GSK-3 promotes the mitochondrial intrinsic apoptosis signaling pathway that is induced by many types of cellular insults, including DNA damage, ER stress, mitochondrial toxins, hypoxia and ischemia, glutamate excitotoxicity, heat shock, oxidative stress, etc. (reviewed in (212)). This leads to loss of mitochondrial integrity and cell destruction (213). Expression of active GSK-3β causes neuronal cell death in neuronal cell lines (214) and cortical neurons (215). In vivo, transgenic overexpression of GSK-3β leads to apoptotic death of neurons in mice (216). Several studies have suggested that GSK-3 inhibition (by lithium) protects from apoptosis (reviewed in (217, 218)). Recently, studies using pharmacological approaches have shown that GSK-3 inhibition attenuates motor neuron cell death in the spinal cord of a mouse model of amyotrophic lateral sclerosis (219), protects dopaminergic neurons from MPTP-induced excitatory toxicity in vivo (220), and promotes axon sprouting and locomotor functional recovery in spinal cord lesioned rats (221). Thus, inhibition of GSK-3 has potential benefits in protecting against neuronal cell death in neurodegenerative conditions.

10. GSK-3, Learning and Memory

Several lines of study support a significant role for GSK-3 in learning and memory. GSK-3β phosphorylation is increased upon induction of long-term potentiation (LTP) in the hippocampus in vivo (162).

Transgenic mice conditionally overexpressing GSK-3β showed impaired spatial memory and LTP in the CA1 and DG, implicating GSK-3β in synaptic plasticity (162, 222). In GSK-3αβ knockin mice (186), NMDA receptor-dependent LTD in the CA1 was converted to a slow onset, LTP-like response (223). LTP deficits can be attenuated and rescued by chronic treatment with lithium (162). GSK-3 inhibitors block the NMDA receptor-dependent induction of LTP (224). In a study employing 57 different Ser/Thr protein kinase inhibitors, suppression of only one protein kinase, GSK-3, affected LTD (160). GSK-3β heterozygote mice have been shown to have retrograde amnesia in the reconsolidation test of contextual fear conditioning as well as in water maze tests, suggesting that GSK-3β plays role in memory reconsolidation (225). By contrast, GSK-3α KO mice showed impaired fear conditioning and locomotor memory performance (142). GSK-3 modulates structural plasticity such as growth cone formation and synaptogenesis (as discussed above).

11. GSK-3 and Brain Diseases

Dysregulation of GSK-3-substrate-mediated phosphorylation and associated signaling pathways have been implicated in the pathogenesis of psychiatric and neurodegenerative diseases such as schizophrenia, Alzheimer disease, bipolar mood disorder, and ADHD (reviewed in (226–230)).

11.1. Alzheimer Disease

The global prevalence of dementia is estimated to be as high as 24 million, and is predicted to double every 20 years through to 2040 (231). Alzheimer disease (AD) is the leading cause of dementia and is characterized by a progressive decline in cognitive function, which typically begins with deterioration in memory. Accumulation of β-amyloid (Aβ) has been hypothesized to be the primary influence triggering the cascade of pathogenic events leading to tau alterations and neuronal death in AD (reviewed in (232, 233)). Thus, amyloid deposits, formation of neurofibrillary tangles (NFT) and neuronal death in selective brain regions are the chief pathologies found in the brains of AD patients.

(a) GSK-3 is a key protein kinase required for AD-type abnormal phosphorylation of tau, generating paired helical filaments (reviewed in (232)). Overexpression of GSK-3β results in tau hyperphosphorylation and disrupted microtubules in transgenic mice (216).

(b) In vitro studies have shown that GSK-3α modulates APP processing and Aβ production (146).

(c) Both in vitro and in vivo overexpression of GSK-3 promote apoptotic neuronal cell death (214–216).

(d) GSK-3β is involved in memory and synaptic plasticity. Transgenic mice overexpressing GSK-3β exhibit impaired spatial memory and LTP in CA1 and DG (222).

(e) Lithium and other GSK-3 inhibitors reduce tauopathy in vivo (234, 235). Both lithium and valproic acid have been reported to inhibit Aβ peptide production in APP transgenic mice (236). However, the effect of GSK-3 inhibition on APP metabolism and Aβ remain controversial, and a recent study found no effect of lithium on Aβ load or memory deficits, despite the reduction of phospho-tau in a triple transgenic mouse model of AD (237).

(f) Treatment of APP^{SW}–tau^{vlw} transgenic mice with the TZDZ compound, NP12, a novel non-ATP-competitive GSK-3 inhibitor had a beneficial impact on AD pathology, including reduction of tau phosphorylation, amyloid deposition and neuroinflammatory changes, preserved neuronal survival in EC and CA1 hippocampal region, and enhanced memory after treatment (238).

(g) Treatment with an ATP-competitive GSK-3 inhibitor, SB216763, protected against Aβ-induced neuronal damage in an intracerebroventricular Aβ peptide infusion model—but only partially ameliorated neuroinflammation and behavioral deficits (239).

(h) Lithium is used clinically to augment antipsychotic treatment in patients with AD (240).

11.2. Schizophrenia

Schizophrenia is an under-appreciated but devastating illness that has been estimated to affect approximately 1 in 200 people over their lifetime (241). It is a severe psychotic disorder with abnormalities associated with thought process and content, perception, cognition, and affect (242, 243). The clinical symptoms of schizophrenia are divided into broad categories, including positive or negative symptoms and cognitive deficits. The etiology of the disease remains unknown, however a strong genetic component to the illness (over 130 genes have been reported to predispose to schizophrenia), influence of environmental factors, neurodevelopmental defects, as well as dysfunction of different neurotransmitter systems (glutamate, GABA, dopamine, and acetylcholine) have been reported to be involved (244–246). Pathophysiology of the disease includes deficits in neural circuits between the neocortical association area, the limbic system and midbrain thalamic structures, and basal ganglia as well as dysorganization of synaptic connectivity.

Several studies have revealed a relationship between GSK-3 and the pathogenesis of schizophrenia in patients (210, 247–251):

(a) Overexpression of GSK-3β in mice produced manic-like behavior (252).
(b) Two genetic studies reported a direct association between GSK-3 polymorphisms and schizophrenia (251, 253).
(c) Elevations in GSK-3 activity were found in animal models characterized by hyperlocomotion, such as DAT-KO (163) and Disc1L100p (254) mutant mice. Moreover, pharmacological and genetic inhibition of GSK-3 in these models yielded beneficial effects and rescued schizophrenia-like phenotypes in mice (163, 254).
(d) There is evidence of impaired Akt/PKB/GSK-3β signaling in subjects with schizophrenia (reviewed in (227–229, 255)).
(e) The Akt/PKB/GSK-3 pathway has been implicated in the mechanisms of action of several antipsychotic and psychoactive drugs (75, 163, 209–211).
(f) Lithium is used to augment antipsychotic treatment in patients with schizophrenia (256, 257).

11.3. Addiction

GSK-3 has been recognized as an important downstream substrate of dopamine receptor signaling which is linked to addictive behaviors (163, 252, 258). Several studies support a role for GSK-3 and addiction.

(a) Regulation of GSK-3 by the D2 receptor involves inactivation of Akt/PKB in a protein complex including the scaffolding protein β-arrestin2 and PP2A (259).
(b) Genetic or pharmacological inhibition of GSK-3 activity attenuated dopamine-mediated behaviors (163, 254, 260, 261).
(c) Inhibition of GSK-3 activity blocks the development of behavioral sensitization to cocaine (260, 262).
(d) GSK-3β heterozygous null mice showed attenuated behavioral responses to amphetamine (163), whereas GSK-3 (S21A/S9A) knockin mice displayed increased susceptibility to amphetamine-induced hyperactivity (223). Mice overexpressing constitutively active S9A-GSK-3β exhibited hyperactivity (252).
(e) Administration of the psychostimulants amphetamine and cocaine induced time-dependent changes in GSK-3 activity. Increases in phosphorylation of GSK-3β in the striatum were reported at early time-points after acute amphetamine and cocaine (75, 263, 264), and decreases in phosphorylation of GSK-3α and β were reported at later times (163, 260, 264, 265). Rapid increases in GSK-3 phosphorylation were observed

in the nucleus accumbens after an acute injection of cocaine, and tolerance to this increase in GSK-3 phosphorylation developed upon repeated cocaine administration (266).

11.4. Bipolar Disorder

Bipolar disorder afflicts approximately 2% of the population and causes debilitating and life-threatening behaviors if untreated. The defining characteristic of bipolar disorder is uncontrollable and extreme shifts in mood between mania and depression, usually triggered by environmental factors such as stress (267, 268). The pathophysiological mechanisms underlying bipolar disorder are unknown, but may include genetic variations, abnormal development of neural networks, and dysregulation of neurotransmitter, neuroendocrine, and signaling activities (268–270). There is accumulating evidence that GSK-3 plays a significant role in this disorder:

(a) GSK-3 has been implicated in the pathogenesis of bipolar disorder since the discovery that lithium, which is widely used to augment antipsychotic treatment in patients with bipolar mood disorders (reviewed in (271–273)), was found to inhibit GSK-3 (22, 23).

(b) Several effective mood stabilizer drugs inhibit GSK-3 activity (reviewed in (226)).

(c) Treatment with lithium results in changes in GSK-3 phosphorylation (274), as do treatments with serotonin (165) and brain-derived neurotrophic factor (BDNF) (209).

(d) Studies in dentate gyrus (DG)-specific GSK-3β knockdown mice, generated by in vivo GSK-3β shRNA lentivirus administration into the DG, demonstrated an antidepressant-like effect on mouse behavior (275).

(e) GSK-3 knockin mice displayed increased susceptibility to stress-induced depressive-like behaviors (223).

(f) GSK-3β heterozygote mice (163, 276) as well as GSK-3α KO mice (142) exhibit antidepressant-like behaviors.

(g) Treatment with GSK-3 inhibitors generated antidepressant-like behavior in animals (164, 276–279).

(h) The serine phosphorylation level of GSK-3 in human peripheral blood mononuclear cells (PBMCs) was lower in bipolar patients (223). Modification of GSK-3 activity occurs during the treatment of bipolar mania in patients (280).

(i) GSK-3 has been implicated in the mechanism of neuroinflammation (as discussed below), which is recognized as having an important influence on the pathology and treatment of mood disorders, especially depression (281, 282).

(j) Recent genetic studies found a direct association between a GSK-3β polymorphism and lithium treatment in acutely depressed, antidepressant-resistant bipolar disorder, and major depressive patients (283). However, another study failed to find an association of this SNP with the degree of prophylactic lithium response in bipolar disorder (284). Other studies have found several associations between polymorphisms in GSK-3 and depressed patients (285, 286).

11.5. Amyotrophic Lateral Sclerosis and Neuroinflammation

GSK-3 has been reported to play an important role in promoting inflammation in both innate and T- and B-cell-mediated immunity (212, 287).

(a) GSK-3 activity promotes production of several pro-inflammatory cytokines after stimulation of multiple types of Toll-like receptors in human monocytes. These responses were tempered by administration of GSK-3 inhibitors or by inducing GSK-3 deficiency (288).

(b) GSK-3 acts to suppress the anti-inflammatory cytokine IL-10. GSK-3 inhibition increases IL-10 levels in mouse models (288, 289).

(c) In vivo, two GSK-3 inhibitors (lithium and SB216763) rescued approximately 70% of mice from an otherwise 100% lethal inflammatory response to lipopolysaccharide (288).

(d) GSK-3 inhibitors were found to reduce inflammatory cytokine production in mouse primary astrocytes and microglia by greater than 90% (287, 290).

12. Therapeutic Applications of GSK-3 Inhibitors

Remarkably, there have been over 40,000 small molecule derivatives reported in the published literature or patents as having inhibitory activity toward GSK-3. Approximately 8,000 of these have been reported with activity values (reviewed in (291)).

GSK-3 inhibitors can be classified dependent upon their mechanism of GSK-3 inhibition: ATP—or non-ATP-competitive, including substrate—competitive, and alternative mode of inhibition (such as lithium). Most GSK-3 inhibitors developed so far are ATP-competitive, where the compound competes for the ATP-binding site within the kinase domain and often shows off-target effects due to a loss of selectivity (since the ATP-binding site of protein kinases is conserved between diverse protein kinase families—discussed in (292)). Non-ATP-competitive inhibitors are expected to be more specific, as the substrate-binding site is quite distinctive between various protein kinases (as discussed above).

The first widely employed GSK-3 inhibitor was lithium (Li), which inhibits GSK-3 in vivo in mouse and rat brain (274, 293–295), in cultured cells (296); and in human PBMCs (297) by a magnesium-competitive mechanism (298). At physiological concentrations of lithium (1 mM) only ~20% of GSK-3 activity is suppressed in vivo (276), depending on the magnesium concentration used in the kinase assay (293).

Several types of GSK-3 inhibitors have been used beneficially to correct neurological conditions in animals such as bipolar disorder and depression (164, 276–279), Alzheimer disease (234, 294, 295, 299), and schizophrenia (163, 254). Both ARA014418, an ATP-competitive GSK-3 inhibitor that has been reported to cross the blood–brain barrier, and the peptide inhibitor L803mts, which was injected directly into the ventricles, reduce immobility time in the forced swim test (277, 278). The thiadiazolidinone TDZD-8 reduces immobility in a related behavioral assessment, the tail suspension test, and reduces latency to emerge from dark to light areas (164). Lithium attenuates hyperactivity induced by amphetamine (300); this amphetamine-induced behavior is believed to be due to increased dopamine signaling, and is mimicked in DAT-KO (dopamine transporter knockout) mice (163). Furthermore, lithium and several GSK-3 inhibitors, including SB216763, alsterpaullone, 6-BIO and TDZD, reduce hyperactivity in DAT-KO mice (163) as well as in Disc1L100p mice (254). Open field activity is also attenuated in wild-type mice by intraperitoneal injection of lithium (as LiCl), TDZD or AR-A014418 (164, 277).

Because the ATP-binding pockets of GSK-3α and GSK-3β are so similar, inhibitors that target these sites are unlikely to differentiate between the two isoforms (discussed in (301)). Such selectivity might only be achieved by drugs that act at other sites on the kinases, by alternative molecular approaches (e.g., RNAi), or if either the intracellular distribution or the expression is different for the two isoforms (discussed in (301)). Thus, despite many publications inferring that only GSK-3β is affected, all small molecule GSK-3 inhibitors are only useful in assessing the function of total cellular GSK-3 activity. In addition, GSK-3 inhibitors are prone to nonspecific "off-target" effects that become more pronounced as the working concentration of inhibitor increases (302). To offset this concern, experiments should be repeated with several classes of inhibitors since each has distinct off-target effects.

Acknowledgments

This work was supported by CIHR grant MOP 74711 (to JW). We also thank our colleagues, John Roder, Albert Wong, Satish Patel and members of the Woodgett lab for their advice and input.

References

1. Kannan N, Neuwald AF (2004) Evolutionary constraints associated with functional specificity of the CMGC protein kinases MAPK, CDK, GSK, SRPK, DYRK, and CK2alpha. Protein Sci 13(8):2059–2077
2. Embi N, Rylatt DB, Cohen P (1980) Glycogen synthase kinase-3 from rabbit skeletal muscle. Separation from cyclic-AMP-dependent protein kinase and phosphorylase kinase. Eur J Biochem/FEBS 107(2):519–527
3. Rylatt DB, Aitken A, Bilham T, Condon GD, Embi N, Cohen P (1980) Glycogen synthase from rabbit skeletal muscle. Amino acid sequence at the sites phosphorylated by glycogen synthase kinase-3, and extension of the N-terminal sequence containing the site phosphorylated by phosphorylase kinase. Eur J Biochem/FEBS 107(2):529–537
4. Vandenheede JR, Yang SD, Goris J, Merlevede W (1980) ATP x Mg-dependent protein phosphatase from rabbit skeletal muscle. II. Purification of the activating factor and its characterization as a bifunctional protein also displaying synthase kinase activity. J Biol chem 255(24):11768–11774
5. Larner J (1972) Insulin and glycogen synthase. Diabetes 21(2 Suppl):428–438
6. Rosell-Perez M, Larner J (1964) Studies on Udpg: alpha-1.4-glucan alpha-4-glucosyltransferase. Vi. Specificity and structural requirements for the activator of the D form of the dog muscle enzyme. Biochemistry 3:773–778
7. Schlender KK, Wei SH, Villar-Palasi C (1969) UDP-glucose:glycogen alpha-4-glucosyltransferase I kinase activity of purified muscle protein kinase. Cyclic nucleotide specificity. Biochim Biophys Acta 191(2):272–278
8. Roach PJ, DePaoli-Roach AA, Larner J (1978) Ca^{2+}-stimulated phosphorylation of muscle glycogen synthase by phosphorylase b kinase. J Cyclic Nucleotide Res 4(4):245–257
9. Embi N, Rylatt DB, Cohen P (1979) Glycogen synthase kinase-2 and phosphorylase kinase are the same enzyme. Eur J Biochem/FEBS 100(2):339–347
10. DePaoli-Roach AA, Roach PJ, Larner J (1979) Rabbit skeletal muscle phosphorylase kinase. Comparison of glycogen synthase and phosphorylase as substrates. J Biol chem 254(10):4212–4219
11. Soderling TR, Srivastava AK, Bass MA, Khatra BS (1979) Phosphorylation and inactivation of glycogen synthase by phosphorylase kinase. Proc Natl Acad Sci U S A 76(6):2536–2540
12. Parker PJ, Caudwell FB, Cohen P (1983) Glycogen synthase from rabbit skeletal muscle; effect of insulin on the state of phosphorylation of the seven phosphoserine residues in vivo. Eur J Biochem/FEBS 130(1):227–234
13. Cross DA, Alessi DR, Cohen P, Andjelkovich M, Hemmings BA (1995) Inhibition of glycogen synthase kinase-3 by insulin mediated by protein kinase B. Nature 378(6559):785–789
14. Welsh GI, Proud CG (1993) Glycogen synthase kinase-3 is rapidly inactivated in response to insulin and phosphorylates eukaryotic initiation factor eIF-2B. Biochem J 294(Pt 3): 625–629
15. Welsh GI, Miller CM, Loughlin AJ, Price NT, Proud CG (1998) Regulation of eukaryotic initiation factor eIF2B: glycogen synthase kinase-3 phosphorylates a conserved serine which undergoes dephosphorylation in response to insulin. FEBS Lett 421(2):125–130
16. Siegfried E, Perkins LA, Capaci TM, Perrimon N (1990) Putative protein kinase product of the Drosophila segment-polarity gene zeste-white3. Nature 345(6278):825–829
17. Bourouis M, Moore P, Ruel L, Grau Y, Heitzler P, Simpson P (1990) An early embryonic product of the gene shaggy encodes a serine/threonine protein kinase related to the CDC28/cdc2+ subfamily. EMBO J 9(9): 2877–2884
18. Siegfried E, Chou TB, Perrimon N (1992) wingless signaling acts through zeste-white 3, the Drosophila homolog of glycogen synthase kinase-3, to regulate engrailed and establish cell fate. Cell 71(7):1167–1179
19. Pierce SB, Kimelman D (1995) Regulation of Spemann organizer formation by the intracellular kinase Xgsk-3. Development 121(3):755–765
20. He X, Saint-Jeannet JP, Woodgett JR, Varmus HE, Dawid IB (1995) Glycogen synthase kinase-3 and dorsoventral patterning in Xenopus embryos. Nature 374(6523):617–622
21. Kao KR, Masui Y, Elinson RP (1986) Lithium-induced respecification of pattern in *Xenopus laevis* embryos. Nature 322(6077):371–373
22. Klein PS, Melton DA (1996) A molecular mechanism for the effect of lithium on development. Proc Natl Acad Sci U S A 93(16): 8455–8459
23. Stambolic V, Ruel L, Woodgett JR (1996) Lithium inhibits glycogen synthase kinase-3 activity and mimics wingless signalling in intact cells. Curr Biol 6(12):1664–1668
24. Ali A, Hoeflich KP, Woodgett JR (2001) Glycogen synthase kinase-3: properties, functions, and regulation. Chem Rev 101(8): 2527–2540

25. Woodgett JR (1991) cDNA cloning and properties of glycogen synthase kinase-3. Methods Enzymol 200:564–577
26. Ruel L, Bourouis M, Heitzler P, Pantesco V, Simpson P (1993) Drosophila shaggy kinase and rat glycogen synthase kinase-3 have conserved activities and act downstream of Notch. Nature 362(6420):557–560
27. Kassir Y, Rubin-Bejerano I, Mandel-Gutfreund Y (2006) The *Saccharomyces cerevisiae* GSK-3 beta homologs. Curr Drug Targets 7(11): 1455–1465
28. Alon LT, Pietrokovski S, Barkan S, Avrahami L, Kaidanovich-Beilin O, Woodgett JR, Barnea A, Eldar-Finkelman H (2011) Selective loss of glycogen synthase kinase-3alpha in birds reveals distinct roles for GSK-3 isozymes in tau phosphorylation. FEBS Lett 585(8): 1158–1162
29. Hemmings BA, Aitken A, Cohen P, Rymond M, Hofmann F (1982) Phosphorylation of the type-II regulatory subunit of cyclic-AMP-dependent protein kinase by glycogen synthase kinase 3 and glycogen synthase kinase 5. Eur J Biochem/FEBS 127(3):473–481
30. Woodgett JR, Cohen P (1984) Multisite phosphorylation of glycogen synthase. Molecular basis for the substrate specificity of glycogen synthase kinase-3 and casein kinase-II (glycogen synthase kinase-5). Biochim Biophys Acta 788(3):339–347
31. Tung HY, Reed LJ (1989) Purification and characterization of protein phosphatase 1I activating kinase from bovine brain cytosolic and particulate fractions. J Biol Chem 264(5):2985–2990
32. Woodgett JR (1990) Molecular cloning and expression of glycogen synthase kinase-3/factor A. EMBO J 9(8):2431–2438
33. Dajani R, Fraser E, Roe SM, Young N, Good V, Dale TC, Pearl LH (2001) Crystal structure of glycogen synthase kinase 3 beta: structural basis for phosphate-primed substrate specificity and autoinhibition. Cell 105(6):721–732
34. MacAulay K, Doble BW, Patel S, Hansotia T, Sinclair EM, Drucker DJ, Nagy A, Woodgett JR (2007) Glycogen synthase kinase 3alpha-specific regulation of murine hepatic glycogen metabolism. Cell Metab 6(4):329–337
35. Hoeflich KP, Luo J, Rubie EA, Tsao MS, Jin O, Woodgett JR (2000) Requirement for glycogen synthase kinase-3beta in cell survival and NF-kappaB activation. Nature 406(6791):86–90
36. Perez-Costas E, Gandy JC, Melendez-Ferro M, Roberts RC, Bijur GN (2010) Light and electron microscopy study of glycogen synthase kinase-3beta in the mouse brain. PLoS One 5(1):e8911
37. Yao HB, Shaw PC, Wong CC, Wan DC (2002) Expression of glycogen synthase kinase-3 isoforms in mouse tissues and their transcription in the brain. J Chem Neuroanat 23(4):291–297
38. Ferrer I, Barrachina M, Puig B (2002) Glycogen synthase kinase-3 is associated with neuronal and glial hyperphosphorylated tau deposits in Alzheimer's disease, Pick's disease, progressive supranuclear palsy and corticobasal degeneration. Acta Neuropathol 104(6): 583–591
39. Pandey GN, Dwivedi Y, Rizavi HS, Teppen T, Gaszner GL, Roberts RC, Conley RR (2009) GSK-3beta gene expression in human postmortem brain: regional distribution, effects of age and suicide. Neurochem Res 34(2): 274–285
40. Hanks SK, Hunter T (1995) Protein kinases 6. The eukaryotic protein kinase superfamily: kinase (catalytic) domain structure and classification. FASEB J 9(8):576–596
41. Mukai F, Ishiguro K, Sano Y, Fujita SC (2002) Alternative splicing isoform of tau protein kinase I/glycogen synthase kinase 3beta. J Neurochem 81(5):1073–1083
42. Lau KF, Miller CC, Anderton BH, Shaw PC (1999) Expression analysis of glycogen synthase kinase-3 in human tissues. J Pept Res 54(1):85–91
43. Schaffer B, Wiedau-Pazos M, Geschwind DH (2003) Gene structure and alternative splicing of glycogen synthase kinase 3 beta (GSK-3beta) in neural and non-neural tissues. Gene 302(1–2):73–81
44. Kwok JB, Hallupp M, Loy CT, Chan DK, Woo J, Mellick GD, Buchanan DD, Silburn PA, Halliday GM, Schofield PR (2005) GSK3B polymorphisms alter transcription and splicing in Parkinson's disease. Ann Neurol 58(6):829–839
45. Takahashi M, Tomizawa K, Kato R, Sato K, Uchida T, Fujita SC, Imahori K (1994) Localization and developmental changes of tau protein kinase I/glycogen synthase kinase-3 beta in rat brain. J Neurochem 63(1):245–255
46. Leroy K, Brion JP (1999) Developmental expression and localization of glycogen synthase kinase-3beta in rat brain. J Chem Neuroanat 16(4):279–293
47. Wood-Kaczmar A, Kraus M, Ishiguro K, Philpott KL, Gordon-Weeks PR (2009) An alternatively spliced form of glycogen synthase

kinase-3beta is targeted to growing neurites and growth cones. Mol Cell Neurosci 42(3):184–194

48. Bijur GN, Jope RS (2001) Proapoptotic stimuli induce nuclear accumulation of glycogen synthase kinase-3 beta. J Biol Chem 276(40):37436–37442
49. Bijur GN, Jope RS (2003) Glycogen synthase kinase-3 beta is highly activated in nuclei and mitochondria. Neuroreport 14(18): 2415–2419
50. Goold RG, Gordon-Weeks PR (2001) Microtubule-associated protein 1B phosphorylation by glycogen synthase kinase 3beta is induced during PC12 cell differentiation. J Cell Sci 114(Pt 23):4273–4284
51. Stambolic V, Woodgett JR (1994) Mitogen inactivation of glycogen synthase kinase-3 beta in intact cells via serine 9 phosphorylation. Biochem J 303(Pt 3):701–704
52. Sutherland C, Leighton IA, Cohen P (1993) Inactivation of glycogen synthase kinase-3 beta by phosphorylation: new kinase connections in insulin and growth-factor signalling. Biochem J 296(Pt 1):15–19
53. Woodgett JR (1994) Regulation and functions of the glycogen synthase kinase-3 subfamily. Semin Cancer Biol 5(4):269–275
54. Kockeritz L, Doble B, Patel S, Woodgett JR (2006) Glycogen synthase kinase-3 – an overview of an over-achieving protein kinase. Curr Drug Targets 7(11):1377–1388
55. Doble BW, Woodgett JR (2003) GSK-3: tricks of the trade for a multi-tasking kinase. J Cell Sci 116(Pt 7):1175–1186
56. Frame S, Cohen P (2001) GSK3 takes centre stage more than 20 years after its discovery. Biochem J 359(Pt 1):1–16
57. Jope RS, Johnson GV (2004) The glamour and gloom of glycogen synthase kinase-3. Trends Biochem Sci 29(2):95–102
58. McNeill H, Woodgett JR (2010) When pathways collide: collaboration and connivance among signalling proteins in development. Nat Rev Mol Cell Biol 11(6):404–413
59. Hughes K, Nikolakaki E, Plyte SE, Totty NF, Woodgett JR (1993) Modulation of the glycogen synthase kinase-3 family by tyrosine phosphorylation. EMBO J 12(2):803–808
60. Lochhead PA, Kinstrie R, Sibbet G, Rawjee T, Morrice N, Cleghon V (2006) A chaperone-dependent GSK3beta transitional intermediate mediates activation-loop autophosphorylation. Mol Cell 24(4):627–633
61. Sutherland C, Cohen P (1994) The alpha-isoform of glycogen synthase kinase-3 from rabbit skeletal muscle is inactivated by p70 S6 kinase or MAP kinase-activated protein kinase-1 in vitro. FEBS Lett 338(1):37–42
62. Bax B, Carter PS, Lewis C, Guy AR, Bridges A, Tanner R, Pettman G, Mannix C, Culbert AA, Brown MJ, Smith DG, Reith AD (2001) The structure of phosphorylated GSK-3beta complexed with a peptide, FRATtide, that inhibits beta-catenin phosphorylation. Structure 9(12):1143–1152
63. Wang X, Liu XT, Dunn R, Ohl DA, Smith GD (2003) Glycogen synthase kinase-3 regulates mouse oocyte homologue segregation. Mol Reprod Dev 64(1):96–105
64. Cole A, Frame S, Cohen P (2004) Further evidence that the tyrosine phosphorylation of glycogen synthase kinase-3 (GSK3) in mammalian cells is an autophosphorylation event. Biochem J 377(Pt 1):249–255
65. Kim L, Liu J, Kimmel AR (1999) The novel tyrosine kinase ZAK1 activates GSK3 to direct cell fate specification. Cell 99(4):399–408
66. Lesort M, Jope RS, Johnson GV (1999) Insulin transiently increases tau phosphorylation: involvement of glycogen synthase kinase-3beta and Fyn tyrosine kinase. J Neurochem 72(2):576–584
67. Plyte SE, O'Donovan E, Woodgett JR, Harwood AJ (1999) Glycogen synthase kinase-3 (GSK-3) is regulated during Dictyostelium development via the serpentine receptor cAR3. Development 126(2):325–333
68. Kim L, Harwood A, Kimmel AR (2002) Receptor-dependent and tyrosine phosphatase-mediated inhibition of GSK3 regulates cell fate choice. Dev Cell 3(4):523–532
69. Lochhead PA, Sibbet G, Morrice N, Cleghon V (2005) Activation-loop autophosphorylation is mediated by a novel transitional intermediate form of DYRKs. Cell 121(6):925–936
70. Zhang F, Phiel CJ, Spece L, Gurvich N, Klein PS (2003) Inhibitory phosphorylation of glycogen synthase kinase-3 (GSK-3) in response to lithium. Evidence for autoregulation of GSK-3. J Biol Chem 278(35):33067–33077
71. Grimes CA, Jope RS (2001) CREB DNA binding activity is inhibited by glycogen synthase kinase-3 beta and facilitated by lithium. J Neurochem 78(6):1219–1232
72. Frame S, Cohen P, Biondi RM (2001) A common phosphate binding site explains the unique substrate specificity of GSK3 and its inactivation by phosphorylation. Mol Cell 7(6):1321–1327
73. Alessi DR, Andjelkovic M, Caudwell B, Cron P, Morrice N, Cohen P, Hemmings BA (1996) Mechanism of activation of protein kinase B by insulin and IGF-1. EMBO J 15(23):6541–6551

74. Li M, Wang X, Meintzer MK, Laessig T, Birnbaum MJ, Heidenreich KA (2000) Cyclic AMP promotes neuronal survival by phosphorylation of glycogen synthase kinase 3beta. Mol Cell Biol 20(24):9356–9363

75. Svenningsson P, Tzavara ET, Carruthers R, Rachleff I, Wattler S, Nehls M, McKinzie DL, Fienberg AA, Nomikos GG, Greengard P (2003) Diverse psychotomimetics act through a common signaling pathway. Science 302(5649):1412–1415

76. Eldar-Finkelman H, Krebs EG (1997) Phosphorylation of insulin receptor substrate 1 by glycogen synthase kinase 3 impairs insulin action. Proc Natl Acad Sci U S A 94(18):9660–9664

77. Sharfi H, Eldar-Finkelman H (2008) Sequential phosphorylation of insulin receptor substrate-2 by glycogen synthase kinase-3 and c-Jun NH2-terminal kinase plays a role in hepatic insulin signaling. Am J Physiol 294(2):E307–E315

78. Shaw M, Cohen P (1999) Role of protein kinase B and the MAP kinase cascade in mediating the EGF-dependent inhibition of glycogen synthase kinase 3 in Swiss 3T3 cells. FEBS Lett 461(1–2):120–124

79. Brady MJ, Bourbonais FJ, Saltiel AR (1998) The activation of glycogen synthase by insulin switches from kinase inhibition to phosphatase activation during adipogenesis in 3T3-L1 cells. J Biol Chem 273(23):14063–14066

80. Saito Y, Vandenheede JR, Cohen P (1994) The mechanism by which epidermal growth factor inhibits glycogen synthase kinase 3 in A431 cells. Biochem J 303(Pt 1):27–31

81. Armstrong JL, Bonavaud SM, Toole BJ, Yeaman SJ (2001) Regulation of glycogen synthesis by amino acids in cultured human muscle cells. J Biol Chem 276(2):952–956

82. Krause U, Bertrand L, Maisin L, Rosa M, Hue L (2002) Signalling pathways and combinatory effects of insulin and amino acids in isolated rat hepatocytes. Eur J Biochem/FEBS 269(15):3742–3750

83. Terruzzi I, Allibardi S, Bendinelli P, Maroni P, Piccoletti R, Vesco F, Samaja M, Luzi L (2002) Amino acid- and lipid-induced insulin resistance in rat heart: molecular mechanisms. Mol Cell Endocrinol 190(1–2):135–145

84. Fang X, Yu SX, Lu Y, Bast RC Jr, Woodgett JR, Mills GB (2000) Phosphorylation and inactivation of glycogen synthase kinase 3 by protein kinase A. Proc Natl Acad Sci U S A 97(22):11960–11965

85. Tanji C, Yamamoto H, Yorioka N, Kohno N, Kikuchi K, Kikuchi A (2002) A-kinase anchoring protein AKAP220 binds to glycogen synthase kinase-3beta (GSK-3beta) and mediates protein kinase A-dependent inhibition of GSK-3beta. J Biol Chem 277(40): 36955–36961

86. Ballou LM, Tian PY, Lin HY, Jiang YP, Lin RZ (2001) Dual regulation of glycogen synthase kinase-3beta by the alpha1A-adrenergic receptor. J Biol Chem 276(44): 40910–40916

87. Fang X, Yu S, Tanyi JL, Lu Y, Woodgett JR, Mills GB (2002) Convergence of multiple signaling cascades at glycogen synthase kinase 3: Edg receptor-mediated phosphorylation and inactivation by lysophosphatidic acid through a protein kinase C-dependent intracellular pathway. Mol Cell Biol 22(7):2099–2110

88. Goode N, Hughes K, Woodgett JR, Parker PJ (1992) Differential regulation of glycogen synthase kinase-3 beta by protein kinase C isotypes. J Biol Chem 267(24):16878–16882

89. Lee E, Salic A, Kruger R, Heinrich R, Kirschner MW (2003) The roles of APC and Axin derived from experimental and theoretical analysis of the Wnt pathway. PLoS Biol 1(1):E10

90. Benchabane H, Hughes EG, Takacs CM, Baird JR, Ahmed Y (2008) Adenomatous polyposis coli is present near the minimal level required for accurate graded responses to the Wingless morphogen. Development 135(5): 963–971

91. Zeng L, Fagotto F, Zhang T, Hsu W, Vasicek TJ, Perry WL III, Lee JJ, Tilghman SM, Gumbiner BM, Costantini F (1997) The mouse Fused locus encodes Axin, an inhibitor of the Wnt signaling pathway that regulates embryonic axis formation. Cell 90(1):181–192

92. Ikeda S, Kishida S, Yamamoto H, Murai H, Koyama S, Kikuchi A (1998) Axin, a negative regulator of the Wnt signaling pathway, forms a complex with GSK-3beta and beta-catenin and promotes GSK-3beta-dependent phosphorylation of beta-catenin. EMBO J 17(5):1371–1384

93. Hart MJ, de los Santos R, Albert IN, Rubinfeld B, Polakis P (1998) Downregulation of beta-catenin by human Axin and its association with the APC tumor suppressor, beta-catenin and GSK3 beta. Curr Biol 8(10):573–581

94. Amit S, Hatzubai A, Birman Y, Andersen JS, Ben-Shushan E, Mann M, Ben-Neriah Y, Alkalay I (2002) Axin-mediated CKI phosphorylation of beta-catenin at Ser 45: a molecular switch for the Wnt pathway. Genes Dev 16(9):1066–1076

95. Hagen T, Vidal-Puig A (2002) Characterisation of the phosphorylation of beta-catenin at the

GSK-3 priming site Ser45. Biochem Biophys Res Commun 294(2):324–328

96. Hagen T, Di Daniel E, Culbert AA, Reith AD (2002) Expression and characterization of GSK-3 mutants and their effect on beta-catenin phosphorylation in intact cells. J Biol Chem 277(26):23330–23335
97. Liu C, Li Y, Semenov M, Han C, Baeg GH, Tan Y, Zhang Z, Lin X, He X (2002) Control of beta-catenin phosphorylation/degradation by a dual-kinase mechanism. Cell 108(6):837–847
98. Sakanaka C (2002) Phosphorylation and regulation of beta-catenin by casein kinase I epsilon. J Biochem 132(5):697–703
99. Yanagawa S, Matsuda Y, Lee JS, Matsubayashi H, Sese S, Kadowaki T, Ishimoto A (2002) Casein kinase I phosphorylates the Armadillo protein and induces its degradation in Drosophila. EMBO J 21(7):1733–1742
100. Aberle H, Bauer A, Stappert J, Kispert A, Kemler R (1997) beta-catenin is a target for the ubiquitin-proteasome pathway. EMBO J 16(13):3797–3804
101. Miller JR (2002) The Wnts. Genome Biol 3(1):REVIEWS3001
102. Polakis P (2000) Wnt signaling and cancer. Genes Dev 14(15):1837–1851
103. Smalley MJ, Dale TC (1999) Wnt signalling in mammalian development and cancer. Cancer Metastasis Rev 18(2):215–230
104. Mi K, Dolan PJ, Johnson GV (2006) The low density lipoprotein receptor-related protein 6 interacts with glycogen synthase kinase 3 and attenuates activity. J Biol Chem 281(8):4787–4794
105. Zeng X, Huang H, Tamai K, Zhang X, Harada Y, Yokota C, Almeida K, Wang J, Doble B, Woodgett J, Wynshaw-Boris A, Hsieh JC, He X (2008) Initiation of Wnt signaling: control of Wnt coreceptor Lrp6 phosphorylation/activation via frizzled, dishevelled and axin functions. Development 135(2):367–375
106. Niehrs C, Shen J (2010) Regulation of Lrp6 phosphorylation. Cell Mol Life Sci 67(15):2551–2562
107. Bilic J, Huang YL, Davidson G, Zimmermann T, Cruciat CM, Bienz M, Niehrs C (2007) Wnt induces LRP6 signalosomes and promotes dishevelled-dependent LRP6 phosphorylation. Science 316(5831):1619–1622
108. Blitzer JT, Nusse R (2006) A critical role for endocytosis in Wnt signaling. BMC Cell Biol 7:28
109. Yamamoto H, Komekado H, Kikuchi A (2006) Caveolin is necessary for Wnt-3a-dependent internalization of LRP6 and accumulation of beta-catenin. Dev Cell 11(2):213–223
110. Jho E, Lomvardas S, Costantini F (1999) A GSK3beta phosphorylation site in axin modulates interaction with beta-catenin and Tcf-mediated gene expression. Biochem Biophys Res Commun 266(1):28–35
111. Yamamoto H, Kishida S, Kishida M, Ikeda S, Takada S, Kikuchi A (1999) Phosphorylation of axin, a Wnt signal negative regulator, by glycogen synthase kinase-3beta regulates its stability. J Biol Chem 274(16):10681–10684
112. Rubinfeld B, Albert I, Porfiri E, Fiol C, Munemitsu S, Polakis P (1996) Binding of GSK3beta to the APC-beta-catenin complex and regulation of complex assembly. Science 272(5264):1023–1026
113. Doble BW, Patel S, Wood GA, Kockeritz LK, Woodgett JR (2007) Functional redundancy of GSK-3alpha and GSK-3beta in Wnt/beta-catenin signaling shown by using an allelic series of embryonic stem cell lines. Dev Cell 12(6):957–971
114. Kimelman D, Xu W (2006) beta-catenin destruction complex: insights and questions from a structural perspective. Oncogene 25(57):7482–7491
115. Cselenyi CS, Jernigan KK, Tahinci E, Thorne CA, Lee LA, Lee E (2008) LRP6 transduces a canonical Wnt signal independently of Axin degradation by inhibiting GSK3's phosphorylation of beta-catenin. Proc Natl Acad Sci U S A 105(23):8032–8037
116. Piao S, Lee SH, Kim H, Yum S, Stamos JL, Xu Y, Lee SJ, Lee J, Oh S, Han JK, Park BJ, Weis WI, Ha NC (2008) Direct inhibition of GSK3beta by the phosphorylated cytoplasmic domain of LRP6 in Wnt/beta-catenin signaling. PLoS One 3(12):e4046
117. Wu G, Huang H, Garcia Abreu J, He X (2009) Inhibition of GSK3 phosphorylation of beta-catenin via phosphorylated PPPSPXS motifs of Wnt coreceptor LRP6. PLoS One 4(3):e4926
118. Taelman VF, Dobrowolski R, Plouhinec JL, Fuentealba LC, Vorwald PP, Gumper I, Sabatini DD, De Robertis EM (2010) Wnt signaling requires sequestration of glycogen synthase kinase 3 inside multivesicular endosomes. Cell 143(7):1136–1148
119. Thornton TM, Pedraza-Alva G, Deng B, Wood CD, Aronshtam A, Clements JL, Sabio G, Davis RJ, Matthews DE, Doble B, Rincon M (2008) Phosphorylation by p38 MAPK as an alternative pathway for GSK3beta inactivation. Science 320(5876):667–670
120. Cohen P (1982) The role of protein phosphorylation in neural and hormonal control of cellular activity. Nature 296(5858):613–620
121. DePaoli-Roach AA, Ahmad Z, Camici M, Lawrence JC Jr, Roach PJ (1983) Multiple

phosphorylation of rabbit skeletal muscle glycogen synthase. Evidence for interactions among phosphorylation sites and the resolution of electrophoretically distinct forms of the subunit. J Biol chem 258(17): 10702–10709

122. Picton C, Woodgett J, Hemmings B, Cohen P (1982) Multisite phosphorylation of glycogen synthase from rabbit skeletal muscle. Phosphorylation of site 5 by glycogen synthase kinase-5 (casein kinase-II) is a prerequisite for phosphorylation of sites 3 by glycogen synthase kinase-3. FEBS Lett 150(1):191–196
123. ter Haar E, Coll JT, Austen DA, Hsiao HM, Swenson L, Jain J (2001) Structure of GSK3beta reveals a primed phosphorylation mechanism. Nat Struct Biol 8(7):593–596
124. Fiol CJ, Mahrenholz AM, Wang Y, Roeske RW, Roach PJ (1987) Formation of protein kinase recognition sites by covalent modification of the substrate. Molecular mechanism for the synergistic action of casein kinase II and glycogen synthase kinase 3. J Biol Chem 62(29):14042–14048
125. Fiol CJ, Wang A, Roeske RW, Roach PJ (1990) Ordered multisite protein phosphorylation. Analysis of glycogen synthase kinase 3 action using model peptide substrates. J Biol Chem 265(11):6061–6065
126. Cole AR, Causeret F, Yadirgi G, Hastie CJ, McLauchlan H, McManus EJ, Hernandez F, Eickholt BJ, Nikolic M, Sutherland C (2006) Distinct priming kinases contribute to differential regulation of collapsin response mediator proteins by glycogen synthase kinase-3 in vivo. J Biol Chem 281(24):16591–16598
127. Li T, Hawkes C, Qureshi HY, Kar S, Paudel HK (2006) Cyclin-dependent protein kinase 5 primes microtubule-associated protein tau site-specifically for glycogen synthase kinase 3beta. Biochemistry 45(10):3134–3145
128. Noble W, Olm V, Takata K, Casey E, Mary O, Meyerson J, Gaynor K, LaFrancois J, Wang L, Kondo T, Davies P, Burns M, Veeranna, Nixon R, Dickson D, Matsuoka Y, Ahlijanian M, Lau LF, Duff K (2003) Cdk5 is a key factor in tau aggregation and tangle formation in vivo. Neuron 38(4):555–565
129. Sengupta A, Wu Q, Grundke-Iqbal I, Iqbal K, Singh TJ (1997) Potentiation of GSK-3-catalyzed Alzheimer-like phosphorylation of human tau by cdk5. Mol Cell Biochem 167(1–2):99–105
130. Nishimura I, Yang Y, Lu B (2004) PAR-1 kinase plays an initiator role in a temporally ordered phosphorylation process that confers tau toxicity in Drosophila. Cell 116(5): 671–682
131. Singh TJ, Zaidi T, Grundke-Iqbal I, Iqbal K (1995) Modulation of GSK-3-catalyzed phosphorylation of microtubule-associated protein tau by non-proline-dependent protein kinases. FEBS Lett 358(1):4–8
132. Thomas GM, Frame S, Goedert M, Nathke I, Polakis P, Cohen P (1999) A GSK3-binding peptide from FRAT1 selectively inhibits the GSK3-catalysed phosphorylation of axin and beta-catenin. FEBS Lett 458(2):247–251
133. Boyle WJ, Smeal T, Defize LH, Angel P, Woodgett JR, Karin M, Hunter T (1991) Activation of protein kinase C decreases phosphorylation of c-Jun at sites that negatively regulate its DNA-binding activity. Cell 64(3):573–584
134. Saksela K, Makela TP, Hughes K, Woodgett JR, Alitalo K (1992) Activation of protein kinase C increases phosphorylation of the L-myc trans-activator domain at a GSK-3 target site. Oncogene 7(2):347–353
135. Diehl JA, Cheng M, Roussel MF, Sherr CJ (1998) Glycogen synthase kinase-3beta regulates cyclin D1 proteolysis and subcellular localization. Genes Dev 12(22):3499–3511
136. Sears R, Nuckolls F, Haura E, Taya Y, Tamai K, Nevins JR (2000) Multiple Ras-dependent phosphorylation pathways regulate Myc protein stability. Genes Dev 14(19):2501–2514
137. Yost C, Torres M, Miller JR, Huang E, Kimelman D, Moon RT (1996) The axis-inducing activity, stability, and subcellular distribution of beta-catenin is regulated in Xenopus embryos by glycogen synthase kinase 3. Genes Dev 10(12):1443–1454
138. Kang T, Wei Y, Honaker Y, Yamaguchi H, Appella E, Hung MC, Piwnica-Worms H (2008) GSK-3 beta targets Cdc25A for ubiquitin-mediated proteolysis, and GSK-3 beta inactivation correlates with Cdc25A overproduction in human cancers. Cancer Cell 13(1):36–47
139. Ding Q, He X, Hsu JM, Xia W, Chen CT, Li LY, Lee DF, Liu JC, Zhong Q, Wang X, Hung MC (2007) Degradation of Mcl-1 by beta-TrCP mediates glycogen synthase kinase 3-induced tumor suppression and chemosensitization. Mol Cell Biol 27(11):4006–4017
140. Fuentealba LC, Eivers E, Ikeda A, Hurtado C, Kuroda H, Pera EM, De Robertis EM (2007) Integrating patterning signals: Wnt/GSK3 regulates the duration of the BMP/Smad1 signal. Cell 131(5):980–993
141. Kerkela R, Kockeritz L, Macaulay K, Zhou J, Doble BW, Beahm C, Greytak S, Woulfe K, Trivedi CM, Woodgett JR, Epstein JA, Force T, Huggins GS (2008) Deletion of GSK-3beta in mice leads to hypertrophic

cardiomyopathy secondary to cardiomyoblast hyperproliferation. J Clin Invest 118(11): 3609–3618

142. Kaidanovich-Beilin O, Lipina TV, Takao K, van Eede M, Hattori S, Laliberte C, Khan M, Okamoto K, Chambers JW, Fletcher PJ, MacAulay K, Doble BW, Henkelman M, Miyakawa T, Roder J, Woodgett JR (2009) Abnormalities in brain structure and behavior in GSK-3alpha mutant mice. Mol Brain 2:35

143. Liang MH, Chuang DM (2006) Differential roles of glycogen synthase kinase-3 isoforms in the regulation of transcriptional activation. J Biol Chem 281(41):30479–30484

144. Liang MH, Chuang DM (2007) Regulation and function of glycogen synthase kinase-3 isoforms in neuronal survival. J Biol Chem 282(6):3904–3917

145. Wang QM, Park IK, Fiol CJ, Roach PJ, DePaoli-Roach AA (1994) Isoform differences in substrate recognition by glycogen synthase kinases 3 alpha and 3 beta in the phosphorylation of phosphatase inhibitor 2. Biochemistry 33(1):143–147

146. Phiel CJ, Wilson CA, Lee VM, Klein PS (2003) GSK-3alpha regulates production of Alzheimer's disease amyloid-beta peptides. Nature 423(6938):435–439

147. Force T, Woodgett JR (2009) Unique and overlapping functions of GSK-3 isoforms in cell differentiation and proliferation and cardiovascular development. J Biol Chem 284(15):9643–9647

148. Chen L, Salinas GD, Li X (2009) Regulation of serotonin 1B receptor by glycogen synthase kinase-3. Mol Pharmacol 76(6):1150–1161

149. Lucas FR, Goold RG, Gordon-Weeks PR, Salinas PC (1998) Inhibition of GSK-3beta leading to the loss of phosphorylated MAP-1B is an early event in axonal remodelling induced by WNT-7a or lithium. J Cell Sci 111(Pt 10):1351–1361

150. Lucas FR, Salinas PC (1997) WNT-7a induces axonal remodeling and increases synapsin I levels in cerebellar neurons. Dev Biol 192(1):31–44

151. Eickholt BJ, Walsh FS, Doherty P (2002) An inactive pool of GSK-3 at the leading edge of growth cones is implicated in Semaphorin 3A signaling. J Cell Biol 157(2):211–217

152. Packard M, Mathew D, Budnik V (2003) Wnts and TGF beta in synaptogenesis: old friends signalling at new places. Nat Rev 4(2):113–120

153. Guan RJ, Khatra BS, Cohlberg JA (1991) Phosphorylation of bovine neurofilament proteins by protein kinase FA (glycogen synthase kinase 3). J Biol Chem 266(13):8262–8267

154. Berling B, Wille H, Roll B, Mandelkow EM, Garner C, Mandelkow E (1994) Phosphorylation of microtubule-associated proteins MAP2a, b and MAP2c at Ser136 by proline-directed kinases in vivo and in vitro. Eur J Cell Biol 64(1):120–130

155. Guidato S, Tsai LH, Woodgett J, Miller CC (1996) Differential cellular phosphorylation of neurofilament heavy side-arms by glycogen synthase kinase-3 and cyclin-dependent kinase-5. J Neurochem 66(4):1698–1706

156. Garcia-Perez J, Avila J, Diaz-Nido J (1998) Implication of cyclin-dependent kinases and glycogen synthase kinase 3 in the phosphorylation of microtubule-associated protein 1B in developing neuronal cells. J Neurosci Res 52(4):445–452

157. Mackie K, Sorkin BC, Nairn AC, Greengard P, Edelman GM, Cunningham BA (1989) Identification of two protein kinases that phosphorylate the neural cell-adhesion molecule, N-CAM. J Neurosci 9(6):1883–1896

158. Hoshi M, Takashima A, Noguchi K, Murayama M, Sato M, Kondo S, Saitoh Y, Ishiguro K, Hoshino T, Imahori K (1996) Regulation of mitochondrial pyruvate dehydrogenase activity by tau protein kinase I/glycogen synthase kinase 3beta in brain. Proc Natl Acad Sci U S A 93(7):2719–2723

159. Peineau S, Bradley C, Taghibiglou C, Doherty A, Bortolotto ZA, Wang YT, Collingridge GL (2008) The role of GSK-3 in synaptic plasticity. Br J Pharmacol 153(Suppl 1):S428–S437

160. Peineau S, Nicolas CS, Bortolotto ZA, Bhat RV, Ryves WJ, Harwood AJ, Dournaud P, Fitzjohn SM, Collingridge GL (2009) A systematic investigation of the protein kinases involved in NMDA receptor-dependent LTD: evidence for a role of GSK-3 but not other serine/threonine kinases. Mol Brain 2(1):22

161. Zhu LQ, Wang SH, Liu D, Yin YY, Tian Q, Wang XC, Wang Q, Chen JG, Wang JZ (2007) Activation of glycogen synthase kinase-3 inhibits long-term potentiation with synapse-associated impairments. J Neurosci 27(45):12211–12220

162. Hooper C, Markevich V, Plattner F, Killick R, Schofield E, Engel T, Hernandez F, Anderton B, Rosenblum K, Bliss T, Cooke SF, Avila J, Lucas JJ, Giese KP, Stephenson J, Lovestone S (2007) Glycogen synthase kinase-3 inhibition is integral to long-term potentiation. Eur J Neurosci 25(1):81–86

163. Beaulieu JM, Sotnikova TD, Yao WD, Kockeritz L, Woodgett JR, Gainetdinov RR, Caron MG (2004) Lithium antagonizes dopamine-dependent behaviors mediated by an AKT/glycogen synthase kinase 3 signaling

cascade. Proc Natl Acad Sci U S A 101(14): 5099–5104

164. Beaulieu JM, Zhang X, Rodriguiz RM, Sotnikova TD, Cools MJ, Wetsel WC, Gainetdinov RR, Caron MG (2008) Role of GSK3 beta in behavioral abnormalities induced by serotonin deficiency. Proc Natl Acad Sci U S A 105(4):1333–1338
165. Li X, Zhu W, Roh MS, Friedman AB, Rosborough K, Jope RS (2004) In vivo regulation of glycogen synthase kinase-3beta (GSK3beta) by serotonergic activity in mouse brain. Neuropsychopharmacology 29(8):1426–1431
166. Yin L, Wang J, Klein PS, Lazar MA (2006) Nuclear receptor Rev-erbalpha is a critical lithium-sensitive component of the circadian clock. Science 311(5763):1002–1005
167. Iitaka C, Miyazaki K, Akaike T, Ishida N (2005) A role for glycogen synthase kinase-3beta in the mammalian circadian clock. J Biol Chem 280(33):29397–29402
168. Hur EM, Zhou FQ (2010) GSK3 signalling in neural development. Nat Rev 11(8): 539–551
169. Espinosa L, Ingles-Esteve J, Aguilera C, Bigas A (2003) Phosphorylation by glycogen synthase kinase-3 beta down-regulates Notch activity, a link for Notch and Wnt pathways. J Biol Chem 278(34):32227–32235
170. Uemura K, Kuzuya A, Shimozono Y, Aoyagi N, Ando K, Shimohama S, Kinoshita A (2007) GSK3beta activity modifies the localization and function of presenilin 1. J Biol Chem 282(21):15823–15832
171. Jin YH, Kim H, Oh M, Ki H, Kim K (2009) Regulation of Notchl/NICD and Hesl expressions by GSK-3alpha/beta. Mol Cells 27(1):15–19
172. Kim WY, Wang X, Wu Y, Doble BW, Patel S, Woodgett JR, Snider WD (2009) GSK-3 is a master regulator of neural progenitor homeostasis. Nat Neurosci 12(11):1390–1397
173. Wei W, Jin J, Schlisio S, Harper JW, Kaelin WG Jr (2005) The v-Jun point mutation allows c-Jun to escape GSK3-dependent recognition and destruction by the Fbw7 ubiquitin ligase. Cancer cell 8(1):25–33
174. Gregory MA, Qi Y, Hann SR (2003) Phosphorylation by glycogen synthase kinase-3 controls c-myc proteolysis and subnuclear localization. J Biol Chem 278(51):51606–51612
175. Beals CR, Sheridan CM, Turck CW, Gardner P, Crabtree GR (1997) Nuclear export of NF-ATc enhanced by glycogen synthase kinase-3. Science 275(5308):1930–1934
176. Ma YC, Song MR, Park JP, Henry HO HY, Hu L, Kurtev MV, Zieg J, Ma Q, Pfaff SL, Greenberg ME (2008) Regulation of motor neuron specification by phosphorylation of neurogenin 2. Neuron 58(1):65–77
177. Zumbrunn J, Kinoshita K, Hyman AA, Nathke IS (2001) Binding of the adenomatous polyposis coli protein to microtubules increases microtubule stability and is regulated by GSK3 beta phosphorylation. Curr Biol 11(1):44–49
178. Kumar P, Lyle KS, Gierke S, Matov A, Danuser G, Wittmann T (2009) GSK3beta phosphorylation modulates CLASP-microtubule association and lamella microtubule attachment. J Cell Biol 184(6):895–908
179. Yoshimura T, Kawano Y, Arimura N, Kawabata S, Kikuchi A, Kaibuchi K (2005) GSK-3beta regulates phosphorylation of CRMP-2 and neuronal polarity. Cell 120(1):137–149
180. Goold RG, Owen R, Gordon-Weeks PR (1999) Glycogen synthase kinase 3beta phosphorylation of microtubule-associated protein 1B regulates the stability of microtubules in growth cones. J Cell Sci 112(Pt 19): 3373–3384
181. Stoothoff WH, Johnson GV (2005) Tau phosphorylation: physiological and pathological consequences. Biochim Biophys Acta 1739(2–3):280–297
182. van Diepen MT, Parsons M, Downes CP, Leslie NR, Hindges R, Eickholt BJ (2009) MyosinV controls PTEN function and neuronal cell size. Nat Cell Biol 11(10): 1191–1196
183. Morfini G, Szebenyi G, Elluru R, Ratner N, Brady ST (2002) Glycogen synthase kinase 3 phosphorylates kinesin light chains and negatively regulates kinesin-based motility. EMBO J 21(3):281–293
184. Inoki K, Ouyang H, Zhu T, Lindvall C, Wang Y, Zhang X, Yang Q, Bennett C, Harada Y, Stankunas K, Wang CY, He X, MacDougald OA, You M, Williams BO, Guan KL (2006) TSC2 integrates Wnt and energy signals via a coordinated phosphorylation by AMPK and GSK3 to regulate cell growth. Cell 126(5): 955–968
185. Mao Y, Ge X, Frank CL, Madison JM, Koehler AN, Doud MK, Tassa C, Berry EM, Soda T, Singh KK, Biechele T, Petryshen TL, Moon RT, Haggarty SJ, Tsai LH (2009) Disrupted in schizophrenia 1 regulates neuronal progenitor proliferation via modulation of GSK3-beta/beta-catenin signaling. Cell 136(6): 1017–1031
186. McManus EJ, Sakamoto K, Armit LJ, Ronaldson L, Shpiro N, Marquez R, Alessi

DR (2005) Role that phosphorylation of GSK3 plays in insulin and Wnt signalling defined by knockin analysis. EMBO J 24(8):1571–1583

187. Eom TY, Jope RS (2009) Blocked inhibitory serine-phosphorylation of glycogen synthase kinase-3alpha/beta impairs in vivo neural precursor cell proliferation. Biol Psychiatry 66(5):494–502

188. Duman RS (2002) Structural alterations in depression: cellular mechanisms underlying pathology and treatment of mood disorders. *CNS Spectr* 7(2):140-142

189. Jacobs BL, van Praag H, Gage FH (2000) Adult brain neurogenesis and psychiatry: a novel theory of depression. Mol Psychiatry 5(3):262–269

190. Kempermann G (2002) Regulation of adult hippocampal neurogenesis – implications for novel theories of major depression. Bipolar Disord 4(1):17–33

191. Schloesser RJ, Chen G, Manji HK (2007) Neurogenesis and neuroenhancement in the pathophysiology and treatment of bipolar disorder. Int Rev Neurobiol 77:143–178

192. Reif A, Fritzen S, Finger M, Strobel A, Lauer M, Schmitt A, Lesch KP (2006) Neural stem cell proliferation is decreased in schizophrenia, but not in depression. Mol Psychiatry 11(5): 514–522

193. Malberg JE, Eisch AJ, Nestler EJ, Duman RS (2000) Chronic antidepressant treatment increases neurogenesis in adult rat hippocampus. J Neurosci 20(24):9104–9110

194. Czeh B, Michaelis T, Watanabe T, Frahm J, de Biurrun G, van Kampen M, Bartolomucci A, Fuchs E (2001) Stress-induced changes in cerebral metabolites, hippocampal volume, and cell proliferation are prevented by antidepressant treatment with tianeptine. Proc Natl Acad Sci U S A 98(22): 12796–12801

195. Manev H, Uz T, Smalheiser NR, Manev R (2001) Antidepressants alter cell proliferation in the adult brain in vivo and in neural cultures in vitro. Eur J Pharmacol 411(1–2):67–70

196. Banasr M, Soumier A, Hery M, Mocaer E, Daszuta A (2006) Agomelatine, a new antidepressant, induces regional changes in hippocampal neurogenesis. Biol Psychiatry 59(11):1087–1096

197. Warner-Schmidt JL, Duman RS (2007) VEGF is an essential mediator of the neurogenic and behavioral actions of antidepressants. Proc Natl Acad Sci U S A 104(11): 4647–4652

198. Scott BW, Wojtowicz JM, Burnham WM (2000) Neurogenesis in the dentate gyrus of the rat following electroconvulsive shock seizures. Exp Neurol 165(2):231–236

199. Madsen TM, Treschow A, Bengzon J, Bolwig TG, Lindvall O, Tingstrom A (2000) Increased neurogenesis in a model of electroconvulsive therapy. Biol Psychiatry 47(12):1043–1049

200. Segi-Nishida E, Warner-Schmidt JL, Duman RS (2008) Electroconvulsive seizure and VEGF increase the proliferation of neural stem-like cells in rat hippocampus. Proc Natl Acad Sci U S A 105(32):11352–11357

201. Warner-Schmidt JL, Madsen TM, Duman RS (2008) Electroconvulsive seizure restores neurogenesis and hippocampus-dependent fear memory after disruption by irradiation. Eur J Neurosci 27(6):1485–1493

202. Chen G, Rajkowska G, Du F, Seraji-Bozorgzad N, Manji HK (2000) Enhancement of hippo campal neurogenesis by lithium. J Neurochem 75(4):1729–1734

203. Hashimoto R, Senatorov V, Kanai H, Leeds P, Chuang DM (2003) Lithium stimulates progenitor proliferation in cultured brain neurons. Neuroscience 117(1):55–61

204. Shimomura A, Nomura R, Senda T (2003) Lithium inhibits apoptosis of mouse neural progenitor cells. Neuroreport 14(14):1779–1782

205. Dawirs RR, Hildebrandt K, Teuchert-Noodt G (1998) Adult treatment with haloperidol increases dentate granule cell proliferation in the gerbil hippocampus. J Neural Transm 105(2–3):317–327

206. Wakade CG, Mahadik SP, Waller JL, Chiu FC (2002) Atypical neuroleptics stimulate neurogenesis in adult rat brain. J Neurosci Res 69(1):72–79

207. Wang HD, Dunnavant FD, Jarman T, Deutch AY (2004) Effects of antipsychotic drugs on neurogenesis in the forebrain of the adult rat. Neuropsychopharmacology 29(7):1230–1238

208. Newton SS, Duman RS (2007) Neurogenic actions of atypical antipsychotic drugs and therapeutic implications. CNS Drugs 21(9):715–725

209. Mai L, Jope RS, Li X (2002) BDNF-mediated signal transduction is modulated by GSK3beta and mood stabilizing agents. J Neurochem 82(1):75–83

210. Emamian ES, Hall D, Birnbaum MJ, Karayiorgou M, Gogos JA (2004) Convergent evidence for impaired AKT1-GSK3beta signaling in schizophrenia. Nat Genet 36(2):131–137

211. Li X, Rosborough KM, Friedman AB, Zhu W, Roth KA (2007) Regulation of mouse brain

glycogen synthase kinase-3 by atypical antipsychotics. Int J Neuropsychopharmacol 10(1):7–19

212. Beurel E, Jope RS (2006) The paradoxical pro- and anti-apoptotic actions of GSK3 in the intrinsic and extrinsic apoptosis signaling pathways. Prog Neurobiol 79(4):173–189

213. Jin Z, El-Deiry WS (2005) Overview of cell death signaling pathways. Cancer Biol Ther 4(2):139–163

214. Bhat RV, Shanley J, Correll MP, Fieles WE, Keith RA, Scott CW, Lee CM (2000) Regulation and localization of tyrosine216 phosphorylation of glycogen synthase kinase-3beta in cellular and animal models of neuronal degeneration. Proc Natl Acad Sci U S A 97(20):11074–11079

215. Hetman M, Cavanaugh JE, Kimelman D, Xia Z (2000) Role of glycogen synthase kinase-3beta in neuronal apoptosis induced by trophic withdrawal. J Neurosci 20(7):2567–2574

216. Lucas JJ, Hernandez F, Gomez-Ramos P, Moran MA, Hen R, Avila J (2001) Decreased nuclear beta-catenin, tau hyperphosphorylation and neurodegeneration in GSK-3beta conditional transgenic mice. EMBO J 20(1–2):27–39

217. Jope RS (2003) Lithium and GSK-3: one inhibitor, two inhibitory actions, multiple outcomes. Trends Pharmacol Sci 24(9):441–443

218. Chuang DM (2005) The antiapoptotic actions of mood stabilizers: molecular mechanisms and therapeutic potentials. Ann N Y Acad Sci 1053:195–204

219. Koh SH, Kim Y, Kim HY, Hwang S, Lee CH, Kim SH (2007) Inhibition of glycogen synthase kinase-3 suppresses the onset of symptoms and disease progression of G93A-SOD1 mouse model of ALS. Exp Neurol 205(2):336–346

220. Wang W, Yang Y, Ying C, Li W, Ruan H, Zhu X, You Y, Han Y, Chen R, Wang Y, Li M (2007) Inhibition of glycogen synthase kinase-3beta protects dopaminergic neurons from MPTP toxicity. Neuropharmacology 52(8):1678–1684

221. Dill J, Wang H, Zhou F, Li S (2008) Inactivation of glycogen synthase kinase 3 promotes axonal growth and recovery in the CNS. J Neurosci 28(36):8914–8928

222. Hernandez F, Borrell J, Guaza C, Avila J, Lucas JJ (2002) Spatial learning deficit in transgenic mice that conditionally over-express GSK-3beta in the brain but do not form tau filaments. J Neurochem 83(6):1529–1533

223. Polter A, Beurel E, Yang S, Garner R, Song L, Miller CA, Sweatt JD, McMahon L, Bartolucci AA, Li X, Jope RS (2010) Deficiency in the inhibitory serine-phosphorylation of glycogen synthase kinase-3 increases sensitivity to mood disturbances. Neuropsychopharmacology 35(8):1761–1774

224. Peineau S, Taghibiglou C, Bradley C, Wong TP, Liu L, Lu J, Lo E, Wu D, Saule E, Bouschet T, Matthews P, Isaac JT, Bortolotto ZA, Wang YT, Collingridge GL (2007) LTP inhibits LTD in the hippocampus via regulation of GSK3beta. Neuron 53(5):703–717

225. Kimura T, Yamashita S, Nakao S, Park JM, Murayama M, Mizoroki T, Yoshiike Y, Sahara N, Takashima A (2008) GSK-3beta is required for memory reconsolidation in adult brain. PLoS One 3(10):e3540

226. Li X, Jope RS (2010) Is glycogen synthase kinase-3 a central modulator in mood regulation? Neuropsychopharmacology 35(11):2143–2154

227. Koros E, Dorner-Ciossek C (2007) The role of glycogen synthase kinase-3beta in schizophrenia. Drug News Perspect 20(7):437–445

228. Lovestone S, Killick R, Di Forti M, Murray R (2007) Schizophrenia as a GSK-3 dysregulation disorder. Trends Neurosci 30(4):142–149

229. Freyberg Z, Ferrando SJ, Javitch JA (2009) Roles of the Akt/GSK-3 and Wnt signaling pathways in schizophrenia and antipsychotic drug action. Am J Psychiatry 167(4):388–396

230. Takashima A (2009) Drug development targeting the glycogen synthase kinase-3beta (GSK-3beta)-mediated signal transduction pathway: role of GSK-3beta in adult brain. J Pharmacol Sci 109(2):174–178

231. Reitz C, Brayne C, Mayeux R (2011) Epidemiology of Alzheimer disease. Nat Rev Neurol 7(3):137–152

232. Hardy JA, Higgins GA (1992) Alzheimer's disease: the amyloid cascade hypothesis. Science 256(5054):184–185

233. Kosik KS (1992) Cellular aspects of Alzheimer neurofibrillary pathology. Prog Clin Biol Res 379:183–193

234. Noble W, Planel E, Zehr C, Olm V, Meyerson J, Suleman F, Gaynor K, Wang L, LaFrancois J, Feinstein B, Burns M, Krishnamurthy P, Wen Y, Bhat R, Lewis J, Dickson D, Duff K (2005) Inhibition of glycogen synthase kinase-3 by lithium correlates with reduced tauopathy and degeneration in vivo. Proc Natl Acad Sci U S A 102(19):6990–6995

235. Caccamo A, Oddo S, Sugarman MC, Akbari Y, LaFerla FM (2005) Age- and region-dependent alterations in Abeta-degrading enzymes: implications for Abeta-induced disorders. Neurobiol Aging 26(5):645–654

236. Su Y, Ryder J, Li B, Wu X, Fox N, Solenberg P, Brune K, Paul S, Zhou Y, Liu F, Ni B (2004) Lithium, a common drug for bipolar disorder

treatment, regulates amyloid-beta precursor protein processing. Biochemistry 43(22): 6899–6908

237. Caccamo A, Oddo S, Tran LX, LaFerla FM (2007) Lithium reduces tau phosphorylation but not A beta or working memory deficits in a transgenic model with both plaques and tangles. Am J Pathol 170(5):1669–1675

238. Sereno L, Coma M, Rodriguez M, Sanchez-Ferrer P, Sanchez MB, Gich I, Agullo JM, Perez M, Avila J, Guardia-Laguarta C, Clarimon J, Lleo A, Gomez-Isla T (2009) A novel GSK-3beta inhibitor reduces Alzheimer's pathology and rescues neuronal loss in vivo. Neurobiol Dis 35(3):359–367

239. Hu S, Begum AN, Jones MR, Oh MS, Beech WK, Beech BH, Yang F, Chen P, Ubeda OJ, Kim PC, Davies P, Ma Q, Cole GM, Frautschy SA (2009) GSK3 inhibitors show benefits in an Alzheimer's disease (AD) model of neurodegeneration but adverse effects in control animals. Neurobiol Dis 33(2):193–206

240. Zhong J, Lee WH (2007) Lithium: a novel treatment for Alzheimer's disease? Expert Opin Drug Saf 6(4):375–383

241. Perala J, Suvisaari J, Saarni SI, Kuoppasalmi K, Isometsa E, Pirkola S, Partonen T, Tuulio-Henriksson A, Hintikka J, Kieseppa T, Harkanen T, Koskinen S, Lonnqvist J (2007) Lifetime prevalence of psychotic and bipolar I disorders in a general population. Arch Gen Psychiatry 64(1):19–28

242. McGurk SR, Mueser KT (2004) Cognitive functioning, symptoms, and work in supported employment: a review and heuristic model. Schizophr Res 70(2–3):147–173

243. Tamminga CA, Holcomb HH (2005) Phenotype of schizophrenia: a review and formulation. Mol Psychiatry 10(1):27–39

244. Ross CA, Margolis RL, Reading SA, Pletnikov M, Coyle JT (2006) Neurobiology of schizophrenia. Neuron 52(1):139–153

245. Harrison PJ, Weinberger DR (2005) Schizophrenia genes, gene expression, and neuropathology: on the matter of their convergence. Mol Psychiatry 10(1):40–68, image 45

246. Lisman JE, Coyle JT, Green RW, Javitt DC, Benes FM, Heckers S, Grace AA (2008) Circuit-based framework for understanding neurotransmitter and risk gene interactions in schizophrenia. Trends Neurosci 31(5):234–242

247. Kozlovsky N, Belmaker RH, Agam G (2001) Low GSK-3 activity in frontal cortex of schizophrenic patients. Schizophr Res 52(1–2): 101–105

248. Kozlovsky N, Regenold WT, Levine J, Rapoport A, Belmaker RH, Agam G (2004) GSK-3beta in cerebrospinal fluid of schizophrenia patients. J Neural Transm 111(8):1093–1098

249. Kozlovsky N, Shanon-Weickert C, Tomaskovic-Crook E, Kleinman JE, Belmaker RH, Agam G (2004) Reduced GSK-3beta mRNA levels in postmortem dorsolateral prefrontal cortex of schizophrenic patients. J Neural Transm 111(12):1583–1592

250. Yang SD, Yu JS, Lee TT, Yang CC, Ni MH, Yang YY (1995) Dysfunction of protein kinase FA/GSK-3 alpha in lymphocytes of patients with schizophrenic disorder. J Cell Biochem 59(1):108–116

251. Souza RP, Romano-Silva MA, Lieberman JA, Meltzer HY, Wong AH, Kennedy JL (2008) Association study of GSK3 gene polymorphisms with schizophrenia and clozapine response. Psychopharmacology 200(2):177–186

252. Prickaerts J, Moechars D, Cryns K, Lenaerts I, van Craenendonck H, Goris I, Daneels G, Bouwknecht JA, Steckler T (2006) Transgenic mice overexpressing glycogen synthase kinase 3beta: a putative model of hyperactivity and mania. J Neurosci 26(35):9022–9029

253. Benedetti F, Poletti S, Radaelli D, Bernasconi A, Cavallaro R, Falini A, Lorenzi C, Pirovano A, Dallaspezia S, Locatelli C, Scotti G, Smeraldi E (2010) Temporal lobe grey matter volume in schizophrenia is associated with a genetic polymorphism influencing glycogen synthase kinase 3-beta activity. Genes Brain Behav 9(4):365–371

254. Lipina TV, Kaidanovich-Beilin O, Patel S, Wang M, Clapcote SJ, Liu F, Woodgett JR, Roder JC (2010) Genetic and pharmacological evidence for schizophrenia-related Disc1 interaction with GSK-3. Synapse 65(3):234–248

255. Beaulieu JM, Gainetdinov RR, Caron MG (2009) Akt/GSK3 signaling in the action of psychotropic drugs. Annu Rev Pharmacol Toxicol 49:327–347

256. Kang UG, Seo MS, Roh MS, Kim Y, Yoon SC, Kim YS (2004) The effects of clozapine on the GSK-3-mediated signaling pathway. FEBS Lett 560(1–3):115–119

257. Gould TD (2006) Targeting glycogen synthase kinase-3 as an approach to develop novel mood-stabilising medications. Expert Opin Ther Targets 10(3):377–392

258. Grimes CA, Jope RS (2001) The multifaceted roles of glycogen synthase kinase 3beta in cellular signaling. Prog Neurobiol 65(4):391–426

259. Beaulieu JM, Sotnikova TD, Marion S, Lefkowitz RJ, Gainetdinov RR, Caron MG (2005) An Akt/beta-arrestin 2/PP2A signaling complex mediates dopaminergic neurotransmission and behavior. Cell 122(2):261–273

260. Miller JS, Tallarida RJ, Unterwald EM (2009) Cocaine-induced hyperactivity and sensitization are dependent on GSK3. Neuropharmacology 56(8):1116–1123
261. Miller JS, Tallarida RJ, Unterwald EM (2010) Inhibition of GSK3 attenuates dopamine D1 receptor agonist-induced hyperactivity in mice. Brain Res Bull 82(3–4):184–187
262. Xu CM, Wang J, Wu P, Zhu WL, Li QQ, Xue YX, Zhai HF, Shi J, Lu L (2009) Glycogen synthase kinase 3beta in the nucleus accumbens core mediates cocaine-induced behavioral sensitization. J Neurochem 111(6):1357–1368
263. Brami-Cherrier K, Valjent E, Garcia M, Pages C, Hipskind RA, Caboche J (2002) Dopamine induces a PI3-kinase-independent activation of Akt in striatal neurons: a new route to cAMP response element-binding protein phosphorylation. J Neurosci 22(20):8911–8921
264. McGinty JF, Shi XD, Schwendt M, Saylor A, Toda S (2008) Regulation of psychostimulant-induced signaling and gene expression in the striatum. J Neurochem 104(6):1440–1449
265. Shi X, McGinty JF (2007) Repeated amphetamine treatment increases phosphorylation of extracellular signal-regulated kinase, protein kinase B, and cyclase response element-binding protein in the rat striatum. J Neurochem 103(2):706–713
266. Nwaneshiudu CA, Unterwald EM (2010) NK-3 receptor antagonism prevents behavioral sensitization to cocaine: a role of glycogen synthase kinase-3 in the nucleus accumbens. J Neurochem 115(3):635–642
267. Feder A, Alonso A, Tang M, Liriano W, Warner V, Pilowsky D, Barranco E, Wang Y, Verdeli H, Wickramaratne P, Weissman MM (2009) Children of low-income depressed mothers: psychiatric disorders and social adjustment. Depress Anxiety 26(6):513–520
268. Martinowich K, Schloesser RJ, Manji HK (2009) Bipolar disorder: from genes to behavior pathways. J Clin Invest 119(4):726–736
269. Gold PW, Chrousos GP (2002) Organization of the stress system and its dysregulation in melancholic and atypical depression: high vs low CRH/NE states. Mol Psychiatry 7(3):254–275
270. Pittenger C, Duman RS (2008) Stress, depression, and neuroplasticity: a convergence of mechanisms. Neuropsychopharmacology 33(1):88–109
271. Zarate CA Jr, Singh J, Manji HK (2006) Cellular plasticity cascades: targets for the development of novel therapeutics for bipolar disorder. Biol Psychiatry 59(11):1006–1020
272. Gould TD, Manji HK (2005) Glycogen synthase kinase-3: a putative molecular target for lithium mimetic drugs. Neuropsychopharmacology 30(7):1223–1237
273. O'Brien WT, Klein PS (2009) Validating GSK3 as an in vivo target of lithium action. Biochem Soc Trans 37(Pt 5):1133–1138
274. De Sarno P, Li X, Jope RS (2002) Regulation of Akt and glycogen synthase kinase-3 beta phosphorylation by sodium valproate and lithium. Neuropharmacology 43(7):1158–1164
275. Omata N, Chiu CT, Moya PR, Leng Y, Wang Z, Hunsberger JG, Leeds P, Chuang DM (2011) Lentivirally mediated GSK-3beta silencing in the hippocampal dentate gyrus induces antidepressant-like effects in stressed mice. Int J Neuropsychopharmacol 14(5):711–717
276. O'Brien WT, Harper AD, Jove F, Woodgett JR, Maretto S, Piccolo S, Klein PS (2004) Glycogen synthase kinase-3beta haploinsufficiency mimics the behavioral and molecular effects of lithium. J Neurosci 24(30):6791–6798
277. Gould TD, Einat H, Bhat R, Manji HK (2004) AR-A014418, a selective GSK-3 inhibitor, produces antidepressant-like effects in the forced swim test. Int J Neuropsychopharmacol 7(4):387–390
278. Kaidanovich-Beilin O, Milman A, Weizman A, Pick CG, Eldar-Finkelman H (2004) Rapid antidepressive-like activity of specific glycogen synthase kinase-3 inhibitor and its effect on beta-catenin in mouse hippocampus. Biol Psychiatry 55(8):781–784
279. Rosa AO, Kaster MP, Binfare RW, Morales S, Martin-Aparicio E, Navarro-Rico ML, Martinez A, Medina M, Garcia AG, Lopez MG, Rodrigues AL (2008) Antidepressant-like effect of the novel thiadiazolidinone NP031115 in mice. Prog Neuropsychopharmacol Biol Psychiatry 32(6):1549–1556
280. Li X, Liu M, Cai Z, Wang G, Li X (2010) Regulation of glycogen synthase kinase-3 during bipolar mania treatment. Bipolar Disord 12(7):741–752
281. Raison CL, Capuron L, Miller AH (2006) Cytokines sing the blues: inflammation and the pathogenesis of depression. Trends Immunol 27(1):24–31
282. Miller AH, Maletic V, Raison CL (2009) Inflammation and its discontents: the role of cytokines in the pathophysiology of major depression. Biol Psychiatry 65(9):732–741
283. Adli M, Hollinde DL, Stamm T, Wiethoff K, Tsahuridu M, Kirchheiner J, Heinz A, Bauer M (2007) Response to lithium augmentation in depression is associated with the glycogen synthase kinase 3-beta-50T/C single nucleotide polymorphism. Biol Psychiatry 62(11): 1295–1302

284. Szczepankiewicz A, Rybakowski JK, Suwalska A, Skibinska M, Leszczynska-Rodziewicz A, Dmitrzak-Weglarz M, Czerski PM, Hauser J (2006) Association study of the glycogen synthase kinase-3beta gene polymorphism with prophylactic lithium response in bipolar patients. World J Biol Psychiatry 7(3):158–161

285. Tsai SJ, Liou YJ, Hong CJ, Yu YW, Chen TJ (2008) Glycogen synthase kinase-3beta gene is associated with antidepressant treatment response in Chinese major depressive disorder. Pharmacogenomics J 8(6):384–390

286. Inkster B, Nichols TE, Saemann PG, Auer DP, Holsboer F, Muglia P, Matthews PM (2009) Association of GSK3beta polymorphisms with brain structural changes in major depressive disorder. Arch Gen Psychiatry 66(7):721–728

287. Beurel E, Jope RS (2009) Lipopolysaccharide-induced interleukin-6 production is controlled by glycogen synthase kinase-3 and STAT3 in the brain. J Neuroinflammation 6:9

288. Martin M, Rehani K, Jope RS, Michalek SM (2005) Toll-like receptor-mediated cytokine production is differentially regulated by glycogen synthase kinase 3. Nat Immunol 6(8):777–784

289. Hu X, Paik PK, Chen J, Yarilina A, Kockeritz L, Lu TT, Woodgett JR, Ivashkiv LB (2006) IFN-gamma suppresses IL-10 production and synergizes with TLR2 by regulating GSK3 and CREB/AP-1 proteins. Immunity 24(5):563–574

290. Yuskaitis CJ, Jope RS (2009) Glycogen synthase kinase-3 regulates microglial migration, inflammation, and inflammation-induced neurotoxicity. Cell Signal 21(2):264–273

291. Phukan S, Babu VS, Kannoji A, Hariharan R, Balaji VN (2010) GSK3beta: role in therapeutic landscape and development of modulators. Br J Pharmacol 160(1):1–19

292. Eldar-Finkelman H, Licht-Murava A, Pietrokovski S, Eisenstein M (2009) Substrate competitive GSK-3 inhibitors – strategy and implications. Biochim Biophys Acta 1804(3):598–603

293. Gurvich N, Klein PS (2002) Lithium and valproic acid: parallels and contrasts in diverse signaling contexts. Pharmacol Ther 96(1):45–66

294. Huang HC, Klein PS (2006) Multiple roles for glycogen synthase kinase-3 as a drug target in Alzheimer's disease. Curr Drug Targets 7(11):1389–1397

295. Munoz-Montano JR, Moreno FJ, Avila J, Diaz-Nido J (1997) Lithium inhibits Alzheimer's disease-like tau protein phosphorylation in neurons. FEBS Lett 411(2–3): 183–188

296. Chalecka-Franaszek E, Chuang DM (1999) Lithium activates the serine/threonine kinase Akt-1 and suppresses glutamate-induced inhibition of Akt-1 activity in neurons. Proc Natl Acad Sci U S A 96(15):8745–8750

297. Li X, Friedman AB, Zhu W, Wang L, Boswell S, May RS, Davis LL, Jope RS (2007) Lithium regulates glycogen synthase kinase-3beta in human peripheral blood mononuclear cells: implication in the treatment of bipolar disorder. Biol Psychiatry 61(2):216–222

298. Ryves WJ, Harwood AJ (2001) Lithium inhibits glycogen synthase kinase-3 by competition for magnesium. Biochem Biophys Res Commun 280(3):720–725

299. Engel T, Hernandez F, Avila J, Lucas JJ (2006) Full reversal of Alzheimer's disease-like phenotype in a mouse model with conditional overexpression of glycogen synthase kinase-3. J Neurosci 26(19):5083–5090

300. Murphy DL (1977) The behavioral toxicity of monoamine oxidase-inhibiting antidepressants. Adv Pharmacol Chemother 14:71–105

301. Meijer L, Flajolet M, Greengard P (2004) Pharmacological inhibitors of glycogen synthase kinase 3. Trends Pharmacol Sci 25(9):471–480

302. Bain J, McLauchlan H, Elliott M, Cohen P (2003) The specificities of protein kinase inhibitors: an update. Biochem J 371(Pt 1): 199–204

Chapter 10

Cortical Neurons Culture to Study c-Jun N-Terminal Kinase Signaling Pathway

Alessandra Sclip, Xanthi Antoniou, and Tiziana Borsello

Abstract

Excitotoxicity via *N*-methyl-D-aspartate receptors (NMDA-r) is a key mechanism of neurodegeneration following ischemia and other brain pathologies.

We here describe the use of a fast, reliable and easily reproducible in vitro model of excitotoxicity induced with NMDA stimulation of cortical neurons.

Such a protocol can be used to study the intracellular pathways involved in NMDA-induced death. Here we focused on JNK signaling, a pathway that plays a determinant role in excitotoxic death of cortical neurons. Our proposed model can additionally be used to test different drugs designed to prevent excitotoxic neuronal death. As an example we describe the use of the cell permeable specific JNK inhibitor peptide as a neuroprotective agent.

This chapter will give methodological suggestions to help researchers prepare cortical neurons, to develop an in vitro model of neuronal death by stimulating neurons with NMDA, to quantify neuronal death in vitro using LDH assay and propidium staining, to analyze the intracellular pathways involved (JNK pathway) in neuronal death, and to perform protocols of neuroprotection.

Key words: JNK, Neuronal death, NMDA-toxicity, Neuroprotection, Signaling pathway, Cell permeable JNK inhibitor peptide, D-JNKI1

1. Background and Historical Overview

Protein kinases are enzymes that modify other proteins by adding a phosphate group to them, a process that is chemically defined as phosphorylation. This enzymatic process results in functional changes of the target protein by modulating enzyme activity, cellular localization, or interaction with other proteins. Protein kinases are known to regulate the majority of cellular pathways, especially those involved in signal transduction.

The recent completion of the whole human genome sequence revealed that there are 500 protein kinases (1). Among protein

Hideyuki Mukai (ed.), *Protein Kinase Technologies*, Neuromethods, vol. 68,
DOI 10.1007/978-1-61779-824-5_10, © Springer Science+Business Media, LLC 2012

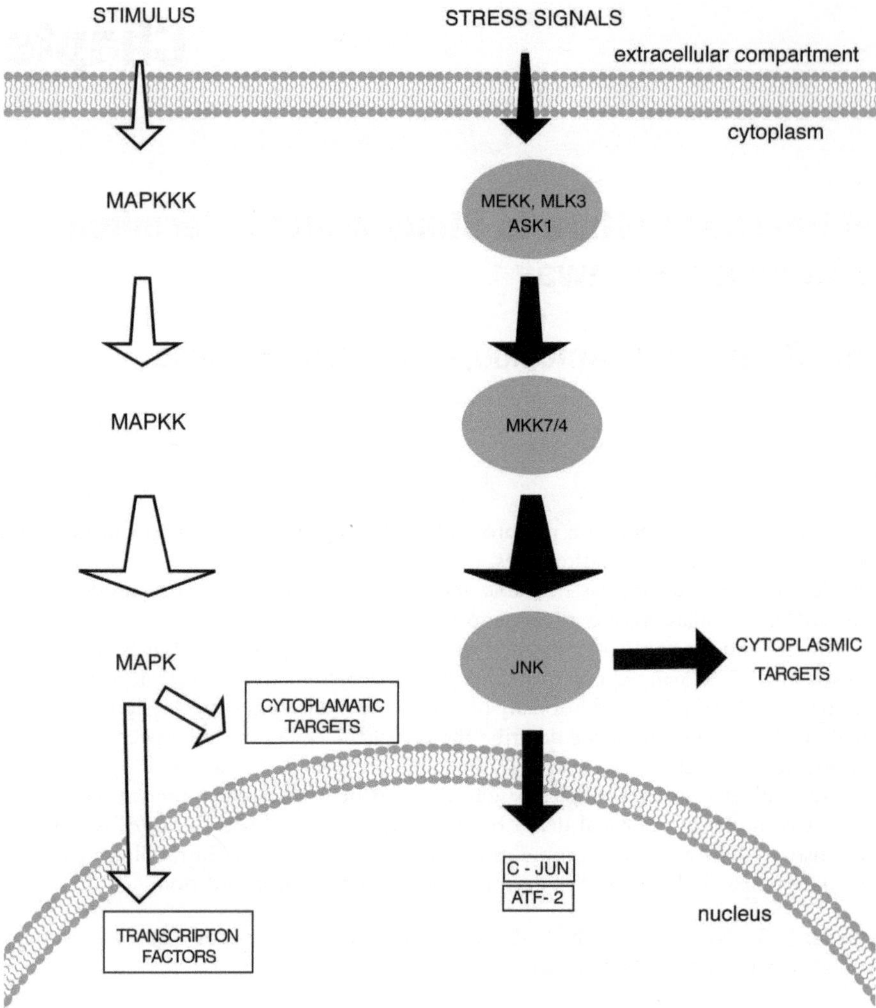

Fig. 1. MAPKs signaling pathway. Schematic representation of MAPKs cascade. The MAPKs cascade transduces and amplifies a signal from the extracellular compartment into the cell. MAPKs are serine/threonine-specific kinases, which are activated by combined phosphorylation on serine/threonine and tyrosine residues from upstream kinases: MAPKKKs activate MAPKK proteins that subsequently phosphorylate MAPKs. In JNK signaling pathway, MEK (MAPKKK) is activated in response to extracellular stimuli and phosphorylates MKK7 or MKK4 (MAPKK) that on their turn activate JNK. JNK itself can phosphorylate different targets localized in the cytoplasm as well as translocate into the nucleus and activate nuclear factor targets (c-Jun and ATF2).

kinases, the MAPK (acronym from: "mitogen-activated protein kinases") are serine/threonine-specific kinases that are activated by combined phosphorylation on serine/threonine and tyrosine residues.

MAPKs family includes extracellular signal-regulated kinase (2), p38, and c-Jun NH_2-terminal kinase (JNK). MAPKs signaling pathway is composed of at least three components, a MAPKKK (MAPKinase kinase kinase), a MAPKK (MAPKinase kinase) and a MAPK (3) (Fig. 1). These enzymes are particularly important since they translate the extracellular changes inside cells; in this manner they regulate diverse cellular-processes ranging from proliferation to apoptosis.

Within the MAPKs family, ERK is an established participant in the regulation of cell growth and differentiation, while p38 and JNK are mainly stress-activated protein kinases.

We will here focus our attention on JNK pathway, which is playing a determinant role in transmitting, amplifying and integrating diverse signals in both physiological and pathological responses in the Central Nervous System (CNS).

JNKs can split into ten isoforms derived from three genes: *Jnk1*, *Jnk2*, *Jnk3* (4). *Jnk3* is found mainly in the brain, while the other two are present in all cells and tissues. Alternate splicing of the gene transcripts results in further molecular diversity. A splice site within the *Jnk1* and two transcripts results in two splice forms; a second alternate splice for all JNK transcripts occurs at the C-terminus of the protein resulting in the 46- and 55-kDa protein isoforms, capable of phosphorylating serine residues 63 and 73 in the N-terminal region of c-Jun, thereby potentiating its transactivation function (5).

JNKs reside in the cytosol where they are activated by phosphorylation in the conserved Thr-Pro-Tyr motif within their activation loop. This dual phosphorylation is mediated by MKK4 and MKK7, two members of the family of dual-specificity kinases referred to as MAPK kinases (MAP2Ks) (1). Concomitantly, JNK activation is regulated by dephosphorylation of some phosphatases MKP-1, M3/6 and MKP7 (2, 6, 7). Another important and most probably key mechanism to control efficiency and specificity of JNKs is mediated by scaffold proteins that bind and sequester JNK in different cellular compartments, and help in minimizing the MAPKs cross-talks (8).

On their turn, activated JNK phosphorylates over 60 substrates ranging from transcription factors (c-Jun, ATF2, Elk1, JunD, JunB, etc., see Table 1) and Histone protein (like H2AX) to cytoskeletal proteins (Keratin8, Tau, SG100 and others) but also kinases like AKT and RSK2, ubiquitin ligase (Itch) as well as scaffold proteins (JIP1, IRS-1, 14-3-3 and ShcA), Bcl2-familly proteins (Bcl-2, Bcl-xl, Bim, Bid, Bad and Bmf), cytosolic proteins involved in cell death (MADD, caspase-3) and receptors (PPAR-gamma1, RAR-alpha, glucocorticoid-receptors, Nur77) (9). Despite JNK has numerous substrates (see Table 1) it is important to note that JNK remains the dominant kinase for c-Jun phosphorylation in vivo.

Stress signals such as *N*-methyl-D-aspartate (NMDA) stimulation, Abeta fragments, hypoxia, reactive oxygen species, ultraviolet radiation, protein synthesis inhibitors can all activate JNK. In particular, excitotoxicity is a stress condition that triggers JNK signal cascade, leading to neuronal death (10).

Excitotoxicity is defined as the excessive activation of glutamate receptors, in which NMDA receptors play a key role, as a

Table 1
JNK substrates

Protein function	Member
Transcription factors	c-Jun, JunB, JunD, JDP2, ATF2, ELK-1, c-Myc, p53, NFAT, Foxo4, Stat3, Stat1, Pax2, TCF-beta
Nuclear hormone receptors	PPAR-gamma, glucocorticoid receptors, RXR, RAR-alpha, Nur77
Scaffold proteins	JIP1, IRS-1, 14-3-3, ShcA
Kinases	Akt, RSK2, MKK4, MKK7
Ubiquitin ligases	Itch
Cytoskeletal proteins and associated proteins	Keratin 8, Tau, SCG100, DXC
Histones	H2AX
Protein implicated in cell death	MADD and caspase-3
Bcl-2 family members	Bcl-2, Bcl-xl, Mcl-1, Bim, Bid, Bad, Bmf

result of their high Ca^{2+} permeability. In cortical neurons exposed to high concentrations of NMDA (100 μM) JNK pathway is tightly connected to extracellular calcium influx and plays a pivotal role in inducing and executing the death pathway (11). Preventing the interactions between JNK and its targets through the synthesized D-retroinverso-JNK inhibitor peptide, D-JNKI1, completely blocks cell death induced by excitotoxicity (12). Moreover, D-JNKI1 shows a strong protective effect in vivo both in transient and permanent middle-cerebral artery occlusion (12). c-Jun and Caspase-3 are the main elements involved in JNK-mediated neuronal death pathway (12).

Here we describe an in vitro model of excitotoxic neuronal death obtained by stimulation of cortical neurons with NMDA. This model is useful to study excitotoxic-induced neuronal death and JNK activation as well as the powerful neuroprotective effect of the JNK inhibitor peptide D-JNKI1.

2. Equipment, Materials, and Setup

2.1. Primary Rat Cortical Neurons

- Poly D-lysine solution: 25 μg/ml poly-D-lysine (P6407, Sigma-Aldrich, St Louis, USA) dissolved in water (Note 1).
- Cortical Neuron Dissociation Medium (CNDM): prepared with 5.8 mM MgCl2, 0.5 mM CaCl2, 3.2 mM Hepes, 0.2 mM NaOH, 30 mM K_2SO_4, 90 mM Na_2SO_4, 0.5 ml Phenol Red in water. pH of the medium is adjusted to 7.4 with NaOH solution

(Note 2). The medium is filtered with 0.22 μm filter and stored at 4°C (Note 3).

- Digestion Medium: 200 U of papain (P3125, Sigma-Aldrich, St Louis, USA) is added to CNDM previously supplemented with 0.01 g of L-cysteine (C6852, Sigma-Aldrich, St Louis, USA) and 0.36% D-glucose (G5400-1 kg, Sigma-Aldrich, St Louis, USA) (Note 4).
- Protease inhibition medium: trypsin inhibitor (T-9253 Sigma-Aldrich, St Louis, USA) in CNDM supplemented with 0.36% glucose.
- Plating medium: is made with Minimum Essential Medium (MEM, 21575-022, Gibco–Invitrogen, Paisley, Scotland, UK) supplemented with 0.36% glucose, 10% fetal bovine serum (SH30070.03, Hy Clone) (Note 5), 0.5 mM glutamine (25030 Gibco–Invitrogen, Paisley, Scotland, UK), 100 U/ml penicillin and 10 μg/ml streptomycin (15140-122, Gibco–Invitrogen, Paisley, Scotland, UK).
- Growth medium is prepared with Neurobasal-A medium (10888-022, Gibco–Invitrogen, Paisley, Scotland, UK) supplemented with 1:50 B27 supplement (17504-044, Gibco–Invitrogen, Paisley, Scotland, UK), 0.5 mM glutamine and 100 U/ml penicillin and 10 μg/ml streptomycin.
- Plastic ware: neurons are plated on 96 well plate (CC3595, Corning incorporated, New York, USA) to perform cell death assays and neuroprotection tests, on 35 mm dishes (35-3001, Falcon, USA) for western blots and on IBIDI chambers (80826 ibitreat, IBIDI, Martinsries, Germany) for immunocytochemistry.

2.2. NMDA Application

- NMDA stock solution (10 mM) is made by dissolving NMDA (M3262, Sigma-Aldrich, St Louis, USA) in water.

2.3. D-JNKI1 Treatments

- D-JNKI1: stock solution (10 mM) stored at –80°C.
- D-JNKI1-FITC: stock solution (10 mM) stored at –80°C.

2.4. Detection of Cell Death In Vitro

- Propidium iodide (PI): stock solution (1 mg/ml or 1.5 mM) stored at 4°C (P-3566, Molecular probes, Burlington, ON, Canada)
- Hoechst 33258 (H3569, Molecular probes, Burlington, ON, Canada)
- Confocal microscope
- LDH assay: CytoTox 96® Non-Radioactive Cytotoxicity Assay (G1781, Promega, Madison, USA) is composed by the substrate mix, assay buffer, LDH positive control, lysis solution (10×) and stop solution. Reconstitute the substrate mix with 12 ml of assay buffer and store it at –20°C. Prepare LDH positive control diluting 2 μl of positive control solution (supplied

by the kit) into 10 ml of PBS and 1% BSA. LDH positive control is prepared fresh for each use

- Plate reader

2.5. Detection of JNK Activation

Sample Preparation

- Lysis buffer: 20 mM Tris-acetate, 0.27 M sucrose, 1 mM EDTA, 1 mM EGTA, 50 mM NaF, 1 mM Sodium orthovanadate, 5 mM Na pyrophosphate, 10 mM Na b glycerophosphate and 1 mM DTT. Store at −20°C.
- Add proteases (1xCPIK, 10634200, Roche) and phosphatases (10030536, Roche) inhibitors and 1% Triton-X100 before use. Store buffer at 4°C all times.

SDS-PAGE

- Bradford (500-0006, BioRad, Munchen, Germany)
- Acrilamide 30% (#EC-890, National Diagnostic ProtoGel, Georgia, USA)
- Running Buffer: Tris-Glycine 10× (161-0771, BioRad, Munchen, Germany) and SDS 0.5% (161-0418, BioRad, Munchen, Germany) in water
- Blotting Buffer: Tris-Glycine 10×, MeOH 12%, SDS 0.1% in water
- Laemmli Buffer (161-0737, BioRad, Munchen, Germany) is supplemented with β-mercaptoethanol and diluted 1:1 with the sample
- TBS-T: Tris 20 mM, NaCl 150 mM, pH 7.6 added with 0.1% Tween 20 (P5927, Sigma-Aldrich, St Louis, USA)
- Antibodies: P-JNK (phosphorylated JNK; sc-6254, Santa Cruz Biotechnology, California, USA), JNK (#9252, Cell Signaling Technology, Massachusetts, USA), a-tubulin (sc-8035, Santa Cruz Biotechnology, California, USA)
- ECL (W1015, Promega, Madison, USA)
- Quantity One software (BioRad, Munchen, Germany)

3. Procedures

3.1. Primary Rat Cortical Neuronal Culture

- One day before dissection all plastic ware supports are pre-coated with Poly-D-lysine for 1 h at 37°C and then washed with water at least 3 times. All pre-coated plates are transferred to the incubator until use (Note 6)
- Prepare all the media before starting dissection (Note 7)
- P2 rats are anesthetized and washed in 70% ethanol. Under a laminar flow hood dissect cortex and collect the tissue into a dish containing the dissection medium

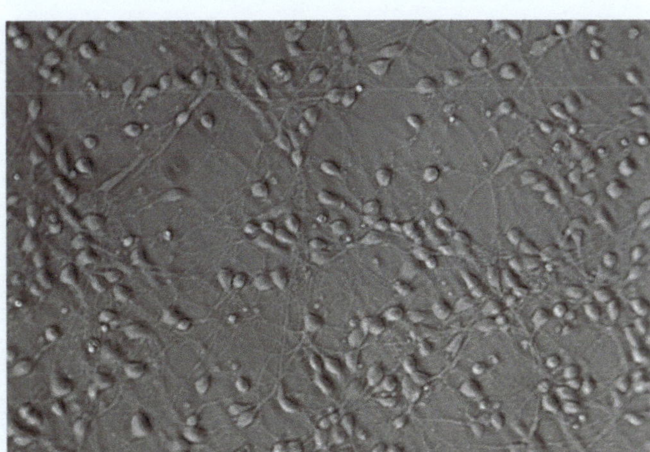

Fig. 2. Primary rat cortical neurons. Phase contrast bright field images of healthy cortical neurons at DIV 12.

- Transfer the dissected tissue into a sterile 15 ml tube containing 10 ml dissection medium and incubate at 37°C for 30 min to allow enzymatic digestion (Note 8)
- Remove the dissection medium and replace it with the protease inhibition medium to block papain digestion. Incubate for 45 min at room temperature
- Remove the medium and add 2 ml of plating medium
- Triturate the tissue with a 2 ml glass pipette taking care to avoid bubbles formation
- Following dissociation centrifuge cells for 2 min at 1,500 rpm (Note 9)
- Resuspend the pellet in 5 ml of plating medium and centrifuge them for 2 min at 1,500 rpm (Note 9)
- Resuspend the pellet in 1 ml of plating medium and then count the number of cells using a Burker chamber (Note 10)
- Plate 70,000 cells into 96 well plates, 700,000 cells into dishes and 80,000 cells in the IBIDI chambers (Note 11)
- After 2 h gently remove all the plating medium and replace it with growth medium
- Change half of the medium every 3–4 days (Note 12)
- Use cells at 12 days in vitro (DIV) for experiments (Note 13) (Fig. 2)

3.2. NMDA Application

- Neurons are treated with NMDA 100 μM for 24 h to induce cell death (Note 14). Activation of the JNK pathway is observed after 30–45 min of treatment.

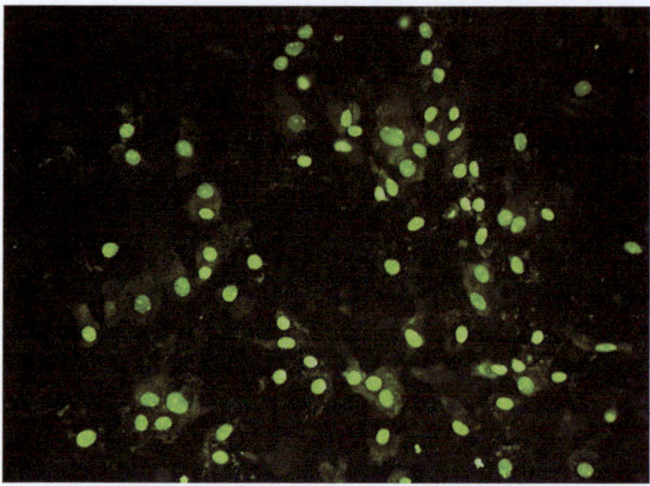

Fig. 3. D-JNKI1 penetrates cortical neurons. FITC labeled D-JNKI1 was added in the medium of cortical neurons and was detected inside cells after 5 min.

3.3. D-JNKI1 Penetration and Treatment

- To detect D-JNKI1 uptake in neurons, D-JNKI1-FITC peptide is added to the medium and fluorescence into cells is detected with a fluorescent microscope (Ex: 488 nm; Em: 525 nm; it fluoresces green) (Fig. 3).
- To test neuroprotection, cells are treated with D-JNKI1 2–4–6 μM 30 min before the toxic stimulus.

3.4. Detection of Cell Death In Vitro

LDH Assay (CytoTox 96® Non-Radioactive Cytotoxicity Assay)

- Induce apoptosis by NMDA as described in Sect. 3.2
- Collect 50 μl of medium from each sample and transfer them into a 96 well flat bottom plate
- Add 50 μl of reconstituted substrate mix to each sample
- Incubate for 30 min at RT in the dark
- After 30 min add 50 μl of stop solution
- Remove bubbles using a syringe needle
- Record the absorbance at 490 nm (Note 15)

Propidium Iodide and Nuclei Staining

- Induce apoptosis by NMDA as described in Sect. 3.2
- Wash gently with fresh medium to eliminate cellular debris and dead neurons
- Dilute the stock solution of PI (final concentration 5 μM) in fresh medium
- Return plates to a 37°C/5% CO_2 incubator for 15 min
- Wash with PBS gently
- Fix cells with 2% formaldehyde in PBS for 20 min at RT

- Incubate fixed cells with Hoechst (1:500 in PBS for 7 min at RT) to stain the nuclei of all cells
- Acquired images of PI staining (Ex: 535 nm; Em: 617 nm; it fluoresces red) and Hoechst staining (Ex: 352 nm; Em: 461 nm) with a confocal microscope

3.5. Detection of JNK Activation

Sample Preparation

- Wash neurons twice with ice cold PBS (10010015, Gibco–Invitrogen, Paisley, Scotland, UK) (Note 16)
- Add 100 μl lysis buffer in each dish
- Incubate for 20 min on ice to avoid protease action
- Lysate cells with a scraper
- Collect the lysate and store it at –80°C

SDS-Page

- Quantify proteins using the Bradford assay (Note 17)
- Prepare 10% SDS polyacrilamide gels
- Prepare samples in leammli buffer (2×) previously supplemented with β-mercaptoethanol
- Denature samples for 7 min at 99°C and load them onto the gel
- Run samples at 50 mA for 1 h at RT
- Transfer protein on a PVDF membrane (162-0177, BioRad) at 400 mA for 2 h in a 4°C room (Note 18)
- Incubate PVDF membrane with 5% no fat milk powder in TBS-T for 1 h at RT to block non-specific sites
- Incubate o/n at 4°C with primary antibodies diluted in 5% no fat milk powder in TBS-T. Suitable dilutions are: P-JNK 1:2,000, JNK 1:1,000, tubulin 1:5,000
- Wash 3 times with TBS-T at RT
- Incubate with secondary antibodies for 1 h at RT. Dilution for secondary antibodies are: P-JNK 1:5,000, JNK 1:2,000, tubulin 1:10,000
- Develop with ECL
- Western blots are quantified by densiometry analysis using Quantity One software

4. Experimental Variables

Note 1. All references to water refer to sterile water (15230-089, Gibco–Invitrogen, Paisley, Scotland, UK).

Note 2. Osmolarity of CNDM must be 292 mOsm.

Note 3. All solutions for cell culture are filtered with 0.22 μm filter and kept sterile.

Note 4. L-cysteine acidifies the medium that turns yellow. Adjust the pH to 7.4 with NaOH 1 N before the use.

Note 5. Fetal bovine serum is inactivated at 56°C for 30 min. Aliquots are stored at −20°C.

Note 6. To ameliorate adhesion of neurons to the plastic support a mixed coating with poli-D-lysine and Laminin can be done. Laminin (23017-015, Invitrogen) is diluted 1:500 and added to the poli-D-lysine solution.

Note 7. All media are stocked at 4°C. To avoid neurons suffering due to thermic shock heat all the media at 37°C before the use.

Note 8. To assure that the enzymatic digestion occurs properly reduce the size of the dissected tissue with a bisturi and collect it in a 15 ml tube containing 10ml of dissociation medium. Gently twist the tube to avoid formation of a pellet and incubate it at 37°C (horizontally). In this way all the tissue is in contact with the medium.

Note 9. If possible centrifuge the samples with a slow acceleration and slow deceleration to preserve neurons.

Note 10. To count cell number, dilute 10 μl of cell suspension in 90 μl of plating medium in a 1.5 ml sterile tube. Add 10 μl of trypan blue. Trypan blue is a viable dye that does not penetrate through the cellular membrane. Only dead cells with a disrupted membrane are permeable to the trypan blue and are labeled in blue. Count only viable cells (not labeled with trypan blue). About 30–50 million neurons are obtained by 4–5 rat pups.

Note 11. To facilitate cell plating dilute neurons at a final concentration of one million cells/ml. Plate 700 μl in each dish, 70 μl in each well of the 96 well plates and 80 μl in the IBIDI chambers. Final volumes are 2 ml for dishes, 200 μl for 96 well plates and 250 μl for ibidi chambers.

Note 12. Remove and replace the medium gently. Neurons are fragile and sensitive to mechanic stress. Avoid prolonged exposure to air.

Note 13. 12 DIV neurons are considered mature and differentiated and are a model for adult neurons.

Note 14. In the conditions described, application of NMDA 100 μM for 24 h resulted in over 90% neuronal death (as reported by Borsello et al. (12)).

Note 15. Remember to add 50 μl LDH positive control and 50 μl of fresh medium as blank. Follow the protocol described in Sect. 3.2. Subtract the blank from all samples value.

Note 16. To obtain a sufficient amount of proteins only neurons plated on dishes are used for Western Blot.

Note 17. To study differences in protein levels in different samples the same amount of protein is loaded on the gel. Protein amount is quantified with Bradford assay.

Note 18. Remember to incubate PVDF membrane with methanol before use.

5. Expected/ Anticipated Results

In the mature brain neuronal death can contribute to the pathogenesis of several neurological disorders including stroke, traumatic brain injury, anoxia, epilepsy, Parkinson's disease and Alzheimer's disease. Thus, the development of new pharmacological strategies to prevent neuronal death is one of the main interests for neuroscientists. In vitro models of neuronal death are extremely useful to study the mechanisms by which neurons die and to test neuroprotective compounds.

To induce neuronal death in vitro, cortical neurons were exposed to NMDA 100 μM for 24 h. Cell death was detected with both LDH and PI protocols. LDH assay measures the content of lactate dehydrogenase released in the medium from degenerating cells. LDH is a cytoplasmic enzyme, which in normal conditions is localized inside cells. During neuronal death, LDH is released in the medium due to plasma membrane breakdown. NMDA stimulation-induced neuronal death resulting in an increase of OD values detected by the LDH kit (Fig. 4). In the same conditions, most of the neurons were positive for PI, a common marker of neuronal death (Fig. 5a). PI is impermeable to plasma membranes

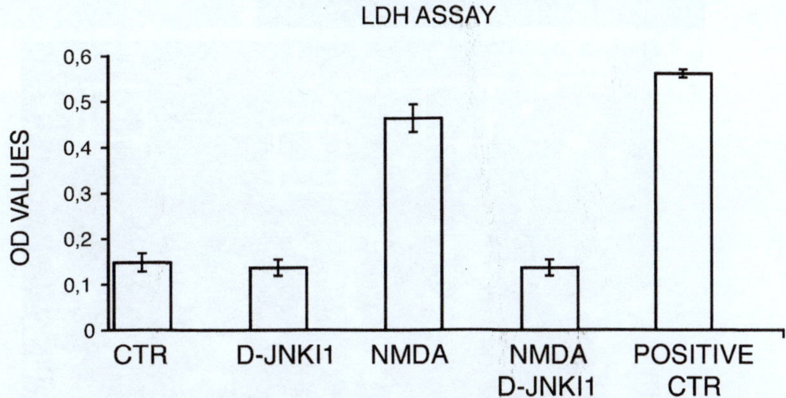

Fig. 4. NMDA induces neuronal death. Typical histogram showing neuronal death after NMDA stimulation of cortical neurons. Death was measured with LDH assay. OD values increased significantly in NMDA stimulated neurons. The 30 min pre-treatment with D-JNKI1 4 μM reduced OD values to control conditions suggesting that JNK inhibition protects neurons from cell death.

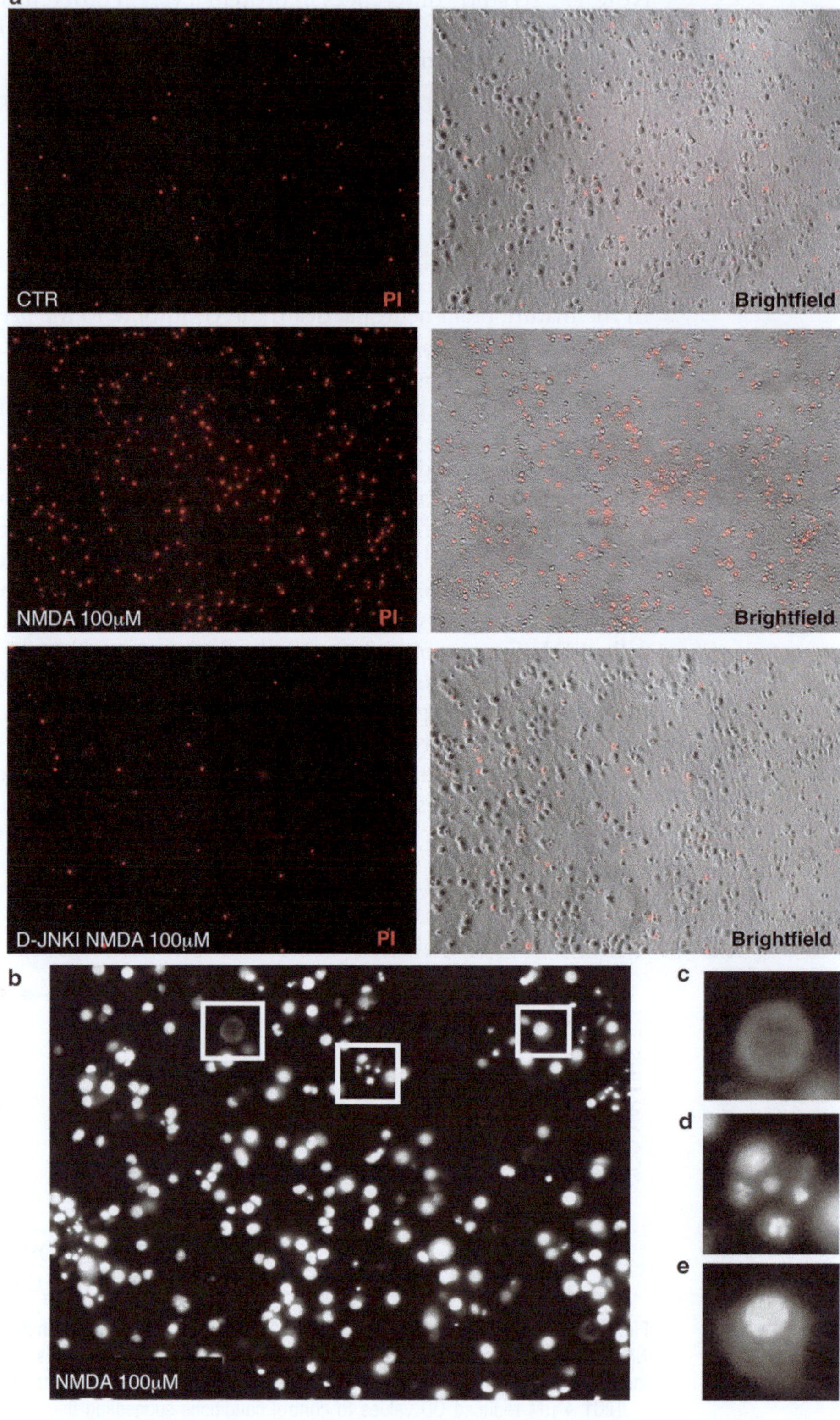
a
CTR
PI
Brightfield
NMDA 100μM
PI
Brightfield
D-JNKI NMDA 100μM
PI
Brightfield
b
NMDA 100μM
c
d
e

and generally excluded from viable cells. When membrane integrity is compromised, like in dead cells, PI can enter cells and bind the DNA. Finally, Hoechst staining (Fig. 5b) showed the presence of ghost neurons (Fig. 5c), of cells characterized by nuclear fragmentation (Fig. 5d) as well as of neurons showing a swelling appearance and nuclear condensation (Fig. 5e). All these phenotypes are hallmarks of neuronal death.

To study the intracellular mechanisms that lead to neuronal death, we performed western blot analysis. We focused on JNK signaling pathway. NMDA stimulation induced an increase in P-JNK, the active form of the enzyme, after 30–45 min, while total JNK levels remained unaffected (Fig. 6). This result suggested that JNK signaling pathway is involved in neuronal death.

Thus, inhibition of JNK pathway can be a strategy to prevent neurodegenaration. To assess this hypothesis, we treated neurons with the most specific JNK inhibitor, D-JNKI1, in order to block JNK activation in response to NMDA stimulation. Treatment with D-JNKI1, by blocking JNK pathway, completely prevented neuronal death induced by a toxic stimulus. Neurons that were

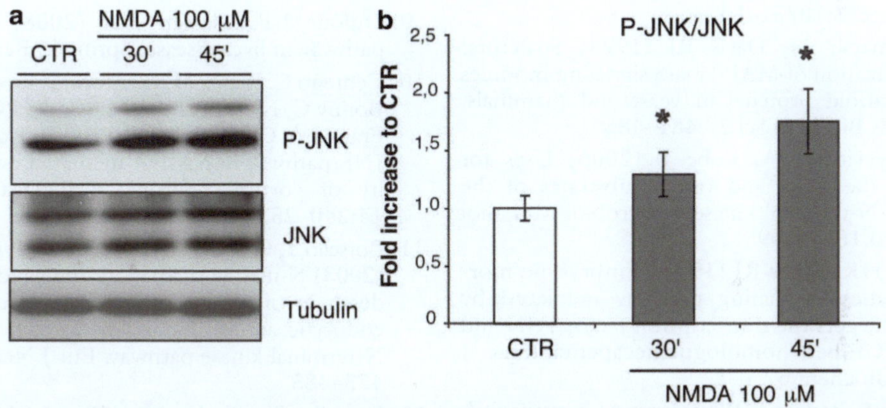

Fig. 6. NMDA induces activation of JNK signaling pathway. (**a**) Representative Western blots showing that NMDA exposure induce activation of JNK. (**b**) Densitometric quantification of the Western blot. The P-JNK (active form)/total JNK levels ratio was increased after 30–45 min of NMDA stimulation suggesting the involvement of JNK in the induction of neuronal death.

Fig. 5. Propidium Iodide (PI) and Hoechst staining to detect neuronal death. Propidium iodide is a marker of cell death and enters only in neurodegenerated neurons. (**a**) Images showing PI staining and the merge with the brightfield in control, NMDA and D-JNKI1/NMDA treated neurons. As expected, NMDA treatment 100 μM for 24 h induces a massive death in cortical neurons as demonstrated by the huge number of neurons positive for PI. In D-JNKI1 pre-treated neurons NMDA did not induce neuronal death and neurons were negative for PI similar to control conditions. (**b**) NMDA treatment induces toxicity and leads to the formation of ghost neurons (**c**), of neurons with nuclear fragmentation (**d**) or swelling appearance and nuclear condensation (**e**) as showed by the Hoechst staining.

pre-treated with D-JNKI1 before NMDA stimulation have LDH levels similar to controls and were negative for PI.

These data show that D-JNKI1 is a potent neuroprotective compound and an interesting tool to develop pharmacological strategies to prevent neurodegeneration.

Acknowledgment

This study was supported by the Marie Curie Industry-Academia Partnerships and Pathways (IAPP) Cpads, San Paolo 2008-2437, CARIPLO 2009-2425.

References

1. Manning AM, Davis RJ (2003) Targeting JNK for therapeutic benefit: from junk to gold? Nat Rev Drug Discov 2:554–565
2. Willoughby EA, Perkins GR, Collins MK, Whitmarsh AJ (2003) The JNK-interacting protein-1 scaffold protein targets MAPK phosphatase-7 to dephosphorylate JNK. J Biol Chem 278:10731–10736
3. Whitmarsh AJ, Davis RJ (1998) Structural organization of MAP-kinase signaling modules by scaffold proteins in yeast and mammals. Trends Biochem Sci 23:481–485
4. Bogoyevitch MA, Kobe B (2006) Uses for JNK: the many and varied substrates of the c-Jun N-terminal kinases. Microbiol Mol Biol Rev 70:1061–1095
5. Sluss HK, Davis RJ (1997) Embryonic morphogenesis signaling pathway mediated by JNK targets the transcription factor JUN and the TGF-beta homologue decapentaplegic. J Cell Biochem 67:1–12
6. Gupta S, Barrett T, Whitmarsh AJ, Cavanagh J, Sluss HK, Derijard B, Davis RJ (1996) Selective interaction of JNK protein kinase isoforms with transcription factors. EMBO J 15:2760–2770
7. Willoughby EA, Collins MK (2005) Dynamic interaction between the dual specificity phosphatase MKP7 and the JNK3 scaffold protein beta-arrestin 2. J Biol Chem 280: 25651–25658
8. Morrison DK, Davis RJ (2003) Regulation of MAP kinase signaling modules by scaffold proteins in mammals. Annu Rev Cell Dev Biol 19:91–118
9. Dufour J-F, Clavien P-A (2008) Signaling pathway in liver disease. Springer, Berlin
10. Centeno C, Repici M, Chatton JY, Riederer BM, Bonny C, Nicod P, Price M, Clarke PG, Papa S, Franzoso G, Borsello T (2007) Role of the JNK pathway in NMDA-mediated excitotoxicity of cortical neurons. Cell Death Differ 14:240–253
11. Borsello T, Croquelois K, Hornung JP, Clarke PG (2003) N-methyl-D-aspartate-triggered neuronal death in organotypic hippocampal cultures is endocytic, autophagic and mediated by the c-Jun N-terminal kinase pathway. Eur J Neurosci 18: 473–485
12. Borsello T, Clarke PG, Hirt L, Vercelli A, Repici M, Schorderet DF, Bogousslavsky J, Bonny C (2003) A peptide inhibitor of c-Jun N-terminal kinase protects against excitotoxicity and cerebral ischemia. Nat Med 9:1180–1186

Chapter 11

MuSK: A Kinase Critical for the Formation and Maintenance of the Neuromuscular Junction

Arnab Barik, Wen-cheng Xiong, and Lin Mei

Abstract

Muscle-specific kinase (MuSK) is a RTK that is specifically expressed in skeletal muscle fibers and critical for the formation and maintenance of the neuromuscular junction (NMJ), a peripheral synapse formed between motoneurons and muscle fibers (1, 2). The acetylcholine receptors (AChRs) are concentrated at the crest of junctional folds on muscle fibers, critical for muscle contraction. Impaired NMJ formation or function inflicts muscular dystrophy. Being large and accessible, this peripheral synapse has served as a classic model of synapse function and synaptogenesis and has contributed a great deal to the understanding of molecular mechanisms of synapse formation in the brain. In this chapter, we will review the structures of MuSK in various vertebrates, its role in NMJ formation and maintenance, possible pathways that have been suggested by recent studies, and how MuSK may be a target in muscular dystrophy.

Key words: Neuromuscular junction, AChR Pre-patterning, Agrin hypothesis, Juxtamembrane domain, MASC, Lrp4

1. Discovery of MuSK

The neuromuscular junction (NMJ) formation is known to be controlled and/or regulated by factors released by motoneurons. A heroic search for such factors by Dr. Jack McMahan and colleagues led to the identification of agrin from the electric organ of *Torpedo californica*, an organ that is homologous to vertebrate skeletal muscles but is more densely innervated. It is a rich source of synapse specific molecules including acetylcholine receptor (AChR), acetylcholinesterase (AChE), and rapsyn (3, 4). Purified agrin was able to stimulate AChR clustering in cultured muscle fibers (5). Per the agrin hypothesis, neural agrin is synthesized in motoneurons and released from motor nerve terminals where it is

Hideyuki Mukai (ed.), *Protein Kinase Technologies*, Neuromethods, vol. 68,
DOI 10.1007/978-1-61779-824-5_11, © Springer Science+Business Media, LLC 2012

deposited into the nascent synaptic basal lamina. Subsequently, agrin interacts with a receptor on the myotube surface to initiate clustering of AChRs and other postsynaptic molecules at the future neuromuscular synapse. Pharmacological studies indicated that agrin-induced AChR clustering requires tyrosine kinase activity (6), tyrosine-phosphorylated proteins are enriched at the NMJ and in the electric organ of *Torpedo californica* (7, 8), and tyrosine-phosphorylation regulates the properties of the AChR (8). These observations suggest a critical role of a tyrosine kinase(s) in NMJ formation or function.

Taking advantage of the electric organ and consensus sequences of previously identified kinases, Burden and colleagues cloned an novel RTK (7). The Torpedo RTK was found to be exclusively expressed at the muscles, closely related to the Trk family of neurotrophin receptors, and the sequence also resembled the Trk family in having a very short C-terminal region beyond the kinase domain (Fig. 1a) Trks are the RTKs for neurotrophic factors (9, 10). In search for Trk-related or novel RTKs, a rat and human ortholog named MuSK (muscle-specific kinase) and mouse ortholog named Nsk2 were identified (11, 12). MuSK is specifically expressed in the muscle cells, whose cytoplasmic domain of MuSK

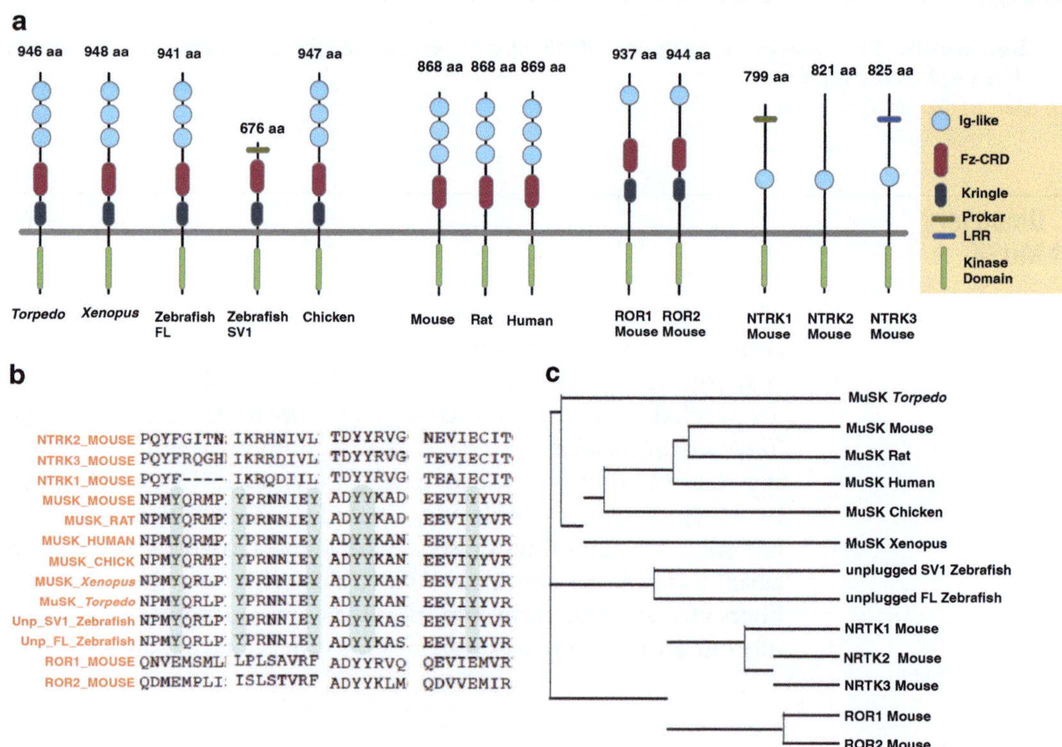

Fig. 1. Structures of MuSK and orthologues. (**a**) Domain structures of MuSK and orthologues. (**b**) Alignment of important tyrosine residues of MuSK, ROR, and Trk. (**c**) Phylogenetic tree showing the clustering relationships of MuSK and related RTKs.

is similar to those of Trk kinase receptors—traditional RTKs for neurotrophic factors including NGF and BDNF. The extracellular domain of MuSK has three immunoglobin (Ig) domains and one cysteine-rich domain (CRD) in rodents and humans; but in Xenopus, zebrafish, and chick, it includes a kringle domain. This topology is different from that of Trks, which has one Ig domain and a leucine rich repeat C-terminal (Lrrct) domain. RTKs that show similar domain structure of MuSK are receptor tyrosine kinase-like orphan receptors (RORs) (Fig. 1a), a family of orphan RTKs that has one Ig domain, one CRD, and one kringle domain. RORs have been implicated in neuronal development, cell migration, and cell polarity (13) and were recently shown to be regulated by Wnt5a as a ligand (14, 15).

2. MuSK Orthlogs and Structures

MuSK orthologs have thus far been cloned in human (11), mouse (12), Xenopus (16), chicken (17), zebrafish (18) (Fig. 1a), but it does not appear to exist in *Drosophila melanogaster*. Although agrin orthologs are present in flies (19), their functions have not been investigated in detail. MuSK in rodents is specifically expressed in early myotomes and developing muscles and at low levels in proliferating myoblasts and upregulated upon differentiation and fusion. While MuSK localizes at the NMJ, it is dramatically down regulated in mature muscles (11, 12).

In zebrafish, "unplugged" mutants with motor axon guidance effects were discovered during a random mutagenesis screen (18). Positional cloning revealed the mutations to be in the kinase domain of the MuSK ortholog unplugged. Alternative splicing of unplugged generates two isoforms, full-length (FL) isoform and unplugged splice variant 1 (SV1) (Fig. 1a), which differ from each other by the presence and absence of three Ig domains, respectively. Each isoform appears to control a different step in synaptogenesis. Unplugged SV1 is transiently expressed during early embryonic development from the tailbud stage up to 48 hpf, while unplugged FL expression starts around the 10-somite stage and persists into adult stages (18, 20, 21). It remains unclear whether such isoforms of MuSK with redundant functions but different structural domains exist in mammals.

A major structural difference between mammalian MuSK and the orthologs found in the lower species, such as Torpedo is the absence of a kringle domain in the extracellular region (Fig. 1a). However, little is known about its function. All MuSK orthologs share the very critical tyrosine residues in their kinase domain (Fig. 1b). Thus, apart from being domain conserved, the kinase domain seems to be functionally conserved across species evolutionarily.

MuSK ectodomain is composed of three Ig domains and a CRD related to those in seven pass transmembrane Frizzled proteins, the receptors for Wnts (22). Fz-CRD domain was shown to be dispensable for agrin-induced MuSK activation and AChR clustering, but critical for co-clustering with rapsyn and AChR co-clustering (23). Evidence suggests possible involvement of Wnt in regulating MuSK function. First, like Frizzled, MuSK also interacts with Disheveled (Dvl), a critical protein in Wnt signaling (24). Second, zebrafish MuSK binds to Wnt11r in a manner that requires the CRD (21) (also discussed below). Third, treatment with Wnt alters AChR clusters in muscle cells (25, 26). It remains unclear whether MuSK may also bind other CRD binding ligands including secreted Frizzled-related proteins (sFRPS) (27), Patched (Ptc) protein, collagen (col18a) and carboxypeptidase Z (CPZ) (28).

3. MuSK is a Component of the Receptor Complex for Agrin

Neural agrin is synthesized exclusively by motoneurons and deposited in the synaptic basal lamina where it is thought to concentrate AChR at junctional folds (29, 30). In agreement with this notion, agrin induces AChR clusters in muscle cells in culture (5, 23, 31). However, the underlying mechanism remain poorly understood (1, 32). Genetic evidence indicates essential roles of agrin and MuSK in NMJ formation (30). MuSK$^{-/-}$ muscle fibers lack AChR clusters prior to or after motor neuron innervation (33). Moreover, agrin could not induce AChR clustering in cultured Musk$^{-/-}$ myotubes.

Agrin dose-dependently stimulate MuSK phosphorylation, with kinetics characteristic of a ligand of the kinase. However, agrin does not bind to its ectodomain suggesting the existence of an intermediate protein between agrin and MuSK, such as myotube-associated specificity component (MASC). This long-sought after receptor of agrin appears to be LRP4 (34, 35), a member of the low-density lipoprotein receptor (LDLR) family. LRP4 contains a large extracellular N-terminal region, a single transmembrane domain and a short C-terminal region without an identifiable catalytic domain (36, 37). Genetic studies indicate that LRP4 is required for NMJ formation, with NMJ defects that resemble those of Musk$^{-/-}$ mutant mice (38). LRP4 is necessary for agrin binding activity to muscle cells, and agrin-induced activation of MuSK and AChR clustering and is sufficient to reconstitute agrin binding activity and MuSK activation in non-muscle cells (34, 35). Moreover, binding of agrin to LRP4 promotes its interaction with MuSK, and its activation. These observations indicate that LRP4 serves as a receptor for agrin to stimulate MuSK (Fig. 2).

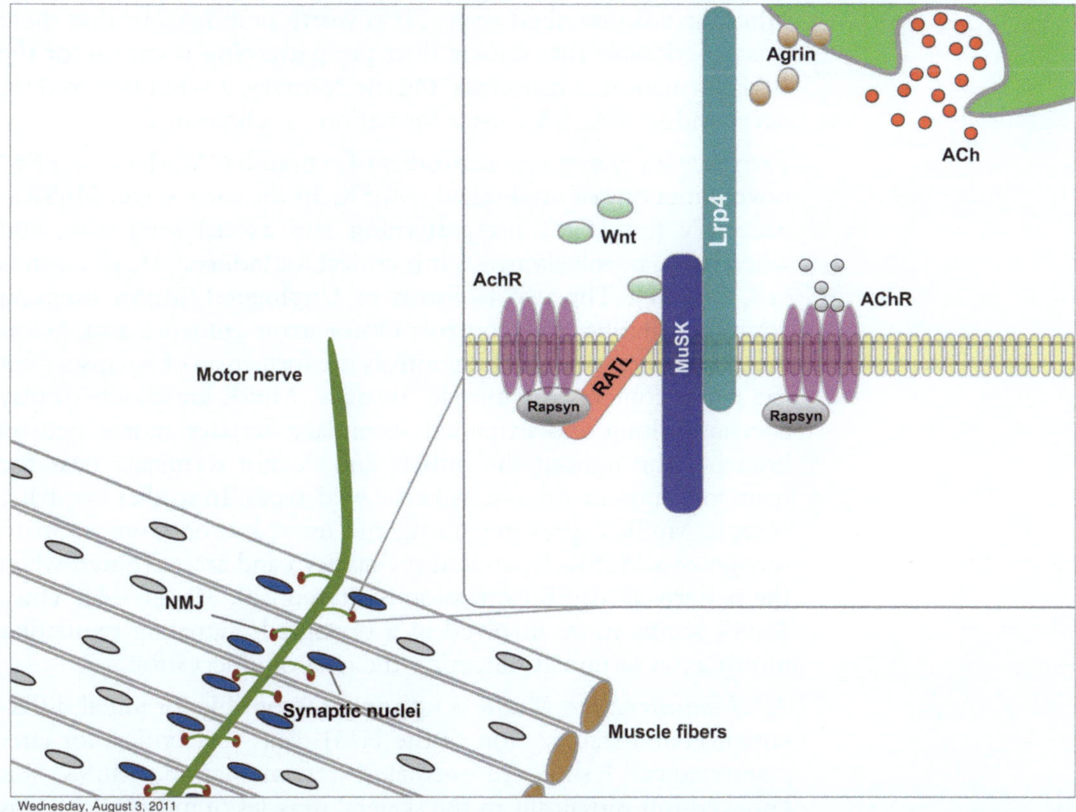

Fig. 2. Schematic diagram showing motor nerve innervating muscle fibers. NMJs are formed in the *middle region* of muscle fibers. Inset depicts an enlarged NMJ.

4. MuSK is the Master Organizer at the NMJ

In addition to AChR clustering, MuSK has the following functions.

Muscle fiber pre-patterning. In addition to AChR clustering, MuSK is required for "pre-patterning" of muscle fibers prior to innervation. AChRs appeared in primitive, small clusters in the central region of muscle fibers in the advance of motor terminal arrival (39–41). This nerve-independent pre-pattern of AChR expression is thought to be involved in deciding where to form eventual NMJs. In a working model, ACh extinguishes AChR clusters in entire muscle fibers whereas agrin from nerves stabilizes AChR clusters beneath the synapse. In MuSK mutant mice, muscle fibers do not form the primitive AChR clusters and thus not pre-patterned; in contrast, muscle fibers remain pre-patterned in agrin mutant mice. This would suggest the existence of agrin-independent ligands that may interact with MuSK directly, including Wnt and other ligands described above. It is worth pointing out that there was no evidence that muscle fiber pre-patterning is critical for the NMJ formation in mammals. On the contrary, it is not required for nerve-induced AChR cluster formation in zebrafish (21).

Presynaptic differentiation. Studies of zebrafish NMJs have revealed novel functions of unplugged/MuSK. In the early stage, MuSK is necessary for AChR pre-patterning and axonal migration, and when nerve terminals arrive, it is critical for induced AChR clusters (18, 20, 21). The two isoforms of Unplugged/MuSK function complementarily: SV1 controls motor axon guidance and AChR pre-patterning, whereas FL controls the formation of synapses after the nerve contacts the muscle. In mice, MuSK knockouts display aberrantly long and extended secondary/tertiary motor neuron branches throughout the muscle and do not terminate near the main intramuscular nerve, as in the wild-types. In studies involving ectopic MuSK expressing transgenic mice, motor axons seem to recognize a MuSK-dependent pre-pattern and are re-routed when the pattern of MuSK expression is altered (20, 39, 41–44). Thus, MuSK seems to be involved in a retrograde signaling regulating motor axon terminals, directing the sites of innervation.

NMJ maintenance. MuSK is not only responsible for initial differentiation and stabilization of the NMJ. But, very critical for later maintenance. Using cre mediated recombination, MuSK was knocked-out mosaically in the skeletal muscles of mice. MuSK was shown to be not only required for initial formation of AChR clusters in embryogenesis but also essential throughout postnatal development to maintain the postsynaptic apparatus and to control axonal growth (45). MuSK inactivation at NMJ of adult muscle causes reduction in AChR density and change in the gross synaptic arborization of the endplate leading finally to the complete loss of AChR and disappearance of the synaptic structure. These observations suggest a critical role of MuSK in NMJ maintenance or function. In agreement, auto-antibodies against MuSK cause myasthenia gravis (MG) (see below).

Formation of synaptic scaffold. A chimera composed of the neurotrophin receptor (trkC kinase) ectodomain and the MuSK cytoplasmic domain can be activated by neurotrophin, but does not induce AChR clustering (46, 47). These results suggest that MuSK is unusual among RTKs in that ligand-dependent activation is insufficient to mediate its effects. In addition to signaling proteins, which will be discussed below, MuSK also interact with a plethora of scaffold proteins based on in vitro studies. Its ectodomain might interact indirectly with rapsyn, a cytoplasmic protein critical for NMJ formation (48), via a hypothetical transmembrane protein, rapsyn-associated transmembrane linking (RATL) molecule (Fig. 2) (49). Being a transmembrane protein able to interact with

MuSK (35, 37), it would be interesting to test if Lrp4 may serve as a candidate for RATL. Other scaffold proteins include ColQ, a protein for AChE enrichment in the synaptic cleft (50), the MAGUK protein MAGI-1c (51) and AChR (52). MuSK has been shown to associate with proteins that may regulate gene expression including 14-3-3γ (53) and with synaptic nuclear envelope 1 (syne-1), a nuclear envelope protein enriched in synaptic nuclei (54). Finally, MuSK interacts with enzymes that control its endocytosis and degradation including NSF (55), and E3 ligases putative Ariadne-like ubiquitin ligase (PAUL) and PDZ domain-containing RING finger 3 (PDZRN3) (56, 57).

5. MuSK Regulation and Intracellular Pathways

The C-terminal kinase domain of MuSK is around ~260 aa and well conserved across the vertebrate species (Fig. 1b). Tyrosine residues in the juxtamembrane domain (Y553) (Fig. 1b), in the activation loop (Y570, Y754, and Y755) (Fig. 1b), and within the kinase domain (Y576 and Y812) (Fig. 1b) are phosphorylated in the activated MuSK (58). Y553 is contained in a NPXY motif (NPMY in MuSK), which is a binding site for PTB domains in adaptor proteins (59, 60) and thus thought to be responsible for the activation of the downstream cascade (23, 58, 61). Y576 is thought to interact with SH2-domain-containing proteins (58). Mutation of the activation loop tyrosine residues results in a failure of agrin to activate MuSK in cultured cells. Mutation of Y553 and Y576, but not Y812, abrogated the ability of Agrin to induce clustering or tyrosine phosphorylation of AChRs in cultured myotubes (23, 58, 61). Remarkably, inclusion of the 13 amino acids from the juxtamembrane region including Y553 is able to rescue NMJ deficits in MuSK null mutant mice (58, 61). Thus, Y553 in the juxtamembrane region appears to be most critical for MuSK function.

In addition, the intracellular domain of MuSK has a consensus binding site (Y831) for p85, the regulatory subunit of phosphatidylinositol-3-kinase. Its C terminus ends with VXV (VGV), a consensus binding site for PDZ domain proteins (23). PDZ proteins play multiple roles in assembly of the postsynaptic apparatus at neuron–neuron synapses (62). However, mutation of these two motifs did not interfere with the capability of MuSK to induce AChR clusters (21).

How MuSK transduces its signals that eventually lead to AChR clustering and NMJ formation/maintenance remains unclear. Many pathways have been implicated (1). First, MuSK may activate Abl and GGT to activate Rho GTPases and Pak1, which may regulate LIM kinases and cofilin, and WASP/Arp2/3 (Fig. 3) (63–66).

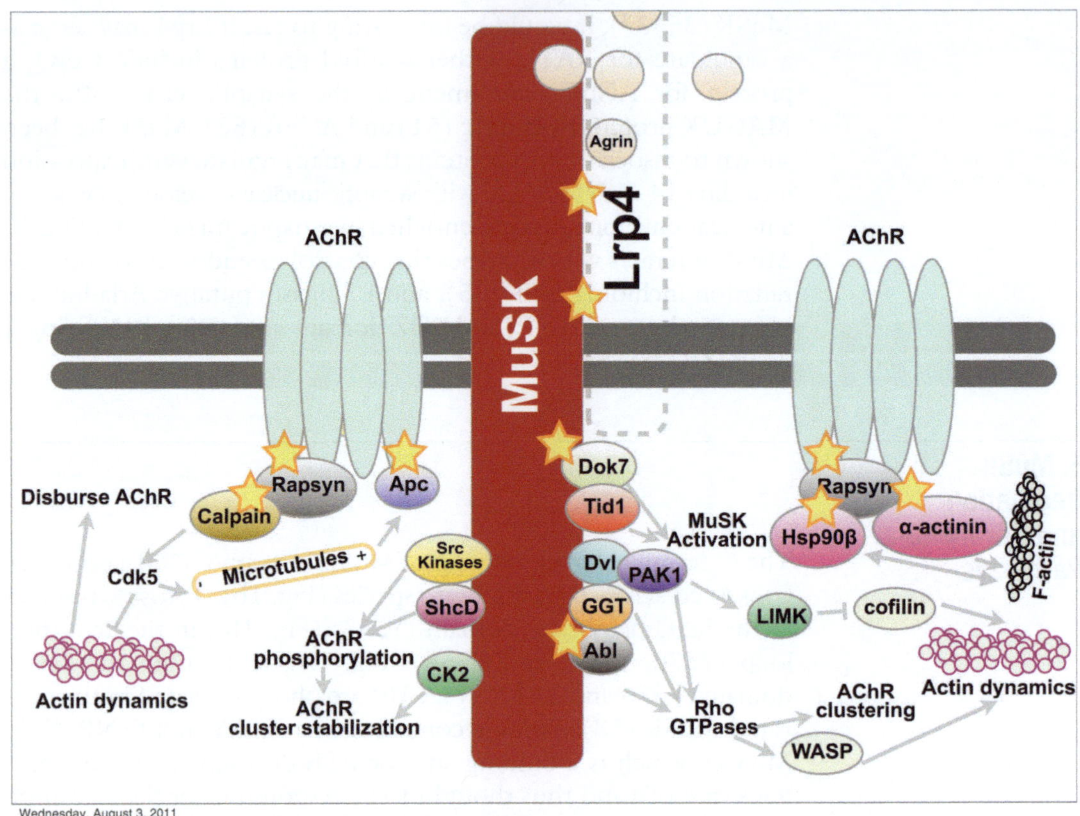

Fig. 3. Agrin/Lrp4/MuSK pathways leading to AChR clustering. See text for details.

Second, it stimulates Src kinases and CK2 (Fig. 3) (67–69), which in turn results into AChR tyrosine phosphorylation and stabilization of AChR clusters. Third, it stimulates the interaction of AChR interaction with rapsyn, a protein critical for AChR clustering and NMJ formation (70) and APC, a protein implicated in microtubule reorganization (71). It also stimulate rapsyn interaction with HSP90β for its stability (Fig. 3) (70), with α-actinin for possible linkage to cytoskeleton (Fig. 3) (72), and calpain to suppress Cdk5 (Fig. 3), a kinase able to disburse AChR clusters (Fig. 3) (40, 73, 74). In the following, we will focus on proteins that directly interact with MuSK (Fig. 3).

Dok7. Docking protein 7 (Dok7) is a member of a family of adaptor proteins, containing N-terminal pleckstrin-homology (PH) and phosphotyrosine-binding (PTB) domains, followed by an extended C-terminal region with multiple sites of tyrosine phosphorylation. Dok-7 appears to regulate MuSK activity and propagate signal to downstream effectors (Fig. 3). Mice deficient of Dok-7 do not form the NMJ (75). The critical Y553, when phosphorylated, binds the PTB domains of Dok7 (75). Dok7 is not only a substrate of MuSK but acts as an activator of MuSK catalytic (kinase) activity (76). MuSK is poorly activated by agrin in myotubes lacking Dok7, and co-transfection of Dok7 and MuSK in non-muscle cells leads to robust MuSK phosphorylation independent of agrin and LRP4 (75, 76). Structural and biochemical analysis of Dok7 revealed a dimeric arrangement of PH–PTB that facilitates *trans*-autophosphorylation of the kinase activation loop of PH–PTB that facilitates trans-autophosphorylation of the kinase activation loop. Downstream of Dok7 may be two adapter proteins: Crk and Crk-L (77). Muscle-specific ablation of Crk and Crk-L leads to NMJ deficits that suggesting their involvement in presynaptic and postsynaptic differentiations (77).

Tid1. Tumorous imaginal discs 1 (also known as Dnaja3) is a protein that interacts with the intracellular domain of MuSK (Fig. 3). It was shown to be necessary for Dok 7 binding to MuSK in response to agrin (78). Tid1 is colocalized with AChR clusters in cultured myotubes and at rodent NMJs. Knockdown of Tid1 in mouse skeletal muscles inhibited spontaneous and agrin-induced AChR clustering in cultured myotubes and resulted in NMJ disassembly.

Abl. Abl is a tyrosine kinase that is activated by agrin (79). It is required for agrin-induced AChR clustering and MuSK tyrosine phosphorylation (79). Abl kinases (Abl1 and Abl2) are localized at the synaptic regions of the developing NMJ. Abl kinase activity appears necessary for agrin-induced AChR clustering and enhancement of MuSK tyrosine phosphorylation (Fig. 3). MuSK and Abl kinases may form a complex after agrin engagement and mutually phosphorylate each other. Abl kinases are known to regulate the activity of Rho GTPases (80, 81), which have been implicated in agrin-induced AChR clustering.

CK2. CK2 (Casein kinase) was shown to interact with MuSK and colocalize at postsynaptic specializations (69). Pharmacological inhibition of CK2 reduced stability of AChRs, though MuSK activity remained unchanged. Muscle-specific CK2β knockout mice develop myasthenic phenotypes due to impaired postsynaptic apparatus (Fig. 3).

GGT. Geranylgeranyltransferase I (GGT) is a zinc metalloenzyme that tethers proteins to the plasma membrane by prenylation (65). Agrin stimulates tyrosine phosphorylation of GGT and increases its activity (Fig. 3) (65). Inhibition of GGT activity or expression attenuates agrin-induced activation of Rho GTPases, AChR clustering, and formation of NMJs in culture (65).

Dvl. Dvl is an adaptor protein that interacts with Frizzled and is critical for Wnt signaling. It interacts with MuSK (24) and Pak1, a downstream kinase of Rho GTPases, which are required for agrin-induced AChR clustering (Fig. 3) (82, 83). Therefore, Dvl may be able to bring Pak1 to the proximity of MuSK (Fig. 3). Disruption of the MuSK-Dvl interaction inhibits agrin-induced AChR clustering (24). Over-expression of the dominant-negative Dvl1 in postsynaptic muscle cells reduces the amplitude of spontaneous synaptic currents at the NMJ, suggesting existence of Dvl downstream signaling involved in presynaptic function. In agreement with this notion, a recent study demonstrates that muscle-specific knockout of β-catenin impairs presynaptic differentiation (43).

ShcD. Src homologous collagen D (ShcD; also known as Shc4) can associate via its PTB domain with phosphorylated MuSK and undergo tyrosine phosphorylation (Fig. 3). ShcD is colocalized with MuSK at the NMJ. Expression of ShcD seemed to be necessary for early tyrosine phosphorylation and clustering of AChRs (84).

6. MuSK in NMJ Diseases

Defects in MuSK is a cause of autosomal recessive congenital myasthenic syndrome (CMS) (85). Frameshift and missense mutations were reported (220insC and V789M) in MuSK in a CMS patient whose endplates are deficient in AChR and MuSK. The mutations may prevent MuSK expression and over-expression of the missense mutant results in decreased end-plate AChRs and aberrant axonal outgrowth in transgenic mice (85, 86).

MG is an autoimmune disorder characterized by weakness and rapid fatigue of skeletal muscles. A majority of MG patients are due to autoantibodies against AChRs. However, approximately 10% of MG patients do not have detectable autoantibodies to AChRs. These cases are generally known as seronegative MG (SNMG) (87). Recent studies suggest that the seronegative patients have antibodies against MuSK (88, 89). Immunization with the extracellular domain of MuSK causes MG in rodents and rabbits (90–93). Passive transfer of IgG from anti-MuSK-positive MG patients causes MG in adult animals (94–96). These observations indicate that MuSK could be a target of autoimmune reaction in SNMG patients.

7. Conclusion

MuSK is involved in almost every aspects of NMJ formation: formation of primitive AChR clusters in advance of innervation (or muscle fiber pre-patterning), agrin-induced AChR clustering, presynapitc differentiation, and maintenance of NMJ. MuSK has a unique combination of domains/motifs in the extracellular and intracellular regions. It does not bind directly to its ligand agrin and thus requires the co-receptor LRP4, which binds agrin and interacts with MuSK at the same time. Recent studies suggest that MuSK may be regulated by non-canonical ligands such as Wnt. In addition, MuSK appears to interact with an increasing number of proteins that may regulate its activity, stability, localization, and signaling. Despite of the fact that MuSK is a RTK, little is known about the identity of its physiological substrates. This glaring gap will have to be filled to better understand the functions and signaling mechanisms of this critical RTK.

References

1. Wu H, Xiong WC, Mei L (2010) To build a synapse: signaling pathways in neuromuscular junction assembly. Development 137:1017–1033
2. Sanes JR, Lichtman JW (1999) Development of the vertebrate neuromuscular junction. Annu Rev Neurosci 22:389–442
3. Nitkin RM, Smith MA, Magill C, Fallon JR, Yao YM, Wallace BG, McMahan UJ (1987) Identification of agrin, a synaptic organizing protein from torpedo electric organ. J Cell Biol 105:2471–2478
4. McMahan UJ (1990) The agrin hypothesis. Cold Spring Harb Symp Quant Biol 55:407–418
5. Reist NE, Werle MJ, McMahan UJ (1992) Agrin released by motor neurons induces the aggregation of acetylcholine receptors at neuromuscular junctions. Neuron 8:865–868
6. Wallace BG (1995) Regulation of the interaction of nicotinic acetylcholine receptors with the cytoskeleton by agrin-activated protein tyrosine kinase. J Cell Biol 128:1121–1129
7. Qu ZC, Moritz E, Huganir RL (1990) Regulation of tyrosine phosphorylation of the nicotinic acetylcholine receptor at the rat neuromuscular junction. Neuron 4:367–378
8. Qu Z, Huganir RL (1994) Comparison of innervation and agrin-induced tyrosine phosphorylation of the nicotinic acetylcholine receptor. J Neurosci 14:6834–6841
9. Barbacid M (1993) Nerve growth factor: a tale of two receptors. Oncogene 8:2033–2042
10. Snider WD (1994) Functions of the neurotrophins during nervous system development: what the knockouts are teaching us. Cell 77:627–638
11. Valenzuela DM, Stitt TN, DiStefano PS, Rojas E, Mattsson K, Compton DL, Nunez L, Park JS, Stark JL, Gies DR et al (1995) Receptor tyrosine kinase specific for the skeletal muscle lineage: expression in embryonic muscle, at the neuromuscular junction, and after injury. Neuron 15:573–584
12. Ganju P, Walls E, Brennan J, Reith AD (1995) Cloning and developmental expression of Nsk2, a novel receptor tyrosine kinase implicated in skeletal myogenesis. Oncogene 11:281–290
13. Green JL, Kuntz SG, Sternberg PW (2008) Ror receptor tyrosine kinases: orphans no more. Trends Cell Biol 18:536–544
14. Mikels AJ, Nusse R (2006) Purified Wnt5a protein activates or inhibits beta-catenin-TCF signaling depending on receptor context. PLoS Biol 4:e115
15. Nomachi A, Nishita M, Inaba D, Enomoto M, Hamasaki M, Minami Y (2008) Receptor tyrosine kinase Ror2 mediates Wnt5a-induced polarized cell migration by activating c-Jun N-terminal kinase via actin-binding protein filamin A. J Biol Chem 283:27973–27981
16. Fu AK, Smith FD, Zhou H, Chu AH, Tsim KW, Peng BH, Ip NY (1999) Xenopus muscle-specific kinase: molecular cloning and prominent expression in neural tissues during early embryonic development. Eur J Neurosci 11:373–382
17. Ip FC, Glass DG, Gies DR, Cheung J, Lai KO, Fu AK, Yancopoulos GD, Ip NY (2000) Cloning and characterization of muscle-specific kinase in chicken. Mol Cell Neurosci 16:661–673
18. Zhang J, Lefebvre JL, Zhao S, Granato M (2004) Zebrafish unplugged reveals a role for muscle-specific kinase homologs in axonal pathway choice. Nat Neurosci 7:1303–1309
19. Husain N, Pellikka M, Hong H, Klimentova T, Choe KM, Clandinin TR, Tepass U (2006) The agrin/perlecan-related protein eyes shut is essential for epithelial lumen formation in the drosophila retina. Dev Cell 11:483–493
20. Jing L, Gordon LR, Shtibin E, Granato M (2010) Temporal and spatial requirements of unplugged/MuSK function during zebrafish neuromuscular development. PLoS One 5:e8843

21. Jing L, Lefebvre JL, Gordon LR, Granato M (2009) Wnt signals organize synaptic prepattern and axon guidance through the zebrafish unplugged/MuSK receptor. Neuron 61: 721–733
22. Wang Y, Macke JP, Abella BS, Andreasson K, Worley P, Gilbert DJ, Copeland NG, Jenkins NA, Nathans J (1996) A large family of putative transmembrane receptors homologous to the product of the drosophila tissue polarity gene frizzled. J Biol Chem 271:4468–4476
23. Zhou H, Glass DJ, Yancopoulos GD, Sanes JR (1999) Distinct domains of MuSK mediate its abilities to induce and to associate with postsynaptic specializations. J Cell Biol 146:1133–1146
24. Luo ZG, Wang Q, Zhou JZ, Wang J, Luo Z, Liu M, He X, Wynshaw-Boris A, Xiong WC, Lu B, Mei L (2002) Regulation of AChR clustering by dishevelled interacting with MuSK and PAK1. Neuron 35:489–505
25. Wang J, Ruan NJ, Qian L, Lei WL, Chen F, Luo ZG (2008) Wnt/beta-catenin signaling suppresses rapsyn expression and inhibits acetylcholine receptor clustering at the neuromuscular junction. J Biol Chem 283:21668–21675
26. Henriquez JP, Webb A, Bence M, Bildsoe H, Sahores M, Hughes SM, Salinas PC (2008) Wnt signaling promotes AChR aggregation at the neuromuscular synapse in collaboration with agrin. Proc Natl Acad Sci USA 105:18812–18817
27. Rattner A, Hsieh JC, Smallwood PM, Gilbert DJ, Copeland NG, Jenkins NA, Nathans J (1997) A family of secreted proteins contains homology to the cysteine-rich ligand-binding domain of frizzled receptors. Proc Natl Acad Sci USA 94:2859–2863
28. Masiakowski P, Yancopoulos GD (1998) The Wnt receptor CRD domain is also found in MuSK and related orphan receptor tyrosine kinases. Curr Biol 8:R407
29. Godfrey EW, Nitkin RM, Wallace BG, Rubin LL, McMahan UJ (1984) Components of torpedo electric organ and muscle that cause aggregation of acetylcholine receptors on cultured muscle cells. J Cell Biol 99:615–627
30. Gautam M, Noakes PG, Moscoso L, Rupp F, Scheller RH, Merlie JP, Sanes JR (1996) Defective neuromuscular synaptogenesis in agrin-deficient mutant mice. Cell 85: 525–535
31. Glass DJ, Bowen DC, Stitt TN, Radziejewski C, Bruno J, Ryan TE, Gies DR, Shah S, Mattsson K, Burden SJ, DiStefano PS, Valenzuela DM, DeChiara TM, Yancopoulos GD (1996) Agrin acts via a MuSK receptor complex. Cell 85:513–523
32. Sealock R, Wray BE, Froehner SC (1984) Ultrastructural localization of the Mr 43,000 protein and the acetylcholine receptor in torpedo postsynaptic membranes using monoclonal antibodies. J Cell Biol 98:2239–2244
33. DeChiara TM, Bowen DC, Valenzuela DM, Simmons MV, Poueymirou WT, Thomas S, Kinetz E, Compton DL, Rojas E, Park JS, Smith C, DiStefano PS, Glass DJ, Burden SJ, Yancopoulos GD (1996) The receptor tyrosine kinase MuSK is required for neuromuscular junction formation in vivo. Cell 85:501–512
34. Zhang B, Luo S, Wang Q, Suzuki T, Xiong WC, Mei L (2008) LRP4 Serves as a coreceptor of agrin. Neuron 60:285–297
35. Kim N, Stiegler AL, Cameron TO, Hallock PT, Gomez AM, Huang JH, Hubbard SR, Dustin ML, Burden SJ (2008) Lrp4 is a receptor for agrin and forms a complex with MuSK. Cell 135:334–342
36. Johnson EB, Hammer RE, Herz J (2005) Abnormal development of the apical ectodermal ridge and polysyndactyly in Megf7-deficient mice. Hum Mol Genet 14:3523–3538
37. Yamaguchi YL, Tanaka SS, Kasa M, Yasuda K, Tam PP, Matsui Y (2006) Expression of low density lipoprotein receptor-related protein 4 (Lrp4) gene in the mouse germ cells. Gene Expr Patterns 6:607–612
38. Weatherbee SD, Anderson KV, Niswander LA (2006) LDL-receptor-related protein 4 is crucial for formation of the neuromuscular junction. Development 133:4993–5000
39. Flanagan-Steet H, Fox MA, Meyer D, Sanes JR (2005) Neuromuscular synapses can form in vivo by incorporation of initially aneural postsynaptic specializations. Development 132:4471–4481
40. Lin W, Dominguez B, Yang J, Aryal P, Brandon EP, Gage FH, Lee KF (2005) Neurotransmitter acetylcholine negatively regulates neuromuscular synapse formation by a Cdk5-dependent mechanism. Neuron 46:569–579
41. Panzer JA, Song Y, Balice-Gordon RJ (2006) In vivo imaging of preferential motor axon outgrowth to and synaptogenesis at prepatterned acetylcholine receptor clusters in embryonic zebrafish skeletal muscle. J Neurosci 26:934–947
42. Kim N, Burden SJ (2008) MuSK controls where motor axons grow and form synapses. Nat Neurosci 11:19–27
43. Lin W, Burgess RW, Dominguez B, Pfaff SL, Sanes JR, Lee KF (2001) Distinct roles of nerve and muscle in postsynaptic differentiation of the neuromuscular synapse. Nature 410:1057–1064

44. Yang X, Arber S, William C, Li L, Tanabe Y, Jessell TM, Birchmeier C, Burden SJ (2001) Patterning of muscle acetylcholine receptor gene expression in the absence of motor innervation. Neuron 30:399–410
45. Hesser BA, Henschel O, Witzemann V (2006) Synapse disassembly and formation of new synapses in postnatal muscle upon conditional inactivation of MuSK. Mol Cell Neurosci 31:470–480
46. Glass DJ, Apel ED, Shah S, Bowen DC, DeChiara TM, Stitt TN, Sanes JR, Yancopoulos GD (1997) Kinase domain of the muscle-specific receptor tyrosine kinase (MuSK) is sufficient for phosphorylation but not clustering of acetylcholine receptors: required role for the MuSK ectodomain? Proc Natl Acad Sci USA 94:8848–8853
47. Jones G, Meier T, Lichtsteiner M, Witzemann V, Sakmann B, Brenner HR (1997) Induction by agrin of ectopic and functional postsynaptic-like membrane in innervated muscle. Proc Natl Acad Sci USA 94:2654–2659
48. Gautam M, Noakes PG, Mudd J, Nichol M, Chu GC, Sanes JR, Merlie JP (1995) Failure of postsynaptic specialization to develop at neuromuscular junctions of rapsyn-deficient mice. Nature 377:232–236
49. Apel ED, Glass DJ, Moscoso LM, Yancopoulos GD, Sanes JR (1997) Rapsyn is required for MuSK signaling and recruits synaptic components to a MuSK-containing scaffold. Neuron 18:623–635
50. Cartaud A, Strochlic L, Guerra M, Blanchard B, Lambergeon M, Krejci E, Cartaud J, Legay C (2004) MuSK is required for anchoring acetylcholinesterase at the neuromuscular junction. J Cell Biol 165:505–515
51. Strochlic L, Cartaud A, Labas V, Hoch W, Rossier J, Cartaud J (2001) MAGI-1c: a synaptic MAGUK interacting with muSK at the vertebrate neuromuscular junction. J Cell Biol 153:1127–1132
52. Fuhrer C, Sugiyama JE, Taylor RG, Hall ZW (1997) Association of muscle-specific kinase MuSK with the acetylcholine receptor in mammalian muscle. EMBO J 16:4951–4960
53. Strochlic L, Cartaud A, Mejat A, Grailhe R, Schaeffer L, Changeux JP, Cartaud J (2004) 14-3-3 Gamma associates with muscle specific kinase and regulates synaptic gene transcription at vertebrate neuromuscular synapse. Proc Natl Acad Sci USA 101:18189–18194
54. Apel ED, Lewis RM, Grady RM, Sanes JR (2000) Syne-1, a dystrophin- and Klarsicht-related protein associated with synaptic nuclei at the neuromuscular junction. J Biol Chem 275:31986–31995
55. Zhu D, Yang Z, Luo Z, Luo S, Xiong WC, Mei L (2008) Muscle-specific receptor tyrosine kinase endocytosis in acetylcholine receptor clustering in response to agrin. J Neurosci 28:1688–1696
56. Bromann PA, Weiner JA, Apel ED, Lewis RM, Sanes JR (2004) A putative ariadne-like E3 ubiquitin ligase (PAUL) that interacts with the muscle-specific kinase (MuSK). Gene Expr Patterns 4:77–84
57. Lu Z, Je HS, Young P, Gross J, Lu B, Feng G (2007) Regulation of synaptic growth and maturation by a synapse-associated E3 ubiquitin ligase at the neuromuscular junction. J Cell Biol 177:1077–1089
58. Watty A, Neubauer G, Dreger M, Zimmer M, Wilm M, Burden SJ (2000) The in vitro and in vivo phosphotyrosine map of activated MuSK. Proc Natl Acad Sci USA 97:4585–4590
59. van der Geer P, Pawson T (1995) The PTB domain: a new protein module implicated in signal transduction. Trends Biochem Sci 20: 277–280
60. Borg JP, Margolis B (1998) Function of PTB domains. Curr Top Microbiol Immunol 228:23–38
61. Herbst R, Burden SJ (2000) The juxtamembrane region of MuSK has a critical role in agrin-mediated signaling. EMBO J 19:67–77
62. Kim E, Sheng M (2004) PDZ domain proteins of synapses. Nat Rev Neurosci 5:771–781
63. Edwards DC, Sanders LC, Bokoch GM, Gill GN (1999) Activation of LIM-kinase by Pak1 couples Rac/Cdc42 GTPase signalling to actin cytoskeletal dynamics. Nat Cell Biol 1:253–259
64. Soosairajah J, Maiti S, Wiggan O, Sarmiere P, Moussi N, Sarcevic B, Sampath R, Bamburg JR, Bernard O (2005) Interplay between components of a novel LIM kinase-slingshot phosphatase complex regulates cofilin. EMBO J 24:473–486
65. Luo ZG, Je HS, Wang Q, Yang F, Dobbins GC, Yang ZH, Xiong WC, Lu B, Mei L (2003) Implication of geranylgeranyltransferase I in synapse formation. Neuron 40:703–717
66. Lee CW, Han J, Bamburg JR, Han L, Lynn R, Zheng JQ (2009) Regulation of acetylcholine receptor clustering by ADF/cofilin-directed vesicular trafficking. Nat Neurosci 12:848–856
67. Smith CL, Mittaud P, Prescott ED, Fuhrer C, Burden SJ (2001) Src, Fyn, and Yes are not required for neuromuscular synapse formation but are necessary for stabilization of agrin-induced clusters of acetylcholine receptors. J Neurosci 21:3151–3160
68. Mohamed AS, Rivas-Plata KA, Kraas JR, Saleh SM, Swope SL (2001) Src-class kinases act

within the agrin/MuSK pathway to regulate acetylcholine receptor phosphorylation, cytoskeletal anchoring, and clustering. J Neurosci 21:3806–3818
69. Cheusova T, Khan MA, Schubert SW, Gavin AC, Buchou T, Jacob G, Sticht H, Allende J, Boldyreff B, Brenner HR, Hashemolhosseini S (2006) Casein kinase 2-dependent serine phosphorylation of MuSK regulates acetylcholine receptor aggregation at the neuromuscular junction. Genes Dev 20:1800–1816
70. Luo S, Zhang B, Dong XP, Tao Y, Ting A, Zhou Z, Meixiong J, Luo J, Chiu FC, Xiong WC, Mei L (2008) HSP90 beta regulates rapsyn turnover and subsequent AChR cluster formation and maintenance. Neuron 60:97–110
71. Wang J, Jing Z, Zhang L, Zhou G, Braun J, Yao Y, Wang ZZ (2003) Regulation of acetylcholine receptor clustering by the tumor suppressor APC. Nat Neurosci 6:1017–1018
72. Dobbins GC, Luo S, Yang Z, Xiong WC, Mei L (2008) Alpha-actinin interacts with rapsyn in agrin-stimulated AChR clustering. Mol Brain 1:18
73. Chen F, Qian L, Yang ZH, Huang Y, Ngo ST, Ruan NJ, Wang J, Schneider C, Noakes PG, Ding YQ, Mei L, Luo ZG (2007) Rapsyn interaction with calpain stabilizes AChR clusters at the neuromuscular junction. Neuron 55:247–260
74. Fu AK, Ip FC, Fu WY, Cheung J, Wang JH, Yung WH, Ip NY (2005) Aberrant motor axon projection, acetylcholine receptor clustering, and neurotransmission in cyclin-dependent kinase 5 null mice. Proc Natl Acad Sci USA 102:15224–15229
75. Okada K, Inoue A, Okada M, Murata Y, Kakuta S, Jigami T, Kubo S, Shiraishi H, Eguchi K, Motomura M et al (2006) The muscle protein Dok-7 is essential for neuromuscular synaptogenesis. Science 312(5781):1802–1805
76. Hamuro J, Higuchi O, Okada K, Ueno M, Iemura S, Natsume T, Spearman H, Beeson D, Yamanashi Y (2008) Mutations causing DOK7 congenital myasthenia ablate functional motifs in Dok-7. J Biol Chem 283:5518–5524
77. Hallock PT, Xu CF, Park TJ, Neubert TA, Curran T, Burden SJ (2010) Dok-7 regulates neuromuscular synapse formation by recruiting Crk and Crk-L. Genes Dev 24:2451–2461
78. Linnoila J, Wang Y, Yao Y, Wang ZZ (2008) A mammalian homolog of drosophila tumorous imaginal discs, Tid1, mediates agrin signaling at the neuromuscular junction. Neuron 60:625–641
79. Finn AJ, Feng G, Pendergast AM (2003) Postsynaptic requirement for Abl kinases in assembly of the neuromuscular junction. Nat Neurosci 6:717–723
80. Weston C, Yee B, Hod E, Prives J (2000) Agrin-induced acetylcholine receptor clustering is mediated by the small guanosine triphosphatases Rac and Cdc42. J Cell Biol 150:205–212
81. Weston CA, Teressa G, Weeks BS, Prives J (2007) Agrin and laminin induce acetylcholine receptor clustering by convergent, Rho GTPase-dependent signaling pathways. J Cell Sci 120:868–875
82. Peng HB, Xie H, Dai Z (1997) Association of cortactin with developing neuromuscular specializations. J Neurocytol 26:637–650
83. Webb BA, Zhou S, Eves R, Shen L, Jia L, Mak AS (2006) Phosphorylation of cortactin by p21-activated kinase. Arch Biochem Biophys 456:183–193
84. Jones N, Hardy WR, Friese MB, Jorgensen C, Smith MJ, Woody NM, Burden SJ, Pawson T (2007) Analysis of a Shc family adaptor protein, ShcD/Shc4, that associates with muscle-specific kinase. Mol Cell Biol 27:4759–4773
85. Engel AG, Sine SM (2005) Current understanding of congenital myasthenic syndromes. Curr Opin Pharmacol 5:308–321
86. Chevessier F, Faraut B, Ravel-Chapuis A, Richard P, Gaudon K, Bauche S, Prioleau C, Herbst R, Goillot E, Ioos C, Azulay JP, Attarian S, Leroy JP, Fournier E, Legay C, Schaeffer L, Koenig J, Fardeau M, Eymard B, Pouget J, Hantai D (2004) MUSK, a new target for mutations causing congenital myasthenic syndrome. Hum Mol Genet 13:3229–3240
87. Vincent A, Li Z, Hart A, Barrett-Jolley R, Yamamoto T, Burges J, Wray D, Byrne N, Molenaar P, Newsom-Davis J (1993) Seronegative myasthenia gravis. Evidence for plasma factor(s) interfering with acetylcholine receptor function. Ann N Y Acad Sci 681:529–538
88. Hoch W, McConville J, Helms S, Newsom-Davis J, Melms A, Vincent A (2001) Autoantibodies to the receptor tyrosine kinase MuSK in patients with myasthenia gravis without acetylcholine receptor antibodies. Nat Med 7:365–368
89. Sanders DB, El-Salem K, Massey JM, McConville J, Vincent A (2003) Clinical aspects of MuSK antibody positive seronegative MG. Neurology 60:1978–1980
90. Shigemoto K, Kubo S, Jie C, Hato N, Abe Y, Ueda N, Kobayashi N, Kameda K, Mominoki K, Miyazawa A, Ishigami A, Matsuda S,

Maruyama N (2008) Myasthenia gravis experimentally induced with muscle-specific kinase. Ann N Y Acad Sci 1132:93–98

91. Xu K, Jha S, Hoch W, Dryer SE (2006) Delayed synapsing muscles are more severely affected in an experimental model of MuSK-induced myasthenia gravis. Neuroscience 143:655–659

92. Jha S, Xu K, Maruta T, Oshima M, Mosier DR, Atassi MZ, Hoch W (2006) Myasthenia gravis induced in mice by immunization with the recombinant extracellular domain of rat muscle-specific kinase (MuSK). J Neuroimmunol 175:107–117

93. Punga AR, Maj M, Lin S, Meinen S, Ruegg MA (2011) MuSK levels differ between adult skeletal muscles and influence postsynaptic plasticity. Eur J Neurosci 33:890–898

94. Cole RN, Reddel SW, Gervasio OL, Phillips WD (2008) Anti-MuSK patient antibodies disrupt the mouse neuromuscular junction. Ann Neurol 63:782–789

95. Cole RN, Ghazanfari N, Ngo ST, Gervasio OL, Reddel SW, Phillips WD (2010) Patient autoantibodies deplete postsynaptic muscle-specific kinase leading to disassembly of the ACh receptor scaffold and myasthenia gravis in mice. J Physiol 588:3217–3229

96. ter Beek WP, Martinez-Martinez P, Losen M, de Baets MH, Wintzen AR, Verschuuren JJ, Niks EH, van Duinen SG, Vincent A, Molenaar PC (2009) The effect of plasma from muscle-specific tyrosine kinase myasthenia patients on regenerating endplates. Am J Pathol 175: 1536–1544

Chapter 12

Assays for Pten-Induced Novel Kinase 1 (PINK1) and Leucine-Rich Repeat Kinase 2 (LRRK2), Kinases Associated with Parkinson's Disease

Alexandra Beilina and Mark R. Cookson

Abstract

PTEN-induced novel kinase 1 (PINK1) and leucine-rich repeat kinase 2 (LRRK2) are two protein kinases associated with recessive and dominant forms of parkinsonism, respectively. Mutations in PINK1 cause loss of protein function whereas mutations in LRRK2 are less easily defined but, in some cases, may cause increased kinase activity. Furthermore, LRRK2 kinase activity is being explored as a therapeutic target in PD. Therefore, in both the cases of PINK1 and LRRK2, measuring kinase activity is important but is complicated by the problem that convincing physiological substrates of the two proteins have not yet been found. In this chapter, we will describe in detail the protocols we use in our lab to measure activity and related functions of these two kinases.

Key words: LRRK2, PINK1, Parkinson's disease, Kinase, Autophosphorylation

1. Introduction

In the past few years, several inherited forms of Parkinson's disease have been identified and the underlying genes cloned (1). Of these, at least two are protein kinases. PTEN-induced novel kinase 1 (PINK1) is a serine/threonine kinase directed to mitochondria where the kinase domain probably faces out towards the cytoplasm (2). Mutations are loss of function and are associated with autosomal recessive early onset parkinsonism (3). Leucine-rich repeat kinase 2 (LRRK2) is a large cytosolic protein with several domains including a kinase domain (4). Mutations are associated with dominantly inherited disease that is clinically and pathologically similar to sporadic Parkinson's disease (5, 6).

Hideyuki Mukai (ed.), *Protein Kinase Technologies*, Neuromethods, vol. 68,
DOI 10.1007/978-1-61779-824-5_12, © Springer Science+Business Media, LLC 2012

Understanding the effects of mutations for both PINK1 and LRRK2 is important for identifying pathological processes involved in the disease and, in both cases, kinase activity measurements are a valuable tool for doing so. Unfortunately, authentic physiological substrates for PINK1 or LRRK2 are not fully established and so kinase activities are usually assayed using surrogate measures. The aim of this chapter is to discuss the methods that are in common usage for PINK1 and LRRK2, with the goal of describing general principles that may be useful to study other kinases.

For the mitochondrial protein PINK1, mutations generally decrease protein stability (7) or kinase activity or both (8). This is consistent with the recessive inheritance of PINK1 mutations. It has been shown that loss of functional PINK1 results in mitochondrial abnormalities in a variety of model organisms (9–19). Although the physiological function of PINK1 is not yet completely understood, some recent evidence implicates PINK1 and a downstream protein ubiquitin E3 ligase, parkin, in mitochondrial turnover via autophagy (8, 10, 20–32). PINK1 can promote the recruitment of parkin to mitochondria in a kinase-dependent fashion, supporting the idea that understanding kinase activity will help us understand disease processes.

Mutations in LRRK2 are found both in the kinase domain and in other regions of the protein. Of the two well-established mutations in the kinase domain, G2019S, which is at the DYG motif that binds divalent metal cations at the N-terminal region of the activation loop, increases kinase activity by about two to threefold (33). However, other mutations in the protein including I2020T at the adjacent residue are not activating to the same extent, meaning that the contribution of kinase activity to mutation mechanisms is not clear. Nonetheless, molecular and pharmacological evidence suggests that inhibition of kinase activity contributes to the neurotoxic effects of mutations (34–36). Therefore, LRRK2 has become an attractive candidate for drug development and kinase activity is used to identify small molecule inhibitors (34, 37, 38).

In this chapter, we will outline our current protocols for measuring kinase activity using truncated protein produced in bacteria (for PINK1) or full length protein prepared by immunopurification from mammalian cells (for LRRK2). The general principle is to incubate purified kinase with the cofactor Mg^{2+} and ATP that is radiolabeled at the gamma-phosphate which is then transferred to a target protein or peptide. The substrates are separated from free radioactivity using either sodium dodecyl sulfate polyacrylamide gel electrophoresis (SDS-PAGE) or by binding to filter paper. Activity can be quantified by relative incorporation of radioactivity using a phosphorimager or a scintillation counter.

As mentioned earlier, physiological substrates for PINK1 or LRRK2 are not known, although several have been nominated. For example, moesin was first proposed as candidate substrates for LRRK2 (39), but it was subsequently shown that LRRK2 is rate-limiting phosphorylation of moesin in cells (40). However, the identification of the threonine site at which moesin is phosphorylated by LRRK2 in vitro led to the development of the artificial peptide substrate LRRKtide, which is now widely used (39–45). The other major assay in use is autophosphorylation, which occurs readily in vitro, and has been mapped to specific sites outside of the kinase domain (46–49), but has not been shown to occur under physiological conditions. Autophosphorylation has also been used to assay PINK1 activity (7), although model substrates such as histone proteins have also been helpful to measure activity (50).

For any kinase activity assay, it is essential to have a negative control of a kinase dead mutant protein. To make kinase dead versions several mutations can be used. There is a critical lysine residue in the ATP binding pocket that interacts with and orients the gamma phosphate of ATP. The motif is VAVK or VAIK in some kinases and the lysine can be mutated to methionine or alanine leading to decreased binding of ATP due to the bulkier side chain. Alternatively, there is a critical aspartate that interacts with Mg^{2+} binding site (DY/FG) that can be mutated to alanine, removing the charge and limiting Mg^{2+} binding. Finally, the aspartate in the enzyme active site, an HRD or YRD motif can also be replaced by alanine. In some cases, we have mutated all three residues to alanine to make a triple kinase dead version. In general, all of the mutants are fully kinase dead when assayed with very pure proteins.

Kinases are also themselves regulated by phosphorylation by kinases in signal transduction pathways. We therefore also include a protocol for metabolic labeling to examine the state of phosphorylation of target proteins, which we have found useful for studying the overall phosphorylation state of LRRK2 in particular.

2. Materials

For each kinase, plasmid containing cDNA of PINK1 or LRRK2 in vectors suitable for protein expression in *Escherichia coli* or in mammalian cells are needed. In general, we have used human cDNA although the methods should work for homologues from other species. Appropriate vectors are available either from the authors or from public repositories such as Addgene (http://www.addgene.org/) where all the vectors from our studies are available. In most cases, it is convenient to "tag" the protein with a fusion protein, which allows for purification on affinity gels, although other labs have reported purification of endogenous LRRK2 using antibodies against the native protein. In this set of protocols, we present methods for purification of GST–PINK1, 2×Myc LRRK2 and 3×FLAG LRRK2 to illustrate the range of methods that we have found useful.

For the kinase activity assays it is essential to have a kinase dead construct, discussed earlier. This is to control for inadvertent co-purification of other kinases with measureable activity, which might otherwise confound analysis.

2.1. Purification and Assay of GST-Tagged PINK1 Kinase Domain

For this protocol, we use an *E. coli* strain optimized for protein expression, such as BL21/DE3. We routinely purchase these are competent cells from New England Biolabs. Routine microbiological media, agar plates, and antibiotics are needed for growth of the bacteria as well as Isopropyl β-D-1-thiogalactopyranoside (IPTG) for induction of protein expression. Purification is achieved using Glutathione Sepharose beads and protein is eluted using glutathione, which provides gentle conditions. Most of these reagents are relatively easy to obtain and we have had good results from several different manufacturers.

For this protocol, and for other kinase assays, we use ATP that has been radiolabeled at the gamma phosphate from Perkin Elmer. Over the years, we have used both ^{32}P-ATP and ^{33}P-ATP. Because ^{33}P is a lower energy beta-emitter compared to ^{32}P, it may be safer, although the amounts used in these protocols are relatively small in any case. Also, 33P has a longer half-life (~25 days vs. ~14 days), which may be helpful if usage is relatively modest. In general, we have used ^{32}P where activity is weak, which is the case for PINK1.

2.2. Autophosphorylation of 2×Myc LRRK2

In this protocol, and in the companion protocol for peptide assays, full length LRRK2 protein is prepared by transfection of plasmids into cell lines and purified using immunoprecipitation. In principle, any mammalian cell line would be suitable for protein production so long as it can be transfected. We have tended to use HEK293 or subclones such as HEK293FT, which can be obtained from ATCC or from Invitrogen, respectively. These cells are very robust, grow quickly, have high transfection rates and can produce large amounts of proteins including LRRK2 without obvious toxic effects. We currently use polyethyleneimine (PEI) for transient transfections because it is reliable, nontoxic, and extremely cheap.

General reagents including protease and phosphatase inhibitors can be obtained from a variety of sources, often as tablets that can be made immediately prior to using a buffer. We have found cell lysis and kinase assay buffers from Cell Signaling technology to be convenient and very reliable and routinely use them. Purification of proteins is achieved using sepharose or agarose gel beads conjugated to the tag of interest, in this case Myc. Again, we find that commercial beads are convenient and currently use EZview™ Red Anti-c-Myc Affinity Gel from Sigma because of the enhanced visibility of the red affinity resin beads makes it is easy to see.

To measure radioactivity incorporated into LRRK2 itself (i.e., autophosphorylation), several additional pieces of equipment are needed and these should be set up within a designated radioactivity area. SDS-PAGE equipment is used for separation of proteins, and specifically we use premade gradient gels that for LRRK2 need to be relatively low percentage acrylamide (i.e., 4–20% gradients) for best separation. We also set up a western blot apparatus and a phosphorimager—the Storm imager from Amersham has been a workhorse for several years but there are other manufacturers who produce similar instruments—and screens.

2.3. LRRKtide Purification by 3×FLAG LRRK2

General reagents for the LRRKtide assays are as for the autophosphorylation assays, except that anti-Flag gel with the M2 monoclonal (Sigma) is used instead of anti-Myc. The LRRKtide peptide, RLGRDKYKTLRQIRQ, can be obtained from Enzo Life sciences or other sources. Measuring of radioactivity requires a scintillation counter and Whatman P81 phosphocellulose filter paper.

2.4. Metabolic Radiolabeling

Metabolic labeling requires 32-P orthophospate, which again can be obtained from Perkin Elmer, and phosphate-free media, DMEM (Invitrogen). Other materials for immunopurification of LRRK2 are as above.

3. Methods

3.1. Purification and Assay of GST-Tagged PINK1 Kinase Domain

Reagents and buffers required:

Name	Supplier, cat#	Ingredients
Lysis buffer A		50 mM Tris–HCl pH 7.5; 50 mM NaCl; 0.5 mM EDTA; 5% glycerol; 1 mM DTT; 0.1% Triton x100; protease inhibitors
Wash buffer B		50 mM Tris–HCl pH 7.5; 300 mM NaCl, 0.5 mM EDTA; 5% glycerol; 1 mM DTT; 0.1% Triton-×100; protease inhibitors
Glutathione elution buffer (GEB)		50 mM Tris–HCl pH 7.5; 10 mM glutathione pH 8
5× kinase buffer		200 mM HEPES pH 7.4; 50 mM $MgCl_2$; 15 mM $MnCl_2$; 5 mM DTT
[γ-^{32}P]ATP	Perkin Elmer, cat# NEG002A	3,000 Ci/mmol, 3.3 μM
Glutathione Sepharose 4B	Amersham, cat# 17075601	
GelCode Blue	Thermo Scientific, cat# 24592	
Protease inhibitors	Roche, cat# 04693124001	

Procedure:

1. Transform expression plasmid pGEX5×-1-PINK1 into BL21/DE3 competent *E. coli*, according to manufacturer's instructions.
2. Plate on LB Agar plates with 100 μg/ml ampicillin and incubate at 37°C overnight.
3. Pick a single colony, inoculate into 10 ml overnight culture in LB with 100 μg/ml ampicillin.
4. Inoculate 200 ml LB ampicillin with 10 ml of overnight culture.
5. Measure optical density at 600 nm (A600) vs. media blank at regular intervals using a UV–Vis spectrophotometer.
6. At A600 between 0.6–0.8 add IPTG to a final concentration of 0.2 mM.
7. Grow bacteria for 2 h at 37°C in a shaking incubator.
8. Pellet cells at 5,000 rpm for 10 min.
9. Resuspend cells completely in ice-cold lysis buffer A: 50 mM Tris–HCl pH 7.5; 50 mM NaCl; 0.5 mM EDTA; 5% glycerol; 1 mM DTT; 0.1% Triton x100; protease inhibitors. Use 10 ml of lysis buffer per 200 ml of total bacterial culture.
10. Sonicate resuspended bacteria cells for 10 min. Remember to keep your samples on ice since sonication causes significant heat.
11. Transfer to the centrifuge tubes and spin to pellet the cellular debris at 12,000 rpm for 15 min at 4°C (the soluble GST proteins should be in the supernatant).
12. While the cells are spinning, prepare the Glutathione Sepharose 4B matrix/resin. Gently mix bottle to resuspend the glutathione resin. Remove 100 μl of glutathione resin and wash it twice with 1 ml of lysis buffer by centrifuging at 500×*g* for 1 min. Decant supernatant carefully. Do not allow the resin to dry out and leave some PBS on top of the resin bed.
13. Add the supernatant from the spin after the sonication to the washed Glutathione Sepharose beads and rotate at 4°C overnight on a rotamixer.
14. Spin down the glutathione resin, to which the protein should be bound, by centrifuging tube at 500×*g* for 1 min. Remove supernatant without disturbing the resin bed.
15. Wash glutathione resin twice with 5 ml Wash buffer B, twice with 5 ml Lysis buffer A, then twice with 50 mM Tris–HCl pH 7.5.
16. After final wash, remove most of the buffer and elute with Glutathione elution buffer (GEB): 50 mM Tris–HCl pH 7.5 with 10 mM glutathione. Volume of GEB should be twice the volume of the glutathione resin, i.e., use 200 μl of GEB for 100 μl of glutathione resin.

17. Incubate the resin with GEB for 15–30 min at 4°C on the rotamixer.
18. Spin down the glutathione resin at 500×*g* for 1 min and collect your supernatant. This is elution 1 and should contain most of the GST–PINK1 protein. Repeat for a 2nd elution.
19. Purity of the eluted proteins should be evaluated by running 5–20 μl of the eluted sample on SDS-PAGE gel.
20. Prepare the 5× kinase buffer: 200 mM HEPES pH 7.4; 50 mM $MgCl_2$; 15 mM $MnCl_2$; 5 mM DTT. Aliquots of the 5× kinase buffer can be stored at −20°C.
21. Dilute 1 μg of purified GST–PINK1 proteins in 50 μl of 1× kinase buffer containing 25 μM ATP, and 10 μCi of [γ-^{32}P] ATP.
22. Incubate the reactions at 30°C for 30 min.
23. Stop the reaction by heating at 95°C for 5 min, and separate mixtures on an SDS–PAGE gel.
24. Stain gels with Gel Code Blue (Pierce) or transfer to PVDF membrane.
25. Dry gel/membrane and expose to X-ray film or phosphoscreen overnight.
26. Read the phosphoscreen on a STORM or equivalent phosphorimager.

3.2. In Vitro Autophosphorylation Kinase Assay with the Immunoprecipitated Full Length 2×Myc–LRRK2

Reagents and buffers required:

Name	Supplier, cat#	Ingredients
1× Lysis buffer	Cell Signaling, cat# 9803	20 mM Tris–HCl pH 7.5; 150 mM NaCl; 1 mM Na_2EDTA; 1 mM EGTA; 1% Triton x100; 2.5 mM sodium pyrophosphate; 1 mM β-glycerophosphate; 1 mM Na_3VO_4; 1 μg/ml leupeptin
Wash buffer		25 mM Tris pH 7.5, 400 mM NaCl, 1% Triton x100
1× kinase buffer	Cell Signaling, cat# 9802	25 mM Tris–HCl pH 7.5; 5 mM β-glycerophosphate; 2 mM (DTT); 0.1 mM Na_3VO_4; 10 mM $MgCl_2$
EZview™ Protein G agarose	Sigma, cat# E3403	

(continued)

Name	Supplier, cat#	Ingredients
EZview™ Red Anti-c-Myc Affinity agarose	Sigma, cat# E6654	
[γ-^{33}P]ATP	Perkin Elmer, cat# NEG602H	3,000 Ci/mmol, 3.3 μM
Myc antibody	Roche, cat# 11667149001	
MJF2 antibody	Epitomics, cat# 35141	
Protease inhibitors	Roche, cat# 04693124001	
Phosphates inhibitors	Thermo scientific, cat# 78426	

Procedure:

1. Transfect HEK293FT cells with LRRK2 plasmids and express protein for 48 h. Lipid-based reagents such as Lipofectamine 2000 or PEI can be used for transfection with equivalent results in our hands.
2. Aspirate media and wash cells with 1×PBS.
3. Resuspend cells in 0.5 ml of lysis buffer per 10 cm dish. Add protease and phosphatase inhibitor cocktail to lysis buffer fresh before use.
4. Incubate lysates on ice for 30 min.
5. Centrifuge lysates at 14,000 rpm for 10 min.
6. During centrifugation, aliquot 20 μl of EZview™ Protein G beads into 1.5 ml tubes and wash them twice with lysis buffer. After removing the supernatant, set the tube with equilibrated red bead pellet on ice.
7. Add centrifuged lysates to the 20 μl of washed EZview™ Protein G beads to preclear lysates from nonspecifically bound protein to the Protein G agarose. Vortex briefly and rotate them for at least 30 min at 4°C.
8. During this preclear step, rinse EZview™ Red Anti-c-Myc Affinity agarose twice with lysis buffer.
9. After preclear, collect EZview™ Protein G beads by centrifugation at 15,000×*g* for 1 min.
10. Carefully transfer the clear lysate supernatant to fresh 1.5 ml tubes containing 20 μl of washed EZview™ Red Anti-c-Myc Affinity agarose. Vortex gently and incubate with gentle mixing for 2 h to overnight by rotating at 4°C to allow c-myc-tagged proteins to bind to the EZview™ Red Anti-c-Myc Affinity agarose.
11. Centrifuge the immunoprecipitated complexes in a refrigerated microcentrifuge at 5,000×*g* for 1 min. Aspirate the supernatant

carefully or remove it with a micropipette. Set tubes with the beads on ice.

12. Wash the bead pellet by adding 1 ml of prechilled wash buffer and mixing beads by inverting a tube several times. Collect beads by centrifugation at 5,000×*g* for 1 min. Remove the supernatant by micropipette or aspirate carefully. Remember to keep tubes with the bead pellet on ice.
13. Repeat washing step 4 times.
14. Prior to kinase assay, rinse the proteins on beads twice with 1 ml of 1× kinase buffer. Carefully remove all kinase buffer and add 30 μl of 1× kinase buffer to each sample.
15. In a separate tube prepare a kinase master mix containing 1× kinase buffer; 100 μM ATP; 40 μCi of [γ-^{32}P]ATP.
16. Add 10 μl of kinase master mix to 30 μl of immunoprecipitated myc proteins. Incubate at 30°C for 30 min by vortexing beads at 1,400 rpm.
17. Stop reaction by adding 10 μl of 5× sample buffer and heating samples at 95°C for 5 min.
18. After 95°C heat vortex samples briefly and centrifuge beads at 14,000 rpm for 1 min.
19. Resolve protein on 4–20% SDS-PAGE gel. It is suggested to load from 10–30 μl of reaction. Be careful not to load beads on gel.
20. Transfer gel to the PVDF membrane, dry the membrane, and expose to film or phosphoscreen. Read the phosphoscreen by STORM.
21. Immunoblot with myc or LRRK2 antibody for total protein levels.

3.3. LRRK2–3× FLAG Purification and LRRKtide Assay

Reagents and buffers required:

Name	Supplier, cat#	Ingredients
1× Lysis buffer	Cell Signaling, cat# 9803	20 mM Tris–HCl pH 7.5; 150 mM NaCl; 1 mM Na_2EDTA; 1 mM EGTA; 1% Triton x100; 2.5 mM sodium pyrophosphate; 1 mM b-glycerophosphate; 1 mM Na_3VO_4; 1 μg/ml leupeptin
Wash buffer		25 mM Tris pH 7.5, 400 mM NaCl, 1% Triton x100
Elution Buffer		1× kinase buffer; 150mM NaCl; and 0.02% Triton x100

(continued)

Name	Supplier, cat#	Ingredients
1× kinase buffer	Cell Signaling, cat# 9802	25 mM Tris–HCl pH 7.5; 5 mM beta-glycerophosphate; 2 mM DTT; 0.1 mM Na_3VO_4; 10 mM $MgCl_2$
STOP solution		0.5 M EDTA with bromophenol blue
EZview™ Protein G agarose	Sigma, cat# E3403	
EZview™ Red Anti-FLAG Affinity agarose	Sigma, cat# F2426	
[γ-^{33}P]ATP	Perkin Elmer, cat# NEG602H	3,000 Ci/mmol, 3.3 μM
3× flag peptide	Sigma, cat# F4799	
LRRKtide	Enzo life sciences, cat# BML-P267	
GelCode Blue	Thermo Scientific, cat# 24592	
Protease inhibitors	Roche, cat# 04693124001	
Phosphates inhibitors	Thermo scientific, cat# 78426	

Procedure:

1. Transfect HEK293FT cells with LRRK2 plasmids and express protein for 48 h as in the protocol above.
2. Aspirate media and wash cells with 1×PBS.
3. Resuspend cells in 1 ml of lysis buffer per 15 cm dish. Add protease and phosphatase inhibitor cocktail to lysis buffer fresh before use.
4. Incubate lysates on ice for 30 min.
5. Centrifuge lysates at 15,000×*g* for 10 min.
6. During centrifugation, aliquot 40 μl of EZview™ Protein G beads into 1.5 ml Eppendorf tubes and wash them twice with lysis buffer. After removing the supernatant, set the tube with equilibrated red bead pellet on ice.
7. Add centrifuged lysates to the 40 μl of washed EZview™ Protein G beads to preclear lysates from nonspecifically bound protein to the Protein G agarose. Vortex briefly and rotate them for at least 30 min at 4°C.
8. During preclear step, rinse EZview™ Red Anti-FLAG Affinity agarose by lysis buffer two times.
9. After preclear, collect Easy view Protein G beads by centrifugation at 15,000 x g for 1 min.
10. Carefully transfer the clear lysate supernatant to the new 1.5 ml tubes containing 40 μl of washed EZview™ Red Anti-FLAG Affinity agarose. Vortex gently and incubate with gentle mixing

for 2 h to overnight by rotating at 4°C to allow FLAG-tagged proteins to bind to the anti-FLAG antibody on the EZview™ Red Anti-FLAG Affinity agarose.

11. Centrifuge the immunoprecipitated complexes in a refrigerated microcentrifuge at 5,000×*g* for 1 min. Aspirate the supernatant carefully or remove it with the micropipette. Set tubes with the beads on ice.
12. Wash the bead pellet by adding 1 ml of prechilled wash buffer (wash buffer: Tris 25 mM pH 7.5, 400 mM NaCl, 1% Triton x100, recommended to store at 4°C) and mixing beads by inverting a tube several times. Collect beads by centrifugation at 5,000×*g* for 1 min. Remove supernatant by micropipette or aspirate carefully. Remember to keep tubes with the bead pellet on ice.
13. Repeat wash step four times.
14. Rinse proteins on beads twice in elution buffer: 1× kinase buffer with 150 mM NaCl; and 0.02% Triton x100. Addition of this small amount triton to the elution buffer is critical for LRRK2 elution and results in higher yield of the recombinant protein.
15. Elute proteins from beads by adding three volumes of elution buffer with 100 μg/ml of 3× flag peptide, and rotating on wheel for 15–20 min at 4°C.
16. Spin down the beads and collect the supernatant with eluted protein. A second elution is recommended since it gives about 50% yield of the first elution.
17. Analyze protein by SDS-PAGE on 4–20% gel followed by GelCode blue staining reagent (with BSA standard). Quantify the LRRK2-3× FLAG concentration for each sample based on BSA standard curve.
18. As a control for elution, keep the eluted beads and include also in the gel analysis: little or no protein should be on the beads if the elution has worked well.

Kinase assay

1. Put the same amount of LRRK2 (total 150–300 ng of purified protein) for each sample in 1.5 ml tube and bring volume up to 50 ul with 1× kinase buffer.
2. Prepare the reaction master mixture of radiolabeled and unlabeled ATP and LRRKtide substrate:

Item	Start conc	End conc in 60 μl rxn	Per rxn
ATP	1 mM	10 μM	0.6 μl
ATP–P32	3.3 μM	50 nM	0.9 μl (=6 μCi)
Peptide substrate	10 mM	200 μM	1,2 μl
Kinase buffer	10×	1×	1 μl
Water			6.3 μl

3. Add 10 μl of master mix to each sample and incubate the reaction at 30°C by shaking at 1,400 rpm.
4. At specific time points (0, 5, 15, 30, 60 min) remove 10 μl aliquot from the reaction and stop the reaction by mixing with 5 μl of 0.5 M EDTA with bromophenol blue.
5. Analyze reaction by spotting 20 μl of reaction mixture on P81 membrane and allow proteins/peptide bind to P81 paper.
6. Allow P81 phosphocellulose filter to dry.
7. Wash four times for 10 min each in 75 mM phosphoric acid.
8. Put dried P81 paper with the bound peptides in the scintillation vials.
9. Add 10 ml of Scintillation counting cocktail to each sample. Mix well to allow the peptide to be dissolved in the cocktail.
10. Monitor radioactivity of the samples using a Scintillation Counter.

3.4. Metabolic Radiolabeling

Reagents and buffers required:

Name	Supplier, cat#	Ingredients
1× Lysis buffer	Cell Signaling, cat# 9803	20 mM Tris–HCl pH 7.5; 150 mM NaCl; 1 mM Na_2EDTA; 1 mM EGTA; 1% Triton x100; 2.5 mM sodium pyrophosphate; 1 mM b-glycerophosphate; 1 mM Na_3VO_4; 1 μg/ml leupeptin
Wash buffer		25 mM Tris pH 7.5, 400 mM NaCl, 1% Triton x100
Phosphate-free DMEM	Invitrogen, cat# 11971	
EZview™ Protein G agarose	Sigma, cat# E3403	
EZview™ Red Anti-FLAG Affinity agarose	Sigma, cat# F2426	
Phosphorus-32 radionuclide disodium phosphate	Perkin Elmer, cat# NEX011001MC	
Protease inhibitors	Roche, cat# 04693124001	
Phosphates inhibitors	Thermo scientific, cat# 78426	

1. HEK293FT cells with LRRK2/PINK1 plasmid DNA using Lipofectamine-2,000 reagent in a 6-well plate. Use 2 μg of plasmid DNA per well and perform transfection-trypsinization protocol for higher transfection efficiency.
2. Thirty six to forty eight hours after transfection wash cells twice with 2 ml of phosphate-free medium. This is an important step for removal of any free phosphate in the medium.
3. Add 0.75 ml phosphate-free DMEM to each well and bring the plate in the radioactive work designated area.
4. Prepare p32-DMEM labeling mix per sample by combining 200 μl of phosphate-free DMEM and 50 μl of 20 mCi/ml of p32-ortophosphate (50 μCi/well). It is necessary to prepare a master mix for multiple samples.
5. Add 0.25 ml of labeling mix to each well and mix gently. Final volume should be 1 ml.
6. Incubate cells for 4 h in a CO_2 incubator designated for radioactive work.
7. While the cells are incubating, prepare lysis and wash buffers and keep them on ice.
8. After incubation of the cells, discard the medium containing 32P by an aspirator.
9. Wash cells once with phosphate-free DMEM and twice with 1 ml of TBS buffer containing phosphatase inhibitors.
10. Add 400 μl of the lysis buffer per well and collect lysates into 1.5 ml tubes.
11. Incubate lysates on ice for 30 min.
12. During incubation, aliquot 30 μl of EZview™ Protein G beads into new 1.5 ml tubes and keep them on ice. Do NOT add antibodies!
13. Centrifuge lysates for 10 min at 12,000 rpm at 4°C.
14. Transfer the supernatants to the tubes containing the EZview™ Protein G beads.
15. Rotate the tubes for 30–60 min at 4°C.
16. During rotation, aliquot 20 μl of EZview™ Red Anti-FLAG Affinity (protein G with flag antibodies). Keep them on ice.
17. Centrifuge the tubes for 1 min at 15,000 rpm at 4°C.
18. Transfer the supernatants to the tubes containing FLAG antibody and the Protein G beads.
19. Rotate the tubes for 2 h at 4°C.
20. Centrifuge the tubes for 2 min at 3,000 rpm at 4°C.
21. Discard the supernatants by aspiration.

22. Add 1 ml of the wash buffer to each tube and mix well by inverting them 5 times.
23. Repeat the washing step 4 times.
24. After washing the beads, discard the supernatants by an aspirator.
25. Add 30 μl of 1×SDS-PAGE sample buffer to the beads and vortex them briefly.
26. Boil the tubes for 5 min at 95°C.
27. Centrifuge the tubes for 1 min at 12,000 rpm at room temperature.
28. Use 20 μl of the supernatants for SDS-PAGE and immunoblotting.
29. After immunoblotting, expose the membrane to phospho-screen and read using a phosphorimager.

3.5. Data Analysis

For autophosphorylation, radioactivity incorporated is normalized to relative amounts of the kinase by western blots or using Coomassie staining, if there is enough protein. For peptide assays, one can measure radioactivity (either as CPM or converted to moles of phosphate) incorporated in a given time or as a rate per minute. These measures tend to be distributed approximately normally and so each point measure can be analyzed using t-tests if there are only two groups or, more commonly, using one-way ANOVA with appropriate post hoc tests if there are more than two groups.

4. Notes

4.1. Purification and Assay of GST-Tagged PINK1 Kinase Domain

1. PINK1 is not a very stable protein and is susceptible to degradation, therefore several tips should be considered during purification:
   (a) Lysis and washing buffers should be prepared in advance and stored at 4°C. Protease inhibitors must be added immediately before purification.
   (b) Sonicate bacterial pellets on ice. Oversonication can lead to overheating, resulting in decrease in protein solubility and co-purification of host proteins. If the sample starts to heat during the sonication, it is suggested to stop and continue sonication after the sample cools down. Avoid frothing. Cell disruption is evidenced by partial clearing of the suspension. Do not allow the tip of the sonicator to touch the sides or bottom of the tube or leave the sonicator on without the tip being submerged.

(c) Perform all purification steps on ice and use a refrigerated centrifuge. It is suggested to elute protein at 4°C.

2. For kinase assay, the PINK1 protein should be used right after the purification since freezing generally results in loss of PINK1 kinase activity.

3. The 5× kinase buffer can be prepared in advance. However, aliquoting of 5× kinase buffer is recommended if many small experiments are to be performed and should be stored at −20°C. Do not re-freeze 5× kinase buffer due to DTT instability.

4. Work in designated radioactivity area equipped with shielding, Geiger counter, waste collection. Remember to follow your institutions rules for working with radioactive materials, and apply ALARA principles when working with high energy beta emitters such as ^{32}P.

5. It is important to remember that the gel and the membrane need to be dried after the transfer of radioactive proteins because even small amounts of water can stop beta radiation from reaching the film or phosphoscreen.

4.2. In Vitro Autophosphorylation Kinase Assay with Immunoprecipitated Full Length 2×Myc–LRRK2

1. It is recommended for LRRK2 protein expression to use HEK293FT (Invitrogen) or HEK293T (ATCC) cell lines over HEK293 cells since these variants of HEK293 cell line contain the SV40 Large T-antigen allowing the replication of transfected plasmids containing the SV40 origin of replication. Further plasmid propagation within these cells lines after cell division results in extended temporal expression of the desired protein and higher protein yield.

2. When you purify recombinant protein from cells remember to:

   (a) Add protease and phosphatase inhibitors to buffers just before the purification. Do not add inhibitors days or weeks before they are used.

   (b) Whenever possible, work at low temperatures, on ice or at 4°C, where protease and phosphotase activity is much lower.

3. To produce high purity protein remember to:

   (a) Preclear lysates with the protein G agarose. This is critical step that helps to reduce nonspecific binding of proteins to the agarose.

   (b) Increase number of washing steps and time for each wash.

4. Prior to performing kinase assay wash your immunoprecipitated proteins on beads with kinase buffer. This helps to remove

chemicals present in the lysis and washing buffers that might potentially inhibit kinase activity of LRRK2.

5. An important consideration for the kinase assay is total ATP concentration. We find that 5–10 μCi radiolabeled ATP and 25 μM unlabeled ATP is optimal. Using only hot ATP means that the ATP concentration is far below the Km of LRRK2 and leads to less total incorporation.
6. It is important to include kinase dead mutant (K1906M/A or D1994A) as a negative control in your experiments. Another helpful control that is LRRK2 G2019S mutant that is twofold more active than wild type.

4.3. LRRK2-3× FLAG Purification and LRRKtide Assay

1. Protein purification tips for LRRK2-3× FLAG proteins are similar to purification of 2×Myc–LRRK2 proteins and can be found in Sect. 4.2. Since LRRKtide assay requires more LRRK2 protein than autophosphorylation assay we suggest purifying LRRK2 from a bigger dish (15 cm dish).
2. We found that LRRK2 elution from flag beads is significantly higher if 0.02% triton concentration is used in elution buffer. We also found that higher than 0.02% triton concentration reduces LRRK2 elution from beads. Two elutions are required to get most of the recombinant LRRK2 from beads.
3. Aliquots of the stocks for 3× flag peptide (5 mg/ml, 20 μl) and LRRKtide peptides (10 mM) can be prepared in advance and stored at −80°C.

4.4. Metabolic Radiolabeling

1. It is critical to use poly-D lysine coated plates for the metabolic labeling experiments since cells are better attached to the plate and multiple media and buffer changes will not result in the cell detachment from the plate.
2. Before you start the metabolic labeling experiments you may need to get the permission from your radiation safety officer to use the designated tissue culture incubator.
3. When you perform metabolic labeling use phosphate free DMEM. Do not use phosphate buffers.
4. It is extremely important to use phosphatase inhibitors in lysis and washing buffers during immunoprecipitation procedure. Absence of phosphatase inhibitors may result in protein dephosphorylation by phosphatases during IP procedure.

Acknowledgments

This research was supported by the Intramural Research Program of the NIH, National Institute on Aging.

References

1. Hardy J (2010) Genetic analysis of pathways to Parkinson disease. Neuron 68:201–206
2. Zhou C, Huang Y, Shao Y et al (2008) The kinase domain of mitochondrial PINK1 faces the cytoplasm. Proc Natl Acad Sci USA 105:12022–12027
3. Valente EM, Abou-Sleiman PM, Caputo V et al (2004) Hereditary early-onset Parkinson's disease caused by mutations in PINK1. Science 304:1158–1160
4. Marin I, Van Egmond WN, Van Haastert PJ (2008) The roco protein family: a functional perspective. FASEB J 22:3103–3110
5. Paisan-Ruiz C, Jain S, Evans EW et al (2004) Cloning of the gene containing mutations that cause PARK8-linked Parkinson's disease. Neuron 44:595–600
6. Zimprich A, Biskup S, Leitner P et al (2004) Mutations in LRRK2 cause autosomal-dominant parkinsonism with pleomorphic pathology. Neuron 44:601–607
7. Beilina A, Van Der Brug M, Ahmad R et al (2005) Mutations in PTEN-induced putative kinase 1 associated with recessive parkinsonism have differential effects on protein stability. Proc Natl Acad Sci USA 102:5703–5708
8. Kawajiri S, Saiki S, Sato S et al (2011) Genetic mutations and functions of PINK1. Trends Pharmacol Sci
9. Clark IE, Dodson MW, Jiang C et al (2006) Drosophila pink1 is required for mitochondrial function and interacts genetically with parkin. Nature 441:1162–1166
10. Dagda RK, Cherra SJ 3rd, Kulich SM et al (2009) Loss of PINK1 function promotes mitophagy through effects on oxidative stress and mitochondrial fission. J Biol Chem 284: 13843–13855
11. Exner N, Treske B, Paquet D et al (2007) Loss-of-function of human PINK1 results in mitochondrial pathology and can be rescued by parkin. J Neurosci 27:12413–12418
12. Gautier CA, Kitada T, Shen J (2008) Loss of PINK1 causes mitochondrial functional defects and increased sensitivity to oxidative stress. Proc Natl Acad Sci USA 105:11364–11369
13. Gegg ME, Cooper JM, Schapira AH et al (2009) Silencing of PINK1 expression affects mitochondrial DNA and oxidative phosphorylation in dopaminergic cells. PLoS One 4:e4756
14. Gispert S, Ricciardi F, Kurz A et al (2009) Parkinson phenotype in aged PINK1-deficient mice is accompanied by progressive mitochondrial dysfunction in absence of neurodegeneration. PLoS One 4:e5777
15. Grunewald A, Gegg ME, Taanman JW et al (2009) Differential effects of PINK1 nonsense and missense mutations on mitochondrial function and morphology. Exp Neurol 219: 266–273
16. Hoepken HH, Gispert S, Morales B et al (2007) Mitochondrial dysfunction, peroxidation damage and changes in glutathione metabolism in PARK6. Neurobiol Dis 25:401–411
17. Morais VA, Verstreken P, Roethig A et al (2009) Parkinson's Disease mutations in PINK1 result in decreased complex I activity and deficient synaptic function. EMBO Mol Med 1:99–111
18. Park J, Lee SB, Lee S et al (2006) Mitochondrial dysfunction in drosophila PINK1 mutants is complemented by parkin. Nature 441:1157–1161
19. Sandebring A, Thomas KJ, Beilina A et al (2009) Mitochondrial alterations in PINK1 deficient cells are influenced by calcineurin-dependent dephosphorylation of dynamin-related protein 1. PLoS One 4:e5701
20. Cui T, Fan C, Gu L et al (2011) Silencing of PINK1 induces mitophagy via mitochondrial permeability transition in dopaminergic MN9D cells. Brain Res 1394:1–13
21. Gegg ME, Cooper JM, Chau KY et al (2010) Mitofusin 1 and mitofusin 2 are ubiquitinated in a PINK1/parkin-dependent manner upon induction of mitophagy. Hum Mol Genet 19:4861–4870
22. Geisler S, Holmstrom KM, Skujat D et al (2010) PINK1/Parkin-mediated mitophagy is dependent on VDAC1 and p62/SQSTM1. Nat Cell Biol 12:119–131
23. Geisler S, Holmstrom KM, Treis A et al (2010) The PINK1/Parkin-mediated mitophagy is compromised by PD-associated mutations. Autophagy 6:871–878
24. Heeman B, Van Den Haute C, Aelvoet SA et al (2011) Depletion of PINK1 affects mitochondrial metabolism, calcium homeostasis and energy maintenance. J Cell Sci 124:1115–1125
25. Jin SM, Lazarou M, Wang C et al (2010) Mitochondrial membrane potential regulates PINK1 import and proteolytic destabilization by PARL. J Cell Biol 191:933–942
26. Kawajiri S, Saiki S, Sato S et al (2010) PINK1 is recruited to mitochondria with parkin and associates with LC3 in mitophagy. FEBS Lett 584:1073–1079
27. Matsuda N, Sato S, Shiba K et al (2010) PINK1 stabilized by mitochondrial depolarization recruits Parkin to damaged mitochondria and activates latent Parkin for mitophagy. J Cell Biol 189:211–221

28. Michiorri S, Gelmetti V, Giarda E et al (2010) The Parkinson-associated protein PINK1 interacts with Beclin1 and promotes autophagy. Cell Death Differ 17:962–974
29. Narendra DP, Jin SM, Tanaka A et al (2010) PINK1 is selectively stabilized on impaired mitochondria to activate Parkin. PLoS Biol 8:e1000298
30. Rakovic A, Grunewald A, Kottwitz J et al (2011) Mutations in PINK1 and Parkin impair ubiquitination of mitofusins in human fibroblasts. PLoS One 6:e16746
31. Rakovic A, Grunewald A, Seibler P et al (2010) Effect of endogenous mutant and wild-type PINK1 on Parkin in fibroblasts from Parkinson disease patients. Hum Mol Genet 19:3124–3137
32. Vives-Bauza C, Zhou C, Huang Y et al (2010) PINK1-dependent recruitment of Parkin to mitochondria in mitophagy. Proc Natl Acad Sci USA 107:378–383
33. Greggio E, Cookson MR (2009) Leucine-rich repeat kinase 2 mutations and Parkinson's disease: three questions. ASN Neuro 1
34. Lee BD, Shin JH, Vankampen J et al (2010) Inhibitors of leucine-rich repeat kinase-2 protect against models of Parkinson's disease. Nat Med 16:998–1000
35. Greggio E, Jain S, Kingsbury A et al (2006) Kinase activity is required for the toxic effects of mutant LRRK2/dardarin. Neurobiol Dis 23:329–341
36. Smith WW, Pei Z, Jiang H et al (2006) Kinase activity of mutant LRRK2 mediates neuronal toxicity. Nat Neurosci 9:1231–1233
37. Deng X, Dzamko N, Prescott A et al (2011) Characterization of a selective inhibitor of the Parkinson's disease kinase LRRK2. Nat Chem Biol 7:203–205
38. Liu M, Poulose S, Schuman E et al (2010) Development of a mechanism-based high-throughput screen assay for leucine-rich repeat kinase 2–discovery of LRRK2 inhibitors. Anal Biochem 404:186–192
39. Jaleel M, Nichols RJ, Deak M et al (2007) LRRK2 Phosphorylates moesin at threonine-558: characterization of how Parkinson's disease mutants affect kinase activity. Biochem J 405:307–317
40. Nichols RJ, Dzamko N, Hutti JE et al (2009) Substrate specificity and inhibitors of LRRK2, a protein kinase mutated in Parkinson's disease. Biochem J 424:47–60
41. Anand VS, Reichling LJ, Lipinski K et al (2009) Investigation of leucine-rich repeat kinase 2: enzymological properties and novel assays. FEBS J 276:466–478
42. Covy JP, Giasson BI (2009) Identification of compounds that inhibit the kinase activity of leucine-rich repeat kinase 2. Biochem Biophys Res Commun 378:473–477
43. Liu M, Dobson B, Glicksman MA et al (2010) Kinetic mechanistic studies of wild-type leucine-rich repeat kinase 2: characterization of the kinase and GTPase activities. Biochemistry 49:2008–2017
44. Lovitt B, Vanderporten EC, Sheng Z et al (2010) Differential effects of divalent manganese and magnesium on the kinase activity of leucine-rich repeat kinase 2 (LRRK2). Biochemistry 49:3092–3100
45. Pedro L, Padros J, Beaudet L et al (2010) Development of a high-throughput AlphaScreen assay measuring full-length LRRK2(G2019S) kinase activity using moesin protein substrate. Anal Biochem 404:45–51
46. Gloeckner CJ, Boldt K, Von Zweydorf F et al (2010) Phosphopeptide analysis reveals two discrete clusters of phosphorylation in the N-terminus and the Roc domain of the Parkinson-disease associated protein kinase LRRK2. J Proteome Res 9:1738–1745
47. Greggio E, Taymans JM, Zhen EY et al (2009) The Parkinson's disease kinase LRRK2 autophosphorylates its GTPase domain at multiple sites. Biochem Biophys Res Commun 389:449–454
48. Kamikawaji S, Ito G, Iwatsubo T (2009) Identification of the autophosphorylation sites of LRRK2. Biochemistry 48:10963–10975
49. Pungaliya PP, Bai Y, Lipinski K et al (2010) Identification and characterization of a leucine-rich repeat kinase 2 (LRRK2) consensus phosphorylation motif. PLoS One 5:e13672
50. Mills RD, Sim CH, Mok SS et al (2008) Biochemical aspects of the neuroprotective mechanism of PTEN-induced kinase-1 (PINK1). J Neurochem 105:18–33

Chapter 13

Imaging PKA Activation Inside Neurons in Brain Slice Preparations

Marina Brito, Elvire Guiot, and Pierre Vincent

Abstract

Cyclic-AMP dependent protein kinase (PKA) is present in most branches of the living kingdom, and as an example in the nervous system where PKA integrates the cellular effects of various neuromodulators. The recent development of fluorescence resonance energy transfer (FRET) biosensors such as AKAR now allows a direct measurement of PKA activation in living cells by simply ratioing the fluorescence emission at the CFP and YFP wavelength. This novel approach provides data with a temporal resolution of a few seconds at the cellular and even sub-cellular level, opening a new possibility for understanding the integration processes in space and time.

Our protocol was optimized to study morphologically intact mature neurons and we describe how simple and cheap wide-field imaging as well as more elaborate two-photon imaging allow real-time monitoring of PKA activation in pyramidal cortical neurons in neonate rodent brain slices. In addition, many practical details presented here also pertain to image analysis in other cellular preparations such as cultured cells. This protocol also applies to the various other CFP–YFP-based FRET biosensors for other kinases or other intracellular signals, and it is likely that this kind of approach will generalize in a near future.

Key words: Protein kinase A, Biosensor, FRET, Time-lapse imaging, Fluorescence, Brain slice preparation

1. Introduction

The signaling pathway of cyclic adenosine monophosphate (cAMP) and protein kinase A (PKA) is involved in virtually all physiological processes. More specifically in the nervous system, the cAMP/PKA cascade plays an important role in several integrated brain functions such as neuronal survival, axonal regeneration, memory and cognitive functions. At the cellular level, the cAMP/PKA cascade is responsible for the modulation of various processes including some specific forms of synaptic plasticity, control of excitability,

Hideyuki Mukai (ed.), *Protein Kinase Technologies*, Neuromethods, vol. 68,
DOI 10.1007/978-1-61779-824-5_13, © Springer Science+Business Media, LLC 2012

regulation of nuclear factors and imprinting of long-term changes. One major target of cAMP is the cAMP-dependent protein kinase (PKA), hence the importance of monitoring its activation. Besides PKA, cAMP also affects the exchange proteins directly activated by cAMP (Epac1 and Epac2) and directly modulates several channels like cyclic-nucleotide gated channels and I_h.

The cAMP/PKA signal, as many other cellular signals, takes place in the complex multi-dimensional space of the living cell. The temporal dimension is obviously a critical one, as the PKA signal rises following a stimulus and then declines over time as the stimulus is removed and/or feedback mechanisms revert the signal toward baseline. The spatial dimension also plays a role since the concentration of signaling molecules decreases with distance. Protein–protein interactions, such as those mediated by AKAPs, maintain all partners of the signaling cascade in close proximity in the so-called signaling microdomains (1). In addition, the geometry of the cell can determine signal integration, with a high surface over volume ratio favoring the accumulation of cAMP and the activation of PKA in thin dendrites versus cell body (2). Temporal and spatial measurements are therefore of critical importance in the understanding of intracellular signaling and direct imaging of the cAMP/PKA signaling cascade therefore seems a very needed approach.

The first attempts to image changes in cAMP concentration in neurons made use of fluorescently labeled PKA (3) to study cAMP diffusion in giant Aplysia neurons (4), then in the intact stomatogastric ganglion of the spiny lobster (5). This first bio-chemical sensor remained however impractical to use in vertebrate neurons because it had to be introduced as a whole reconstituted holoenzyme into the cells of interest, a difficult task for vertebrate neurons (6). A major breakthrough in this field resulted from the creation of fluorescence resonance energy transfer (FRET)-based genetically encoded optical sensors, the first sensors being designed to report calcium (7, 8), eventually leading to sensors for various other biological signals.

Single wavelength sensors have been made which use the intrinsic sensitivity of GFP to pH or chloride, a property amplified by mutating the GFP sequence. They can also have an external sensor grafted to it, such as for the GCamp series of calcium biosensors. However, changes in fluorescence emission measured at a single wavelength can result from various artifacts, technical (fluctuations in illumination intensity, focus drift) or related to the cell (changes in cell volume, hence sensor concentration in the imaged volume). Ratiometric quantification can reduce these artifacts: the measurement is done simultaneously at two wavelength where the intensity change relative to the biological signal of interest goes in opposite directions, whereas artifactual changes affect the two wavelength by a same factor, canceled in the ratio. This quantification method typically applies to most FRET biosensors.

FRET biosensors are commonly constituted of a domain sensitive to the biological signal of interest sandwiched between a pair of fluorophores, usually CFP and YFP or their close derivatives (9). The signal triggers a conformational change of the biosensor, modifying the FRET from CFP to YFP, thus changing the CFP and YFP fluorescence in opposite directions. The last decade has seen the creation of a number of biosensors based on this design for biological signals ranging from the detection of the neurotransmitter glutamate (10, 11) to the monitoring of apoptosis (12). After the original CFP–YFP pair, new fluorophores are progressively being used, resulting in improved ratio changes (13–15). Many biosensors have been created on this scheme, and concerning the cAMP/PKA cascade, although this will not be discussed any further in this chapter, there are various sensors which indicate changes in cAMP concentration. We will now focus on one series of biosensors specifically developed to monitor PKA-mediated phosphorylation events.

A-kinase activity reporter (AKAR) is a recombinant protein composed of a phosphoamino acid binding domain and PKA-specific substrate fused between CFP and YFP. When phosphorylated by PKA, intramolecular binding of the substrate by the phosphoamino acid binding domain drives a conformational reorganization, leading to an increase in FRET between CFP and YFP. Variants with increasingly better signals were produced over the years and are reviewed in (16). Briefly, AKAR2.2 has a CFP as donor and a citrine as an acceptor and exhibits a ~25% ratio change upon PKA phosphorylation (17). AKAR3 has a circularly permuted Venus (cpV E172) as an acceptor with ~40% ratio change (18). Finally, replacing CFP with cerulean led to AKAR4, with ratio changes up to 68% (19).

Although commonly dubbed "PKA sensors," one must keep in mind that the phosphorylation level of the biosensor results from the equilibrium between PKA-dependent phosphorylation and dephosphorylation performed by various phosphatases. An increase in AKAR signal can therefore result either from an activation of PKA, or from an inhibition of phosphatases in the context of a tonic PKA activity. This has been observed experimentally in neurons in brain slices where Gs-coupled receptors, as well as phosphatase inhibition, led to an increase in AKAR phosphorylation level (20). Since the ratio recovered slowly after agonist removal (20), one can assume that the kinetics of phosphatases-mediated dephosphorylation was much slower than PKA-mediated phosphorylation, and as long as this condition is verified, one can assume that AKAR mainly reports PKA activity.

AKAR thus opened the possibility to quantify the relative amplitude of the PKA signal obtained in different situations. For example, we measured the response to different neuropeptides, reporting a relative order of potency of the receptors CRF1 ≅ PAC1 > VPAC1 in their ability to activate the cAMP/PKA signaling cascade in pyramidal cortical neurons (21). AKAR imaging also provided key experimental data in the temporal and spatial dimensions: the comparison of AKAR responses kinetics (indicating PKA activation in the bulk cytosol) with PKA effect measured electrophysiologically (reporting PKA activation underneath the membrane) demonstrated that neurons have a sub-membrane domain where full PKA activation occurs in ~30 s, whereas it takes 2.5 min (5× longer) to reach the maximum phosphorylation level in the bulk cytosol (20). The potential of optical biosensor for spatial resolution was later used in combination with two-photon microscopy and revealed that cAMP accumulates in dendrites of small diameter while its concentration fades away in the bulk cytosol, an effect which most likely results from a higher surface over volume ratio in small dendrites, favoring cAMP concentration buildup (22). Although this had been predicted theoretically (2), imaging provided a clear experimental confirmation of the importance of cell geometry on signal integration.

The spatial dimension can also be access using targeting domains to target the probe toward a specific sub-cellular compartment. AKAR2.2 fused with a nuclear localization signal at its C-terminal was thus used to monitor the kinetics of PKA activation inside the nucleus of neurons in brain slices, and reported a much slower response than in the cytosol (20).

As described above, the spatial organization of the cell of interest directly determines signal integration processes. The biological preparation is therefore an important point, particularly in the case of neurons. Our choice has been to study mature neurons in their most physiological contex, i.e., in brain slice preparations made from rodent neonates. This preparation, derived from the work of electrophysiologists, has the advantage of providing morphologically intact differenciated neurons, which keep functional network connections and bear endogenous receptors. This chapter describes the details of AKAR imaging in brain slices, emphasizing some technical points like image acquisition and image analysis. Most informations presented here also pertain to imaging in any other preparation including cell cultures, and potentially to many other dual-emission biosensors.

2. Materials

2.1. Solutions

Cutting solution contains: 125 mM NaCl, 0.4 mM $CaCl_2$, 1 mM $MgCl_2$, 1.25 mM NaH_2PO_4, 26 mM $NaHCO_3$ and 25 mM glucose, saturated with 5% CO_2 and 95% O_2 (see Note 1).

Culture medium is made of 50% Minimum Essential Medium, 50% Hanks' Balanced Salt Solution, 6.5 g/l glucose, penicillin–streptomycin (Invitrogen).

Recording solution contains: 125 mM NaCl, 2 mM $CaCl_2$, 1 mM $MgCl_2$, 1.25 mM NaH_2PO_4, 26 mM $NaHCO_3$ and 25 mM glucose, saturated with 5% CO_2 and 95% O_2.

A 10× stock solution of 1.25 M NaCl, 12.5 mM NaH_2PO_4, 260 mM $NaHCO_3$ is made weekly and kept at 4°C. On the day of the experiment, 1× solution is prepared by diluting the stock and adding $CaCl_2$, $MgCl_2$ and glucose to the final concentration indicated above.

2.2. Brain Slice Preparation

Brain slices are prepared from young rodents, rats at post-natal age P10–P14, or mice at P8–P12 (see Note 2). Brain slices are cut using a vibrating blade microtome VT1200S (Leica).

Brain slices are kept on a Millicell-CM membrane PICM0RG50 (Millipore).

2.3. Biosensor and Viral Vector

The PKA-sensitive AKAR biosensors were provided by Jin Zhang, Johns Hopkins School of Medicine, Baltimore, USA.

A control version of the AKAR2.2 sensor was made with a T391A point mutation in the phosphorylation site and kindly provided by Jin Zhang.

The viral vector used in our experiments is derived from the Sindbis virus (23) and provided by Invitrogen, San Diego, CA. The coding sequence of the biosensor was sub-cloned into the viral vector pSinRep5. Viral particles are produced following the guidelines published by Invitrogen.

2.4. Wide-Field Imaging

The imaging setup uses an upright microscope (Olympus BX51WI) equipped with a 20× 0.5 NA, a 40× 0.8NA or 60× 0.9 NA water-immersion objective with a piezoelectric device (P-721 PIFOC, Physik Instrumente GmbH) to remotely control image focus. Epifluorescence is excited using a halogen light source, a shutter (Uniblitz, Vincent and Associates) and a D436/20 filter for excitation (see Note 3). A 450 dclp dichroic mirror separates the emission from the excitation. The emitted light is filtered by alternating the emission filters, HQ480/40 for CFP and D535/40 for YFP, with a filter wheel (Sutter Instruments, Novato, CA, USA). Filters were obtained from Chroma Technology (Brattleboro, VT, USA) or Semrock (Rochester, NY, USA). Images are recorded with a low noise CCD camera (ORCA-AG, Hamamatsu).

Images for shading correction are acquired using a 100 μM solution of Coumarin 343 in a 10 mM sodium phosphate buffer at pH 7, stored in small aliquots kept at –20°C (see Note 4).

2.5. Two-Photon Imaging

Two-photon images were obtained with a custom-built two-photon laser scanning microscope based on an Olympus BX51WI upright microscope with a 60× 0.9 NA water-immersion objective and a Ti:sapphire laser (MaiTai HP; Spectra Physics, Mountain View, CA) tuned to 850 nm wavelength and 12 mW power for CFP excitation. Galvanometric scanners (model 6210, Cambridge Technology, Cambridge, MA) were used for raster scanning, and a piezo-driven objective scanner (P-721 PIFOC, Physik Instrumente GmbH) was used to control the depth of the focal plane in the preparation. A two-photon emission filter was used to reject residual excitation light (E700 SP, Chroma Technology, Brattleboro, VT, USA). A fluorescence cube containing 479/40 and 542/50 emission filters and a 506 nm dichroic beamsplitter (FF01-479/40, FF01-542/50 and FF506-Di02-25 × 36 Brightline Filters, Semrock, Rochester, NY, USA) was used for the orthogonal separation of the two fluorescence signals. Two imaging channels (H9305 photomultipliers, Hamamatsu) were used for simultaneous detection of the two types of fluorescence emission.

2.6. Image Acquisition

The wide-field imaging setup is controlled using iVision-Mac software (BioVision technologies, Exton, USA) and custom scripts. The two-photon imaging system is controlled by MPscope software (24).

2.7. Image Analysis

Image analysis is performed using Igor software (Wavemetrics, USA) and custom-written functions.

3. Methods

3.1. Brain Slice Preparation

The animal is killed by decapitation, following the guidelines of our institution. The brain is quickly removed and placed in ice-cold cutting solution. After cutting, brain slices are left 1 h to recover in cutting solution. The Millicell membrane is placed into a 35 mm petri dish containing 1 ml of culture medium and incubated for 1 h at 35°C (Note 5) in 5% CO_2 atmosphere. Two to three brain slices are carefully placed onto the Millicell membrane and all cutting solution is gently removed. The slices are then left for 1 h in the incubator.

Infection: 1–3 μl of virus suspension is gently spread over the surface of the brain slice with a pipette. The dish is then kept overnight in the incubator.

The next day, brain slices are detached from the Millicell membrane and transferred into a 100 ml beaker containing recording solution gassed with 95% O_2/5% CO_2. Slices are left 1 h in this solution, a time needed to recover a pH/metabolic equilibrium (see Note 6).

The imaging chamber on the microscope is continuously perfused with the recording solution saturated with 95% O_2/5% CO_2 at a rate of 2 ml/min. Smoothly switching between different reservoirs allows for changing the bathing solution to a solution containing drugs, without mechanically disturbing the preparation.

3.2. Wide-Field Imaging

Images (one for each CFP and YFP channel) are acquired with an exposure duration of 0.2–1 s, depending on the preparation expression level, illumination power and camera sensitivity. Image pairs are recorded at time intervals from 5 s to several minutes, depending on the duration of the biological event to be monitored (see Note 3). Besides lateral drift of the brain slice, which can be easily corrected off-line (see below), focus drift must be corrected in real time.

3.3. Two-Photon Imaging

CFP and YFP signals were recorded simultaneously during the scanning of the preparation by the laser beam. The 512 × 512 pixel images were acquired at a rate of 1.6 frames/s. Vertical stacks of ~50 images were recorded with a 0.5 μm increment step.

For both wide-field and two-photon imaging, a total number of 100 data points with negligible photobleaching over 1–2 h duration is common practice.

3.4. Image Analysis

Images saved by the acquisition software are immediately processed by the analysis program and the resulting ratio measurement and pseudocolor images are displayed in real time (Fig. 1). Image analysis comprises a series of steps, based on the description made for ratiometric calcium imaging (25):

- The camera offset is subtracted from each image
- Intensity values are divided by the exposure time
- Each image is divided by the shading image of the corresponding imaging channel (see Note 4)
- An optional background value is then subtracted from the image. Since the autofluorescence of the neurons is usually low compared to the fluorescence of the biosensor, and since this value cannot be determined objectively, this parameter is usually left null
- Images are corrected for preparation movements, typically horizontal translations of several microns (Note 7) (Fig. 1C)
- Images are corrected for misregistration, usually by <2 pixels (Note 8) (Fig. 1D)
- The ratio is calculated pixel by pixel as Ratio = (YFP)/(CFP), with YFP and CFP the images at the two wavelength, after the corrections described above
- Images are displayed in pseudocolor, coding the ratio value in hue and the fluorescence in intensity (see Note 9) (Fig. 1A)

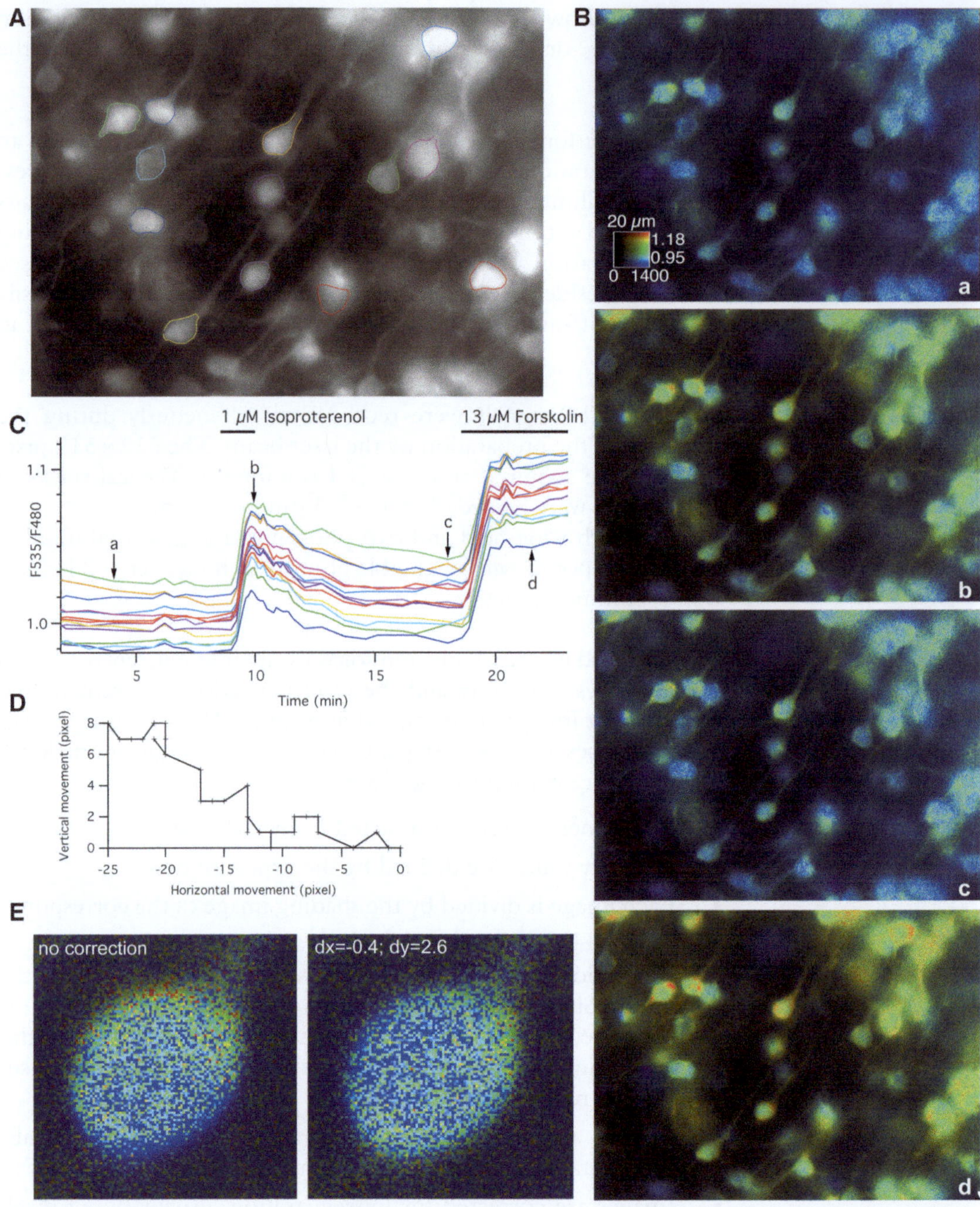

Fig. 1. Monitoring a PKA response resulting from the activation of β1-adrenergic receptors in the cortex using wide-field AKAR2.2 imaging. Brain slice expressing AKAR2.2 viewed in wide-field microscopy. (**A**) Fluorescence image at the 535 emission wavelength. (**B**) Pseudocolor images in control condition (a), in the presence of 1 μM Isoproterenol (b), after recovery (c) and in the presence of 13 μM Forskolin (d). (**C**) Ratio values during the course of the recording, measured over the regions of interest of the same color and drawn on (**A**). *Arrows* indicate the time corresponding to the images in (**B**). (**D**) Horizontal correction of the movement of the brain slice. (**E**) Magnification of a single cell body before (*left*) and after (*right*) correction of misregistration between the F535 and F480 channels. Misregistered ratio images display the characteristic "rainbow" appearance, with one edge with low ratio while the opposite edge has high ratio. The correction is calculated at the 1/5th pixel resolution.

- A calibration scale is added to the image: since pseudocolor images represent two dimensions (ratio and intensity), the calibration becomes a square, showing the intensity of the pixel (in counts per pixel per second) on the horizontal axis and the ratio on the vertical axis. The size of this calibration square also defines the size scale of the image (Fig. 1Ba)
- Regions of interest drawn on the image are used to calculate the average ratio as an average weighted by the intensity (see Note 10) (Fig. 1A)
- There are sometimes non-linear deformation within the brain slice resulting in movements which cannot be corrected by a global translation. Fine local adjustments of the regions of interest compensate for these small distortions

3.5. Controls

CFP and YFP are far from ideal chromophores and both exhibit some pH and environment sensitivities. It is therefore quite useful to check that ratio changes recorded in an experiment are not an artifactual effect. A control biosensor mutated in the phosphorylation site must report no ratio change in the same situation as the experiment and thus demonstrates that the ratio signal indeed reflects a phosphorylation event. The T391A mutant of AKAR2.2 is used systematically to check for a lack of ratio response in the experimental conditions of interest (Fig. 2).

3.6. Two-Photon Imaging

Two-photon image acquisition and image analysis is performed similarly as wide-field imaging (Fig. 3) with the following differences:

- Laser scanning provides images which are intrinsically registered, so there is no need for mis-registration correction
- For each data point, a stack of images is recorded on a thickness of up to 50 μm with 0.5 μm interval. Corrections for preparation movements are applied to the X, Y *and* Z dimensions, thereby correcting for possible focus drift

4. Notes

1. Solutions for cutting: there are various recipes for the cutting buffer, which would maintain the neurons in better shape, containing kynurenate, lower sodium, etc... This may depend on the brain region and should be tested and optimized for each brain region.
2. Age of the animals: although brain slices from young animals are more transparent and have a higher infection rate, both conditions favorable for imaging, it is still possible to do imaging on slices from adult rodents.

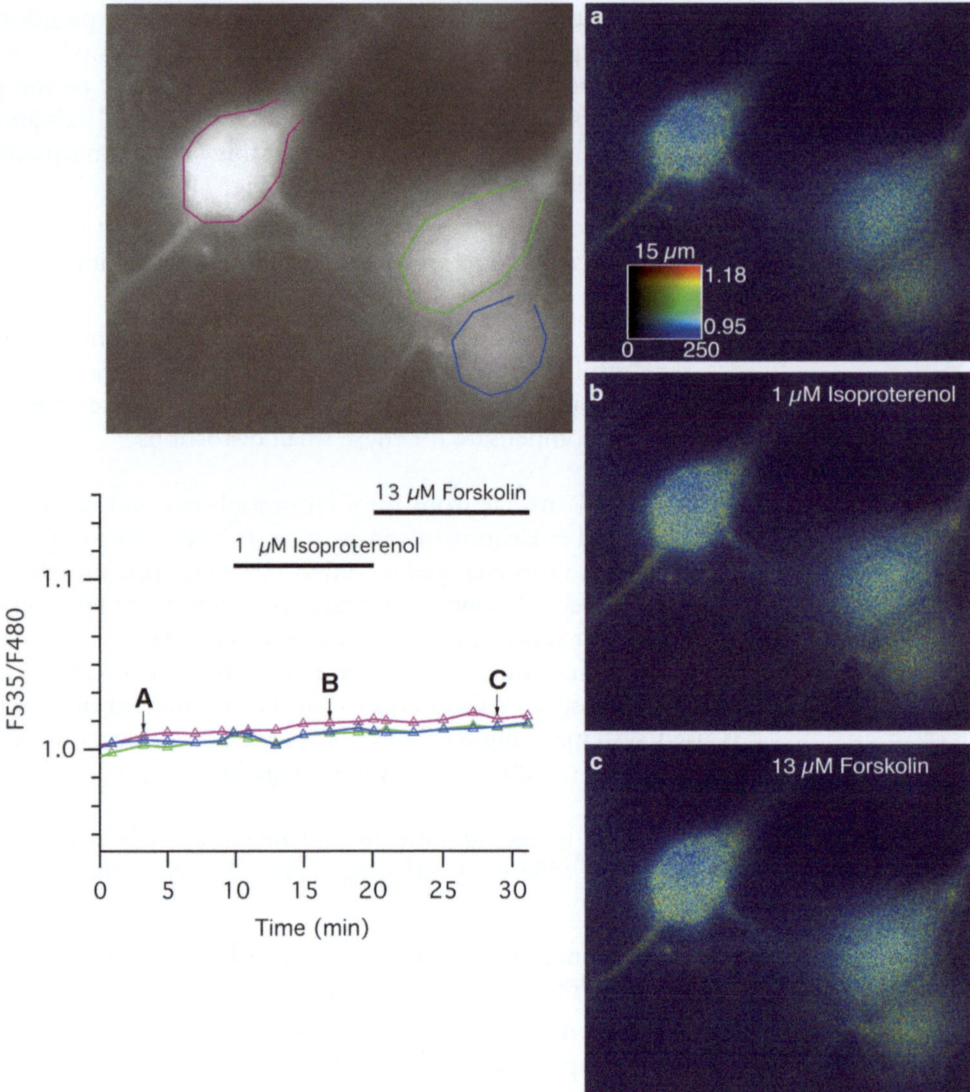

Fig. 2. Lack of response with the T391A mutant of AKAR2.2. Brain slice expressing AKAR2.2mut (T391A) viewed in wide-field microscopy. (**A**) Fluorescence image at the 535 emission wavelength. (**B**) Pseudocolor images in control condition (a), in the presence of 1 μM Isoproterenol (b), and in the presence of 13 μM Forskolin (c). (**C**) Ratio values during the course of the recording, measured over the regions of interest of the same color and drawn on (**A**). *Arrows* indicate the time corresponding to the images in (**B**).

3. Bleaching of the biosensor may dramatically limit the experiment. Unnecessary illumination of the biosensor should be avoided and here are some common causes of useless damage done to the fluorophores.
   - Too high illumination power: epifluorescence imaging is performed quite well with a 100 W halogen light source, sufficient to distinguish the cells by eye and very stable for quantitative imaging. Too strong illumination such as arc

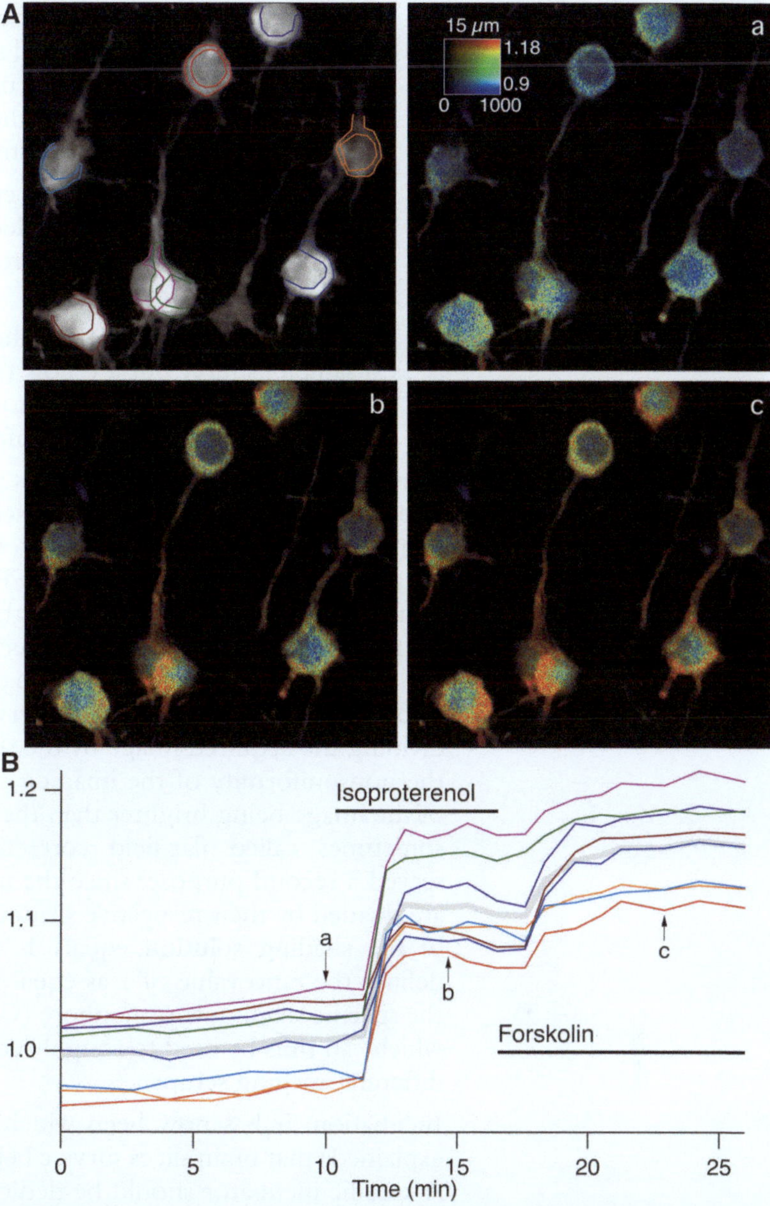

Fig. 3. AKAR2.2 response measured using two-photon microscopy. Brain slice expressing AKAR2.2 viewed in two-photon microscopy. (**A**) *Top left*: vertical projection of the image stack (20 μm thick, 0.5 μm interval) at the 535 emission wavelength. Please note that the projection "mixes" cells from different focal planes, which are actually well separated in the image stack. Pseudocolor images in control condition (a), in the presence of 1 μM Isoproterenol (b) and in the presence of 13 μM Forskolin (c). (**B**) Ratio values during the course of the recording, measured over the regions of interest of the same color and drawn on (**A**). Regions of interest exclude the nucleus, which responds on different kinetics. Measurements are obtained from a "slice" of 4 μm thickness centered on the cell body of the neuron. The *gray line* represents the average of all cells in this experiment. *Arrows* indicate the time corresponding to the images in (**B**).

lamps, if used improperly (i.e., without correct attenuation), can bleach the fluorophores and suppress the ratio change. An indication of this phenomenon is a rapid drift of the ratio baseline as a function of the cumulated illumination duration, associated with very small ratio responses.

- Bad illumination control: the epifluorescence light path must be electronically controlled by a fast shutter, only allowing illumination of the preparation when acquiring images.
- Too strong transillumination while looking at the biological sample: simply put a yellow tainted glass in the transillumination light path in order to prevent bleaching by the blue component of the transillumination light.

4. Shading correction: shading images are images of a uniformly fluorescent object, such as fluorescent solution placed in a hemocytometer. Coumarin 343 (100 μM) is convenient because it emits in both CFP and YFP channels. Since the non-uniformity of the field may be slightly different for the two channels, a shading image is acquired for each detection channel. Data images are divided by the shading image of the corresponding wavelength, which serves two purposes. First, dividing the acquired image by the shading image corrects for the non-uniformity of the imaging system, usually the center of the image being brighter than the edges. This correction is sometimes called flat-field correction. Shading correction served a second purpose: since the images from each channel are divided by their respective shading, by definition, the ratio of the shading solution equals 1. Shading correction thus defines the ratio value of 1 as equal to the ratio measured for the specific shading chromophore (Coumarin 343 in our case) which can thus be used to roughly compare ratio values across different imaging setups.
5. Incubation: it has now been widely reported (although not explained) that brain slices survive better at 35°C than at 37°C. A specific incubator should be dedicated to this temperature. Infections also succeeded at room temperature when performed directly inside the 100 ml beaker: the solution level is adjusted so that the slices are held on a mesh close to the surface, at the interface with air just for the duration of the infection (30 min), then later submerged again and kept overnight, with continuous gassing with 95% O_2/5% CO_2.
6. Baseline drift: drifts of the baseline ratio are sometimes observed during the course of the experiment. This effect can also be observed with AKARmut, showing an effect independent of phosphorylation events, possibly reflecting changes in chromophore properties. Since this effect was reduced when the slices were pre-incubated in the recording medium for ~1 h, it

is suspected to result from changes in pH or chloride concentration, which eventually stabilize in the recording solution.

7. Preparation movement: during the course of the experiment, the biological object often moves as a result of mechanical instability of the microscope or intrinsic movements of the biological object. A brain slice typically moves by a few tens of microns, horizontally and vertically. While the vertical movement must be corrected by adjusting the focus of the microscope, horizontal translation are easy to correct by software, translating the acquired image until it is registered with a reference image. Both image registration and translation are calculated automatically using built-in Igor Pro routines (26).
8. Image registration: since all microscope somehow exhibit some level of chromatic aberration (objective, filter thickness and wedge effect) it is often necessary to register the images from the two channels at the sub-pixel level. This correction is usually determined once and for all for an imaging system and applied to all images acquired on this system.
9. Intensity-modulated display: the color of each pixel is coded in tridimensional space and we use the "hue–saturation-value" coding, representing the ratio with a hue from blue to red, and the fluorescence level with the intensity (25). Saturation is set to 100%. This results in more "natural looking" images, with the most intense regions of the image appearing bright, and the image gradually fading into darker hues and eventually black as the fluorescence intensity in the original image declines toward background. This display mode thus avoids the artificial thresholding of images which show arbitrary sharp edges.
10. Ratio average: the ratio is usually evaluated as a weighted average: each pixel in the region of interest is weighted by the intensity of the fluorescence (25). As a consequence, bright pixels, corresponding to regions of the image containing a large amount of biosensor will contribute more to the average than dim pixels where the biosensor is scarce. This has the advantage that background pixels accidentally contained within a region of interest will be canceled-out in the average. When fluorescence intensity is close to background, the ratio converges toward "0/0" and values can randomly reach arbitrary values from $-\infty$ to $+\infty$. To remove these infinite values, all ratio values are cropped from $0.1 * R_{min}$ to $10 * R_{max}$.

References

1. Tasken K, Aandahl EM (2004) Localized effects of cAMP mediated by distinct routes of protein kinase A. Physiol Rev 84:137–167
2. Neves SR et al (2008) Cell shape and negative links in regulatory motifs together control spatial information flow in signaling networks. Cell 133:666–680. doi:10.1016/j.cell.2008.04.025
3. Adams SR, Harootunian AT, Buechler YJ, Taylor SS, Tsien RY (1991) Fluorescence ratio imaging of cyclic AMP in single cells. Nature 349:694–697. doi:10.1038/349694a0

4. Bacskai BJ et al (1993) Spatially resolved dynamics of cAMP and protein kinase A subunits in Aplysia sensory neurons. Science 260:222–226
5. Hempel CM, Vincent P, Adams SR, Tsien RY, Selverston AI (1996) Spatio-temporal dynamics of cAMP signals in an intact neural circuit. Nature 384:166–169
6. Vincent P, Brusciano D (2001) Cyclic AMP imaging in neurones in brain slice preparations. J Neurosci Methods 108:189–198
7. Miyawaki A et al (1997) Fluorescent indicators for Ca^{2+} based on green fluorescent proteins and calmodulin. Nature 388:882–887
8. Romoser VA, Hinkle PM, Persechini A (1997) Detection in living cells of Ca^{2+}-dependent changes in the fluorescence emission of an indicator composed of two green fluorescent protein variants linked by a calmodulin-binding sequence. A new class of fluorescent indicators. J Biol Chem 272:13270–13274
9. Zhang J, Campbell RE, Ting AY, Tsien RY (2002) Creating new fluorescent probes for cell biology. Nat Rev Mol Cell Biol 3:906–918
10. Dulla C et al (2008) Imaging of glutamate in brain slices using FRET sensors. J Neurosci Methods 168:306–319. doi:10.1016/j.jneumeth.2007.10.017
11. Hires SA, Zhu Y, Tsien RY (2008) Optical measurement of synaptic glutamate spillover and reuptake by linker optimized glutamate-sensitive fluorescent reporters. Proc Natl Acad Sci U S A 105:4411–4416. doi:10.1073/pnas.0712008105
12. Laxman B et al (2002) Noninvasive real-time imaging of apoptosis. Proc Natl Acad Sci U S A 99:16551–16555. doi:10.1073/pnas.252644499
13. Ai HW, Hazelwood KL, Davidson MW, Campbell RE (2008) Fluorescent protein FRET pairs for ratiometric imaging of dual biosensors. Nat Methods 5:401–403. doi:10.1038/nmeth.1207
14. van der Krogt GN, Ogink J, Ponsioen B, Jalink K (2008) A comparison of donor-acceptor pairs for genetically encoded FRET sensors: application to the Epac cAMP sensor as an example. PLoS One 3:e1916. doi:10.1371/journal.pone.0001916
15. Klarenbeek JB, Goedhart J, Hink MA, Gadella TW, Jalink K (2011) A mTurquoise-based cAMP sensor for both FLIM and ratiometric read-out has improved dynamic range. PLoS One 6:e19170. doi:10.1371/journal.pone.0019170
16. Vincent P, Gervasi N, Zhang J (2008) Real-time monitoring of cyclic nucleotide signaling in neurons using genetically encoded FRET probes. Brain Cell Biol 36:3–17. doi:10.1007/s11068-008-9035-6
17. Zhang J, Hupfeld CJ, Taylor SS, Olefsky JM, Tsien RY (2005) Insulin disrupts beta-adrenergic signalling to protein kinase A in adipocytes. Nature 437:569–573. doi:10.1038/nature04140
18. Allen MD, Zhang J (2006) Subcellular dynamics of protein kinase A activity visualized by FRET-based reporters. Biochem Biophys Res Commun 348:716–721
19. Depry C, Allen MD, Zhang J (2010) Visualization of PKA activity in plasma membrane microdomains. Mol Biosyst. doi:10.1039/c0mb00079e
20. Gervasi N et al (2007) Dynamics of protein kinase A signaling at the membrane, in the cytosol, and in the nucleus of neurons in mouse brain slices. J Neurosci 27:2744–2750. doi:10.1523/JNEUROSCI.5352-06.2007
21. Hu E et al (2011) VIP, CRF, and PACAP act at distinct receptors to elicit different cAMP/PKA dynamics in the neocortex. Cereb Cortex 21:708–718. doi:10.1093/cercor/bhq143
22. Castro LR et al (2010) Type 4 phosphodiesterase plays different integrating roles in different cellular domains in pyramidal cortical neurons. J Neurosci 30:6143–6151. doi:10.1523/JNEUROSCI.5851-09.2010
23. Ehrengruber MU et al (1999) Recombinant Semliki Forest virus and Sindbis virus efficiently infect neurons in hippocampal slice cultures. Proc Natl Acad Sci U S A 96:7041–7046
24. Nguyen QT, Tsai PS, Kleinfeld D (2006) MPScope: a versatile software suite for multiphoton microscopy. J Neurosci Methods 156:351–359. doi:10.1016/j.jneumeth.2006.03.001
25. Tsien RY, Harootunian AT (1990) Practical design criteria for a dynamic ratio imaging system. Cell Calcium 11:93–109
26. Thevenaz P, Ruttimann UE, Unser M (1998) A pyramid approach to subpixel registration based on intensity. IEEE Trans Image Process 7:27–41. doi:10.1109/83.650848

Chapter 14

Imaging Oscillations of Protein Kinase C Activity in Cells

Maya T. Kunkel and Alexandra C. Newton

Abstract

The protein kinase C (PKC) family contains ten members that share a catalytic core but differ in the composition of their regulatory modules. The conventional PKCs are a subfamily whose regulatory domains bind to Ca^{2+} and to the lipid second messenger diacylglycerol and thus they are activated by coincidence detection of Ca^{2+} and diacylglycerol at the plasma membrane. In HeLa cells, oscillations in Ca^{2+} evoke oscillations in membrane PKC activity. The only method to date that offers the time resolution to observe these rapid changes in PKC activity is the utilization of a genetically encoded membrane-targeted PKC activity reporter, MyrPalm-CKAR (C Kinase Activity Reporter). CKAR is a fluorescence resonance energy transfer (FRET)-based, modular protein that undergoes a conformational change upon phosphorylation resulting in a change in FRET, thereby serving as a direct readout of cellular kinase activity. As a genetically encoded reporter, it is easily introduced into cells and can be targeted to distinct intracellular compartments through the addition of short targeting sequences. MyrPalm-CKAR is targeted to plasma membranes through the amino-terminal addition of seven amino acids that encode a sequence that is myristoylated and palmitoylated in the cell. This chapter details the method of utilizing MyrPalm-CKAR to monitor acute changes in PKC signaling at the plasma membrane that are a consequence of acute changes in Ca^{2+} levels.

Key words: PKC, FRET, Kinase activity reporter, CKAR, Phorbol esters, Diacylglycerol, Ca^{2+}, Live cell imaging

1. Background and Historical Overview

The PKC family of isozymes is perched on a branch of the kinase tree that contains the AGC kinases (1). The ten members of the PKC family have in common a conserved catalytic core but differ in the composition of their regulatory modules (reviewed in (2)). It is these differences in their regulatory modules that control their regulation by lipid second messengers and Ca^{2+}. Conventional isozymes (PKC α, β, and γ) are regulated by Ca^{2+} via a C2 domain and by diacylglycerol via tandem C1 domains; novel isozymes (PKC δ, ε, θ, η) are regulated by diacylglycerol, and atypical PKC

Hideyuki Mukai (ed.), *Protein Kinase Technologies*, Neuromethods, vol. 68,
DOI 10.1007/978-1-61779-824-5_14, © Springer Science+Business Media, LLC 2012

isozymes (ζ, ι/λ) are regulated by scaffold interactions via a PB1 domain.

PKC isozymes transduce an abundance of signals and play central, and sometimes opposing, roles in cell survival and apoptotic pathways (2, 3). Misregulation of PKC leads to pathogenic states, notably cancer (4). PKC isozymes are highly expressed in the brain, where an abundance of substrates have been identified (5). Expression of the conventional isozyme PKC γ and an alternatively transcribed form of the atypical PKC ζ consisting only of its catalytic domain, PKM ζ, are confined to brain, suggesting a unique role of PKC signaling in brain (6, 7). A number of naturally occurring mutations in PKC γ have been identified in patients with spinocerebellar ataxia type 14, resulting in reduced protein stability and signaling output (8, 9). PKM ζ has been proposed to play a key role in learning and memory (10), although this work remains controversial because experiments to date have depended on the use of pharmacological inhibitors that do not inhibit PKM ζ in cells (11). Another important isozyme in the brain is PKC α, an isozyme that is required for long term depression via a mechanism that depends on scaffolding via its C-terminal PDZ ligand (12).

The advent of genetically encoded kinase activity reporters has provided previously unattainable insight into the spatiotemporal dynamics of PKC activity in cells (13, 14). Targeting the reporter to discrete intracellular locations by fusion of appropriate targeting sequences has allowed visualization of the rate, amplitude, and duration of PKC signaling at the plasma membrane, Golgi membranes, mitochondria, cytosol, and nucleus (15). Furthermore, the generation of an isozyme-specific reporter, δCKAR, has enabled dissection of the role of PKCδ in second messenger signaling (16).

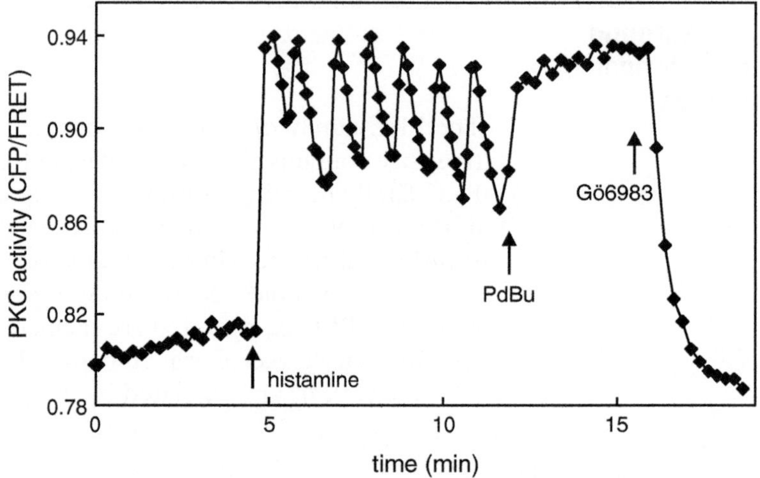

Fig. 1. The CFP/FRET ratio from MyrPalm-CKAR oscillates after stimulation of HeLa cells with 10 μM histamine. Subsequent addition of 200 nM phorbol-12,13-dibutyrate (PdBu) maximally activates PKCs and the FRET change from CKAR. The FRET ratio from CKAR reverses once PKCs are inhibited by addition of the PKC inhibitor Gö6983.

The ability to monitor PKC activity at specific cellular locations has revealed a unique signature of activity depending on cellular location. Relevant to this chapter, the activity of conventional PKC isozymes dominates at the plasma membrane, where under appropriate conditions it oscillates in phase with Ca^{2+} oscillations (see Fig. 1) (13). Here, we describe how to measure oscillatory PKC activity in response to agonist stimulation.

2. Materials

1. Sterile glass-bottom culture dishes (MatTek Corporation)
2. HeLa cells
3. FuGENE 6 transfection reagent (Roche)
4. DNA encoding MyrPalm-CKAR (distributed by Addgene)
5. Hanks' Balanced Salt Solution (HBSS) containing 1 mM Ca^{2+}
6. Axiovert 200 M microscope (Zeiss)
7. MicroMAX 512BFT CCD camera (Roper Scientific)
8. Lambda 10-2 filterwheel shutter controller (Sutter)
9. Metafluor software (Molecular Devices)
10. 40×/1.3 NA oil-immersion objective (Zeiss)
11. Immersion oil, Type DF (Cargille Labs)
12. Filters from Chroma Technology:
    (a) CFP: 420/20 nm excitation, 450 nm dichroic, 475/40 nm emission
    (b) YFP: 495/10 nm excitation, 505 nm dichroic, 535/25 nm emission
    (c) 10% neutral density filter
13. histamine (Sigma)
14. phorbol-12,13-dibutyrate (Calbiochem)
15. Gö6983 (Calbiochem)
16. Excel (Microsoft)

3. Methods

Day 1:

Plate HeLa cells on a sterile glass-bottomed tissue culture dish.

Day 2:

Transfect HeLa cells that are between 50 and 80% confluent with 1 μg of MyrPalm-CKAR DNA using FuGENE 6 according to the manufacturer's protocol.

Day 3:

1. Turn on the microscope, camera, filterwheel controller and lamp.
2. Launch Metafluor software and allow CCD camera to cool (approximately 15 min).
3. Remove media from the cells to prepare the dish for the imaging experiment.
4. Rinse cells one time in HBSS, image in 2 ml HBSS.
5. Clean the 40× oil-immersion objective and apply one drop of oil.
6. Place the imaging dish on the microscope stage and secure it (we use clay) to prevent subtle movements of the dish during the addition of agonists/inhibitors over the course of the experiment.
7. Focus on the cells.
8. The following details were established on the above setup to allow for FRET imaging of CFP- and YFP-based FRET experiments:
   (a) Acquire all fluorescent images through a 10% neutral density filter.
   (b) Perform all experiments at room temperature.
   (c) Set up a protocol within Metafluor to acquire CFP, FRET, YFP, and CFP/FRET with the parameters listed below.
      - Acquire CFP for 200 ms using 420/20 nm excitation, 450 nm dichroic, 475/40 nm emission filters.
      - Acquire FRET for 200 ms using 420/20 nm excitation, 450 nm dichroic, 535/25 nm emission filters.
      - Acquire YFP for 100 ms using 495/10 nm excitation, 505 nm dichroic, 535/25 nm emission filters (this channel is acquired to monitor against photobleaching of the FRET acceptor).
      - Find cells expressing approximately 1 μM of MyrPalm-CKAR (too much or too little will not yield optimal results). In our setup, this was calibrated to be between 400 and 800 intensity units in the CFP channel and 1,000–2,000 units in the YFP channel.
   (d) With the above parameters, one can acquire a series of images every 7 s without photobleaching on this setup.
9. Define regions for each cell within the Metafluor software. MyrPalm-CKAR targets the reporter to plasma membranes so we choose a region from the top of the cell and make certain to avoid the edges which may move in or out of the region within the course of the experiment (see Fig. 2).

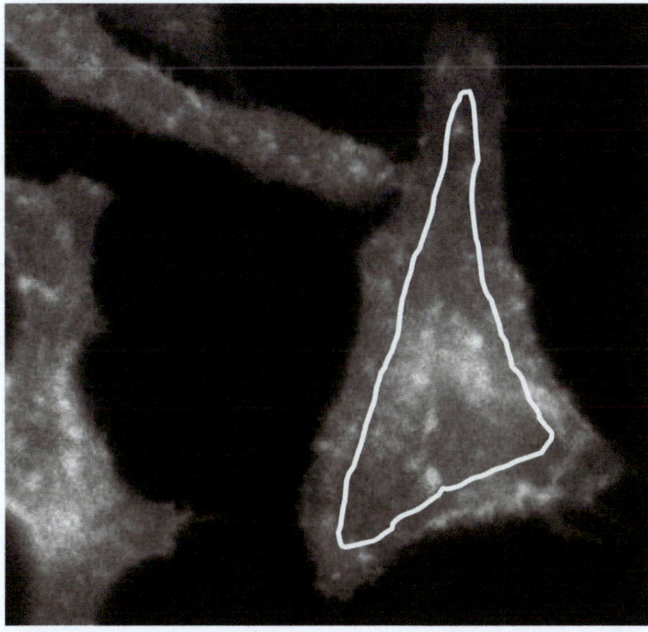

Fig. 2. The fluorescent signal from MyrPalm-CKAR expressed in HeLa cells is distributed throughout plasma membranes of the cells. Shown drawn over the cell image in *white* is the region defined in Metafluor for analysis of FRET ratio changes from MyrPalm-CKAR.

10. Save a log file of the data from within the software.
11. Importantly, save the images within the software as these can be re-analyzed after the experiment if necessary.
12. Begin the experiment acquiring data every 7 s until a constant baseline is established.
13. Once a steady baseline is established, stimulate cells with 10 μM histamine by removing 500 μl of HBSS from the dish, mixing this with the histamine to be added, then adding the mixture back drop-wise to the dish between time points.
14. Monitor the changes in the CFP/FRET ratio following histamine. Oscillations in Ca^{2+} translate to oscillations in membrane PKC activity in HeLa cells; this is observed by oscillations in the CFP/FRET ratio from MyrPalm-CKAR (see Fig. 1).
15. One can add a PKC inhibitor (e.g., Gö6983) to halt/reverse changes in the CFP/FRET ratio from CKAR (Fig. 1).
16. When imaging is complete, close the log file and the saved images file within Metafluor.
17. Data analysis can be performed using an Excel spreadsheet, i.e., open the log file from within Excel to view the data saved. The file will contain time as the first column, followed by four columns from each region analyzed: CFP intensity, FRET intensity, YFP intensity, CFP/FRET ratio calculation.

4. Notes

1. We acquire all of our data at room temperature. The cells are healthy in HBSS for the duration of our experiments (under an hour); furthermore, the lower temperature allows one to visualize rapid changes with better time resolution.
2. Metafluor software is our software of choice for time-lapse FRET experiments, as the software performs a real-time analysis of each region during acquisition. That is, the software quantifies the intensity within the user-defined regions from the CFP, FRET, and YFP channels, and then plots these intensities, as well as the CFP/FRET ratio, after each time point acquired.
3. A CFP/FRET ratio (versus the more common FRET/CFP ratio) is suggested as CKAR displays FRET in the unphosphorylated state and undergoes a decrease in FRET upon phosphorylation by PKC (this is true for all FHA2 domain-based kinase activity reporters such as BKAR and DKAR) (17, 18). Plotting CFP/FRET results in an increase in the ratio when there is an increase in kinase activity.
4. Cells expressing too little or too much reporter will not yield optimal results. Each distinct setup needs to be calibrated for expression levels of reporter. We performed such calibration by determining the approximate concentration of reporter in the cell based on fluorescence intensity and then aimed for imaging approximately 1 μM reporter expressed within a cell (see (19)).
5. Each distinct setup should be tested to determine how much light the reporter can be exposed to (as determined by acquisition times and frequency of acquisition) without photobleaching.
6. If one does not observe changes in the CFP/FRET ratio following histamine stimulation, one can test the setup by adding 200 nM phorbol-12,13-dibutyrate to the cells while imaging; this will robustly stimulate PKC activity at the plasma membrane by binding to its regulatory domains with high affinity.
7. Background levels should be subtracted from each channel post-acquisition in references within Metafluor, after the levels throughout the course of the experiment have been established.

Acknowledgments

This work was supported by NIH GM 43154. We thank Corina Antal, Noel Warfel and Alyssa Wu-Zhang for helpful comments.

References

1. Manning G, Whyte DB, Martinez R, Hunter T, Sudarsanam S (2002) The protein kinase complement of the human genome. Science 298:1912–1934
2. Newton AC (2010) Protein kinase C: poised to signal. Am J Physiol Endocrinol Metab 298:E395–E402
3. Rosse C, Linch M, Kermorgant S, Cameron AJ, Boeckeler K, Parker PJ (2010) PKC and the control of localized signal dynamics. Nat Rev Mol Cell Biol 11:103–112
4. Griner EM, Kazanietz MG (2007) Protein kinase C and other diacylglycerol effectors in cancer. Nat Rev Cancer 7:281–294
5. Bignon E, Ogita K, Kishimoto A, Nishizuka Y (1994) The protein kinase C family for neuronal signaling. Annu Rev Neurosci 17:551–567
6. Coussens L, Parker PJ, Rhee L, Yang-Feng TL, Chen E, Waterfield MD, Francke U, Ullrich A (1986) Multiple, distinct forms of bovine and human protein kinase C suggest diversity in cellular signaling pathways. Science 233: 859–866
7. Hernandez AI, Blace N, Crary JF, Serrano PA, Leitges M, Libien JM, Weinstein G, Tcherapanov A, Sacktor TC (2003) Protein kinase M{zeta} synthesis from a brain mRNA encoding an independent protein kinase C{zeta} catalytic domain: implications for the molecular mechanism of memory. J Biol Chem 278: 40305–40316
8. Yamamoto K, Seki T, Adachi N, Takahashi T, Tanaka S, Hide I, Saito N, Sakai N (2010) Mutant protein kinase C gamma that causes spinocerebellar ataxia type 14 (SCA14) is selectively degraded by autophagy. Genes Cells 15: 425–438
9. Verbeek DS, Goedhart J, Bruinsma L, Sinke RJ, Reits EA (2008) PKC gamma mutations in spinocerebellar ataxia type 14 affect C1 domain accessibility and kinase activity leading to aberrant MAPK signaling. J Cell Sci 121:2339–2349
10. Ling DS, Benardo LS, Serrano PA, Blace N, Kelly MT, Crary JF, Sacktor TC (2002) Protein kinase Mzeta is necessary and sufficient for LTP maintenance. Nat Neurosci 5:295–296
11. Wu-Zhang AX, Schramm CL, Nabavi S, Malinow R, Newton AC. Cellular pharmacology of protein kinase Mζ (PKMζ) contrasts with its In Vitro profile: implications for PKMζ as a mediator of memory. J Biol Chem. 2012 (in press).
12. Leitges M, Kovac J, Plomann M, Linden DJ (2004) A unique PDZ ligand in PKCalpha confers induction of cerebellar long-term synaptic depression. Neuron 44:585–594
13. Violin JD, Zhang J, Tsien RY, Newton AC (2003) A genetically encoded fluorescent reporter reveals oscillatory phosphorylation by protein kinase C. J Cell Biol 161:899–909
14. Gallegos LL, Newton AC (2008) Spatiotemporal dynamics of lipid signaling: protein kinase C as a paradigm. IUBMB Life 60:782–789
15. Gallegos LL, Kunkel MT, Newton AC (2006) Targeting protein kinase C activity reporter to discrete intracellular regions reveals spatiotemporal differences in agonist-dependent signaling. J Biol Chem 281:30947–30956
16. Kajimoto T, Sawamura S, Tohyama Y, Mori U, Newton AC (2010) Protein kinase C δ-specific activity reporter reveals agonist-evoked nuclear activity controlled by Src family of kinases. J Biol Chem 285(53):41896–41910
17. Kunkel MT, Ni Q, Tsien RY, Zhang J, Newton AC (2005) Spatio-temporal dynamics of protein kinase B/Akt signaling revealed by a genetically encoded fluorescent reporter. J Biol Chem 280:5581–5587
18. Kunkel MT, Toker A, Tsien RY, Newton AC (2007) Calcium-dependent regulation of protein kinase D revealed by a genetically encoded kinase activity reporter. J Biol Chem 282:6733–6742
19. Kunkel MT, Newton AC (2009) Spatiotemporal dynamics of kinase signaling visualized by targeted reporters. Curr Protoc Chem Biol 1:17–28

Chapter 15

Diacylglycerol Kinase (DGK) as a Regulator of PKC

Yasuhito Shirai and Naoaki Saito

Abstract

Diacylglycerol kinase (DGK) is a lipid kinase but not protein kinase. However, the enzyme regulates many protein kinases including PKC, PI4P-Kinase, and mTOR by reducing the amount of diacylglycerol and producing phosphatidic acid. To date, ten mammalian DGKs have been cloned and divided into five groups. Interestingly, most subtypes show high expression in the brain, specifically in distinct regions, suggesting that each subtype has unique and important function in the brain. Indeed, neural functions of some DGK subtypes have recently been reported. Here, we summarize enzymatic properties and neural functions of DGKs, and introduce useful information and methods to investigate DGK function in the brain.

Key words: Diacylglycerol kinase (DGK), Diacylglycerol (DG), Phosphatidic acid (PA), Protein kinase C (PKC), Phosphorylation, Lipid, Neurite branching, Neurite extension, Spine formation, Synaptic transmission, Memory, Bipolar disorder, Lithium,

1. Background and Overview

1.1. History and Structure of Diacylglycerol Kinase

Diacylglycerol kinase (DGK) is the enzyme which phosphorylates diacylglycerol (DG) and produces phosphatidic acid (PA) (1–5), resulting in attenuation of protein kinase C (PKC) activity. PA is also an activator for many other kinases including phosphatidylinositol 4-phosphate 5-kinase (6, 7), PKC ζ (8), and mammalian target of rapamycin (mTOR) (9). Therefore, DGK is thought to be one of the key enzymes in the lipid signal transduction. Indeed, recent researches using DGK KO mice clearly demonstrated DGK's importance in the immune system (10, 11), pathophysiological roles in the heart (12), insulin resistance in diabetes (13), and in the brain (12, 14).

DGKα has been cloned from pig brain by Kanoh and his colleagues in 1990 (15). Since then, nine mammalian subtypes have so far been cloned (16–24) and categorized into five groups depending on their primary structures (1–5). Figure 1 represents domain

Hideyuki Mukai (ed.), *Protein Kinase Technologies*, Neuromethods, vol. 68,
DOI 10.1007/978-1-61779-824-5_15, © Springer Science+Business Media, LLC 2012

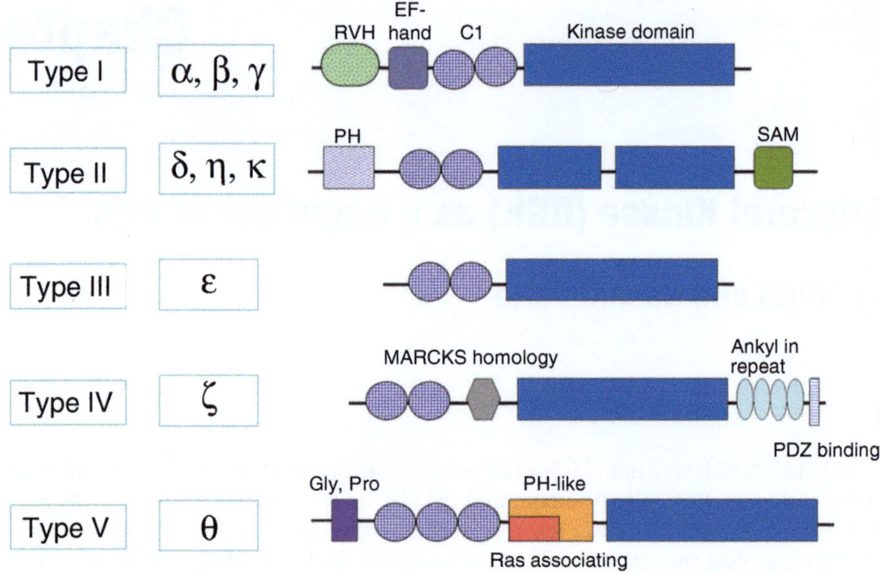

Fig. 1. Mammalian DGK subtypes.

structure of ten mammalian DGK subtypes. All DGKs, except for DGKθ which has three, have two cysteine-rich regions (C1A and C1B domain) in the regulatory domain at N terminus. Although DGK's C1 domains are homologous to PKC, only those of DGKβ and γ bind to DG (25). All DGKs have catalytic domain at C terminus, but catalytic domains of Type II DGKs (δ, η, and κ) are separated. In addition to these domains, they have different structures depending on their groups. Type I DGKs, DGKα, β, and γ, have EF-hand motif and recoverin homology (RVH) domain. Type II DGKs, DGKδ, η, and κ, have pleckstrin homology (PH) domain instead of these two domains at N terminus. DGKδ and η have sterile alpha motif (SAM) domain at C terminus. Type III DGK, DGKε, has only C1 domain. Type IV DGKs, DGK ζ and ι, have myristoylated alanine rich PKC substrate phosphorylation site like region (MARCKS homology domain) at N terminus and four ankyrin repeats and PDZ-binding domain at C terminus. Finally, Type V DGK, DGKθ, has proline and glycine rich domain and PH domain with overlapping Ras associating domain. In addition, many splice variants are reported (1, 4).

1.2. Regulation of Enzyme

Although enzymatic characteristics of all DGK subtypes have not been investigated, activities of some DGKs are regulated by ionic detergents and phospholipids (26, 27). For example, activities of DGKα and ζ depend on deoxycholate or cholic acid, so that a detergent, octyl glucoside, is used for the DGK assay as described below. But dependency of phospholipids seems to be subtype-specific (Table 1). The activity of purified DGKα is remarkably

Table 1
Localization and enzymatic properties of DGK subtypes

	Species	MW(KDa)	Localization	Enzymatic property	References
α	Pig	80	Thymus, spleen >> kidney, brain	Ca^{2+}, PS dependent	(15, 28, 35)
	Rat	80	Thymus, spleen >> brain, kidney (oligodendrocytes not neuron)	R59022 sensitive activated by PIP_3, $PI(3,4)P_2$	
β	Rat	90	Brain (CP > hip, cortex, olf) > adrenal gland > small intestine	Ca^{2+}, PS dependent activated by $PI(4,5)P_2$ R59022 sensitive	(14, 16, 29, 32)
γ	Rat	88	Brain (Cb, hip) >>>> other tissues	Ca^{2+}, PS dependent	(17, 29, 30)
	Human		Retina >> brain	R59022 sensitive	
δ	Human	130	skeletal muscle > testis, colon	Ca^{2+}, PS independent	(19)
η	Hamster	130–140	Testis > brain, lung, spleen > heart	Ca^{2+} insensitive	(20)
κ	Human	142	Testis > spleen, pracenta	Ca^{2+} insensitive	(24)
ε	Human	64	Testis > ovary >> skeletal muscle	Ca^{2+} insensitive	(18)
	Rat	64	Retina > brain > heart	Selectively for arachidonoyl DG	
ζ	Human	104	Brain > skeltal muscle, heart, pancreas	Ca^{2+} insensitive	(21, 31, 33)
	Rat	104	Thymus > brain(Cb, hip, olf) >> small intestine	Mg^{2+}, PS dependent	
ι	Human	130	Retina > brain (hip, CP, cortex, C*b*, D*g*)	Ca^{2+} insensitive	(23, 34)
θ	Human	110	Brain (C*b*, hip) >> small intestine, liver, etc.	Ca^{2+} insensitive	(22)

Cb cerebellum, *CP* caudate putamen, *Dg* dandate gyrus, *DG* diacylglycerol, *hip* hippocampus, *Olf* olfactory bulb, *PS* phosphatidylserine

enhanced by phosphatidylcholine but not by phosphatidylinositol (26), while DGKζ is activated by phosphatidylinositol and phosphatidylserine, but not so remarkably by phosphatidylcholine (27). DGKα is activated by PIP_3, while DGKβ is activated by PIP_2 (27). Moreover, divalent cations including calcium and magnesium are also required (31, 35). Specifically, type I DGKs are believed to depend on calcium because they have EF hands and RVH domains.

Indeed, calcium dependency of DGKα has been clearly shown in vitro (35). However, in vitro calcium-dependent activities of DGKβ and γ have not been reported, although their EF-hands bind to calcium. The calcium dependency of DGKβ and γ might be masked because the detergent was used for the assay or phospholipids used was not suitable. At this moment, information of enzymatic properties of each DGK subtype is not enough. Precise enzymatic characterization should be analyzed using purified proteins under optimized conditions.

Some DGKs show the translocation to the plasma membrane in response to several stimulations (35–42). For example, DGKγ is translocated from the cytoplasm to the plasma membrane by calcium, phorbol ester, and purinergic receptor stimulations (36). DGKα is also translocated to the plasma membrane by calcium (35), purinergic receptor stimulation (36), and T-cell receptor stimulation (37). Moreover, DGKδ translocates to the plasma membrane in response to phorbol ester (38). These are coincident with the fact that DG is produced on the plasma membrane and DGKs work there. In addition to the plasma membrane, DGKα is accumulated at the Golgi complex in the case of arachidonic acid and vitamin E stimulations (36, 39). Furthermore, it was reported that DGKζ is translocated from the nucleus to the cytoplasm (40), and DGKθ and γ are localized in the nucleus (41, 42), consistent with the fluctuation of DG and PA contents in the nucleus (43–45). These findings indicate that translocation is the key regulator to define where and how long DGK works, and suggest that DGKs are important for the DG/PA metabolism in many organelles including the plasma membrane, nucleus, and Golgi complex.

The translocation and activation of some DGK are regulated by phosphorylation. We have revealed that DGKγ is subtype-specifically phosphorylated by PKCγ at Ser-776 and Ser-779 upon the purinergic stimulation, resulting in upregulation of kinase activity of DGKγ (46). Membrane translocation of DGKδ is regulated by phosphorylation at Ser-22 and Ser-26 by cPKC (47), and nuclear export of DGKζ is dependent on the phosphorylation at MARCKS homology domain by PKCα (48). Not only serine phosphorylation but also tyrosine phosphorylation is important: membrane translocation of DGKα is dependent on phosphorylation at Tyr-334 (Tyr-335 in mouse) by Src family tyrosine phosphorylation (39, 49, 50). These results indicate that phosphorylation is one of the important regulation mechanisms for spatial regulation of diacylglycerol signaling.

1.3. Localization and Function of DGKs in the Brain

Most subtypes of DGKs are abundant in the brain (Table 1). Above all, high level mRNA of DGKβ, γ, ζ, ι, and θ are detected in the neurons. In situ hybridization revealed that DGKβ is expressed in the caudate putamen, accumbens nucleus, and hippocampus (16). Indeed, protein expression of DGKβ in these regions was reported

(14, 29). DGKβ is not at the birth but its expression is rapidly increased from day 14 to day 28 (29), and is localized on the plasma membrane of perisynaptic sites (29, 32), suggesting its importance in the neuronal network. DGKγ, which is predominantly localized in Purkinje cells and hippocampus, is present at birth and then gradually increased (17, 29). DGKζ is highly expressed in thymus and the brain but with substantial levels in skeletal muscle, heart, and pancreas (21, 33). This subtype is abundant in cerebellum, hippocampus, and olfactory bulb in the brain (33). Interestingly, DGKζ is detected in the nucleus in the hippocampal neurons but it is translocated to the cytoplasm by ischemia or kainite stimulation (34, 51). mRNA of DGKι is abundantly detected in the retina and brain (23), and in situ hybridization showed its high expression in hippocampus and dentate gyrus (52). mRNA expression of DGKθ is the highest in the cerebellum and hippocampus (22). Protein level of DGKη is found predominantly in the brain, although distribution of mRNA is different (20), but precise localization in the brain has not been reported. Finally, it is noteworthy that DGKα is detected in oligodentrocytes, although the subtype is enriched in thymus, not in the brain (15, 53).

Neural function of DGKs has not been fully understood. The most plausible function is to regulate synaptic function by modulating morphology of neuron and/or synaptic transmission. We showed that the primary cultured hippocampal neurons from DGKβ KO mice had less branches and spines compared to the wild type, and its long-term potentiation in the hippocampal CA1 region was reduced, causing impairment of cognitive functions including spatial and long-term memories in Y-maze and Morris water-maze tests (14). Furthermore, the KO mice showed impairment of emotion which was normalized by lithium treatment (54). The importance of DGKβ in the memory and emotion fits to localization of DGKβ in hippocampus and caudate putamen, and its developmental changes (29). In addition, the control of splicing of the enzyme, which generates non-membrane bound variants differing at the C terminus, is associated with mood disorders (55), supporting its important role in spine formation and emotion. Indeed, the splice variant form did not induce branches and spines (14). Similarly, a mutation in DGKη was reported to correlate with bipolar disorder (56) and increase in its mRNA level has been reported in some patients with bipolar disorder and schizophrenia (57), although it is still controversial (58). In addition to DGKβ, DGKζ regulates morphological change of neurons such as neurite outgrowth and spine formation via PDZ-binding regions (59, 60). Specifically, it has a role to maintenance of spine density (60). DGKι and DGKε are likely to modulate synaptic activity, although they did not affect on the neuron specific morphological change. DGKι is shown to regulate presynaptic release during mGluR-dependent LTD (61), and DGKε KO mice showed an increases resistance to electroconvulsive and faster recovery than wild type (62).

In addition to these functions, an important role of DGKζ in hypoxic response has been suggested (12, 63). It may be correlated to the DGKζ function in the heart (12). The subtype is also expected to be involved in leptin receptor signaling because it associates with leptin (64). However, detailed mechanisms are still unclear. Physiological functions of DGKα, γ, and θ in the brain are still unknown. Further examination would be necessary.

2. Protocol

In this section, we would like to introduce methods used for to investigate neural function of DGKβ. They would be helpful to investigate functions of other DGK subtypes in the brain.

2.1. Kinase Assay

To determine the kinase activity, appropriate volumes of the homogenate samples or purified protein were subjected to octyl glucoside mixed-micelle assay (35). A 2 mM 1-Steroyl-2-arachidonoyl *sn*-glycerol was used as a substrate. The reaction mixture was 50 mM Mops (pH 7.2), 50 mM octyl glucoside (Calbiochem), 100 mM NaCl, 1 mM dithiothreitol, 20 mM NaF, 2.1 mM $CaCl_2$, 2.0 mM EGTA, 0.8 mM EDTA, 10 mM $MgCl_2$, 6.7 mM phosphatidylserine (Avanti Polar Lipids) and 1 μM [(-^{32}P]ATP (10,000 cpm/nmol; ICN).The radioactivity of phosphatidyl acid was separated on a 20-cm Silica Gel 60 (Merck) thin layer chromatography plate using a chloroform:methanol:acetic acid (65:15:5) solution and detected by BAS 2000 (Fujix, Tokyo, Japan).

2.2. Measurement of PA and DG Content

Membrane PA in SH-SY5Y cells was measured based on the methods by Aragones et al. (65). The 5×10^5 SH-SY5Y cells were lipofected with GFP-DGKβ or GFP and cultured for 2 days. The cells were pre-incubated in phosphate-free medium for 30 min and then incubated with [^{32}P]-monosodium phosphate (100 μCi/ml) for additional 2 h. Thereafter, the cells were harvested and lysed in 20 mM Tris–HCl (pH 7.5) containing 1 mM $MgCl_2$, 1 mM EGTA, 1 mM PMSF, 20 μg/ml leupeptin. After sonication and centrifugation at 800×*g*, lipids including PA were extracted from the pellet, and separated by TLC. Spot of [^{32}P]-PA was measured by BAS2500 (FUJIFILM, Tokyo, Japan). To measure membrane DG, membrane lipids were extracted from the SH-SY5Y cells normally cultured. The amount of DG was determined by its conversion into [^{32}P]-PA by *Escherichia coli* DGK in the presence of [γ-^{32}P]-ATP. DGK assay was performed as described above.

To measure PA and DG in the brain from WT and DGKβKO mice (14), mice were anesthetized by the intraperitoneal injection of Nembutal and perfused with 0.1 M phosphate buffered saline. The brain was removed and a section (5 mm × 2 mm × 1 mm thick) was cut out from the hippocampus and cerebellum region of WT and KO mice. The sections were sonicated (30 s, input 5 by Tomi Seiko Co. Ltd.'s) in 100 μl of PBS (-). Protein concentration of the homogenate was measured by Bradford methods. DG and PA were extracted from the homogenates using chloroform and methanol as described previously (42). A half of chloroform phase was dried up and subjected to measurement of DG using *E. coli* DGK as described above. The half was subjected for measurement of PA using enzymatic reaction of lipase as described previously (65). 1-Steraroyl-2-arachidonoyl-sn-glycerol (1–100 pmol) and egg PA (1–40 pmol) were used for making standard carve, respectively.

2.3. Northern Blotting

Total RNA was extracted from several tissues of adult rats by guanidine thiocyanate/phenol/chloroform extraction. Poly(A)+RNA was isolated by chromatography on an oligo(dT)-cellulose column and each poly(A) + RNA sample (5 μg per lane) was denatured with formaldehyde and size-separated by agarose gel electrophoresis. The RNA was transferred and fixed to a nylon membrane (Nytran, Schleicher & Schuell) and hybridized with a ^{32}P-labeled 1.2-kb EcoRI fragment of pNDGKl cDNA for DGKβ (16). Hybridization and washing were performed under high stringency. Autoradiography was at –80°C for 3 days.

2.4. In Situ Hybridization

Cryostat sections of adult rat brain were hybridized with $0.5-1.0 \times 10^6$ cpm per slide of the same cDNA probe labeled with a-^{35}S[thio]dATP as used for Northern analysis and washed (16). After exposure to Hyperfilm-βmax (Amersham) for 2–3 weeks, the sections were dipped in Kodak NTB2 emulsion and exposed for 3 months.

2.5. Immunohistochemistry

Mice were deeply anesthetized by the intraperitoneal injection of Nembutal (50 mg/kg) (14, 29). The mice were perfused with 0.9% NaCl through the left ventricle at a flow rate of 5–50 ml/min and then perfused with 250 ml of 0.1 M phosphate buffer (PB, pH 7.4) containing 4% PFA and 0.2% picric acid at 4°C. The brain was removed and immersed for 48 h in the same fixative. After washing with several changes of PB containing 30% sucrose for at least 2 days at 4°C, serial coronal sections of 20 μm in thickness were cut on a cryostat. These sections were immersed directly in PBS-T (0.1–0.3% Triton X-100) for at least 4 days at 4°C before use. The following steps were carried out at 25°C, unless otherwise specified. The frontal sections were washed for 5 min with PBS-T between each step. The frontal sections were pre-incubated with 0.3% H_2O_2 for 20 min to inactivate endogenous peroxidase activity, with 5% NGS for 20 min to block non-specific binding sites and then with 0.1% phenylhydrazine to inactivate endogenous peroxidase activity for 20 min. The sections were then incubated with the antisera against DGKβ (1:2,000) in PBS-T for 72 h at 4°C. After washing with PBS-T, the sections were incubated for an additional 2 h with biotinylated goat anti-mouse IgG (1:1,000) (Vector, Burlingame, CA), and then for 1.5 h with avidin–biotin–peroxidase complex (1:1,000) (Vector). After rinsing three times, the reaction product was visualized with 0.01% 3.3 diaminobenzidine (DAB, Sigma, CA, USA), and 1% nickel ammonium sulfate in 0.05 M Tris–HCl (pH 7.6) with 0.0003% H_2O_2. Finally, the stained sections were mounted on gelatin-coated glass slides, dehydrated by a graded series of ethanol and covered with Entellan (Merck, Whitehouse Station, NJ, USA) for observation, and then photographed under a light microscope (Carl Zeiss, Esslingen, Germany).

2.6. Primary Culture of Mouse Hippocampal Neurons

Fetuses were removed on embryonic days 17–18 from mice anesthetized by intraperitoneal injection of Nembutal (Abbott Laboratories, Abbott Park, IL, USA). Hippocampi were dissected and placed in Ca^{2+}- and Mg^{2+}-free HEPES-buffered Hanks salt solution (HHSS) at pH 7.45 (Invitrogen). Primary culturing of hippocampal neurons was carried out using Nerve Cell Culture System (SUMITOMO BAKELITE Co. Ltd., Tokyo, Japan). Briefly, hippocampal neurons were dissociated using dissociation solution and cultured using the glial-conditioned medium, in a glass-bottomed culture dish (Matek Corp, USA). Half of the medium was exchanged every 3 days with fresh medium.

2.7. Adenovirus Infection to the Primary Cultured Hippocampal Neurons

After culturing for 3, 10, or 15 days, adenoviruses NSE-tTA, TetOp-GFP, or TetOP-GFP-DGKβ (14) were applied to a dish culturing hippocampal neurons. After 1 h incubation, the medium was washed well and cultured for a further 48 h. After fixation with 4% PFA and 0.2% picric acid at 4°C and washing with PBS-T, the fluorescence of GFP was monitored under confocal microscopy.

2.8. Immunocytochemistry of the Primary Cultured Hippocampal Neurons

The hippocampal neurons dissected from DGKβKO mice (14) or their litter mates WT mice were cultured for 3, 10, or 15 days, and fixed with 4% PFA and 0.2% picric acid at 4°C. After washing with PBS containing 0.03% triton X-100 (PBS-T), the cells were tritonized with 0.3% triton X in PBS(−) for 20 min and then blocked with 5% normal goat serum (NGS) for 1 h. The neurons were incubated with anti-MAP-2 antibody (1:2,000) for 16 h at 4°C. After rinsing three times, the neurons were visualized with Alexa 488-labeled goat anti-mouse IgG (1:1,000), followed observation under confocal microscope.

2.9. Confocal Microscopic Analysis and Image Processing

The fluorescence of the GFP or Alxa 488 was observed with a confocal laser scanning fluorescent microscope (LSM 510 invert, Carl Zeiss, Jena, Germany) at 488-nm argon excitation using a 515–535-nm band pass barrier filter. The images were recorded as TIFF files.

To count the number of neurites, branches, and spines, the image was analyzed with Neurolucida and Nurolucida Explorer software (MBF Bioscience, Tokyo, Japan).

2.10. Electrophysiology

Brains were rapidly removed from ether-anesthetized mice and the hippocampi were dissected out. Transverse hippocampal slices (400 μm thick), prepared using a vibratome (microslicer DTK-1000), were incubated for 2 h in continuously oxygenized (95% O_2, 5% CO_2) artificial cerebrospinal fluid (ACSF) containing 126 mM NaCl, 5 mM KCl, 26 mM $NaHCO_3$, 1.3 mM $MgSO_4 \cdot 7H_2O$, 1.26 mM KH_2PO_4, 2.4 mM $CaCl_2 \cdot 2H_2O$ and 1.8% glucose at room temperature (28°C). After a 2-h recovery period, a slice was transferred to an interface recording chamber and perfused at a flow rate of 2 ml/min with ACSF warmed to 34°C. Field excitatory postsynaptic potentials (fEPSPs) were evoked by a 0.05 Hz test stimulus through a bipolar stimulating electrode placed on the Schaffer collateral/commissural pathway and recorded from the stratum radiatum of CA1 using a glass electrode filled with 3 M NaCl. A single-electrode amplifier (CEZ-3100, Nihon Kohden, Tokyo, Japan) was used to record the responses, and the maximal value of the initial fEPSPs slope was collected and averaged every 1 min (3 traces) using an A/D converter (PowerLab 200, AD Instruments, Castle Hill, Australia) and a personal computer. After a stable baseline was obtained, high frequency stimulation (HFS) of 100 Hz with a 1-s duration was applied twice with a 10-s interval and test stimuli were continued for the indicated periods (14).

2.11. Golgi Staining

The Golgi staining is a useful method to visualize spines in vivo (14, 54). The hippocampal region was cut out from the brains fixed as described above, and further immersed in 30% sucrose for 2–3 days. The tissue block was placed in 2% potassium dichromate for 2 days at 4°C and then in 2% silver nitrate solution for 2 days at 4°C in the dark. The block was cut into 60 μm thick sliced into distilled water. Finally, the sections were mounted onto slides, dried for 10 min, and dehydrated through 95% alcohol, 100% alcohol, clear in xylene.

2.12. Electromicroscopic Analysis

The anesthetized DGKβKO and WT mice were perfused with 0.9% NaCl through the left ventricle at a flow rate of 5–50 ml/min and then perfused with 250 ml of 0.1 M phosphate buffer (PB, pH 7.4) containing 2% PFA and 2% glutaraldehyde at 4°C. The dissected brain was sliced into sections of 1 mm thickness and 4 mm square, and immersed in the same fixative. The slice was further fixed in the 2% osmium tetroxide in 0.1 M PB, dehydrated through a graded series of ethanol and embedded in Epon (quetol-812). The ultrathin sections were mounted on the mesh and stained with 2% uranyl acetate and Reynolds solution, then observed and photographed with a JEOL JEM1200EX electron microscope.

2.13. Y-Maze Test

Spontaneous alternation behavior in a Y-maze was assessed as a spatial working memory task. The Y-maze apparatus consisted of three identical arms (length 40 × width 10 × height 12 cm). Each mouse was placed at the end of one fixed arm and allowed to move freely through the maze during an 8-min session. The sequence of arm entries was recorded manually. An alternation was defined as entering each of the three arms consecutively. The maximum number of alternations was thus the total number of arms entered minus two, and the percentage of alternations was calculated as (actual alternations/maximum alternations) × 100. The total number of arms entered during the session was also recorded.

2.14. Morris Water-Maze Test

A circular pool (diameter 120 × height 45 cm) was filled to a depth of 30 cm with water (21–23°C). Four equally spaced points around the edge of the pool were designated as four starting positions. A hidden platform (diameter 10 cm) was set 0.5 cm below the surface of the water in a fixed position. Mice were placed in the water facing the wall and trained with four trials per day for 5 days. In each trial, the starting position was changed, and the mouse swam until it found the platform, or after 60 s was guided to the platform; the mouse was then placed on the platform for 15 s before being picked up. Twenty-four hours after the last training trial the mice were given a prove test without the platform. In this test, each mouse was placed in the pool once and allowed to search for 60 s. Mean distance from the original location of the platform and the time spent in the quadrant where the platform had been was recorded using a video camera-based Ethovision XT system (Noldus, Wageningen, The Netherlands).

References

1. Kanoh H, Yamada K, Sakane F (2002) Diacylglycerol kinases: emerging downstream regulators in cell signaling systems. J Biochem 131:629–633
2. Topham MK, Prescott SM (1999) Mammalian diacylglycerol kinases, a family of lipid kinases with signaling functions. J Biol Chem 274: 11447–11450
3. van Bitterswijk WJ, Houssa B (2000) Properties and functions of diacylglycerol kinases. Cell Signal 12:595–605
4. Sakane F, Imai S, Kai M, Yasuda S, Kanoh H (2007) Diacylglycerol kinases: why so many of them? Biochim Biophys Acta 1771:793–806
5. Merida I, Avila-Flores A, Merino E (2008) Diacylglycerol kinases: at the hub of cell signaling. Biochem J 409:1–18
6. Jones DR, Sanjuan MA, Merida I (2000) Type Ialpha phosphatidylinositol 4-phosphate 5-kinase is a putative target for increased intracellular phosphatidic acid. FEBS Lett 476: 160–165
7. Luo B, Prescott SM, Topham MK (2004) Diacylglycerol kinase zeta regulates phosphatidylinositol 4-phosphate 5-kinase Ialpha by a novel mechanism. Cell Signal 16:891–897
8. Limatola C, Schaap D, Moolenaar WH et al (1994) Phosphatidic acid activation of protein kinase C-zeta overexpressed in COS cells: comparison with other protein kinase C isotypes and other acidic lipids. Biochem J 304:1001–1008
9. Fang Y, Vilella-Bach M, Bachmann R et al (2001) Phosphatidic acid-mediated mitogenic activation of mTOR signaling. Science 294:1942–1945
10. ZhaY MR, Ho AW et al (2006) T cell anergy is reversed by active Ras and is regulated by diacylglycerol kinase-α. Nat Immunol 11:1166–1173

11. Olenchock BA, Guo R, Carpenter JH et al (2006) Distruption of diacylglycerol metabolism impairs the induction of T cell anergy. Nat Immunol 11:1174–1181
12. Goto K, Nakano T, Hozumi Y (2006) Diacylglycerol kinase and animal models: the pathophysiological roles in the brain and heart. Adv Enzyme Regul 46:192–202
13. Chibalian AV, Leng Y, Vieira E, Krook A, Bjomholm M et al (2008) Down regulation of diacylglycerol kinase delta contributes to hyperglycemia-induced insulin resistance. Cell 132:375–386
14. Shirai Y, Kouzuki K, Kakefuda K et al (2010) Essential role of neuron-enriched diacylglycerol kinase (DGK), DGKβ in neurite spine formation, contributing to cognitive function. PLoS One 5:e11602
15. Sakane F, Yamada K, Kanoh H et al (2000) Porcine diacylglycerol kinase sequence has zinc finger and E-F hand motifs. Nature 334: 345–348
16. Goto K, Kondo H (1993) Molecular cloning and expression of a 90-kDa diacylglycerol kinase that predominantly localizes in neurons. Proc Natl Acad Sci U S A 90:7598–7602
17. Goto K, Funayama M, Kondo H (1994) Cloning and expression of a cytosckeleton-associated diacylglycerol kinase that is dominantly expressed in cerebellum. Proc Natl Acad Sci U S A 91:13042–13046
18. Tang W, Bunting M, Zimmerman GA et al (1996) Molecular cloning of a novel human diacylglycerol kinase highly selective for arachidonate-containing substrates. J Biol Chem 271:10237–10241
19. Sakane F, Imai S, Kai M et al (1996) Molecular cloning of a novel diacylglycerol kinase isozyme with a pleckstrin homology domain and a C-terminal tail similar to those of the EPH family of protein-tyrosine kinases. J Biol Chem 271:8394–8401
20. Klauck TM, Xu X, Mousseau B et al (1996) Cloning and characterization of a glucocorticoid-induced diacylglycerol kinase. J Biol Chem 271:19781–19788
21. Bunting M, Tang W, Zimmerman G et al (1996) Molecular cloning and characterization of a novel human diacylglycerol kinase ζ. J Biol Chem 271:10230–10236
22. Houssa B, Schaap D, van der Wal J et al (1997) Cloning of a novel human diacylglycerol kinase (DGKη) containing three cysteine-rich domains, a prorine-rich region, and a pleckstrin homology domain with an overlapping Ras-associating domain. J Biol Chem 272:10422–10428
23. Ding L, Traer E, McIntyre TM et al (1998) The cloning and characterization of a novel human diacylglycerol kinase, DGKι. J Biol Chem 273:32746–32752
24. Imai S, Kai M, Yasuda S et al (2005) Identification and characterization of a novel human type II diacylglycerol kinase, DGKκ. J Biol Chem 280:39870–39881
25. Shindo M, Irie K, Masuda A et al (2003) Synthesis and phorbol ester binding of the cysteine-rich domains of diacylglycerol kinase (DGK) isozymes. J Biol Chem 278:18448–18454
26. Kanoh H, Kondoh H, Ono T (1983) Diacylglycerol kinase from pig brain. J Biol Chem 258:1767–1774
27. Cipres A, Carrasco S, Merino E et al (2003) Regulation of diacylglycerol kinase a by phosphoinositide 3-kinase lipid products. J Biol Chem 278:35629–35635
28. Goto K, Watanabe M, Kondo H et al (1992) Gene cloning, sequence, expression and in situ localization of 80 kDa diacylglycerol kinase specific to oligodendrocyte of rat brain. Mol Brain Res 16:75–87
29. Adachi N, Oyasu M, Taniguchi T et al (2005) Immunocytochemical localization of a neuron-specific diacylglycerol kinase β and γ in the developing rat brain. Mol Brain Res 139:288–299
30. Kai M, Sakane F, Imai S et al (1994) Molecular cloning of a diacylglycerol kinase isozyme predominantly expressed in human retina with a truncated and inactive enzyme expressed in most other human cells. J Biol Chem 269: 18492–18498
31. Kato M, Takenawa T (1990) Purification and characterization of membrane-bound and cytosolic forms of diacylglycerol kinase from rat brain. J Biol Chem 265:794–800
32. Hozumi Y, Fukuya M, Adachi N et al (2008) Diacylglycerol kinase β accumulates on the perisynaptic site of medium spiny neurons in the striatum. Eur J Neurosci 28:2409–2422
33. Goto K, Kondo H (1996) A 104 kDa diacylglycerol kinase containing ankyrin-like repeats localize in the cell nucleus. Proc Natl Acad Sci U S A 93:11196–11201
34. Saino-Saito S, Hozumi Y, Goto K (2011) Excitotoxicity by kainate-induced seizure diacylglycerol kinase ζ to shuttle from the nucleus to the cytoplasm in hippocampal neurons. Neurosci Lett 494:185–189
35. Sakane F, Yamada K, Imai S et al (1991) Porcine 80-kDa diacylglycerol kinse is a calcium-binding and calcium/phospholipid-dependent enzyme and undergoes calcium-dependent translocation. J Biol Chem 266:7096–7100

36. Shirai Y, Segawa S, Kuriyama M (2000) Subtype-specific translocation of diacylglycerol kinase α and γ and its correlation with protein kinase C. J Biol Chem 275:24760–24766
37. Sanjuán MA, Pradet-Balade B, Jones DR et al (2003) T cell activation in vivo targets diacylglycerol kinase to the membrane: a novel mechanism for Ras attenuation. J Immunol 170: 2877–2883
38. Imai S, Sakane F, Kanoh H (2002) Phorbol ester-regulated oligomerization of diacylglycerol kinase δ linked to its phosphorylation and translocation. J Biol Chem 277:35323–35332
39. Fukunaga-Takenaka R, Shirai Y, Yagi K et al (2005) Importance of chroman ring and tyrosine phosphorylation in the subtype-specific translocation and activation of diacylglycerol kinase alpha by d-α-tocopherol. Genes Cells 10:311–319
40. Topham M, Bunting M, Zimmerman GA et al (1998) Protein kinase C regulates the nuclear localization of diacylglycerol kinase-ζ. Nature 394:697–700, Immunol. 2003; 170: 2877–2883
41. Tabellini G, Bortul R, Santi S et al (2003) Diacylglycerol kinase-θ is localized in the speckle domains of the nucleus. Exp Cell Res 287:143–154
42. Matsubara T, Shirai Y, Miyasaka K et al (2006) Nuclear transportation of diacylglycerol kinase γ and its possible function in the nucleus. J Biol Chem 281:6152–6164
43. Martelli AM, Tabellini G, Bortul R et al (2000) Enhanced nuclear diacylglycerol kinase activity in response to a mitogenic stimulation of quiescent Swiss 3T3 cells with insulin-like growth factor I. Cancer Res 60:815–821
44. Irvine R (2000) Nuclear lipid signaling. Sci STKE 2000(48):RE1
45. Martelli AM, Fala F, Faenza I et al (2004) Metabolism and signaling activities of nuclear lipids. Cell Mol Life Sci 61:1143–1156
46. Yamaguchi Y, Shirai Y, Matsubara T et al (2006) Phosphorylation and up-regulation of diacylglycerol kinase γ via its interaction with protein kinase C γ. J Biol Chem 281:31627–31637
47. Imai S, Kai M, Yamada K et al (2004) The plasma membrane translocation of diacylglycerol kinase d1 is negatively regulated by conventional protein kinase C-dependent phosphorylation at Ser-22 and Ser-26 within the pleckstrin homology domain. Biochem J 382:957–966
48. Luo B, Prescott SM, Topham MK (2003) Association of diacylglycerol kinase ζ with protein kinase C a: spatial regulation of diacylglycerol signaling. J Cell Biol 160:929–937
49. Baldanzi G, Cutrupi S, Chianale F et al (2007) Diacylglycerol kinase-α phosphorylation by Src on Y335 is required for activation, membrane recruitment and Hgf-induced cell motility. Oncogene 27:942–956
50. Merino E, Avila-Flores A, Shirai Y et al (2008) Lck-dependent tyrosine phosphorylation of diacylglycerol kinase α regulates its membrane association in T cells. J Immunol 180: 5805–5815
51. Ali H, Nakano T, Saino-Saito S et al (2004) Selective translocation of diacylglycerol kinase ζ in hippocampal neurons under transient forebrain ischemia. Neurosci Lett 372:190–195
52. Sommer W, Arlinde C, Caberlotto L et al (2001) Differential expression of diacylglycerol kinase iota and L18A mRNAs in the brains of alcohol-preferring AA and alcohol-avoiding ANA rats. Mol Psychiatry 6:103–108
53. Goto K, Kondo H (1996) Heterogeneity of diacylglycerol kinase in terms of molecular structure, biochemical characteristics and gene expression localization in the brain. J Lipid Mediat Cell Signal 14(1–3):251–257
54. Kakefuda K, Oyagi A, Ishisaka M et al (2010) Diacylglycerol kinase b knockout mice exhibit lithium-sensitive behavioral abnormalities. PLoS One 5:e13447
55. Caricasole A, Bettini E, Sala C et al (2002) Molecular cloning and characterization of the human diacylglycerol kinase β (DGKβ) gene. J Biol Chem 277:4790–4796
56. Baum AE, Akula N, Cabanero M et al (2008) A genome-wide association study implicates diacylglycerol kinase eta (DGKH) and several other genes in the etiology of bipolar disorder. Mol Psychiatry 13:197–207
57. Moya PR, Murphy DL, McMahon FJ et al (2010) Increased gene expression of diacylglycerol kinase eta in bipolar disorder. Int J Neuropsychopharmacol 13:1127–1128
58. Tesli M, Kahler AK, Andreassen BK (2009) No association between DGKH and bipolar disorder in a Scandinavian case-control sample. Psychiatr Genet 19:269–272
59. Yakubchyk Y, Abramovici H, Maillet JC et al (2005) Regulation of neurite outgrowth in N1E-115 cells through PDZ-mediated recruitment of diacylglycerol kinase ζ. Mol Cell Biol 25:7289–7302
60. Kim K, Yang J, Zhong XP et al (2009) Synaptic removal of diacylglycerol by DGKζ and PSD-95 regulates dendritic spine maintenance. EMBO J 28:1170–1179
61. Yang J, Seo J, Nair R et al (2011) DGKι regulates presynaptic release during mGluR-dependent LTD. EMBO J 30:165–180
62. Rodriguez de Turco EB, Tang W, Topham MK et al (2001) Diacylglycerol kinase ε regulates

seizure susceptibility and long-term potentiation through arachidonoyl–inositol lipid signaling. Proc Natl Acd Sci U S A 98:4740–4745

63. Aragones J, Jones DR, Martin S et al (2001) Evidence for the involvement of diacylglycerol kinase in the activation of hypoxia-inducible transcription factor 1 by low oxygen tension. J Biol Chem 276:10548–10555

64. Liu Z, Chang GQ, Leibowitz SF (2001) Diacylglycerol kinase zeta in hypothalamus interacts with long form leptin receptor. Relation to dietary fat and body weight regulation. J Biol Chem 276:5900–5907

65. Morita S, Ueda K, Kitagawa S (2009) Enzymatic measurement of phosphatidic acid in cultured cells. J Lipid Res 50:1945–1952

Chapter 16

Trafficking of Trk Receptors

Daniel Bodmer and Rejji Kuruvilla

Abstract

Endocytosis and trafficking of Trk receptors are central to their functions in neuronal development. Neurotrophin binding to Trk receptors induces endocytosis of Trk receptors as a ligand–receptor complex in axon terminals. This endocytic ligand–receptor complex, termed the signaling endosome, is retrogradely transported to cell bodies of developing neurons to regulate transcriptional programs that promote neuronal survival. Compartmentalized cultures are an essential tool for studying the directional trafficking of Trk receptors in neurons. Compartmentalized cultures allow the separation of cell bodies and distal axons into distinct fluid environments. This enables the researcher to add neurotrophins or other factors exclusively to the distal axons or cell bodies of neurons. Here we describe two related compartmentalized culture methods, the antibody feeding assay and cell surface biotinylation assay, to study the directional transport of Trk receptors in neurons. Selective addition of antibody or biotin to compartmentalized cultures allows researchers to specifically label Trk receptors in the distal axons or cell bodies of neurons, and follow their trafficking to other cellular compartments using standard detection methods.

Key words: Neurotrophin, Nerve growth factor, Trk receptor, Endocytosis, Axonal transport, Compartmentalized cultures, Signaling endosomes

1. Background and Historical Overview

Receptor-mediated endocytosis modulates the signaling and function of growth factors. In many cases, receptor-mediated endocytosis functions to down-regulate signal transduction from receptors on the plasma membrane. However, receptor-mediated endocytosis can also lead to the spatial propagation of signal transduction via the trafficking of endosomal ligand–receptor complexes. One of the best examples of this process is found in the peripheral nervous system where the target-derived growth factor, nerve growth factor (NGF), is endocytosed in developing axon tips and retrogradely transported to the neuronal soma to control transcription and other cellular processes.

Hideyuki Mukai (ed.), *Protein Kinase Technologies*, Neuromethods, vol. 68,
DOI 10.1007/978-1-61779-824-5_16, © Springer Science+Business Media, LLC 2012

NGF is the founding member of the neurotrophin family of growth factors, which consists of NGF, brain-derived neurotrophic factor (BDNF), neurotrophin-3 (NT-3), and neurotrophin-4/5 (NT-4/5). Neurotrophins have been best characterized in the peripheral nervous system where they are secreted from sympathetic and sensory target tissues, such as the salivary glands, eyes, skin, and muscle. As developing axons project into these target tissues, secreted neurotrophins bind their cognate receptor tyrosine kinase, TrkA (NGF), TrkB (BDNF, NT4/5), or TrkC (NT-3), present on the axon tips (1). Ligand binding to Trk receptors induces dimerization, autophosphorylation of tyrosine residues, activation of signaling pathways, and receptor-mediated endocytosis (2). Many of these processes are common to other receptor tyrosine kinases and may occur in any cell type. One of the unique aspects of neurotrophin–Trk signaling is that although neurotrophins bind their cognate Trk receptor on developing axons, they must control transcription and other cellular processes long distances away in the neuronal soma to promote neuronal survival and other functions (3).

What is the mechanism by which target-derived NGF signals retrogradely to neuronal soma to support survival? Although many models for NGF retrograde transport have been proposed, the "signaling endosome" model has garnered the most empirical evidence and is likely to be the primary mechanism (4). In the signaling endosome model, NGF–TrkA complexes are endocytosed in distal axons and retrogradely transported along microtubules to the neuronal soma (3). In support of this model, NGF injected into the targets of sympathetic and sensory neurons is retrogradely transported to the neuronal soma, indicating that the ligand is directly transported in vivo (5–7). Phosphorylated Trk receptors are also retrogradely transported in vivo, and this process is dependent on endogenous ligand present in the target tissues (5, 8). Finally, exogenous application of NGF to distal axons leads to the appearance of phosphorylated TrkA and activated downstream signaling molecules in the neuronal soma (9–14). Together, these experiments provide evidence that both NGF and TrkA are retrogradely transported in vivo.

Numerous lines of evidence also support the idea that NGF and TrkA are retrogradely transported as a ligand–receptor complex that is endocytosed in distal axons. First, NGF is able to induce the endocytosis of TrkA receptors (15). Preventing endocytosis, by covalently cross-linking NGF to beads or disrupting the function of endocytic effectors, such as dynamin or pincher, blocks the retrograde activation of downstream signaling effectors and neuronal survival (11, 16, 17). Immunostaining data indicate that NGF and TrkA colocalize in early endosomes in both axons and cell bodies (9). Co-precipitation studies indicate that neurotrophins and Trk receptors are transported in association (13, 18). Thus, NGF and TrkA are retrogradely transported as a complex and this process is dependent on endocytosis. Interestingly, biochemical characterization of sciatic nerve chamber preparations suggests that NGF–TrkA complexes in early endosomes maintain association with activated signaling proteins during retrograde transport (9). These and other findings suggest that NGF–TrkA continues to signal from an endocytic platform, in a manner similar to the way signals are generated from the plasma membrane. Receptor-mediated endocytosis may lead to the spatial propagation of signal transduction from the plasma membrane of projecting axons throughout the entire neuron. In fact, NGF–TrkA-containing signaling endosomes can be retrogradely transported from nerve terminals to the dendrites of sympathetic neurons (19).

Many of the above results would not have been possible without the invention of a unique culture method that separates the distal axons and cell bodies of neurons. This compartmentalized culture method, also known as a Campenot chamber as it was pioneered by the studies of Robert Campenot in the 1970s, has been an essential tool for investigations of retrograde transport and neurotrophin–Trk biology (20, 21). In compartmentalized cultures, the cell bodies and proximal axons are cultured in a separate fluid compartment from distal axons. The compartmentalized culture system is a powerful tool in neuronal cell biology because (1) one may add neurotrophins or other factors exclusively to the cell bodies or distal axons of cultured neurons and (2) one may isolate pure axonal fractions for biochemical or molecular characterization.

Compartmentalized cultures were used to characterize the spatial aspects of neurotrophin function. For example, by adding NGF exclusively to distal axons, researchers demonstrated that axonal NGF locally controls axon growth (20), and is able to retrogradely promote neuronal survival (17, 22). Other studies have extended the uses of compartmentalized culture systems to characterize the retrograde transport of NGF. They have been used to define the kinetics of NGF retrograde transport (23), visualize the retrograde transport of single NGF dimers (24), and characterize the cellular mechanisms and signaling pathways that underlie the process (23). The compartmentalized culture system also provides an excellent system for spatial analysis of activation and transport of Trk receptor tyrosine kinases. For example, compartmentalized cultures were used to demonstrate the appearance of phosphorylated TrkA after application of NGF to distal axons (13). Pharmacological studies in compartmentalized cultures showed that the catalytic activity of TrkA is required in both distal axons and cell bodies for the retrograde activation of TrkA, activation of downstream signaling effectors, and neuronal survival (10–12, 17, 18). Pharmacological inhibition of cytoplasmic effectors in cell bodies or distal axons of compartmentalized cultures facilitated the spatial characterization of signaling pathways necessary to support TrkA retrograde transport and neuronal survival (10).

Compartmentalized cultures may also be used to study the anterograde transport of TrkA. A recent study uncovered a positive feedback mechanism, wherein NGF on distal axons actually promotes the anterograde transport of its own receptor, TrkA (25).

The compartmentalized culture system is a truly versatile tool and may be adapted to study the anterograde or retrograde trafficking of any receptor of interest. The following chapter will provide an overview of two related techniques used to characterize the signaling and trafficking of Trk receptor tyrosine kinases in compartmentalized cultures; the antibody feeding assay, and the cell surface biotinylation assay.

2. Equipment and Materials

2.1. Compartmentalized Cultures

A compartmentalized culture is prepared with a Teflon chamber, two glass syringes filled with silicone vacuum grease (Dow Corning, DC-976), 20 and 25 gauge needles, 90° angle hemostats, dissection light microscope, DMEM–methylcellulose (1 gm methylcellulose dissolved in 100 mL DMEM), and collagen-coated culture dishes. Teflon chambers, which exist in a variety of designs, may be purchased from vendors (Tyler Research) or built in a university machine shop. We use a smaller chamber design (Fig. 1a, Camp10) for most applications and a larger chamber design (Fig. 1b) when more cellular material is needed. To clean the Teflon chambers, they are placed in sulfuric acid for 24 h, thoroughly rinsed in deionized-H_2O, boiled for 30 min, and dry-autoclaved. The two glass syringes are filled with silicone grease using a 10 mL plastic

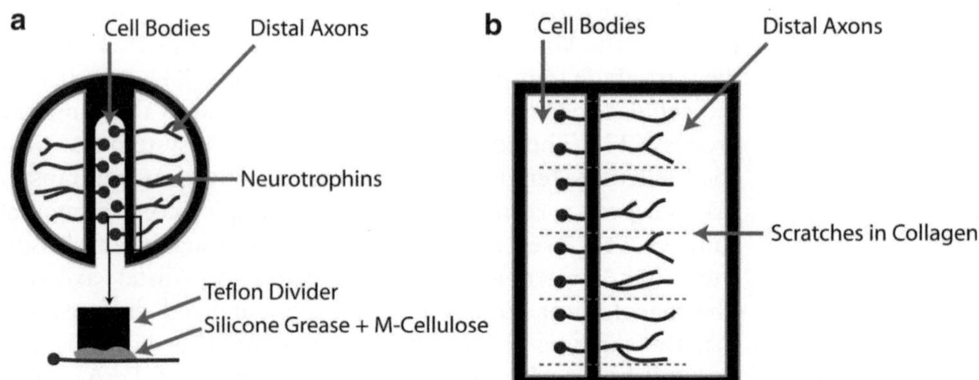

Fig. 1. Schematic of compartmentalized cultures. (**a**) Standard compartmentalized culture design with cell bodies residing in the central compartment and axons projecting to the distal compartments, after 5–9 days in vitro (d.i.v.). The liquid impermeable barrier is formed from a Teflon Divider and a layer of silicone grease and methylcellulose. (**b**) The biochemistry compartmentalized culture design accommodates more neurons for biochemical analyses and requires 9–14 d.i.v. for axons to project into the distal compartment. The scratches in the collagen substratum guide axon growth, perpendicular to the barrier.

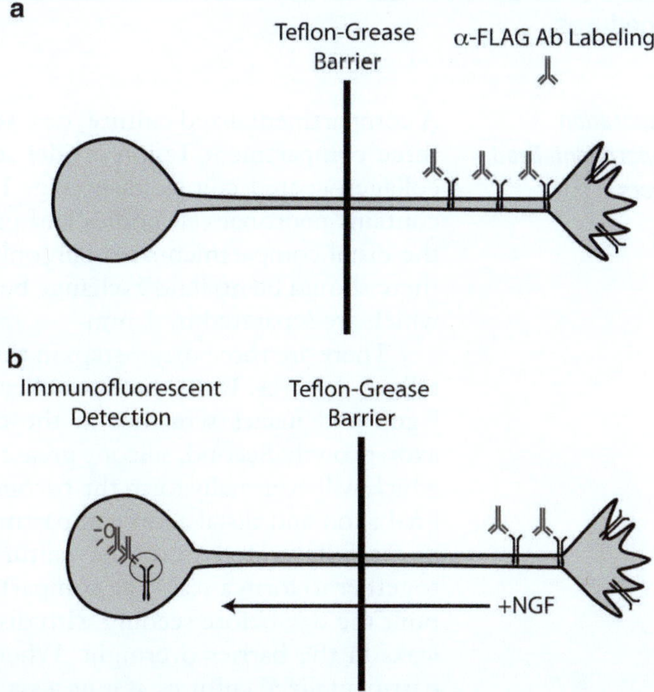

Fig. 2. Antibody feeding assay in compartmentalized cultures. (**a**) The distal axon compartment is incubated with antibody to label epitope-tagged TrkA receptors exclusively in the distal axons. (**b**) Retrograde transport is induced by adding NGF to distal axons and the researcher may detect retrogradely transported TrkA by immunostaining with fluorescently labeled secondary antibody.

syringe and autoclaved. Each glass syringe is equipped with either a 20 or 25 gauge needle that is cut with a pair of wire cutters to 1 cm in length. When cutting the needles with wire cutters, it is important to leave a large bore in the end of the needle.

2.2. Antibody Feeding Assay

An antibody feeding assay is performed to image the endocytosis and transport of Trk receptor tyrosine kinases in compartmentalized cultures. Antibody feeding assays require projecting compartmentalized neuronal cultures expressing a FLAG-tagged Trk receptor, an anti-FLAG antibody (M2, 4.2 μg/mL; Sigma), fixing solution (4% paraformaldehyde/1× PBS), blocking solution (0.1% Triton X-100/1%BSA/PBS), and anti-mouse secondary antibody.

2.3. Cell Surface Biotinylation Assay

A cell surface biotinylation assay allows the researcher to detect the endocytosis and transport of endogenous Trk receptor tyrosine kinases in compartmentalized cultures. Cell surface biotinylation assays require projecting compartmentalized cultures, a reversible membrane-impermeable form of biotin (EZ-Link NHS-S-S-biotin, 1.5 mg/mL in 1× PBS; Pierce Chemical), glycine (Sigma), iodoacetamide (Sigma), glutathione (Sigma), and neutravidin-agarose beads (Pierce Chemical), RIPA solution (50 mM Tris–HCl, 150 mM NaCl, 1 mM EDTA, 1% NP-40, 0.25% deoxycholate).

3. Procedures

3.1. Preparation of Compartmentalized Chambers

A compartmentalized culture, or Campenot chamber, consists of a three-compartment Teflon divider sealed with silicone grease to a collagen-coated culture dish (Fig. 1a). The central compartment contains neuronal cell bodies and proximal axon segments, while the distal compartments contain only axons. If prepared correctly, there should be no fluid exchange between the two compartments, which are separated by 1 mm.

There are three major steps in the preparation of compartmentalized cultures. First, the researcher coats culture dishes with collagen and makes scratches in the collagen substratum to guide axon growth. Second, silicone grease is added to the Teflon divider, which will eventually form the barrier between the cell body/proximal axon and distal axon compartment. Third, the Teflon divider is carefully seated onto the culture dish and they are pressed together to form a seal. The compartmentalized cultures should be built the day before seeding with dissociated neurons to check for leaks in the barrier overnight. When first learning to build compartmentalized cultures, it is necessary to build more compartmentalized cultures than are needed for your experiments because some will leak. Specific neuronal populations are routinely cultured in the compartmentalized culture system. The most common are sympathetic neurons derived from the superior cervical ganglia and sensory neurons derived from the dorsal root ganglia.

First, the researcher coats the desired number of tissue culture dishes with collagen. The tissue culture dishes are initially coated with collagen, at a final concentration of 0.71 μg/mL in 0.1% acetic acid solution, under a sterile tissue culture hood. Normally, the compartmentalized culture will be seated directly onto the tissue culture dish. However, for immunocytochemical assays the compartmentalized culture will be seated onto a glass slide within the culture dish. Seating the compartmentalized culture on a glass slide will allow the researcher to subsequently mount the neurons for microscopic analysis. To guide axon growth from the central compartment into the distal axon compartment, parallel scratches are made into the dried collagen substrate using a pin rake tool. Removing the collagen substratum directs axon growth between the scratches, which run perpendicular to the Teflon grease barrier (Fig. 1b). This ensures that a high percentage of neurons successfully project their axons into the distal compartment.

Following preparation of the collagen-coated tissue culture dishes, the researcher traces the bottom of the Teflon divider with a smooth line of silicone grease where the divider will contact the tissue culture dish. This layer of silicone grease will act as a liquid impermeable barrier between the compartments; therefore, it must be applied in a continuous and even line to form an effective seal.

To obtain a consistent application of silicone grease, the researcher should grip the Teflon divider with hemostats and place the divider under a light microscope. When applying the grease, visually confirm that there are no breaks in the line of silicone grease. To apply the silicone grease, the researcher uses sterilized glass pipettes equipped with needles that are cut to approximately 1 cm in length. An autoclaved glass pipette equipped with a 25-gauge needle should be used to apply the layer of silicone grease where the axons will penetrate the barrier and a 20-gauge needle should be used for the rest of the divider. The 25-gauge needle is used where the axons cross the barrier in order to create a thinner layer of silicone grease, which will enhance axon penetration into the distal axon compartment. After applying the silicone grease, the researcher is ready to place the Teflon divider onto the collagen-coated tissue culture dish.

The barrier is formed by seating the Teflon divider coated with silicone grease onto the culture dish with a drop of DMEM-methylcellulose in between (Fig. 1a). The drop (45 μL) of DMEM–methylcellulose is placed onto the collagen culture dish over the entire region where the axons will cross into the distal chamber. The region is clearly demarcated by the axon-guiding scratches laid down in step one. A layer of media is necessary for the axons to grow through the silicone grease barrier; therefore it is important that its covers the entire length of the barrier to promote robust axonal projections. The next step is to seat the Teflon divider onto the culture dish and form a seal. The Teflon divider is oriented upside down with the silicone grease facing up. Then the researcher places the culture dish onto the Teflon divider, with the drop of DMEM–methylcellulose aligned to the central compartment, and gently taps the culture dish down with a pipette tip until a continuous seal is formed between the two surfaces. Finding the right amount of force is important. Applying too little force results in a permeable barrier and applying too much force results in sparse or absent axonal projections. After allowing the compartmentalized culture to set at 37°C for at least 4 h, media should be added to the distal axon compartment for 12 h to check for any leaks in the seal. Any leaking compartmentalized cultures should be discarded. With the remaining chambers, the researcher blocks the mouth of the center compartment with silicone grease to prepare the center compartment for seeding with dissociated neurons.

Dissociated sympathetic or sensory neurons (~6,000) are plated into the center compartment of the compartmentalized chamber. The neurons are cultured in an artificial neurotrophin gradient to promote the projection of axons into the distal axon compartment. Media containing 10 ng/mL NGF is added to the center compartment and 50–100 ng/mL NGF is added to the distal compartment. The anti-mitotic agent, cytosine-arabinofuranoside (AraC, 50 μM), is added to the central compartment for 48–72 h to eliminate glial cells and obtain a pure culture of neurons. After 5–9 days in culture, with media changes every 48 h, axons will project through the barrier into the distal axon compartment. To detect NGF-dependent activation of TrkA signaling or retrograde transport, it is first necessary to deprive the cultures of NGF. To prevent neuronal apoptosis during NGF deprivation, the general caspase inhibitor boc-aspartyl(*O*-methyl)-fluoromethylketone (BAF, 50 μM, MP Biomedicals) is added to the cell body compartment.

Biochemistry compartmentalized cultures, which are built with a larger Teflon divider design, accommodate approximately 30,000 neurons. These cultures are preferable for assays that require biochemical analysis of target proteins, such as cell surface biotinylation assays. Although the protocol is similar, there are a few differences in the preparation of biochemistry compartmentalized cultures. The larger Teflon dividers are seated onto 6 cm tissue culture dishes instead of 3.5 cm dishes. To cover the longer length of the axon-crossing barrier, four successive groups of scratches are made into the collagen surface with the pin rake and 100 μL of DMEM–methylcellulose is used. Finally, the researcher may use a 22-gauge needle, instead of a 25-gauge needle, for the application of silicone grease to the Teflon divider where the axons will penetrate. This may result in a more reliable seal in the liquid impermeable barrier between the cell body/proximal axon and distal axon compartment. Biochemistry compartmentalized cultures are more difficult to prepare without a leaking barrier due to their larger size. Therefore, approximately twice as many chambers should be prepared than are needed for the desired experiment. It may be necessary to ensure there are no leaks across the barrier by the addition of fluorescently labeled microspheres (Fluospheres, Invitrogen) to the distal axon compartment and confirming the lack of diffusion to the cell body compartment. Biochemistry compartmentalized cultures require 11–14 days in vitro to project axons into the distal axon compartment.

3.2. Antibody Feeding Assay in Compartmentalized Cultures

Antibody feeding assays are a well-established immunocytochemical method for following the trafficking of cell surface receptors (26). This assay can be easily adapted to follow the directional transport of Trk receptors in compartmentalized cultures. By adding antibody to the appropriate compartment, the researcher can live label epitope-tagged receptors on distal axons or proximal axons/cell bodies and follow their transport (Fig. 2a). The immunofluorescent detection of receptors that were labeled in a specific cellular compartment provides a method for analyzing receptor trafficking and may be used in a variety of experimental paradigms. We will focus on the retrograde and anterograde transport of NGF–TrkA, although these assays are applicable to other ligand–receptor combinations of interest.

First, compartmentalized cultures are established that express an epitope-tagged TrkA receptor. This may be accomplished by the transfection of neurons with FLAG-TrkA or the use of FLAG-TrkA knock-in mice (19). We have previously used a FLAG-tagged chimeric Trk receptor that has the intracellular and transmembrane domain of TrkA and the extracellular domain of TrkB. This results in a chimeric receptor that responds to a different ligand (BDNF) that may limit potential complications due to transactivation with endogenous TrkA receptors (25). For structure–function studies, it is important to eliminate the possibility of compensation from endogenous TrkA receptors. To avoid this complication, the researcher may either generate chimeric receptors, transfect FLAG-TrkA into *TrkA*$^{-/-}$ neurons, or use neurons derived from FLAG-TrkA knock-in mice.

To measure the retrograde transport of TrkA, the researcher exclusively labels the surface TrkA receptors in distal axons by incubating the distal compartments under live cell conditions with anti-FLAG antibody (M2 Sigma 4.2 μg/mL in DMEM without serum or Fab fragments) at 4°C (Fig. 2a). Controlling the temperature allows the researcher to gate the endocytosis and transport of cell surface receptors. At 4°C there is minimal endocytosis or movement of cell surface receptors; therefore, a stable baseline of surface TrkA receptors is labeled with antibody. After washing off the excess antibody, moving the cultures to 37°C will permit the endocytosis of TrkA receptors upon stimulation. Addition of NGF (100 ng/mL in DMEM) to the distal axons induces the retrograde transport of TrkA that was labeled with antibody on the surface of the axons. NGF–TrkA retrograde transport reaches a steady state at approximately 8 h (27); therefore, the researcher may perform analysis after 8 h for maximal detection. Standard immunocytochemical detection of the FLAG antibodies with fluorophore-labeled secondary antibodies enables the researcher to follow the movement of TrkA receptors that were labeled in the distal axons to the proximal axons and cell bodies (Fig. 2b).

Researchers may also use an antibody feeding assay to measure the anterograde transport of Trk receptors via transcytosis (25). Transcytosis is the process by which newly synthesized receptors are first transported to the plasma membrane of neuronal cell bodies and are then endocytosed and transported in the anterograde direction to distal axons. To detect anterograde trafficking of TrkA, the researcher simply adds anti-FLAG antibody (M2 Sigma 4.2 μg/mL in DMEM without serum or Fab fragments) to the cell body/proximal axon compartment. This allows the researcher to label receptors present on the surface of neuronal cell bodies and proximal axons. By following the transport of labeled receptors to the distal axons via immunocytochemistry, this method was recently used to detect the transport of TrkA receptors via transcytosis (25). TrkA labeled on the surface of neuronal cell bodies and proximal axons is detectable in the distal axons within 4 h, uncovering an unexpected mechanism underlying the transport of newly synthesized TrkA to axons (25). Interestingly, addition of NGF to the distal axons enhances the transcytosis of TrkA. This suggests that a positive feedback loop exists by which neurons that receive NGF may enhance their sensitivity to the target-derived growth factor by increasing the transport of TrkA to axons. Similar antibody feeding assay experiments may be adapted to test the anterograde transport of other neuronal receptors via transcytosis. To date, only a few neuronal receptors have been described to undergo transcytosis, namely L1/neuron–glia cell adhesion molecule (28) and type 1 cannabinoid receptor CB1R (29). However, it is possible that transcytosis may be a relatively common mechanism of anterograde transport in neurons.

3.3. Cell Surface Biotinylation Assay in Compartmentalized Chambers

The cell surface biotinylation assay is a well-established biochemical method to follow the trafficking of endogenous receptors in neurons (30). This assay can be easily adapted to follow the directional trafficking of Trk receptors in compartmentalized cultures. The researcher may add biotin to the appropriate compartment and label surface receptors on the proximal axons/cell bodies or distal axons of neurons (Fig. 3a). Subsequently, collecting the cellular lysate in each compartment will allow the detection of transported TrkA by neutravidin precipitation and immunoblot analysis (Fig. 3b). The cell surface biotinylation assay offers the advantages of following endogenous receptor trafficking with minimal adulteration. Due to its small size (244.31 Da), biotin is unlikely to affect the natural function of labeled proteins. However, this method of TrkA detection is less sensitive than the antibody feeding assay. Therefore, it may be necessary to use the biochemistry compartmentalized cultures to obtain more cellular material for analysis.

The researcher may label the surface TrkA receptors exclusively in the distal axons by adding biotin solution (EZ-Link NHS-S-S-biotin, 1.5 mg/mL in 1× PBS/2 mM $CaCl_2$/1 mM $MgCl_2$) to the distal axon compartment for 30 min at 4°C (Fig. 3a). Chemical addition of NHS-Biotin to primary amines (lysines and amino termini) effectively labels TrkA and other proteins present on the plasma membrane. Again, reducing the temperature limits the endocytosis and transport of cell surface receptors during the biotinylation step. After washing off the excess biotin (3×, 10 min, PBS/2 mM $CaCl_2$/1 mM $MgCl_2$), adding NGF (100 ng/mL in DMEM) to the distal axons at 37°C will induce retrograde transport of TrkA. Subsequently, neutravidin precipitation of cell lysates (60 min precipitation @ 4°C, of cell lysates that were centrifuged to remove cell debris in RIPA solution; 50 mM Tris–HCl, 150 mM NaCl, 1 mM EDTA, 1% NP-40, 0.25% deoxycholate) collected from the cell bodies/proximal axon and axon compartments will

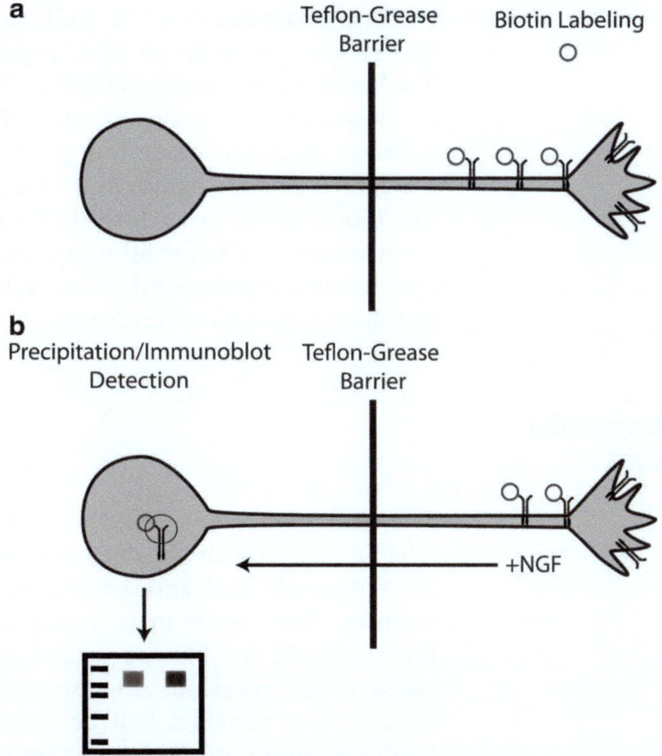

Fig. 3. Cell surface biotinylation assay in compartmentalized cultures. (**a**) The distal axon compartment is incubated with biotin to label surface receptors exclusively in the distal axons. (**b**) Retrograde transport of TrkA is induced by adding NGF to distal axons. Subsequently, the researcher may detect retrogradely transported TrkA by a neutravidin precipitation and immunoblotting for TrkA.

isolate the biotin-labeled proteins in a spatially selective manner. Finally, immunoblotting for TrkA provides a reliable method for the detection of retrogradely transported TrkA present in the cell body lysates (Fig. 3b). Biotinylated TrkA will appear on a 7.5% SDS-PAGE gel at approximately 140 kDa, which is representative of its glycosylated and biochemically mature form. For a quantitative measure of transported TrkA, the researcher may employ densitometric quantification of the immunoblot signal. An important advantage of cell surface biotinylation assays is that they provide an opportunity to assess the activation of downstream signaling pathways in parallel with TrkA retrograde transport. Immunoblotting for NGF–TrkA signaling effectors, such as phospho-Akt, phospho-Erk1/2, or phospho-CREB, enhances reliability by measuring multiple facets of retrograde signaling. Furthermore, due to the unprejudiced labeling of surface receptors with biotin this method also allows the researcher to assess the transport of other receptors, such as p75, in parallel with TrkA. This method was employed to demonstrate that TrkA and NGF are transported as a complex, by adding radiolabeled ^{125}I-NGF to distal axons and combining a neutravidin precipitation of retrogradely transported biotinylated TrkA with autoradiography (13).

Researchers may also use the cell surface biotinylation assay to measure the anterograde transport of Trk receptors via transcytosis. To detect anterograde trafficking of TrkA, the researcher adds the biotin solution exclusively to the cell body/proximal axon compartment (4°C for 30 min). Then, a neutravidin precipitation and immunoblotting of distal axon lysates will allow the researcher to follow transcytosed receptors.

4. Experimental Variables

The antibody feeding assay and cell surface biotinylation assay are reliable methods for the detection and measurement of Trk receptor retrograde and anterograde transport in compartmentalized cultures. These assays may be adapted to a variety of culture conditions, genetic modifications, and pharmacological treatments to assess spatial signaling downstream of Trk receptors. The basic experimental variables will be the addition of control media or NGF containing media to the distal axons, which mimics the presentation of NGF in vivo. The researcher can then test the requirements for specific cytoplasmic effectors in TrkA retrograde transport by genetic or pharmacological manipulation.

For example, expression of dominant negative dynamin (dynamin1K44A) blocks the ability of axon-applied NGF to retrogradely promote neuronal survival in compartmentalized cultures, demonstrating the requirement for endocytosis in retrograde signaling (17). Expression of dominant negative dynamin1 also blocks the anterograde transport of TrkA via transcytosis, indicating that anterograde transport of TrkA from the membrane is an endocytic process and not due to lateral diffusion of TrkA on the membrane (25). Genetic means may also be used to identify components of the vesicular transport mechanism. For example, impairing dynein activity by overexpression of dynamitin blocks Trk retrograde signaling (31). Expression of dominant negative Rab-11, a monomeric G protein that regulates endocytic recycling, blocks the anterograde transport of TrkA via transcytosis (25).

Adding pharmacological agents to the proximal axons/cell bodies or distal axons enables the researcher to locally inhibit targets of interest and characterize the location of their function. Pharmacological inhibition of TrkA kinase activity in compartmentalized cultures has been used to show that TrkA kinase activity is required in the cell bodies for retrograde activation of downstream signaling pathways and neuronal survival (10, 11, 17).

Recently, a chemical genetic approach has been described to inhibit Trk kinase activity, employing the small molecule inhibitor, NMPP1. This small molecule specifically inhibits genetically modified Trk receptors, with phenylalanine-to-alanine substitutions in the ATP binding pocket. Knock-in mice harboring a knock-in *TrkA* allele (*TrkA*^{F592A}) has been generated to allow specific, acute, and reversible inhibition of TrkA kinase activity in vivo and in vitro (32). Addition of NMPP1 to the distal axons or proximal axons/cell bodies of neurons expressing TrkA^{F592A} may be done to address the spatial requirements for TrkA kinase activity in compartmentalized cultures. Pharmacological agents may also be used to test the requirements of downstream signaling pathways for TrkA transport. NGF–TrkA signaling activates three major pathways: PI3K, MAPK, and PLCγ. By adding inhibitors of signaling effectors in these pathways, the researcher can exert spatial control of TrkA signaling. For example, addition of PI3K inhibitors, wortmannin or LY294002, to the distal axons of compartmentalized cultures blocks NGF–TrkA retrograde transport and neuronal survival (10). Recently, we uncovered a PLCγ/calcineurin pathway that is important for TrkA endocytosis (33). Addition of calcineurin inhibitors to the distal axons of compartmentalized cultures blocks NGF–TrkA retrograde signaling (Ascano, Bodmer, and Kuruvilla, unpublished results) presumably by blocking TrkA endocytosis in distal axons.

5. Typical/ Anticipated Results

5.1. Anterograde Trk Transport Monitored by an Antibody-Feeding Assay

A typical experiment assessing TrkA anterograde transport using an antibody-feeding assay is presented in Fig. 4. In this study, sympathetic neurons expressing a FLAG tagged TrkB:A chimeric receptor that has the extracellular domain of TrkB and the transmembrane and intracellular domains of TrkA were grown in compartmentalized cultures. Sympathetic neurons do not normally express TrkB (34), but neurons expressing chimeric receptors showed signaling responses to the TrkB ligand, BDNF, in a manner that is similar to endogenous TrkA receptors stimulated with NGF (25). Live cell immunocytochemistry was performed using an antibody directed against the extracellular FLAG epitope of chimeric Trk receptors and surface receptors were labeled exclusively in cell body compartment. After 4 h, the researchers could detect labeled receptors in the distal axons of compartmentalized cultures with secondary antibody demonstrating direct evidence for Trk transcytosis. Addition of neurotrophin to the distal axon compartment significantly enhanced the transcytosis of FLAG-Trk receptors.

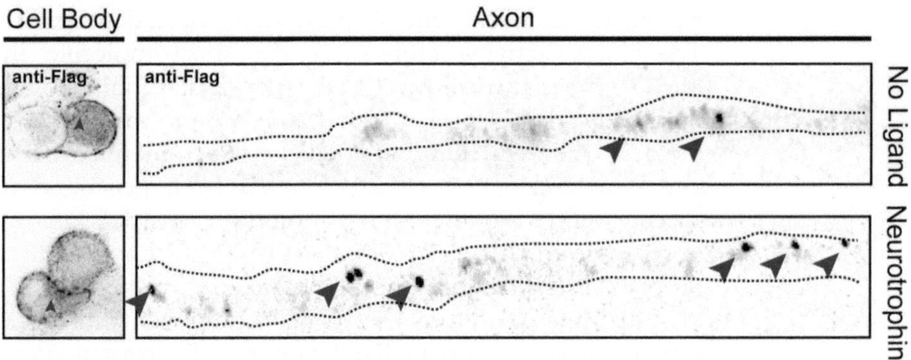

Fig. 4. Representative antibody feeding assay in compartmentalized cultures. Surface Flag-Trk receptors were labeled with anti-Flag antibody in the cell body compartment. Four hours after labeling, the distal axon and cell body compartment were immunostained to detect transported Trk receptors. The arrows mark transcytosed Trk receptors that were originally labeled on the surface of the cell bodies. Neurotrophin treatment increases the amount of transcytosed Trk present in the distal axons. From Ascano et al. (25).

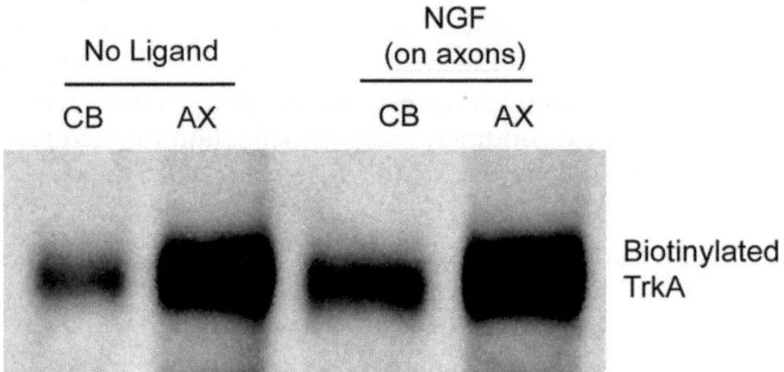

Fig. 5. Representative cell surface biotinylation experiment in compartmentalized cultures. Surface receptors in the distal axons were labeled with biotin and the distal axons were treated with NGF to induce retrograde transport. After 8 h, cellular lysates from the distal axon (AX) and cell body (CB) compartments were subjected to neutravidin precipitation and TrkA immunoblot. NGF treatment increases the amount of retrogradely transported TrkA present in the cell bodies.

5.2. Retrograde Trk Transport Detected by a Cell Surface Biotinylation Assay

A representative experiment to assess the retrograde transport of endogenous TrkA receptors using a cell surface biotinylation assay is depicted in Fig. 5. In this study, membrane proteins were selectively biotinylated in distal axon compartments of compartmentalized cultures. Neutravidin precipitation followed by Western blot analysis of lysates collected from the cell body and distal axon compartments was used to detect the presence of retrogradely transported TrkA receptors in cell bodies. As expected, treating the distal axons with NGF for 8 h increases the retrograde transport of TrkA.

6. Troubleshooting

1. The neurons may not project into the distal axon chamber of compartmentalized cultures. In this event, the researcher should try to press down on the culture dish with less force when forming a seal with the Teflon divider. Alternatively, the problem of axons not projecting may also arise when the drop of DMEM–methylcellulose does not fully cover the region where the axons should be crossing into the distal axon compartment.
2. Leaking chambers when you plate them on glass coverslips. Allow the collagen to completely dry before making the chambers, do not make the "axon guiding" scratches into the collagen surface, and use more force when forming the seal.
3. Antibody binding to epitope tagged-Trk receptors may induce autoactivation. This is a result of the antibody-dependent formation of receptor dimers that undergo autophosphorylation. It is important that the researcher ensures that addition of antibody does not lead to the activation of tagged Trk receptors. To control for autoactivation, test for the activation of downstream signaling pathways after antibody treatment. In addition, it is wise to use a functional test such as testing for neuronal survival with the addition of antibody in the absence of NGF.
4. It is important to ensure that pharmacological agents added exclusively to the cell body or distal axon compartments do not diffuse to the other compartment. Generally, the best method for testing the spatial effects of pharmacological agents is to use an appropriate phosphospecific antibody to probe cell body or distal axon lysates to verify spatially restricted pharmacological inhibition.
5. Low levels of transported TrkA are detected by immunocytochemistry or precipitation/immunoblotting. The researcher may modify the protocol to continuously label the surface TrkA receptors, instead of pulse labeling, before detecting transported TrkA. By adding biotin or antibody to the desired cellular compartment for 4–8 h without washing, the researcher can continuously label surface receptors on the distal axons or cell bodies and greatly enhance the Trk signal.

7. Conclusion

The compartmentalized culture system has been a fundamental tool for answering biological questions where it is important to distinguish between the axons and cell bodies of neurons. In particular, this system has revolutionized our understanding of NGF–TrkA transport. To specifically study directional trafficking of cell surface receptors in neurons we have outlined two methods: the antibody feeding assay and the cell surface biotinylation assay. The antibody feeding assay offers an opportunity to visually detect the anterograde and retrograde transport of epitope-tagged receptors of interest. This allows the researcher to perform colocalization studies as well as to assess the specific location of retrogradely transported receptors within the cell body compartment. The cell surface biotinylation assays offers an opportunity to quantitative and biochemically measure the retrograde and anterograde transport of endogenous receptors in compartmentalized cultures. One major advantage of this system is that the researcher may assess directional transport in neurons with minimal adulteration to the natural system. A second advantage is its versatility, in that one may analyze multiple facets of directional transport and signaling in parallel.

References

1. Huang EJ, Reichardt LF (2001) Neurotrophins: roles in neuronal development and function. Annu Rev Neurosci 24:677–736
2. Huang EJ, Reichardt LF (2003) Trk receptors: roles in neuronal signal transduction. Annu Rev Biochem 72:609–642
3. Zweifel LS, Kuruvilla R, Ginty DD (2005) Functions and mechanisms of retrograde neurotrophin signalling. Nat Rev Neurosci 6: 615–625
4. Howe CL, Mobley WC (2004) Signaling endosome hypothesis: a cellular mechanism for long distance communication. J Neurobiol 58: 207–216
5. Ehlers MD, Kaplan DR, Price DL, Koliatsos VE (1995) NGF-stimulated retrograde transport of TrkA in the mammalian nervous system. J Cell Biol 130:149–156
6. Hendry IA, Stockel K, Thoenen H, Iversen LL (1974) The retrograde axonal transport of nerve growth factor. Brain Res 68:103–121
7. Yip HK, Johnson EM Jr (1986) Comparative dynamics of retrograde transport of nerve growth factor and horseradish peroxidase in rat lumbar dorsal root ganglia. J Neurocytol 15: 789–798
8. Bhattacharyya A, Watson F, Bradlee T, Pomeroy S, Stiles C, Segal R (1997) Trk receptors function as rapid retrograde signal carriers in the adult nervous system. J Neurosci 17: 7007–7016
9. Delcroix JD, Valletta JS, Wu C, Hunt SJ, Kowal AS, Mobley WC (2003) NGF signaling in sensory neurons: evidence that early endosomes carry NGF retrograde signals. Neuron 39: 69–84
10. Kuruvilla R, Ye H, Ginty DD (2000) Spatially and functionally distinct roles of the PI3-K effector pathway during NGF signaling in sympathetic neurons. Neuron 27:499–512
11. Riccio A, Pierchala B, Ciarallo C, Ginty DD (1997) An NGF–TrkA-mediated retrograde signal to transcription factor CREB in sympathetic neurons. Science 227:1097–1100
12. Senger DL, Campenot RB (1997) Rapid retrograde tyrosine phosphorylation of TrkA and other proteins in rat sympathetic neurons in compartmented cultures. J Cell Biol 138: 411–421
13. Tsui-Pierchala BA, Ginty DD (1999) Characterization of an NGF-P-TrkA retrograde signaling complex and age-dependent regulation of TrkA phosphorylation in sympathetic neurons. J Neurosci 19:8207–8218
14. Watson FL, Heerssen HM, Bhattacharyya A, Klesse L, Lin MZ, Segal RA (2001) Neurotrophins use the Erk5 pathway to mediate a retrograde survival response. Nat Neurosci 4:981–988
15. Grimes ML, Zhou J, Beattle EC, Yuen EC, Hall DE, Valletta JS, Topp KS, LaVail JH, Bunnett NW, Mobley WC (1996) Endocytosis of activated TrkA: evidence that nerve growth factor induces formation of signaling endosomes. J Neurosci 16(24):7950–7964
16. Valdez G, Akmentin W, Philippidou P, Kuruvilla R, Ginty DD, Halegoua S (2005) Pincher-mediated macroendocytosis underlies

retrograde signaling by neurotrophin receptors. J Neurosci 25:5236–5247

17. Ye H, Kuruvilla R, Zweifel LS, Ginty DD (2003) Evidence in support of signaling endosome-based retrograde survival of sympathetic neurons. Neuron 39:57–68
18. Watson FL, Heerssen HM, Moheban DB, Lin MZ, Sauvageot CM, Bhattacharyya A, Pomeroy SL, Segal RA (1999) Rapid nuclear responses to target-derived neurotrophins require retrograde transport of ligand–receptor complexes. J Neurosci 19:7889–7900
19. Sharma N, Deppmann CD, Harrington AW, St Hillaire C, Chen ZY, Lee FS, Ginty DD (2010) Long-distance control of synapse assembly by target-derived NGF. Neuron 67:422–434
20. Campenot RB (1977) Local control of neurite development by nerve growth factor. Proc Natl Acad Sci USA 74:4516–4519
21. Campenot RB (1979) Independent control of local environment of somas and neurites. Meth Enzymol 58:302–307
22. MacInnis BL, Campenot RB (2002) Retrograde support of neuronal survival without retrograde transport of nerve growth factor. Science 295:1536–1539
23. Ure DR, Campenot RB (1997) Retrograde transport and steady-state distribution of ^{125}I-nerve growth factor in rat sympathetic neurons in compartmentalized cultures. J Neurosci 17:1282–1290
24. Cui B, Wu C, Chen L, Ramirez A, Bearer EL, Li WP, Mobley WC, Chu S (2007) One at a time, live tracking of NGF axonal transport using quantum dots. Proc Natl Acad Sci USA 104:13666–13671
25. Ascano M, Richmond A, Borden P, Kuruvilla R (2009) Axonal targeting of Trk receptors via transcytosis regulates sensitivity to neurotrophin responses. J Neurosci 29:11674–11685
26. Cao TT, Deacon HW, Reczek D, Bretscher A, von Zastrow M (1999) A kinase-regulated PDZ-domain interaction controls endocytic sorting of the beta2-adrenergic receptor. Nature 401:286–290
27. Claude P, Hawrot E, Dunis DA, Campenot RB (1982) Binding, internalization, and retrograde transport of ^{125}I-nerve growth factor in cultured rat sympathetic neurons. J Neurosci 2:431–442
28. Wisco D, Anderson ED, Chang MC, Norden C, Boiko T, Folsch H, Winckler B (2003) Uncovering multiple axonal targeting pathways in hippocampal neurons. J Cell Biol 162: 1317–1328
29. Leterrier C, Laine J, Darmon M, Boudin H, Rossier J, Lenkei Z (2006) Constitutive activation drives compartment-selective endocytosis and axonal targeting of type 1 cannabinoid receptors. J Neurosci 26:3141–3153
30. Kuruvilla R, Zweifel LS, Glebova NO, Lonze BE, Valdez G, Ye H, Ginty DD (2004) A neurotrophin signaling cascade coordinates sympathetic neuron development through differential control of TrkA trafficking and retrograde signaling. Cell 118:243–255
31. Heerssen HM, Pazyra MF, Segal RA (2004) Dynein motors transport activated Trks to promote survival of target-dependent neurons. Nat Neurosci 7:596–604
32. Chen X, Ye H, Kuruvilla R, Ramanan N, Scangos KW, Zhang C, Johnson NM, England PM, Shokat KM, Ginty DD (2005) A chemical-genetic approach to studying neurotrophin signaling. Neuron 46:13–21
33. Bodmer D, Ascano M, Kuruvilla R (2011) Isoform-specific dephosphorylation of Dynamin1 by calcineurin couples neurotrophin receptor endocytosis to axonal growth. Neuron 70:1085–1099
34. Atwal JK, Massie B, Miller FD, Kaplan DR (2000) The TrkB-Shc site signals neuronal survival and local axon growth via MEK and P13-kinase. Neuron 27:265–277

Chapter 17

Mammalian Target of Rapamycin

Lukasz J. Swiech*, Malgorzata Urbanska*, Matylda Macias, Agnieszka Skalecka, and Jacek Jaworski

Abstract

Mammalian target of rapamycin (mTOR) is a serine/threonine protein kinase known to merge extracellular instructions with information about cellular metabolic resources and control the rate of anabolic and catabolic processes accordingly. In neurons, mTOR has been implicated in several aspects of development and physiology. The major problems with studying mTOR functions in neurons are caused by the restricted knowledge of downstream mTOR effectors and relatively poor tools for studying this particular kinase. In this chapter, we describe the materials and methods routinely used in our laboratory to study mTOR function and activity and the binding partners of mTOR in neurons.

Key words: mTOR, mTOR inhibitors, Nervous system, Cell culture, Genetic modification, Protein–protein interactions, Proximity ligation assay, Bio pull-down

1. Background

Mammalian target of rapamycin (mTOR) is a serine/threonine protein kinase known to merge extracellular instructions with information about cellular metabolic resources and control the rate of anabolic and catabolic processes accordingly (1, 2). mTOR forms two functionally different complexes in mammalian cells: mTORC1 and mTORC2 (3, 4). These complexes share some protein components, but their distinctive activities are defined by their unique components: Raptor, Rictor and mSin1. mTORC1 contains Raptor and is the actual target of rapamycin, a macrolide that has been isolated from *Streptomyces hygroscopicus* (3). mTORC2 contains Rictor and mSin1 and is considered resistant to this drug ((4), but see (5)).

*Swiech and Urbanska contributed equally to this work.

Hideyuki Mukai (ed.), *Protein Kinase Technologies*, Neuromethods, vol. 68,
DOI 10.1007/978-1-61779-824-5_17, © Springer Science+Business Media, LLC 2012

mTORC1 activation is a well-studied phenomenon. Hormones, trophic factors, mitogens, and amino acids activate mTORC1-dependent processes (6, 7), whereas energy or amino acid deficiency decreases it (6, 7). Consequently, the application of neurotrophins (e.g., brain-derived neurotrophic factor (BDNF)], hormones (e.g., insulin), and amino acids is widely used for the experimental induction of mTOR in neurons by physiological means (see section "Pharmacological Modulation of mTOR Activity"). In neuronal cells, mTORC1 is also modulated by neurotransmitters, such as glutamate (via ionotropic and metabotropic receptors) (8, 9), dopamine (10), and noradrenaline (11). Indirect evidence also shows that activation of muscarinic acetylcholine receptors leads to activation of the mTORC1 pathway (12). In contrast, knowledge of the conditions needed for mTORC2 activation is very limited, even in nonneuronal cells (13, 14).

Studies in different model systems (e.g., yeast, fruit flies, and mammalian nonneuronal cells) strongly suggest the involvement of mTORC1 in the regulation of cellular processes, such as translation, transcription, autophagy, lipid metabolism, mitochondrial function, microtubule dynamics, and membrane trafficking (15, 16). In contrast, mTORC2 has been implicated in the control of actin dynamics and activity of several AGC (cyclic adenosine monophosphate-dependent protein kinases A, cyclic guanosine monophosphate-dependent protein kinases G, and phospholipid-dependent protein kinases C) family kinases (e.g., Akt, SGK1, PKCα) (4, 17–19). In neurons, mTOR is mostly studied as a part of mTORC1 in the context of protein translation or degradation (20–22). Consequently, assays to study mTORC1 activity in neurons rely on monitoring the phosphorylation status of eukaryotic initiation factor 4E binding proteins (4E-BPs) or p70S6 kinase (p70S6K). Moreover, p70S6K downstream effectors, such as ribosomal protein S6 (rpS6), eukaryotic initiation factor 4B (eIF4B), and eukaryotic elongation factor-2 (eEF2; an indirect effector), are often used (see Sect. 3.2). 4E-BPs and p70S6K are the most extensively studied mTORC1 targets, the biological role of which is the control of translation (23, 24). mTOR promotes protein translation by inhibiting translation suppressor 4E-BP and activating the protein synthesis inducer p70S6K (1, 2). Active p70S6K, in turn, phosphorylates several targets important for the further regulation of translation initiation and elongation (25). Our recent research (26), however, demonstrated the importance of mTORC1 activities unrelated to protein translation in neurons. We showed that CLIP-170, which is controlled by mTORC1, regulates microtubule-actin cross-talk during dendritic arbor development. However, studies of "unconventional" mTOR roles in neurons are quite rare, and further advances require the identification of novel interacting partners and methods to reproducibly follow such interactions. In Sect. 3.3, we provide a description of the methods we found useful in this context.

In neurons, mTOR has been implicated in several aspects of development and physiology (20–22, 27). Neuronal differentiation, axon elongation and directional movements, dendritogenesis, and spinogenesis have all been shown to depend on mTORC1 (21). Long-term synaptic plasticity and learning and memory can be influenced by rapamycin or overactivation of the mTOR signaling pathway (20–22). Relatively recent research linked changes in mTORC1 activity in the hypothalamus and superchiasmatic nucleus with the regulation of feeding behavior (28) and alterations in circadian rhythm in rodents in response to environmental cues (29), respectively. In contrast to extensive knowledge about mTORC1, revealing the neuronal functions of mTORC2 has just begun. Siuta et al. (30) reported that mice that lack Rictor in neuronal cells display cortical hypodopaminergia and schizophrenia-like behavior, most likely attributable to dysregulation of Akt activity. Some hints also come from research on dendritic arbor development in *Drosophila*, showing the importance of dTORC2 for the establishment of proper boundaries between neurons with nonoverlapping dendritic fields (i.e., tilling) (31).

Important aspects of mTOR-related brain research are tools that modulate mTOR function in neuronal cells. Rapamycin and its derivatives (Rapalogs; e.g., CCI-779 and RAD0001) are very specific inhibitors of mTOR activity and therefore are widely employed for neuronal cell culture studies or in biochemical preparations (e.g., synaptosomes). The in vivo use of these drugs, however, has been quite limited, mostly because of poor penetrance through the blood–brain barrier and the consequent need for long-term application of high drug doses (32). This problem can be overcome by brain injections of Rapalogs (28, 33, 34). This strategy, although laborious, provides better control over drug application with regard to both concentration and time. These aspects are quite important because the effects of rapamycin change with time and drug concentration. Short-term rapamycin application (from minutes to 1 h) results in the inhibition of selected mTORC1 targets (e.g., p70S6K), whereas long-term application (up to a few hours) and higher doses further extend the group of inhibited mTORC1 effectors (e.g., 4E-BP1) (35). Chronic rapamycin application eventually leads to the inhibition of both mTORCs (5), at least in selected cell types. Some mTORC1 targets, however, are never inhibited by Rapalogs and can be targeted only with the use of new-generation adenosine triphosphate (ATP)-competitive inhibitors of mTOR (e.g., Torin-1, Ku-0063794, PP242, and WYE-354) (35), but these compounds do not discriminate between mTORCs. Therefore, the most straightforward way to fully distinguish mTORC1 and mTORC2 functions is either by knockout (KO) or knockdown of Raptor or Rictor (or mSin1), respectively. In neurons, an RNA interference (RNAi) approach is thus far the only option for mTORC1 because Raptor KOs, similar to mTOR

Table 1
Genetic modification used to study mTOR pathway in neuronal cells

Type of modification	Effect on mTOR pathway	References
Upstream mTOR		
PTEN KO[a]	↑, Increased P-S6	(32, 40–42)
TSC1 KO[a]	↑, Increased P-S6	(43–47)
TSC2 KO[a]	↑, Increased P-S6	(48, 49)
TSC1 RNAi	↑, Not defined in neurons; expected increase in mTOR activity	(45, 50)
TSC2 RNAi	↑, Increased P-S6	(43, 46, 51)
PI3K-CA (p110[a]; p110-CAAX)	↑, Increased P-S6	(52, 53) and Fig. 1
AMPK KO[a]	↑, Not defined in neurons; expected increase in mTOR activity	(50)
Akt-CA (myr-Akt)	↑, Increased P-S6	(52–54)
CA-Rheb (Q64L; S16H)	↑, Increased P-S6	(49, 54, 55)
DN-Rheb (D60K)	↓, Not defined in neurons; expected decrease in mTOR activity	(55)
FKBP12 KO[a]	↑, Increased P-mTOR, P-S6K(389), and interaction between mTOR and Raptor	(56)
mTOR complexes		
Rictor KO	↓, Decreased P-Akt (473)	(30)
mTOR RNAi	↓, Decreased activity of eEF2; inhibition of BDNF-induced S6K1 phosphorylation	(52, 57, 58)
Downstream mTOR		
4E-BP2 KO	↑, Increased eIF4E and eif4G association	(36, 39)
S6K1 KO	↓, Increased level of EF1A	(37, 38)
S6K2 KO	↓, Decreased basal levels of P-S6; increased level of EF1A	(37, 38)
S6K1CA (ΔCT[T389E])	↑, Increased local translation	(55)
S6K1-DN (T229A)	↓, Not defined in neurons; expected decrease in protein translation	(55)
S6K RNAi	↓, Not defined in neurons; expected decrease in protein translation	(52)
4E-BP1(2A); 4E-BP1(5A)	↓, Not defined in neurons; expected decrease in protein translation	(52, 55)

[a]Note that several of these mice are conditional KOs, and the cited studies need to be verified for further genotype details

KOs, are embryonically lethal, and conditional brain KO of this protein has not been described. Some neuronal functions of mTORC1 were studied, however, with the use of KO of its downstream effectors 4EBP2 and p70S6K (36–39) (Table 1). Mice that lack Rictor exclusively in the brain, in contrast, are available but have not been extensively used for the thorough dissection of the molecular functions of mTORC2. An alternative approach to downregulate mTORC2 functions in neurons is to knockdown Rictor or mSin1 using an RNAi approach.

The genetic approach is also used to study the neuronal effects of mTOR upregulation (Table 1). The most often used models are mice that lack either phosphatase and tensin homolog deleted on chromosome 10 (PTEN) or components of the tuberous sclerosis complex (TSC1 or TSC2) (Table 1). Recently, mice that lack 12-kDa FK506-binding protein (FKBP12), a protein needed for inhibitory rapamycin effects, have been used as a model to study the effects of mTOR upregulation (56). In our laboratory, for in vitro culture research, we usually use overexpression of constitutively active mutants of PI3K or Akt, which act upstream of mTOR (26, 52). The overexpression of constitutively active Ras homolog enriched in brain (Rheb) is also an option (Table 1). Alternatively, short-hairpin RNA (shRNA)-mediated knockdown of PTEN, TSC1/TSC2, or dominant-negative mutants of TSC2 can be used (Table 1). Such tools can also be used in vivo with the use of appropriate viral vectors. However, these approaches suffer from changes in the activity of several proteins unrelated to mTOR, and the results need to be interpreted very carefully.

This brief introduction shows that mTOR is important for understanding neuronal physiology and pathology, but the tools that have been developed thus far are still far from optimal. In the following sections, we discuss the materials and methods we routinely use to study mTOR function and activation and its binding partners in neurons.

2. Materials and Equipment

2.1. Manipulating mTOR Activity in Neurons

2.1.1. Culturing Primary Hippocampal Neurons

Plasticware and glassware. Plastic noncoated culture dishes (e.g., TPP); 13–24 mm borosilicate glass coverslips (VWR); Biocoat™ poly-D-lysine Cellware dishes (BD Biosciences).

Coating. Laminin (Roche); poly-L-lysine (Sigma); 0.1 M borate buffer (pH 8.5, filter-sterilized, stored at 4°C; 1.24 g boric acid, 1.90 g Borax in 500 ml distilled water); coating buffer (freshly prepared; 37.5 μg/ml poly-L-lysine, 2.5 μg/ml Laminin in 0.1 M borate buffer).

Dissection reagents. 10× Hanks Buffer (Sigma); 2.5% Trypsin (Invitrogen); Hank's Balanced Salt Solution (HBSS; filter-sterilized, stored at 4°C; 1× Hanks Buffer, 10 mM HEPES [pH 7.3], 1% penicillin–streptomycin solution).

Media and supplements. Neurobasal™ (Invitrogen); B27 Supplement (Invitrogen; aliquoted and kept frozen); penicillin–streptomycin solution (Sigma; 10,000 units penicillin/10 mg streptomycin/ml; aliquoted and kept frozen), glutamine (Sigma; 200 mM solution; aliquoted and kept frozen), glutamate (Sigma; 10 mM solution in water, filter-sterilized, kept at 4°C); neuron plating medium (freshly prepared; 2% B27, 0.5 mM glutamine, 12.5 μM glutamate, 1% penicillin–streptomycin in Neurobasal™).

Equipment. Stereoscopic microscope, forceps (#4 and #5), CO_2 incubator, laminar flow cabinet.

2.1.2. Modulation of mTOR Activity

Pharmacological Modulation of mTOR Activity

Drugs. Rapamycin (Calbiochem or LC Laboratories; rapamycin for in vitro use: 10 mM in dimethylsulfoxide [DMSO], stored at −20°C; rapamycin stock for in vivo use: 75 mg of rapamycin in 1 ml of absolute ethanol, stored at −20°C; rapamycin working solution for in vivo use: 5 mg/ml in 5% polyethyleneglycol 400 and 5% Tween 80, freshly prepared prior to injection); Ku-0063794 (Chemdea; 30 mM in DMSO, stored at −20°C, protected from light); insulin (Sigma; 10 mg/ml in 25 mM HEPES [pH 8.2], stored at 4°C); BDNF (Sigma or Alomone Labs; 50 ng/μl in water, aliquoted and stored at −20°C, repetitive thawing and refreezing should be avoided).

Genetic Modulation of mTOR Activity

Plasmid DNA. Purified plasmid DNA for neuronal transfections should be of good quality (prepared with Qiagen Mini, Midi, or Maxi Plasmid Kit or equivalent; use of Endofree Plasmid Maxi Kit is unnecessary).

Transfection reagents. Lipofectamine 2000™ (Invitrogen); Combimag (OZ Biosciences); Amaxa® Rat Neuron Nucleofector® Kit (Lonza).

Transfection equipment. Magnetic plate (OZ Biosciences); Amaxa Nucleofector® Device (Lonza) with 0.2 cm electroporation cuvettes.

2.1.3. Image Analysis and Morphometry of Dendritic Tree

Equipment. Fluorescence or confocal microscope equipped with 20× and 40× objectives.

Software. ImageJ 1.44 (http://rsbweb.nih.gov/ij/); NeuronJ plug-in (59); Sholl Analysis plugin v.1.0. (http://biology.ucsd.edu/labs/ghosh/software).

2.2. mTOR Pathway Activity Analysis in Neurons

Biological material. Hippocampal or cortical neurons in culture or brain tissue (lysates or floating cryostat sections; see Sect. 4.2d).

Antibodies. For details regarding primary antibodies, see Table 2; anti-Mouse IgG IRDye 800CW, anti-Rabbit IgG IRDye 680 (LI-COR), anti-Rabbit IgG horseradish peroxidase (HRP), anti-Mouse IgG HRP (Jackson ImmunoResearch).

Cell lysis reagents. 5× Laemlli sample buffer (0.375 M Tris [pH 6.8], 30% glycerol, 6% sodium dodecyl sulfate [SDS], 0.25 M dithiothreitol, 20% β-mercaptoethanol, 0.06% bromophenol blue); western blot lysis buffer (20 mM Tris [pH 6.8], 135 mM NaCl, 2 mM ethylenediaminetetraacetic acid [EDTA], 1% Triton X-100, 10% glycerol, 0.5 mM dithiothreitol [DTT], protease [Roche]

Table 2
Primary antibodies applicable to mTOR studies

Antibody	Western blot	Cell-Western	IF	IP
P S235/S236-S6 #4856 Cell Signaling Technology	1:1,500	1:200	1:150	–
S6 #2217 Cell Signaling Technology	1:1,000	1:200	1:100	–
P S473-Akt #4060 Cell Signaling Technology	1:1,000	1:200	1:100	4 μg/800 μl of lysate
Akt #2920 Cell Signaling Technology	1:1,000	1:100	1:50	4 μg/800 μl of lysate
P S65-4EBP1 #9451 Cell Signaling Technology	1:200	Not tested	–	–
4EBP1 #9452 Cell Signaling Technology	1:1,000	Not tested	–	4 μg/800 μl of lysate
P S422-eIF4B #3591 Cell Signaling Technology	1:500	Not tested	–	–
P T389-S6K1 #9206 Cell Signaling Technology	1:500	Not tested	–	–
S6K1 #2708 Cell Signaling Technology	1:1,000	Not tested	–	–
P S636/S639-IRS1 #2388 Cell Signaling Technology	1:1,000	Not tested	–	–
Tubulin #T5168 Sigma	1:5,000–1:20,000	1:2,000	–	–
mTOR #2972 Cell Signaling Technology	1:1,000	Not tested	–	4 μg/800 μl of lysate
mTOR #4517 Cell Signaling Technology	1:1,000	Not tested	–	–
mTOR #2983 Cell Signaling Technology	1:1,000	Not tested	1:100	–
mTOR #OP97 Calbiochem	1:200	Not tested	–	4 μg/800 μl of lysate
Rictor #81538 Santa Cruz Biotechnology	1:1,000	Not tested	–	4 μg/800 μl of lysate
Raptor #81537 Santa Cruz Biotechnology	1:1,000	Not tested	–	4 μg/800 μl of lysate

and phosphatase [Sigma] inhibitors); tissue homogenization buffer (6 mM TRIS [pH 8.0], 150 mM $NaCl_2$, 0.5 mM $CaCl_2$, 1 mM $MgCl_2$, 0.3% 3-[(3-Cholamidopropyl)dimethylammonio]-1-propanesulfonate [CHAPS], protease [Roche] and phosphatase [Sigma] inhibitors).

Western blot analysis. Polyacrylamide gel (60); 10× TRIS-glycine buffer (0.25 M Tris [pH 8.4], 2 M glycine, 10% SDS); Towbin buffer (25 mM Tris [pH 8.3], 20% methanol, 0.192 M glycine with 0.03% SDS); Tris-buffered saline (TBS)-T (50 mM Tris [pH 7.5]; 150 mM NaCl, 0.1% Tween20); blocking buffer (5% bovine serum albumin [BSA] or 5% nonfat dry milk in TBS-T); chemiluminescent substrate solution (ECL; 1.25 mM luminol, 200 nM cumaric acid, 0.01%, H_2O_2 in 100 mM Tris [pH 8.5]); X-ray films.

Immunofluorescence on cultured cells. Phosphate-buffered saline (PBS; 0.137 M NaCl, 2.7 mM KCl, 4.3 mM Na_2HPO_4, 1.4 mM KH_2PO_4 [pH 7.4]); 4% paraformaldehyde (PFA) solution (4% PFA, 4% sucrose in PBS [pH 7.4]); Immunofluorescence Blocking Buffer I (1× PBS, 5% normal serum [compatible with secondary antibody], 0.3% Triton X-100); Immunofluorescence Antibody Dilution Buffer I (1X PBS, 1% BSA, 0.3% Triton X-100).

Immunofluorescence on brain sections. 1×PBS (pH 7.4); 0.1 M phosphate buffer (PB; 0.1 M Na_2HPO_4, 0.1 M KH_2PO_4 [pH 7.4]); 4% paraformaldehyde in 0.1 M PB; 30% sucrose in 0.1 M PB; heptane; Jung tissue-freezing medium (Leica Microsystems); antifreeze medium (150 g sucrose, 300 ml ethylene glycol, 500 ml 0.05 M PB [pH 7.4], 0.1 M sodium fluoride); PBS-T (1× PBS, 0.2% Triton X-100); Immunofluorescence Blocking Buffer II (PBS-T, 5% normal serum); Immunofluorescence Antibody Dilution Buffer II (PBS-T, 1% normal serum [compatible with secondary antibody]).

Equipment. Infrared Odyssey Imaging System; cryostat.

2.3. Analysis mTOR Interactions

2.3.1. Immunoprecipitation of mTOR Kinase

Biological material. Cortical neurons in culture ($\sim 6 \times 10^7$) or brain tissue.

Antibodies. mTOR antibody #2972 (Cell Signaling Technology).

Buffers. TBS buffer (20 mM Tris [pH 7.5], 100 mM NaCl); immunoprecipitation lysis buffer (10 mM Tris [pH 7.5], 1 mM $MgCl_2$, 0.5 mM $CaCl_{2,}$ 1% BSA, 0.5% CHAPS, protease inhibitors [Roche], phosphatase inhibitor cocktails 1 and 2 [Sigma]); immunoprecipitation wash buffer (10 mM Tris [pH 7.5], 1 mM $MgCl_2$, 0.5 mM $CaCl_{2,}$ 1% BSA, protease and phosphatase inhibitors).

Other reagents. Protein G Sepharose™ (Amersham).

2.3.2. In Vitro Phosphorylation by mTOR Kinase

Recombinant proteins. mTOR (1362-end, active; Millipore); purified protein of interest.

Radioisotopes. [γ-^{33}P]ATP (Hartmann Analytic).

Buffers. 10× reaction buffer (filter-sterilized, stored at room temperature; 450 mM HEPES [pH 7.5], 9 mM EGTA, 0.09% Tween-20); 10× dilution buffer (filter-sterilized, stored at room temperature; 450 mM HEPES [pH 7.5], 9 mM EGTA, 0.1% Tween-20); [γ-^{33}P]ATP solution (freshly prepared, kept on ice; 15 mM $MnCl_2$, 0.5 mM ATP with [γ-^{33}P]ATP 250–400 cpm/pmol).

Equipment. PhosphoImager™ (Molecular Dynamics).

2.3.3. Co-Precipitation of mTOR Protein Kinase Partners Using Streptavidin Resin

Biological material. HEK293 cells; cultured cortical neurons; brain tissue.

Plasmids. pBirA (61); pbioGFP-mTOR, pbioGFP-βGAL (26).

Transfection reagents. Lipofectamine2000™ (Invitrogen).

Buffers. All buffers should be filtered and stored at room temperature; TBS buffer (20 mM Tris [pH 7.5], 100 mM NaCl); bioIP blocking buffer (20 mM Tris [pH 7.5], 150 mM KCl, 20% glycerol, 0.2 μg/μl BSA); bioIP lysis buffer (20 mM Tris [pH 7.5], 150 mM KCl, 1% Triton X-100, protease inhibitors [Roche] and phosphatase inhibitor cocktails 1 and 2 [Sigma]); bioIP wash buffer A (20 mM Tris [pH 7.5], 500 mM KCl, 0.1% Triton X-100); bioIP wash buffer B (20 mM Tris [pH 7.5], 150 mM KCl); bioIP wash buffer C (20 mM HEPES [pH 7.5], 150 mM NaCl); bioIP Neuron lysis buffer (20 mM HEPES [pH 7.5], 150 mM NaCl, 1 mM EDTA, 0.3% CHAPS, protease, and phosphatase inhibitors).

Other reagents. Dynabeads-280 (Invitrogen).

2.3.4. Proximity Ligation Assay

Biological material. Neuronal cells cultured on glass coverslips.

Antibodies. mTOR antibody (see Table 2) against protein of interest suitable for immunofluorescence (produced in species different from mTOR antibody); complementary proximity ligation assay (PLA) probes, recognizing chosen primary antibodies (e.g., Duolink II anti-mouse minus [Olink Biosciences], Duolink II anti-rabbit plus [Olink Biosciences]); fixatives and buffers (4% PFA solution [2.2]); GDB (stored frozen; 30 mM phosphate buffer [pH 7.4], 0.2% gelatin, 0.5% Triton X-100, 450 mM NaCl); PLA wash buffer A (stored at 4°C, used at room temperature; 10 mM Tris [pH 7.4], 150 mM NaCl, 0.05% Tween-20); PLA wash buffer B (stored at 4°C, used at room temperature; 0.2 M Tris [pH 7.5], 100 mM NaCl).

Detection reagents. Duolink II Detection kit (Olink Biosciences).

3. Procedures

3.1. Manipulating mTOR Activity in Neurons

The following section describes the protocols we routinely use to modulate mTOR signaling and study neuronal development and plasticity in vitro. The dissociated cultures of different types of neurons are a very convenient model for such studies, the major advantages of which are easy maintenance (Sect. 3.1.1), simplicity of manipulation (Sect. 3.1.2), and morphological assessment (Sect. 3.1.3).

3.1.1. Culturing Primary Hippocampal Neurons

Coating. This step is required only when further analysis requires microscopy and therefore cells are grown on glass coverslips. Coverslips are first cleaned by a 48 h bath in 68% nitric acid and subsequently subjected to four to five 1 h washes with distilled water (see Sect. 4.1.1a). Washed coverslips are then sterilized for 3 h at 180°C and then stored in sterile Petri dishes until needed. Two days prior to dissection, the coverslips are treated overnight at room temperature with coating buffer. Next, excess coating buffer is washed out by five 45 min washes with Milli-Q water. Coverslips are left with the last wash overnight. Water is changed to Neuron Plating Medium before beginning of dissection.

Tissue dissection and cell seeding. The protocol for the preparation of primary hippocampal and cortical neuronal cultures presented herein is consistent with one published previously by Banker and Goslin (62). Neuronal cells are obtained from embryonic day 18 (E18) to E20 rat embryos (a similar procedure works well for E16-E18 mice). Embryos are extracted from the euthanized pregnant female and decapitated, and the heads are placed in ice-cold HBSS. The brains are then extracted from the skulls with forceps (#4 and #5) and transferred to another change of ice-cold HBSS. Further steps are performed with the aid of a stereoscopic microscope in a Petri dish filled with ice-cold HBSS and #5 forceps. The hemispheres are separated from each other and the brainstem and cleared of meninges. The hippocampi are then very carefully dissected and placed in a 15 ml conical tube filled with ice-cold HBSS. Cortices that remain after hippocampal dissection can be used for further cell isolation using an analogous protocol after removing the brain steam residuals and olfactory bulbs (see Sect. 4.1.1b). Isolated hippocampi are washed three times with 10 ml ice-cold HBSS and dissociated by 15 min incubation with trypsin in HBSS (4 ml HBSS with the addition of 10 μl 2.5% trypsin per hippocampus) at 37°C. After trypsinization, the hippocampi are very carefully washed three times to remove any traces of trypsin with HBSS preheated to 37°C and dissociated in warm HBSS (see Sect. 4.1.1c). We usually dissociate cells obtained from 10–12 embryos in 1 ml HBSS using 1 ml pipette tips and dilute dissociated cells up to 4 ml (but see Sect. 4.1.1d). Cells are plated at a density of 250 cells/mm^2 and cultured at 37°C and 5% CO_2 for up to 3 weeks (see Sect. 4.1.1e–g).

3.1.2. Modulation of mTOR Activity

Pharmacological Modulation of mTOR Activity

The modulation of mTOR kinase activity can be easily achieved in cultured neurons by treatment with pharmacological reagents. As already mentioned, the most commonly used inhibitor is rapamycin. The concentrations found in the literature for in vitro studies are quite diverse (10 nM to 1 μM). For cultured neurons, we typically

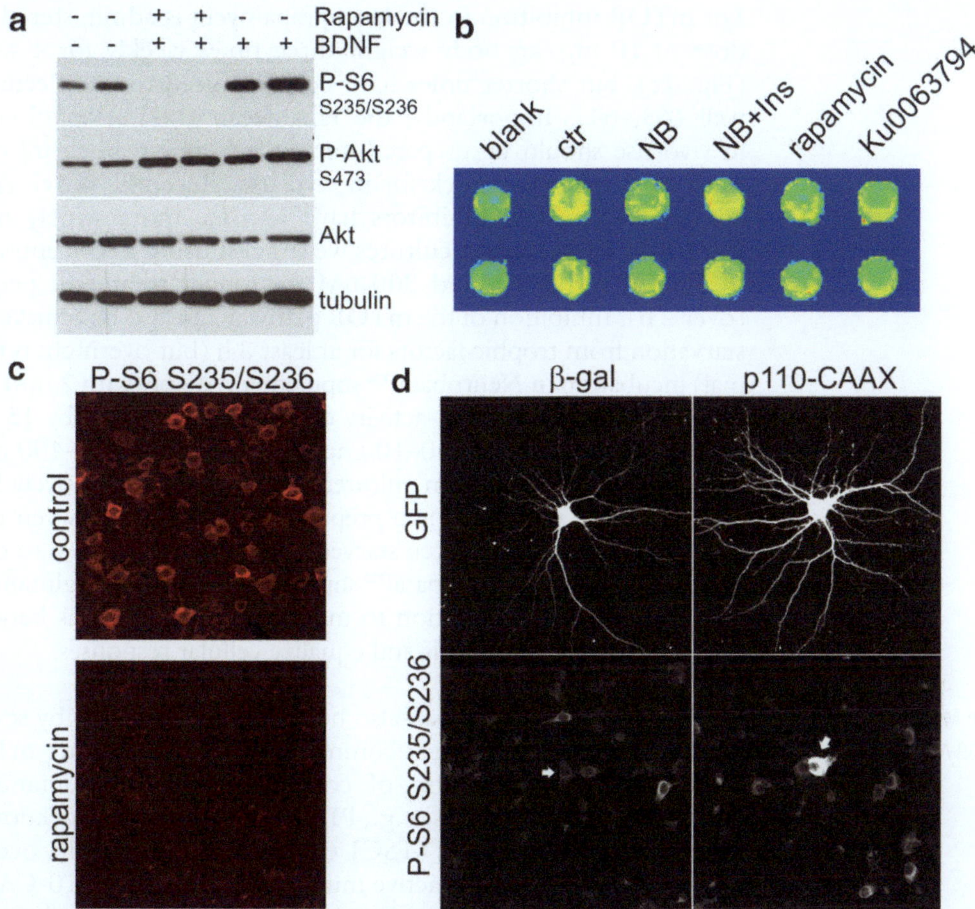

Fig. 1. mTOR activity can be manipulated in neurons using pharmacological and genetic approaches. (**a**) Western blot analysis of mTOR activity, assessed by phosphorylation levels of S6 (P-S6 S235/236) and Akt, in neurons cultured in vitro. Cortical neurons at day in vitro (DIV) 10 were stimulated with 50 ng/ml BDNF for 15 min with or without 2 h pretreatment with 20 nM rapamycin. (**b**) In-cell western analysis of mTOR activity, assessed by phosphorylation levels of S6 (P-S6 S235/236). Cortical neurons at DIV7 were either starved in Neurobasal for 3 h (NB) and subsequently treated with 400 nM insulin (NB + Ins) for 15 min or treated for 1 h with 100 nM rapamycin or 300 nM Ku0063794. For better visualization of intensity changes image from scanner was converted in Zen2009 software (Carl Zeiss MicroImaging GmbH) to a heat map (**c**) Analysis of mTOR activity in the brain sections of control or rapamycin-treated rats, assessed by immunofluorescence of P-S6 (S235/236). Rapamycin was administered intraperitoneally at a dose of 10 mg/kg body weight three times weekly for 4 weeks. Representative confocal images of cortical region are presented (**d**) Analysis of mTOR activity in cultured neurons transfected with control plasmid (β-gal) or constitutively active PI3K (p110CAAX), assessed by immunofluorescence of P-S6 (S235/236). Rat hippocampal neurons were transfected, with respective plasmids at DIV7 for 5 days with the use of Lipofectamine2000™. Plasmid coding GFP was cotransfected to visualize cell morphology. *Arrows* indicate transfected cell.

use 1–24 h treatment with 10–100 nM rapamycin (Fig. 1a, b). Cultured neurons can also be treated with rapamycin for longer times (up to 1 week) without drastic effects on neuronal viability (52). In contrast to cultured neurons, the in vivo inhibition of brain mTOR by intraperitoneal drug injections requires the chronic application of high doses of rapamycin or its derivatives (32).

For mTOR inhibition in the brain, rapamycin is administered at a dose of 10 mg/kg body weight three times weekly for 4 weeks (Fig. 1c), but shorter times have been reported to be effective as well (63, 64). Importantly, the rapamycin working solution for in vivo use should be prepared immediately before each injection from the rapamycin stock for in vivo use. Recently, novel, ATP-competitive mTOR inhibitors have become more widely used. For Ku-0063794 use in cultures we suggest using a concentration of 30 nM for 1–2 h and 300 nM for longer treatment periods (over 2 h). Inhibition of the mTOR pathway can also be achieved by starvation from trophic factors for at least 3 h (but overnight is optimal) incubation in Neurobasal™ supplemented only with 2 mM glutamine. However, mTOR activity can also be increased by 15 min incubation with BDNF (50–100 ng/ml) or insulin (200–400 nM). Notably, mTOR activity in cultured neurons can be relatively high and vary from preparation to preparation as well as between cells. Therefore, neurons are often starved of trophic factors (3 h to overnight incubation in Neurobasal™ supplemented only with glutamine) prior to stimulant application to maximize the differences between control and stimulated cells and equalize cellular responses.

Genetic Modulation of mTOR Activity

mTOR pathway activity can also be changed genetically by several means (Table 1). The most common ways to upregulate mTOR activity are overexpression of constitutively active mutants of mTOR upstream inducers (e.g., PI3K, Akt, Rheb) or knockdown/knockout of its inhibitors (TSC1 or TSC2). For example, overexpression of constitutively active mutant class I PI3K (p110-CAAX) in developing neurons efficiently increases the phosphorylation of S6 (S235/S236) and Akt (S473) and results in increased dendritic arborization (Fig. 1d). The easiest way to obtain the opposite effect (i.e., genetically inhibit mTOR) relies on the RNAi-based silencing of the expression of mTOR or unique components of its complexes. We transfect cultured neurons at the desired age for 2–4 days with pSuper (65) plasmids that encode shRNAs directed against mRNA of these proteins. The hallmarks of an efficient decrease in mTOR activity in developing neurons are a reduction in dendritic arbor complexity (52) and decrease in either basal or trophic factor-induced levels of P-S6 (S235/S236) and P-Akt (S473).

Neuronal cultures can be successfully transfected with a variety of methods with different efficacy. Optimal efficiency does not always mean the highest possible efficiency. Although high efficiency is desired for biochemical studies, it is very inconvenient for morphological assessment. We typically use a calcium phosphate method (52) or Lipofectamine2000™ (66) when low transfection efficiency is needed (Fig. 2). Both approaches can be used to genetically modify neurons of different ages, but some adjustments to the Lipofectamine2000™ protocol provided herein might be needed if young neurons are transfected. Some promoters used to

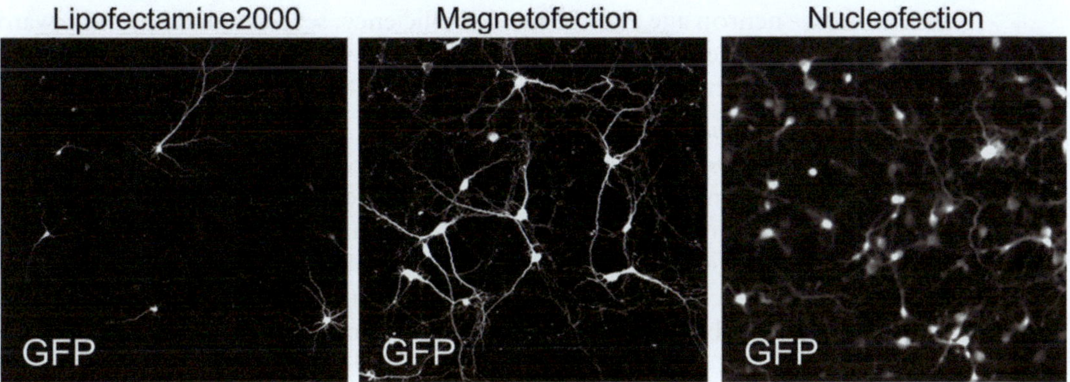

Fig. 2. Comparison of efficacy of lipofection, magnetofection, and nucleofection. Plasmid coding GFP was introduced to neurons with the use of: Lipofectamine2000 at DIV7 for 3 days (*left panel*), magnetofection at DIV7 for 3 days (*middle panel*), or nucleofection at DIV0 for 3 days (*right panel*). Representative images obtained by confocal (lipofection and magnetofection) or fluorescence microscopy (nucleofection) are presented.

drive the expression of exogenous genes, although widely considered constitutive in cell lines, cannot work efficiently in mature neurons or are induced by neuronal activity (67). When high gene transfer efficiency is needed (e.g., for protein or mRNA isolation), one option is the construction of a viral vector (68). However, the generation of viruses is time-consuming and requires appropriate biosafety precautions. As an alternative, two physical methods can be used: magnetofection and nucleofection (Fig. 2). An advantage of the first method is the use of already growing neurons of any age. In contrast, only freshly isolated neurons can be used for efficient nucleofection, but this technique usually assures higher efficiency. Below we discuss all three approaches in detail.

Lipofectamine 2000 transfection. The following protocol is suitable for transfecting cells grown in one well of a 24-well plate, a format typically used in our laboratory. Nevertheless, it can be easily adjusted for other types of plates. The first step is preparing separate mixtures of DNA (0.9 μg total DNA; we tested combinations of up to four plasmids without a drastic decrease in cotransfection efficiency) and Lipofectamine2000™ (1.65 μl) with Neurobasal™ (50 μl each; see Sect. 4.1.2a). After 5 min at room temperature, both mixtures are combined and left at room temperature for another 20–40 min. Meanwhile, fresh medium that contains Neurobasal™/1×B27/0.5 mM glutamine is prepared (0.5 ml per well) and preheated to 37°C. Prior to transfection, fresh medium is added to the neurons and mixed with one already present in the well, and immediately half of the mixed medium is collected and saved for the time of transfection. Next, the complexes of DNA and Lipofectamine2000™ are added dropwise to the cells, and the dish is returned to the CO_2 incubator for 2–4 h (depending on neuron age and the needed efficiency; see Sect. 4.1.2b). Afterward, the cells should be washed twice with warm Neurobasal™ and returned to the saved medium (see Sect. 4.1.2c).

Magnetofection. Similar to Lipofectamine2000™ transfection, the magnetofection protocol is adjusted for the transfection of cells grown in one well of a 24-well plate. The first step is the preparation of separate mixtures of DNA (0.9 μg total DNA) and Lipofectamine2000™ (1.65 μl) with Neurobasal™ (50 μl each; see Sect. 4.1.2d). Additionally, a tube containing 0.9 microliters Combimag™ is needed. After 5 min at room temperature, the DNA and Lipofectamine2000™ solutions are mixed and immediately transferred to the tube that contains Combimag™ for 30 min incubation at room temperature. Prior to transfection, the conditioned medium is changed (and saved for transfection time) to plain, prewarmed, pH-neutral Neurobasal™. Next, DNA/Lipofectamine2000™/Combimag™ mix is added dropwise to the cells, and the culture dish is placed on the magnetic plate in the CO_2 incubator for 15 min. Importantly, the dish should fit the plate well and should be placed perfectly horizontally in the incubator. Afterward, the dish that still remains on the magnetic plate is removed from the incubator, the cells are washed twice with warm Neurobasal™, and the conditioned media collected before transfection is added again. Finally, the dish can be removed from the magnetic plate and returned to the CO_2 incubator.

Nucleofection with Amaxa system. The following protocol is suitable for a single experimental variant (i.e., two wells of a 12-well plate) of transfection of rat cortical neurons. During the first use of Amaxa® Rat Neuron Nucleofector® Kit, Supplement 1 should be added to the Rat Neuron Nucleofection Solution (heretofore referred to as Nucleofection Solution). Such a mixture can be stored at 4°C up to 3 months. Prior to transfection (i.e., during the dissection procedure; see Sect. 3.1.1), the Nucleofection Solution should be brought to room temperature. At the same time, a precoated dish should be filled with Neuron Plating Medium (1 ml per well). An additional 0.5 ml of medium per variant (two wells) should be prepared and equilibrated in a CO_2 incubator in a partially unscrewed conical tube. Next, 2×10^6 freshly isolated rat embryonic cortical neurons (see Sects. 3.1.1 and 4.1.1b) are centrifuged at 800×g at room temperature and resuspended in 100 μl of the Nucleofection Solution. Next, 3 μg of the DNA are added (see Sect. 4.1.2e), and the cells are transferred to a nucleofection cuvette, which should be immediately placed on the Nucleofector® Cuvette Holder, followed by application of the G-013 program. Immediately afterward, 0.5 ml of the equilibrated Neuron Plating Medium is added to the cuvette, and the cells are gently transferred with a pipette (supplied with the kit) to a new 1.5 ml tube. Finally, 250 μl of the cell suspension per well is seeded on a prepared

12-well plate (see above), and the neurons are placed in the CO_2 incubator. The medium should be changed for the Neuron Plating Medium 2 and 24 h after nucleofection (Sect. 4.1.2f).

3.1.3. Image Analysis and Morphometry of Dendritic Tree

One of the important outcomes of manipulation of the mTOR signaling pathway and mTOR binding proteins in cultured neurons is the pronounced change of neuronal morphology (21, 26, 52). Morphological aspects, such as the structure and geometry of dendritic arbors, can be used as a readout of functional neuronal changes. To assess the morphology of dendritic trees, one can focus on versatile parameters that describe two main aspects: its complexity and expansiveness (Fig. 3). For example, the number of dendritic ends, a very basic and often used parameter, describes the complexity of the dendritic arbor. Its limitation, however, is its negligence of dendritic length and spatial distribution. However, total dendritic length (i.e., the sum of the lengths of all dendrites of a particular neuron) describes the expansiveness of the dendritic

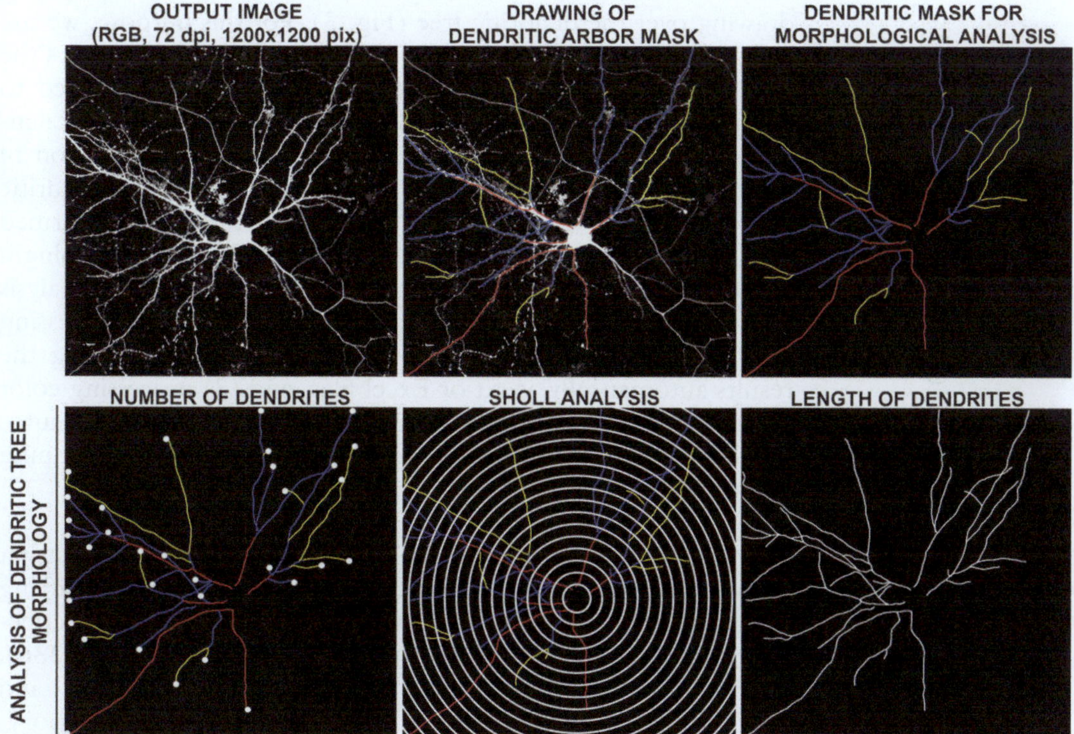

Fig. 3. Morphometry of dendritic tree. Morphological analysis of the dendritic tree can be proceeded on images obtained with a 20× or 40× objective, using the ImageJ 1.42 software (http://rsbweb.nih.gov/ij/) with the NeuronJ plugin (59), which facilitates the tracing of neurites. Using NeuronJ dendrites are traced to obtain dendritic arbor mask, which is next used for automatic calculation of length of individual dendritic segments as well as their sum. Masks can be also used for automatic Sholl analysis by ImageJ supplemented with modified Sholl plugin as well as manual counting of the number of dendritic ends.

field but tells nothing about its complexity. A third very often used analysis, Sholl analysis (69), provides information about both the complexity and expansiveness of dendritic arbors. Sholl analysis measures the number of dendrites that cross concentric circles at various radial distances from the cell soma, but it does not provide information regarding the ultimate number of dendrites. All three parameters, however, increase with more active mTOR and decrease with mTOR inhibition (26, 52). Notably, several additional approaches can be used to describe the morphology of dendritic arbors (e.g., dendrite branching index, number of primary dendrites, contribution of dendrites of different orders to total dendritic length, average length of dendritic segment, number of branch points, etc.) (26, 70) and should be chosen to optimally illustrate the observed phenotype. Principally, however, no single approach fully describes changes in dendritic morphology. Therefore, several parameters should be used.

Morphometric analysis of dendritic trees should be performed on images of neurons obtained with a minimal resolution of 512×512 pixels using a 20× or 40× objective of a fluorescent or confocal microscope. After image acquisition, the first step is mask drawing over the dendritic tree (Fig. 3). For this purpose, we use ImageJ software with the NeuronJ plug-in (59), which allows the semi-manual tracing of dendrites. Such masks are then used to manually count the number of dendritic tips, automatically calculate the total length of dendrites, and assess the contribution of dendrites of different orders to total dendritic length. Dendritic arbor masks can also be used for Sholl analysis, which is performed with ImageJ software supplemented with the Sholl Analysis plugin v.1.0. We recently tailored an original plug-in for mask analysis (26, 70). Additional improvements include (1) the batch processing of images from a given directory, (2) the option of saving the results automatically as txt or Excel files, and (3) processing color images to allow their conversion to 8-bit grayscale images automatically, which previously had to be performed manually running an original plug-in.

3.2. mTOR Pathway Activity Analysis in Neurons

An important aspect of mTOR studies is the range of tools that assess changes in mTOR activity. A standard approach is verifying the phosphorylation status of known mTOR downstream effectors, including S6K1 (T389), eIF4B (S422), S6 (S235/236), 4E-BP1 (S65), and Akt (S473) in the case of mTORC1 and mTORC2, respectively. Despite phospho-Akt, PKCα phosphorylation at S657 can be used as an indicator (71), but this particular modification still awaits more thorough experimental confirmation. For neuronal studies, we found that checking the phosphorylation status of S6 (S235/S236) and Akt (S473) was the most convenient because commercially available antibodies work very well for immunofluorescence and western blot analysis on rat and mouse neuronal samples (Fig. 1). Basal levels of P-S6 and P-Akt in neuronal cultures differ between cells. Therefore, to equalize mTOR activity in cells and maximize the differences between variants, starving the cells prior to either pharmacological treatment or fixation after transfection might be important. Considering the use of samples of neurons that are treated with an mTOR inhibitor or growth factors as controls may also be important (see section "Pharmacological Modulation of mTOR Activity"). For in vivo studies, animals should be habituated to the experimental procedure prior to testing whenever the analysis of the mTOR pathway is considered.

Western blot. To obtain protein extracts of cultured neurons, cells can be lysed directly in 1× Laemmli sample buffer if the cells were seeded equally. If the measurement of total protein level is needed, then using western blot lysis buffer, collecting the sample for measurements, and adding 5× Laemlli sample buffer to a final 1× concentration are better. To extract brain proteins, the proper part of the tissue should be dissected and mechanically homogenized (e.g., with the use of a PRO200 Homogenizer; Pro Scientific) in tissue homogenization buffer. The homogenate is then centrifuged at 1,400×*g* for 10 min at 4°C to remove any tissue debris. Finally, 5× Laemlli sample buffer is added to the supernatant to a final 1× concentration. Part of the supernatant should be saved prior to the addition of buffer for protein concentration measurements. Samples should be boiled for 5–10 min at 95°C. The proteins are then separated by SDS-polyacrylamide gel electrophoresis (SDS-PAGE; 5–10 μg of proteins/line) (60) and electrotransferred to a nitrocellulose membrane in Towbin buffer that contains 0.03% SDS for 1 h at 400 mA. The membrane is then blocked for 1 h at room temperature in 5% nonfat dry milk in TBS-T (all of the subsequent steps require gentle agitation). Afterward, the membrane should be washed four times for 5 min with TBS-T to remove residual milk and incubated overnight, optimally with gentle shaking, at 4°C with an appropriate primary antibody diluted in 5% BSA in TBS-T (refer to Table 2 for appropriate antibody concentrations; see Sect. 4.2a). The next day, the membrane is washed six times for 5–10 min each in TBS-T and incubated for 1 h with an appropriate secondary antibody conjugated with horseradish peroxide (e.g., from Jackson ImmunoResearch; 1:10,000), diluted in TBS-T that contains 5% nonfat dry milk. Finally, the membrane is washed three times with TBS-T, incubated for 1 min with ECL, and immediately exposed to X-ray film.

As an alternative to the chemiluminescent method described above, the Infrared Odyssey Imaging System, which utilizes fluorescently labeled secondary antibodies, can be used. An advantage of this method is the linear measurement of a signal that allows a more accurate quantitative analysis of the results. Moreover, the simultaneous recording of two epitopes is possible using two secondary antibodies, conjugated with fluorophores of different emission spectra. This is very convenient because measurements of phospho- and total protein levels can be performed without the need for membrane stripping when primary antibodies are obtained from different species (e.g., rabbit α-P-Akt and mouse α-Akt). The protocol for fluorescence detection based on western blot is almost identical to the chemiluminescence-based one, but DyLight Dye secondary antibodies (see Sect. 4.2b) diluted 1:10,000 in 5% nonfat dry milk in TBS-T must be used instead of HRP-conjugated antibodies (1 h incubation at room temperature). Afterward, the membranes are washed three times in TBS-T for 5 min and dried (Sect. 4.2c). The membrane images are collected with the Infrared Odyssey Imaging System.

In-Cell Western. This technique allows for high-throughput phosphorylation analysis (Fig. 1b). For this purpose, cortical neurons are plated on 96-well plates (60,000 cells per well in 300 μl medium). After the experimental procedures (e.g., pharmacological treatment, small-interfering RNA [siRNA] transfection) are performed at the needed age, the cells are fixed with ice-cold 4% PFA solution for 20 min. The cells are then washed five times with PBS that contains 0.1% Triton X-100 for 5 min and blocked in 5% nonfat dry milk in PBS-T for 1 h at room temperature. Subsequently, the cells should be washed four times for 5 min with PBS to remove residual milk and incubated overnight at 4°C with primary antibodies diluted in Immunofluorescence Antibody Dilution Buffer. The next day, the cells should be washed five times for 5 min with PBS-T and incubated for 1 h at room temperature with DyLight Dye secondary antibodies diluted 1:2,000 in 5% nonfat dry milk in PBS-T while avoiding exposure to light. Afterward, the cells are washed five times for 5 min with PBS-T and dried, and plate images are collected with the Infrared Odyssey Imaging System.

Immunofluorescence staining and image analysis. In contrast to the methods described above, immunofluorescence can be used for the analysis of protein phosphorylation levels in individual cells (e.g., transfected with the gene of interest or shRNA; Fig. 1). For the immunofluorescence detection of mTOR pathway proteins in cultured neurons, cells grown on glass coverslips are fixed with ice-cold 4% PFA solution for 10–15 min. After fixation, the samples are rinsed three times with PBS and incubated with Immunofluorescence Blocking Buffer I at room temperature for at least 1 h. The coverslips are then placed in a humidified chamber overnight at 4–8°C on one drop (e.g., 40 μl per 13 mm coverslip) of the primary antibodies diluted in Immunofluorescence Antibody Dilution Buffer I placed on parafilm attached to the bottom of the chamber. After incubation with the primary antibodies, the cells are washed three times with PBS for 5–10 min each and placed again in a humidified chamber for 1 h at room temperature on one drop of fluorescent dye-conjugated secondary antibodies (e.g., Alexa Fluor®; 1:200) in the Immunofluorescence Antibody Dilution Buffer. This incubation is followed by three 5 min PBS washes. Finally, each coverslip is dipped for a few seconds in the distilled water to wash out the PBS and mounted on a microscope slide with the use of antifade medium (e.g., VECTASHIELD® Mounting Media or ProLong® Gold antifade reagent). The edges of the coverslip can be sealed with nail polish for long-term storage.

The immunofluorescence detection of activity of brain mTOR pathway proteins can be performed on 40 μm free-floating brain sections (Fig. 1c; see Sect. 4.2d) according to a procedure published elsewhere (72). In the first step, the sections are rinsed three times with PBS-T and incubated with Immunofluorescence Blocking Buffer II at room temperature for at least 1 h. The sections are then incubated with a primary antibody diluted in Immunofluorescence Antibody Dilution Buffer II overnight at 4–8°C. Afterward, the sections are washed three times for 5–10 min each with PBS-T and incubated with a fluorescent dye-conjugated secondary antibody (e.g., Alexa Fluor®; 1:500) in Immunofluorescene Antibody Dilution Buffer II for 1 h at room temperature. This incubation is followed by three PBS-T washes. Finally, the sections are placed on a microscope slide, dried for 5–10 min in the dark, and covered with a coverslip placed on one drop of an antifade medium (e.g., VECTASHIELD® Mounting Media or ProLong® Gold antifade reagent). The edges of the coverslip can be sealed with nail polish for long-term storage.

For the quantitative analysis of fluorescence, the images should be collected with a confocal microscope. Keeping the microscope settings (e.g., laser power, photomultiplier sensitivity, etc.), which influence the brightness of the acquired picture, constant between the cells used for further analysis is important. Fluorescence intensity measurements from the acquired pictures can be performed with the use of almost any image analysis software, but we usually use Metamorph® 7.0 or ImageJ. If three-dimensional pictures are analyzed, then we use Imaris® 6.2.0 (Bitplane A.G). In most cases, we measure the signal from the cell body because of convenience, but this can be done for axons, dendrites, or even dendritic spines.

3.3. Analysis of mTOR Interactions

The analysis of mTOR interactions can be used to assess the activity of mTOR pathways. For example, the amount of Raptor bound to mTOR provides information about mTORC1 activity (56). The methods we propose herein are very useful in the search for novel neuronal mTOR partners and elucidating the functional effects of mTOR activation (26). Below we provide a description of the methods that allow the study of biochemical preparations (Sects. 3.3.1–3.3.3) and individual cells (Sect. 3.3.4).

3.3.1. Immunoprecipitation of mTOR Kinase

A protein extract for mTOR immunoprecipitation can be obtained from either in vitro cultured cortical neurons or brain tissue. In the first case, cultured cells are harvested in TBS and spun down at 1000×*g* at 4°C for 10 min, and the pellet is homogenized in the immunoprecipitation lysis buffer by pipetting (see Sect. 4.3.1). To obtain the protein extract from a specific brain region, tissue is mechanically homogenized (we usually use a PRO200 Homogenizer; Pro Scientific) in the immunoprecipitation lysis buffer. The lysate is then left on ice for 20 min and centrifuged at 18,000×*g* at 4°C for 30 min to remove cellular debris. Meanwhile, Protein G Sepharose™ (20 μl/sample) should be prepared by incubation with (1) mTOR antibody (4 μg/sample) or (2) rabbit IgG (4 μg/sample; negative control) in 0.8 ml immunoprecipitation lysis buffer for 1 h at 4°C with rotation, followed by centrifugation of the resin at 400×*g* for 5 min at 4°C. The protein extract and antibodies conjugated with Protein G Sepharose™ are mixed and incubated overnight at 4°C with constant rotation. Afterward, the resins are spun down at 400×*g* for 5 min at 4°C and washed extensively four times with immunoprecipitation lysis buffer. Such mTOR complexes attached to the resin can be further used for different types of analysis or for in vitro phosphorylation assays. For western blot analysis (Fig. 4a), the samples are simply resuspended in 1× Laemmli sample buffer. For mass spectrometry, the samples should be diluted in 20 mM ammonium bicarbonate.

3.3.2. In Vitro Phosphorylation by mTOR Kinase

As an mTOR substrate for this protocol, purified proteins, soluble or attached to the resin, can be used. Their phosphorylation by mTOR kinase is examined by incubating with [γ-^{32}P]ATP in the presence of either immunoprecipitated mTOR or a commercially available active fragment of the kinase. For a single reaction, 2.5 μl

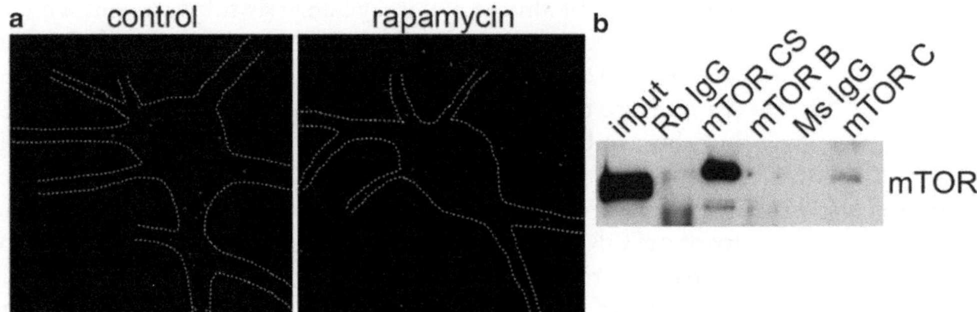

Fig. 4. Techniques for mTOR interactions analysis. (**a**) Results of in situ PLA detection of protein complexes containing mTOR and CLIP-170 (*red dots*) in hippocampal neurons incubated for 1 h with 100 nM rapamycin or vehicle. *Dashed lines* correspond to cell boundaries. (**b**) Western blot analysis of mTOR immunoprecipitation from cortical neurons lysates. mTOR was immunoprecipited with different antibodies against mTOR (Cell Signaling Technology #2972, CS; Bethyl Laboratories #A300-503A, B; Calbiochem #OP97, C). Western blot was performed with the use of antibody from Cell Signaling Technology #2972.

of 10× reaction buffer and the protein of interest (50 μg per tube are diluted in Milli-Q water to a final volume of 5 μl) in a test tube (Sect. 4.3.2a). mTOR substrate (Millipore) should be used as a positive control. One extra tube with the protein of interest should also be prepared as a negative control (without mTOR kinase). Next, 2.5 μl (25 μg) of an active mTOR fragment (1362-end) and 5 μl [γ-^{33}P]ATP are added to the test tube, and the reaction proceeds for 20 min at 30°C. The reaction is stopped by the addition of 5× Leamlli sample buffer to the sample and boiling. The proteins are then separated by SDS-PAGE, and the gel is dried and analyzed with PhosphoImager (Sect. 4.3.2b).

3.3.3. Coprecipitation of mTOR Kinase Protein Partners Using Streptavidin Resin

Preparation of the mTOR resin. To produce biotinylated proteins, we transfect HEK293 (~10^8 cells/transfection) with bioGFP-mTOR or bioGFP-β-galactosidase (negative control) (26) together with BirA-encoding plasmid (2:1 ratio) using Lipofectamine2000™ according to the manufacturer's protocol. Two days after transfection, the cells are harvested, washed with TBS, and lysed with bioIP lysis buffer (see Sect. 4.3.1). The cell lysates are then centrifuged for 10 min at 14,000×*g*. Meanwhile, Dynabeads-280 (20 μl/transfection) are incubated with bioIP blocking buffer for 30 min at room temperature. Dynabeads-280 are then incubated for 1 h at 4°C with the supernatant obtained after cell lysate centrifugation. Afterward, the beads are washed three times with bioIP lysis buffer, three times with bioIP wash buffer A, and three times with bioIP wash buffer B. Although short-term storage (up to 24 h) of the obtained mTOR resin is possible, proceeding as soon as possible is highly recommended.

Identification of mTOR kinase protein partners. The neuronal protein extract is obtained from either cultured cortical neurons or brain tissue. In the first case, the cells are washed with TBS, harvested, and lysed with co-immunoprecipitation neuronal lysis buffer. The cell lysate is then centrifuged for 10 min at 14,000×*g*, and the supernatant is mixed with Dynabeads-280 resin that contains biotinylated proteins (see above) and left overnight at 4°C. Afterward, the beads are washed four times with bioIP neuronal lysis buffer and twice with bioIP wash buffer C. Finally, the beads are resuspended in 20 mM ammonium bicarbonate, pH 7.5, for mass-spec analysis or in 1× Laemlli sample buffer for western blotting. In experimental variants that involve mTOR inhibition, the neurons should be incubated with 100 nM rapamycin 1 h prior to lysis. The drug should also be present in the respective cell extracts.

3.3.4. Proximity Ligation Assay

The PLA allows for the detection and quantification of protein interactions in individual cells (73, 74), and we successfully applied this technique to analyze mTOR interactions in neurons (Fig. 4a) (26). Interactions between proteins are detected with modified secondary antibodies conjugated with complementary DNA oligonucleotides. Annealing and subsequent ligation of the oligonucleotides form a circular DNA that serves as a template for rolling-circle polymerase chain reaction. The reaction product is visualized by fluorescent in situ hybridization with fluorescently labeled oligonucleotides (73, 74). For the PLA, neurons grown on glass coverslips can be fixed and permeabilized with a standard protocol, which works for primary antibodies against the proteins of interest (see Sect. 4.3.3a). The fixed cells are then washed three times with PLA wash buffer A (see Sect. 4.3.3b, c). Incubation of the cells with primary antibodies is performed in GDB buffer overnight at 4°C in a humidified chamber (see Sect. 4.3.3d). The next day, the cells are washed three times for 10 min each with PLA wash buffer A and incubated with PLA probes diluted in the buffer provided by the manufacturer (Olink) for 60 min at 37°C (see Sect. 4.3.3e). Unhybridized probes are then washed out by two changes of PLA wash buffer A (5 min each). Ligation and amplification reactions are then performed using the Duolink II Detection kit according to the manufacturer's protocol. After amplification, the cells are washed twice for 10 min with 1× PLA wash buffer B. Finally, the coverslips are dipped in 0.1× PLA wash Buffer B and left to dry at room temperature in the dark (see Sect. 4.3.3f). The coverslips are then mounted on microscope slides with VectaShield Mounting Medium or Duolink II Mounting Medium. Specimens prepared this way can be frozen at –20°C for months. The PLA results can be analyzed by either fluorescent or confocal microscopy. The number of PLA puncta can be quantified with any image analysis software.

4. Notes

4.1. Manipulating mTOR Activity in Neurons

4.1.1. Culturing Primary Hippocampal Neurons

(a) The optimal way for coverslip cleaning is placing them on the porcelain racks (Thomas Scientific). Nitric acid used for coverslips cleaning may be reused up to five times. (b) The preparation of cortical neurons differs from the dissection of hippocampal neurons only slightly, mostly because of the larger amounts of tissue. Before trypsinization, cortices should be washed three times with 20 ml of cold HBSS (instead of 10 ml). The cortices are trypsinized for 20–30 min in 8 ml HBSS with the use of 20 μl trypsin per cortex and washed three times with 20 ml of prewarmed HBSS. After dissociation with a pipette, the cells are suspended in 10 ml of HBSS and plated at a density of 625 cells/mm^2. (c) The most crucial steps of the neuronal culture preparation are tissue trypsinization and dissociation. To avoid cell loss, the pellet obtained after trypsinization needs to be washed very carefully to prevent premature dissociation and cell loss. During dissociation, low pressure should be applied, and air bubbling should be avoided. (d) If you expect a low cell number because of cell loss during preparation, or fewer embryos are used for the preparation, then dilute the dissociated cells to a lower final volume so that the culturing medium will not be unnecessarily diluted with HBSS during cell seeding. (e) Seeding cells at too low or too high of a density should be avoided because this usually results in poor culture quality. For higher densities, culture media that contain serum and cytostatics (e.g., AraC) are often used. (f) The culturing medium does not have to be replaced for up to 3 weeks. When changing the medium is necessary, remove half of the medium volume and replace it with a fresh portion without glutamate. (g) The most frequent reasons for poor quality of hippocampal and neuronal cultures are stressed animals, bad coating conditions, and a bad lot of B27 Supplement. Usually researchers do not have control over the first parameter, but signs of poor conditions in dams are a lower number of embryos and any signs of embryo resorption. Such cases should be reported to the animal facility. In the case of a bad coating or bad lot of B27, usually a lower density of cells that survive until DIV1 is observed. In such situations, we suggest changing the coating reagents to new ones and testing two to three different lots of B27. Keeping track of the lot numbers of reagents used for each preparation is recommended.

4.1.2. Genetic Modulation of mTOR Activity

(a) Buying 0.75 ml vials of Lipofectamine2000™ and using relatively fresh Lipofectamine2000™ for neuronal transfection are important because oxidation of the reagent leads to extreme toxicity. (b) Depending on the age of transfected neurons and transfection efficiency, some adjustments to the provided Lipofectamine2000™ protocol may be needed. In principle, shorter transfection decreases efficacy but can result in better morphology and survival, especially when younger neurons are transfected (<DIV3). (c) Special care should be taken during the final washings after Lipofectamine2000™ transfection because the cells are extremely vulnerable to even brief drying. (d) Ensure a neutral pH of Neurobasal™, in which the transfection reagents are prepared; this is crucial for the success of magnetofection. (e) Nucleofection enables obtaining very high transfection efficiency, but one must be aware that not all plasmids will be expressed in a high content of cells. We observed over 80% transfection efficiency with shRNA-coding or GFP-coding plasmids. Larger plasmids may "nucleofect" less efficiently (our unpublished observations), and the efficacy of gene transfer in such cases needs to be defined prior to the experiment. (f) Nucleofection is not well tolerated by neurons, and we do not recommend culturing cells for longer than 72 h post-nucleofection. The cotransfection of plasmids that encode anti-apoptotic proteins (e.g., Bcl-2) (75) to increase the cell survival rate is a solution that works well. However, it is not always accepted by all researchers.

4.2. mTOR Pathway Activity Analysis in Neurons

(a) Most antibodies for western blot can be reused. They can be stored either at 4°C with the addition of sodium azide or frozen at −20°C. (b) For weaker antibodies, using DyLight 800 Dye conjugated antibodies for the Odyssey system is preferred because no background is observed from a membrane at the 800 nm channel. (c) Dried membranes (after incubation with DyLight Dye secondary antibodies and washing) can be stored for weeks in the dark before collecting images with an Infrared Odyssey Imaging System. (d) To obtain material for the immunofluorescence of brain sections, animals should be anesthetized with a lethal dose of anesthetic (e.g., sodium pentobarbital, 80 mg/kg, intraperitoneal) and perfused transcardially with 200 ml PBS, pH 7.4, for 2–3 min and subsequently with 400–500 ml of ice-cold fixative (4% paraformaldehyde in 0.1 M PB) for an additional 20 min. Afterward, the brain is removed and postfixed in the fixative for 1.5 h at room temperature. Next, the tissue should be cryoprotected in 30% sucrose in 0.1 M PB at 4°C (until the brain sinks) and frozen with precooled heptane (approximately −30°C). Such tissue can be stored at −80°C and sectioned at the researcher's convenience. For sectioning, the tissue is first brought to −20°C and mounted on a cryostat tissue holder with the use of Jung tissue-freezing medium (Leica Microsystems). Free-floating sections (40 μm) should be collected from the antifreeze medium and stored at −20°C until needed.

4.3. Analysis of mTOR Interactions

4.3.1. Immunoprecipitation of mTOR Kinase and Coprecipitation of mTOR Kinase Protein Partners Using Streptavidin Resin

An important aspect of these procedures is the optimization of experimental conditions depending on the purpose of the experiment, which should involve the proper choice of the type and concentration of detergent (e.g., Triton X-100 or NP-40 disrupts the interaction between mTOR and Raptor) (3) and buffer composition, pH, and salt concentration. For example, PBS has been suggested to decrease mTOR kinase activity.

4.3.2. In Vitro Phosphorylation by mTOR Kinase

(a) Although in the case of CLIP-170, approximately 50 μg of protein per in vitro phosphorylation reaction works well for our laboratory (26), we suggest testing several different amounts of target protein in preliminary experiments. (b) Although PhospoImager is a very convenient equipment, the results of in vitro phosphorylation assays can be visualized simply by the exposure of a radioactive gel to X-ray film.

4.3.3. Proximity Ligation Assay

(a) Although we found that fixation with 4% PFA solution (see Sect. 2.2) is a preferable method for PLA, methanol also usually works well. (b) All PLA buffers should be prewarmed to room temperature before use. The incubation chamber should be preheated to 37°C prior to the reaction. All washing steps should be performed at room temperature. (c) At each step, ensure that no excess amount of wash buffer remains on the coverslip. (d) The best positive control for PLA is the one that uses an antibody that recognizes the known interacting protein. Omitting one or both primary antibodies may serve as a negative control. (e) A ligation step longer than 30 min leads to a small increase in signal number. By modifying the duration of the amplification reaction, one can change the signal size (i.e., times shorter than 90 min lead to smaller signals, and longer times cause larger dots). (f) If additional staining of cells is needed, this can be performed after the final PLA wash.

Acknowledgments

This work was supported by FP7 grant# 223276 ("NeuroGSK3"), FP7 grant# 229676 ("HealthProt"), and ERA-NET-NEURON/03/2010 (co-financed by the National Centre for Research and Development).

References

1. Schmelzle T, Hall MN (2000) TOR, a central controller of cell growth. Cell 103:253–262
2. Harris TE, Lawrence JC Jr (2003) TOR signaling. Sci STKE 2003:re15
3. Kim DH, Sarbassov DD, Ali SM, King JE, Latek RR, Erdjument-Bromage H, Tempst P, Sabatini DM (2002) mTOR interacts with raptor to form a nutrient-sensitive complex that signals to the cell growth machinery. Cell 110:163–175
4. Sarbassov DD, Ali SM, Kim DH, Guertin DA, Latek RR, Erdjument-Bromage H, Tempst P, Sabatini DM (2004) Rictor, a novel binding partner of mTOR, defines a rapamycin-insensitive and raptor-independent pathway that regulates the cytoskeleton. Curr Biol 14: 1296–1302
5. Sarbassov DD, Ali SM, Sengupta S, Sheen JH, Hsu PP, Bagley AF, Markhard AL, Sabatini DM (2006) Prolonged rapamycin treatment inhibits mTORC2 assembly and Akt/PKB. Mol Cell 22:159–168
6. Hay N, Sonenberg N (2004) Upstream and downstream of mTOR. Genes Dev 18: 1926–1945
7. Sengupta S, Peterson TR, Sabatini DM (2010) Regulation of the mTOR complex 1 pathway by nutrients, growth factors, and stress. Mol Cell 40:310–322
8. Hou L, Klann E (2004) Activation of the phosphoinositide 3-kinase-Akt-mammalian target of rapamycin signaling pathway is required for metabotropic glutamate receptor-dependent long-term depression. J Neurosci 24:6352–6361
9. Lenz G, Avruch J (2005) Glutamatergic regulation of the p70S6 kinase in primary mouse neurons. J Biol Chem 280:38121–38124
10. Santini E, Heiman M, Greengard P, Valjent E, Fisone G (2009) Inhibition of mTOR signaling in Parkinson's disease prevents L-DOPA-induced dyskinesia. Sci Signal 2:ra36
11. Chenal J, Pellerin L (2007) Noradrenaline enhances the expression of the neuronal monocarboxylate transporter MCT2 by translational activation via stimulation of PI3K/Akt and the mTOR/S6K pathway. J Neurochem 102: 389–397
12. Slack BE, Blusztajn JK (2008) Differential regulation of mTOR-dependent S6 phosphorylation by muscarinic acetylcholine receptor subtypes. J Cell Biochem 104:1818–1831
13. Zinzalla V, Stracka D, Oppliger W, Hall MN (2011) Activation of mTORC2 by association with the ribosome. Cell 144:757–768
14. Chen CH, Shaikenov T, Peterson TR, Aimbetov R, Bissenbaev AK, Lee SW, Wu J, Lin HK, dos Sarbassov D (2011) ER stress inhibits mTORC2 and Akt signaling through GSK-3beta-mediated phosphorylation of rictor. Sci Signal 4:ra10
15. Xie MW, Jin F, Hwang H, Hwang S, Anand V, Duncan MC, Huang J (2005) Insights into TOR function and rapamycin response: chemical genomic profiling by using a high-density

cell array method. Proc Natl Acad Sci U S A 102:7215–7220

16. Guertin DA, Guntur KV, Bell GW, Thoreen CC, Sabatini DM (2006) Functional genomics identifies TOR-regulated genes that control growth and division. Curr Biol 16:958–970
17. Garcia-Martinez JM, Alessi DR (2008) mTOR complex 2 (mTORC2) controls hydrophobic motif phosphorylation and activation of serum- and glucocorticoid-induced protein kinase 1 (SGK1). Biochem J 416:375–385
18. Sarbassov DD, Guertin DA, Ali SM, Sabatini DM (2005) Phosphorylation and regulation of Akt/PKB by the rictor-mTOR complex. Science 307:1098–1101
19. Jacinto E, Loewith R, Schmidt A, Lin S, Ruegg MA, Hall A, Hall MN (2004) Mammalian TOR complex 2 controls the actin cytoskeleton and is rapamycin insensitive. Nat Cell Biol 6:1122–1128
20. Hoeffer CA, Klann E (2010) mTOR signaling: at the crossroads of plasticity, memory and disease. Trends Neurosci 33:67–75
21. Swiech L, Perycz M, Malik A, Jaworski J (2008) Role of mTOR in physiology and pathology of the nervous system. Biochim Biophys Acta 1784:116–132
22. Garelick MG, Kennedy BK (2011) TOR on the brain. Exp Gerontol 46:155–163
23. Burnett PE, Barrow RK, Cohen NA, Snyder SH, Sabatini DM (1998) RAFT1 phosphorylation of the translational regulators p70 S6 kinase and 4E-BP1. Proc Natl Acad Sci U S A 95:1432–1437
24. Hara K, Yonezawa K, Kozlowski MT, Sugimoto T, Andrabi K, Weng QP, Kasuga M, Nishimoto I, Avruch J (1997) Regulation of eIF-4E BP1 phosphorylation by mTOR. J Biol Chem 272:26457–26463
25. Ruvinsky I, Meyuhas O (2006) Ribosomal protein S6 phosphorylation: from protein synthesis to cell size. Trends Biochem Sci 31:342–348
26. Swiech L, Blazejczyk M, Urbanska M, Pietruszka P, Dortland BR, Malik AR, Wulf PS, Hoogenraad CC, Jaworski J (2011) CLIP-170 and IQGAP1 cooperatively regulate dendrite morphology. J Neurosci 31:4555–4568
27. Jaworski J, Sheng M (2006) The growing role of mTOR in neuronal development and plasticity. Mol Neurobiol 34:205–219
28. Cota D, Proulx K, Smith KA, Kozma SC, Thomas G, Woods SC, Seeley RJ (2006) Hypothalamic mTOR signaling regulates food intake. Science 312:927–930
29. Cao R, Li A, Cho HY, Lee B, Obrietan K (2010) Mammalian target of rapamycin signaling modulates photic entrainment of the suprachiasmatic circadian clock. J Neurosci 30: 6302–6314
30. Siuta MA, Robertson SD, Kocalis H, Saunders C, Gresch PJ, Khatri V, Shiota C, Kennedy JP, Lindsley CW, Daws LC, Polley DB, Veenstra-Vanderweele J, Stanwood GD, Magnuson MA, Niswender KD, Galli A (2010) Dysregulation of the norepinephrine transporter sustains cortical hypodopaminergia and schizophrenia-like behaviors in neuronal rictor null mice. PLoS Biol 8:e1000393
31. Koike-Kumagai M, Yasunaga K, Morikawa R, Kanamori T, Emoto K (2009) The target of rapamycin complex 2 controls dendritic tiling of Drosophila sensory neurons through the Tricornered kinase signalling pathway. EMBO J 28:3879–3892
32. Kwon CH, Zhu X, Zhang J, Baker SJ (2003) mTor is required for hypertrophy of Pten-deficient neuronal soma in vivo. Proc Natl Acad Sci U S A 100:12923–12928
33. Tischmeyer W, Schicknick H, Kraus M, Seidenbecher CI, Staak S, Scheich H, Gundelfinger ED (2003) Rapamycin-sensitive signalling in long-term consolidation of auditory cortex-dependent memory. Eur J Neurosci 18:942–950
34. Parsons RG, Gafford GM, Helmstetter FJ (2006) Translational control via the mammalian target of rapamycin pathway is critical for the formation and stability of long-term fear memory in amygdala neurons. J Neurosci 26:12977–12983
35. Dowling RJ, Topisirovic I, Fonseca BD, Sonenberg N (2010) Dissecting the role of mTOR: lessons from mTOR inhibitors. Biochim Biophys Acta 1804:433–439
36. Banko JL, Poulin F, Hou L, DeMaria CT, Sonenberg N, Klann E (2005) The translation repressor 4E-BP2 is critical for eIF4F complex formation, synaptic plasticity, and memory in the hippocampus. J Neurosci 25:9581–9590
37. Antion MD, Merhav M, Hoeffer CA, Reis G, Kozma SC, Thomas G, Schuman EM, Rosenblum K, Klann E (2008) Removal of S6K1 and S6K2 leads to divergent alterations in learning, memory, and synaptic plasticity. Learn Mem 15:29–38
38. Antion MD, Hou L, Wong H, Hoeffer CA, Klann E (2008) mGluR-dependent long-term depression is associated with increased phosphorylation of S6 and synthesis of elongation factor 1A but remains expressed in S6K-deficient mice. Mol Cell Biol 28:2996–3007
39. Banko JL, Hou L, Poulin F, Sonenberg N, Klann E (2006) Regulation of eukaryotic

initiation factor 4E by converging signaling pathways during metabotropic glutamate receptor-dependent long-term depression. J Neurosci 26:2167–2173

40. Kwon CH, Luikart BW, Powell CM, Zhou J, Matheny SA, Zhang W, Li Y, Baker SJ, Parada LF (2006) Pten regulates neuronal arborization and social interaction in mice. Neuron 50: 377–388
41. Chow DK, Groszer M, Pribadi M, Machniki M, Carmichael ST, Liu X, Trachtenberg JT (2009) Laminar and compartmental regulation of dendritic growth in mature cortex. Nat Neurosci 12:116–118
42. Park KK, Liu K, Hu Y, Smith PD, Wang C, Cai B, Xu B, Connolly L, Kramvis I, Sahin M, He Z (2008) Promoting axon regeneration in the adult CNS by modulation of the PTEN/mTOR pathway. Science 322:963–966
43. Tavazoie SF, Alvarez VA, Ridenour DA, Kwiatkowski DJ, Sabatini BL (2005) Regulation of neuronal morphology and function by the tumor suppressors Tsc1 and Tsc2. Nat Neurosci 8:1727–1734
44. Goorden SM, van Woerden GM, van der Weerd L, Cheadle JP, Elgersma Y (2007) Cognitive deficits in Tsc1+/- mice in the absence of cerebral lesions and seizures. Ann Neurol 62:648–655
45. Feliciano DM, Su T, Lopez J, Platel JC, Bordey A (2011) Single-cell Tsc1 knockout during corticogenesis generates tuber-like lesions and reduces seizure threshold in mice. J Clin Invest 121(4):1596–1607
46. Choi YJ, Di Nardo A, Kramvis I, Meikle L, Kwiatkowski DJ, Sahin M, He X (2008) Tuberous sclerosis complex proteins control axon formation. Genes Dev 22:2485–2495
47. Wang Y, Greenwood JS, Calcagnotto ME, Kirsch HE, Barbaro NM, Baraban SC (2007) Neocortical hyperexcitability in a human case of tuberous sclerosis complex and mice lacking neuronal expression of TSC1. Ann Neurol 61:139–152
48. Ehninger D, Han S, Shilyansky C, Zhou Y, Li W, Kwiatkowski DJ, Ramesh V, Silva AJ (2008) Reversal of learning deficits in a Tsc2+/- mouse model of tuberous sclerosis. Nat Med 14: 843–848
49. Nie D, Di Nardo A, Han JM, Baharanyi H, Kramvis I, Huynh T, Dabora S, Codeluppi S, Pandolfi PP, Pasquale EB, Sahin M (2010) Tsc2-Rheb signaling regulates EphA-mediated axon guidance. Nat Neurosci 13:163–172
50. Williams T, Courchet J, Viollet B, Brenman JE, Polleux F (2011) AMP-activated protein kinase (AMPK) activity is not required for neuronal development but regulates axogenesis during metabolic stress. Proc Natl Acad Sci U S A 108(14):5849–5854
51. Di Nardo A, Kramvis I, Cho N, Sadowski A, Meikle L, Kwiatkowski DJ, Sahin M (2009) Tuberous sclerosis complex activity is required to control neuronal stress responses in an mTOR-dependent manner. J Neurosci 29:5926–5937
52. Jaworski J, Spangler S, Seeburg DP, Hoogenraad CC, Sheng M (2005) Control of dendritic arborization by the phosphoinositide-3′-kinase-Akt-mammalian target of rapamycin pathway. J Neurosci 25:11300–11312
53. Kumar V, Zhang MX, Swank MW, Kunz J, Wu GY (2005) Regulation of dendritic morphogenesis by Ras-PI3K-Akt-mTOR and Ras-MAPK signaling pathways. J Neurosci 25: 11288–11299
54. Kim SR, Chen X, Oo TF, Kareva T, Yarygina O, Wang C, During M, Kholodilov N, Burke RE (2011) Dopaminergic pathway reconstruction by Akt/Rheb-induced axon regeneration. Ann Neurol 70(1):110–120
55. Morita T, Sobue K (2009) Specification of neuronal polarity regulated by local translation of CRMP2 and tau via the mTOR-p70S6K pathway. J Biol Chem 284(40):27734–27745
56. Hoeffer CA, Tang W, Wong H, Santillan A, Patterson RJ, Martinez LA, Tejada-Simon MV, Paylor R, Hamilton SL, Klann E (2008) Removal of FKBP12 enhances mTOR-Raptor interactions. LTP, memory, and perseverative/repetitive behavior. Neuron 60:832–845
57. Inamura N, Nawa H, Takei N (2005) Enhancement of translation elongation in neurons by brain-derived neurotrophic factor: implications for mammalian target of rapamycin signaling. J Neurochem 95:1438–1445
58. Takei N, Inamura N, Kawamura M, Namba H, Hara K, Yonezawa K, Nawa H (2004) Brain-derived neurotrophic factor induces mammalian target of rapamycin-dependent local activation of translation machinery and protein synthesis in neuronal dendrites. J Neurosci 24:9760–9769
59. Meijering E, Jacob M, Sarria JC, Steiner P, Hirling H, Unser M (2004) Design and validation of a tool for neurite tracing and analysis in fluorescence microscopy images. Cytometry A 58:167–176
60. Sambrook J, Fritsch EF, Maniatis T (1989) Molecular cloning: a laboratory manual, 2nd edn. Cold Spring Harbor Laboratory, Cold Spring Harbor, NY
61. de Boer E, Rodriguez P, Bonte E, Krijgsveld J, Katsantoni E, Heck A, Grosveld F, Strouboulis J (2003) Efficient biotinylation and single-step purification of tagged transcription factors in mammalian cells and transgenic mice. Proc Natl Acad Sci U S A 100:7480–7485

62. Banker G, Goslin K (1988) Developments in neuronal cell culture. Nature 336:185–186
63. Zeng LH, McDaniel S, Rensing NR, Wong M (2010) Regulation of cell death and epileptogenesis by the mammalian target of rapamycin (mTOR): a double-edged sword? Cell Cycle 9:2281–2285
64. Zeng LH, Rensing NR, Wong M (2009) The mammalian target of rapamycin signaling pathway mediates epileptogenesis in a model of temporal lobe epilepsy. J Neurosci 29: 6964–6972
65. Brummelkamp TR, Bernards R, Agami R (2002) A system for stable expression of short interfering RNAs in mammalian cells. Science 296:550–553
66. Jaworski J, Kapitein LC, Gouveia SM, Dortland BR, Wulf PS, Grigoriev I, Camera P, Spangler SA, Di Stefano P, Demmers J, Krugers H, Defilippi P, Akhmanova A, Hoogenraad CC (2009) Dynamic microtubules regulate dendritic spine morphology and synaptic plasticity. Neuron 61:85–100
67. Konopka W, Duniec K, Mioduszewska B, Proszynski T, Jaworski J, Kaczmarek L (2005) hCMV and Tet promoters for inducible gene expression in rat neurons in vitro and in vivo. Neurobiol Dis 19:283–292
68. Cid-Arregui A, García-Carrancá A (2000) Viral vectors: basic science and gene therapy. Eaton Pub, Natick, MA
69. Sholl DA (1953) Dendritic organization in the neurons of the visual and motor cortices of the cat. J Anat 87:387–406
70. Perycz M, Urbanska AS, Krawczyk PS, Parobczak K, Jaworski J (2011) Zipcode Binding Protein 1 regulates the development of dendritic arbors in hippocampal neurons. J Neurosci 31:5271–5285
71. Ikenoue T, Inoki K, Yang Q, Zhou X, Guan KL (2008) Essential function of TORC2 in PKC and Akt turn motif phosphorylation, maturation and signalling. EMBO J 27:1919–1931
72. Skup M, Dwornik A, Macias M, Sulejczak D, Wiater M, Czarkowska-Bauch J (2002) Long-term locomotor training up-regulates TrkB(FL) receptor-like proteins, brain-derived neurotrophic factor, and neurotrophin 4 with different topographies of expression in oligodendroglia and neurons in the spinal cord. Exp Neurol 176:289–307
73. Soderberg O, Gullberg M, Jarvius M, Ridderstrale K, Leuchowius KJ, Jarvius J, Wester K, Hydbring P, Bahram F, Larsson LG, Landegren U (2006) Direct observation of individual endogenous protein complexes in situ by proximity ligation. Nat Methods 3: 995–1000
74. Weibrecht I, Leuchowius KJ, Clausson CM, Conze T, Jarvius M, Howell WM, Kamali-Moghaddam M, Soderberg O (2010) Proximity ligation assays: a recent addition to the proteomics toolbox. Expert Rev Proteomics 7:401–409
75. Szatmari E, Kalita KB, Kharebava G, Hetman M (2007) Role of kinase suppressor of Ras-1 in neuronal survival signaling by extracellular signal-regulated kinase 1/2. J Neurosci 27: 11389–11400

Chapter 18

Protein kinase G (PKG): Involvement in Promoting Neural Cell Survival, Proliferation, Synaptogenesis, and Synaptic Plasticity and the Use of New Ultrasensitive Capillary-Electrophoresis-Based Methodologies for Measuring PKG Expression and Molecular Actions

Ronald R. Fiscus and Mary G. Johlfs

Abstract

Cyclic GMP-dependent protein kinase (protein kinase G, PKG) plays an important role as a key protein kinase mediating the neuroprotective effects of nitric oxide (NO) at low, physiological levels (10 pM–10 nM) and by the natriuretic peptides, atrial natriuretic peptide (ANP) and brain (B-type) natriuretic peptide (BNP). The NO-, ANP-, and BNP-induced stimulation of PKG kinase (serine/threonine-phosphorylating) activity promotes the survival of neural cells, preventing or minimizing both spontaneous apoptosis and the apoptosis induced by neurotoxins [e.g., minimizing the apoptotic cell death caused by high/toxic levels of NO (>100 nM NO)]. PKG is also involved in promoting/regulating cell proliferation and migration of neural progenitor cells and neural cancer cells, synaptogenesis (promoting formation of filopodia and regulating the guidance of growth cones), synaptic plasticity [contributing to memory consolidation, including long-term potentiation (LTP) in hippocampus and long-term depression (LTD) in cerebellum], nociception, and circadian rhythm. Studies from our laboratory suggest that the pro-growth and pro-survival actions of PKG, specifically the PKG-Iα isoform, involve the downstream phosphorylation of specific target proteins, including BAD (an apoptosis-regulating protein), CREB (a transcription factor involved in memory consolidation and neuroprotection), c-Src (a tyrosine kinase that promotes cell proliferation and survival in both normal and cancerous cells), and vasodilator-stimulated phosphoprotein (VASP, a protein that regulates actin filament formation and focal adhesions). This chapter highlights the use of capillary electrophoresis (CE), coupled with either LED-induced fluorescence or chemoluminescence detection, for providing ultrasensitive, highly quantitative measurements of (1) apoptotic DNA fragmentation, (2) RT-PCR products (determining mRNA expression levels) in a multiplexed analysis, and (3) levels of protein expression and site-specific phosphorylation. The ultrasensitive measurement of protein expression/phosphorylation levels utilizes the recently developed NanoPro100 system [ProteinSimple (originally named Cell Biosciences, Inc.), Santa Clara, CA, USA], an automated CE-chemoluminescence-based immunoquantification instrument, which provides exceedingly high sensitivity (e.g., femtogram quantities of protein) and much better phosphoprotein resolving power and quantification, compared with conventional Western blot analysis. These novel methodologies can now provide neuroscientists with valuable new tools for studying the expression and phosphorylation of lower abundance proteins, such as PKG, which in the past have been difficult to measure because of the relatively low-level sensitivity and inadequate quantification of conventional techniques. These new CE-based methodologies also allow the

Hideyuki Mukai (ed.), *Protein Kinase Technologies*, Neuromethods, vol. 68,
DOI 10.1007/978-1-61779-824-5_18,

accurate quantification of apoptotic DNA fragmentation, multiplexed mRNA levels, and protein expression/phosphorylation levels in very small biological samples, potentially containing fewer than 1,000 mammalian cells, thus ideal for studies of primary cell cultures and isolated stem/progenitor cells.

Key words: Protein kinase G, Cyclic GMP, Nitric oxide, Neuroprotection, Neural cell proliferation, Protein expression, Capillary electrophoresis, Apoptosis, Natriuretic peptide

1. Introduction

1.1. General Background and the Role of PKG in Regulating Multiple Neural Functions (Neural Cell Survival and Proliferation, Synaptic Plasticity, Synaptogenesis/Axon Guidance, Nociception, and Circadian Rhythm)

Detailed descriptions of the biochemistry and the many biological roles of cyclic GMP (cGMP)-dependent protein kinase (protein kinase G, PKG) in various types of mammalian tissues and cells have been given in a number of excellent reviews (1–4). There are also several exceptional reviews and book chapters that focus on the specific role and molecular actions of PKG in multiple neural systems, determined in studies utilizing both pharmacological inhibitors and genetic knockout (ablation of PKG expression) to define the role of PKG (5–9). These reviews/book chapters, along with specialized reviews from our laboratory describing the molecular actions and physiological/pathological roles of PKG in the cardiovascular and neural systems, have documented the important role of PKG in regulating blood pressure and blood flow dynamics, in facilitating cytoprotection of cardiovascular cells (e.g., cardiac muscle and vascular endothelial cells) and in promoting the survival of many types of neural cells (especially, protecting neural cells against neurotoxin-induced apoptosis) (10–14). Our studies have suggested that nitric oxide (NO), the natriuretic peptides [atrial natriuretic peptide (ANP) and brain (or B-type) natriuretic peptide (BNP)] and the secreted form of amyloid precursor protein (sAPP) all depend on the kinase (serine/threonine-phosphorylating) activity of PKG in mediating their neuroprotective effects, potentially playing a key role in preventing or at least minimizing the pathological progression of various neurodegenerative diseases, such as Alzheimer's disease, amyotrophic lateral sclerosis (ALS), HIV dementia, and Parkinson's disease (11, 12, 15–19).

Early immunohistochemical studies of PKG in the brain by Suzanne Lohmann, Ulrich Walter, Paul Greengard, and others at Yale University in the late 1970s and early 1980s had suggested that PKG was localized almost exclusively to the Purkinje cells in the cerebellum, with the staining of PKG in other regions of the brain occurring only within the vascular smooth muscle cells (VSMCs) of blood vessels (20). However, other studies by Marjorie Ariano, also in the early 1980s, indicated the presence of immunostaining for PKG as well as other components of the cGMP signaling pathway, i.e., soluble guanylyl cyclase (sGC) and cGMP itself, in the caudate–putamen (21). More recent studies by the laboratories of Franz Hofmann and Robert Feil in Germany now

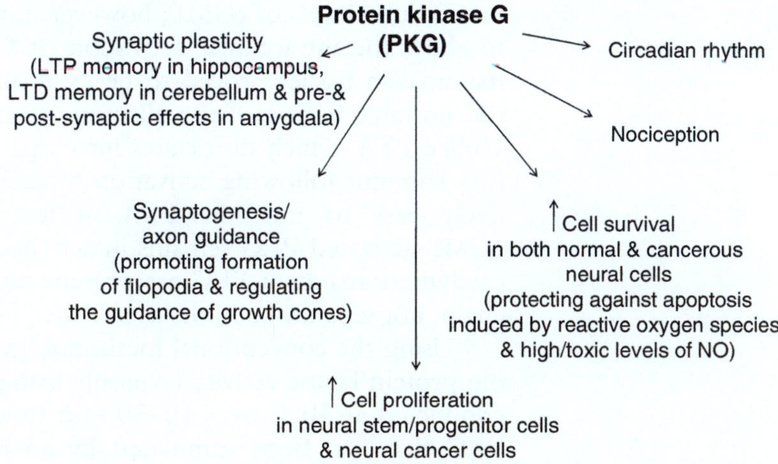

Fig. 1. Model illustrating the multiple physiological actions of PKG in neural cells.

indicate that PKG has a wider distribution in many neural cells throughout the brain (22), suggesting that PKG may play a wider role in regulating brain function than previously recognized.

Figure 1 illustrates the many physiological functions in neural systems that are thought to be mediated (or regulated) by PKG, including (1) synaptic plasticity (LTP memory in hippocampus, LTD memory in cerebellum, and pre- and postsynaptic alterations in lateral amygdala synapses) (6–8, 23–25), (2) synaptogenesis/axon guidance, promoting the formation of filopodia and regulating the guidance of growth cones (2, 6), (3) increased cell proliferation in neural stem/progenitor cells (2, 6, 26–29) as well as neural cancer cells (29, 30), (4) increased cell survival of both normal neural cells (15, 16, 25) and cancerous neural cells (18, 30), with a special role of PKG in protecting the normal neural cells against the toxic/pro-apoptotic effects of neurotoxin, such as reactive oxygen species (ROSs) and high/toxic-level NO (11, 12, 17, 19), (5) nociception (2, 31, 32), and (6) circadian rhythm (2). Further descriptions of these neural functions of PKG are given in several specialized reviews and book chapters (5–9).

1.2. Activation of PKG by cGMP, Nitric Oxide, and Natriuretic Peptides

PKG possesses some properties that are similar to the closely related and more intensely studied cyclic AMP (cAMP)-dependent protein kinase (protein kinase A, PKA), including catalyzing the phosphorylation of some of the same proteins as PKA (e.g., histones, CREB and BAD) (1–4, 6–14, 18). In early studies using in vitro experiments with mixed histones or histone H2B as the substrate, the kinase activity of purified PKG was found to be stimulated by the addition of cGMP, suggesting that this cyclic nucleotide may serve as the allosteric activator of PKG, analogous to the activation of PKA by cAMP. Thus, early on, it was proposed that the kinase activity of PKG within mammalian cells may be regulated by the intracellular levels of cGMP; however, early experiments designed to show the intracellular activation of PKG by cGMP in intact mammalian tissues proved to be exceedingly difficult because of the unstable nature of the allosteric stimulation by cGMP (33). Unlike PKA, which dissociates into separate catalytic and regulatory subunits following activation by cAMP (thus making it relatively easy to trap the PKA in its active form) (34), the cGMP-activated PKG remains intact (because the regulatory and catalytic domains of PKG are present on a common polypeptide chain, not separate subunits as in PKA (1–3, 10, 13)).

Using the conventional methodology of the time for measuring protein kinase activity (typically using assays with incubations conducted at 30°C over 10–30 min time periods), the activated PKG that had been stimulated by an elevation in intracellular cGMP levels within mammalian cells would rapidly inactivate during the kinase reaction (in part, because of the dissociation of cGMP from PKG), causing a substantial underestimation of the real PKG kinase activity that had existed within the intact cells at the time of tissue freezing (followed by homogenization) or cell lysis (10, 13, 33). To solve this problem, a specialized cold (0°C)-temperature kinase assay technique, coupled with an abbreviated assay incubation time (2.5 min), was developed by Ronald Fiscus between 1981 and 1983 while a Postdoctoral Fellow in the Steven Mayer laboratory at the University of California, San Diego (UCSD), designed to minimize the dissociation of cGMP from PKG during the assay procedure, thus preserving the PKG activation state as it existed within the intact cells of the biological sample (33). Also, because of the lower temperature and shortened incubation time, this kinase assay assured that the measurements were made during the "initial velocity" of the kinase reaction. Typical "end-point" protein kinase assays that use 10–30 min incubations times and 30°C temperature may not be measuring the initial velocity of the kinase reaction, potentially leading to an underestimation of the true kinase catalytic activity. By using the novel cold-temperature/short-incubation methodology, PKG kinase activity in airway smooth muscle cells was shown to be significantly elevated by exposure to the NO donor sodium nitroprusside (SNP), thus providing early evidence that the NO-induced relaxation of airway smooth muscle involved the stimulation of the intracellular kinase activity of PKG (33).

Further studies by Ronald Fiscus, while a Senior Postdoctoral Fellow in the Ferid Murad laboratory at Stanford University School of Medicine, utilized the cold-temperature/short-incubation PKG-kinase-assay methodology to study the effects of NO on PKG activation in VSMCs of blood vessels (35). The resulting data showed that both exogenous NO (provided by adding sodium nitroprusside) and endogenous NO [then called EDRF (endothelium-derived relaxant factor) released from endothelial cells stimulated by acetylcholine] significantly elevated the PKG kinase activity in the VSMCs of aortic strips, providing early evidence that PKG was indeed stimulated by NO in vascular tissue and that PKG may be the key protein kinase mediating NO-induced vasodilation.

Interestingly, even basal release of EDRF (endogenous NO) from unstimulated endothelial cells, now recognized to generate local concentrations of NO in the "picomolar" range (see discussion below), was found to be sufficient to significantly elevate the intracellular PKG kinase activity in nearby VSMCs, representing a 45% elevation of PKG kinase activity above the control values in blood vessels lacking endothelium (35). These data from 1983, showing that exogenous NO and endogenous EDRF caused identical stimulation of intracellular PKG kinase activity in VSMCs, provided early evidence that EDRF was very similar (and quite possibly identical) to NO, which was later confirmed in 1987 by the laboratories of Louis Ignarro and Salvador Moncada (using techniques other than PKG stimulation) (36, 37). Because NO at basally released levels (i.e., picomolar concentrations) was sufficient to significantly stimulate intracellular PKG activation in the early studies (35), it became clear that the NO-induced stimulation of PKG kinase activity in mammalian cells represented an exquisitely sensitive cell signaling (signal transduction) pathway, which could be used as a biomarker for the presence of physiological (picomolar) levels of NO in mammalian tissues and cell cultures.

Further studies by Ronald Fiscus at Stanford University School of Medicine, conducted using the cold-temperature/short-incubation PKG kinase assay, showed that exposure of blood vessels (both with or without endothelium) to atrial natriuretic peptide (ANP, called atriopeptin in early studies) caused concentration-dependent increases in the kinase activity of PKG in VSMCs, similar to the PKG activation caused by exposure to NO (38). These data provided the early evidence that ANP (in an endothelium-independent manner) uses PKG activation in VSMCs to mediate its vasodilatory effects in blood vessels. The data also suggested that the many other actions of ANP, such as neuroprotection (specifically, the protection of neural cells against the stressful condition of trophic factor withdrawal or the protection against the toxic/pro-apoptotic effects of high/toxic-level NO) are likely mediated by ANP-induced elevation of PKG kinase activity in neural cells (17, 19).

1.3. NO at Different Concentrations Activate Different Molecular Mechanisms, Which Mediate the Many Biological Effects of NO in Mammalian Cells

The first step in the ultrasensitive effects of NO in vascular tissue (i.e., stimulation of PKG kinase activity and downstream vasodilation) involves the binding of NO to the heme moiety of sGC, resulting in increased synthesis of cGMP from GTP, based in part on the many early studies conducted in the laboratories of Ferid Murad and Louis Ignarro (36, 39). Further studies throughout the 1990s and early 2000 by John Garthwaite's laboratory, focusing on the actions of NO in the brain, had suggested that the concentration of NO needed for activation of sGC was in the range of 1–100 nM, with the calculated EC_{50} values for NO-induced sGC activation becoming lower and lower (from 60 nM to ~1 nM) as the methodology for the measurements improved (40). Interestingly, the activation kinetics from these studies resulted in a Hill constant equal to 2.1, suggesting that the NO-induced activation of sGC involves the binding of NO at more than one site and that the multiple-site binding of NO behaves in a cooperative manner (40).

Recent studies from Michael Marletta's laboratory have suggested that full activation of sGC involves a two-stage process, with the first stage involving binding of NO to the heme moiety of sGC, which occurs at picomolar concentrations of NO (resulting in low level, approximately fourfold increase in sGC activity), followed by binding of additional NO (when NO is present at higher, nanomolar concentrations) to one or more cysteine residues, causing full activation of sGC (~200-fold increase in sGC activity) (41). Within intact mammalian cells, this NO–sGC relationship would translate into small, but biologically important (two- to fourfold) increases in intracellular concentrations of cGMP when mammalian cells are exposed to NO in the picomolar range, and much larger (5- to 200-fold) increases in intracellular cGMP levels when these same cells are exposed to NO at higher, nanomolar concentrations.

Figure 2 illustrates the concentration dependence of the many biological actions of NO in neural cells, which we are presenting in a style inspired by the many impressive NO-concentration-dependence models published by David Wink and his colleagues at the National Cancer Institute, National Institutes of Health (42, 43). Our model in Fig. 2 incorporates the unique NO–sGC relationship described in the two paragraphs above and also includes recent evidence that NO at concentrations as low as 10 pM (i.e., 0.01 nM, 100 times lower than the previously thought threshold concentrations for NO-induced activation of sGC) is capable of consistently stimulating increases in the intracellular cGMP levels in mammalian cells, determined using both a cell-based FRET analysis system (44) and a new non-FRET-fluorescence-based analysis system utilizing the new δ-FlincG biosensor for intracellular cGMP (45). The new δ-FlincG methodology, based on using a truncated PKG-Iα isoform (containing the high-affinity cGMP binding sites of PKG-Iα) coupled with circular permutated enhanced green fluorescent protein (EGFP), was developed by the laboratory of Wolfgang Dostmann, University of Vermont,

Fig. 2. (continued) cycase (sGC), with subsequent elevation of cyclic GMP (cGMP) levels and downstream stimulation of protein kinase G (PKG). At the lower range of physiological levels (0.01–1 nM), NO activates sGC via NO binding to the heme moiety of sGC, resulting in low-level activation and small (but functionally important) increases in cGMP levels in neural cells, which would selectively stimulate the PKG-Iα isoform. At higher physiological levels (and possible including the lower pathological levels) (i.e., 1–50 nM), NO causes high-level activation of sGC, resulting in large (5- to 200-fold) increases in cGMP levels and downstream stimulation of additional PKG isoforms, i.e., PKG-Iβ and PKG-II. These higher levels of cGMP can also regulate other signaling proteins, including protein kinase A (PKA), rod and olfactory cyclic-nucleotide-gated (CNG) cation channels, and various phosphodiesterases.

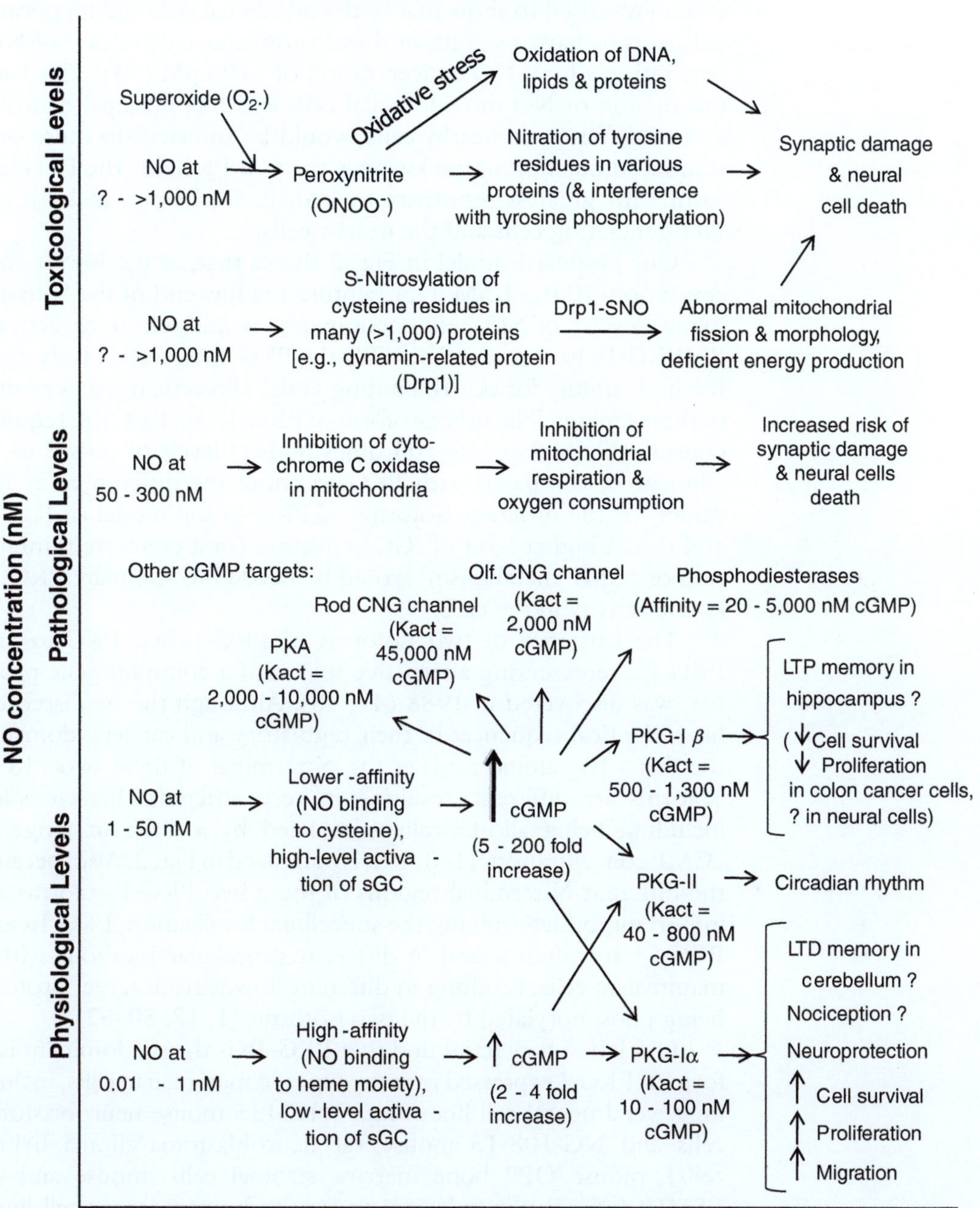

Fig. 2. Model showing the concentration dependence of the many biological effects of nitric oxide (NO) in neural cells. Synaptic damage and neuronal cell death can be caused by NO when present at higher, pathological and toxicological levels (i.e., high nanomolar to micromolar concentrations). These high levels of NO can occur during certain pathologies (e.g., Alzheimer's disease, brain ischemia during a stroke, inflammation, and some cancers) or during experimental studies involving the addition of NO donors to biological tissues or cultured cells. At these higher concentrations, NO induces the S-nitrosylation of cysteine residues and, in combination with superoxide, the nitration of tyrosine residues, both posttranslational modifications causing altered protein function and potentially contributing to the neural pathologies. In contrast, NO at physiological levels (0.01–50 nM) mediates many of its biological effects via the activation of soluble guanylyl

and provides an exquisitely sensitive method for accurately measuring intracellular levels of cGMP (46). The cell-based FRET analysis system was used to show that both endothelial cells and hippocampal neurons have a significant basal (unstimulated) release of NO, generating a local NO concentration of ~100 pM (44). This basal production of NO in endothelial cells and hippocampal neurons, and its diffusion to nearby cells, would be sufficient to cause substantial stimulation of the kinase activity of PKG-Iα (the PKG isoform with greatest sensitivity to cGMP, see below) in both the NO-generating cells and the nearby cells.

Our proposed model in Fig. 2 shows that, at the lowest concentrations (0.01–1 nM, representing the low end of the "physiological levels" of NO), endogenous NO would selectively activate the PKG-Iα isoform (and possibly the PKG-II isoform), because of the high affinity for cGMP binding at the allosteric regulatory sites of these kinases. The other isoform of PKG-I, i.e., PKG-Iβ, requires considerably higher (10–100 times higher) levels of cGMP to be stimulated, compared with PKG-Iα (note the differences in K_{act} values for the different isoforms of PKG in the model of Fig. 2) and thus a higher level of sGC activation (or a concurrent inhibition of cGMP metabolism) would be needed to stimulate PKG-Iβ kinase activity in the cells.

The existence of two isoforms of PKG-I, i.e., PKG-Iα and PKG-Iβ, representing alternative splices of a common gene product, was discovered in 1988 (47, 48). Although the two isoforms have identical sequences in their regulatory and catalytic domains, the first ~100 amino acids at the N-terminal of these two PKG-I isoforms are different, resulting in very different characteristics, including being allosterically stimulated by a different range of cGMP concentrations (1–3, 49), emphasized in Fig. 2. Also, because the different N-terminal regions of these two PKG-I isoforms are important for determining the subcellular localization, PKG-Iα and PKG-Iβ are often found in different subcellular locations within mammalian cells, resulting in different downstream target proteins being phosphorylated by the two isoforms (1, 18, 50–52).

Our laboratory has found that PKG-Iα is the predominant isoform of PKG-I expressed in many types of mammalian cells, including several neural cell lines (e.g., N1E-115 mouse neuroblastoma cells and NG-108-15 mouse/rat neuroblastoma-glioma hybrid cells), mouse OP9 bone marrow stromal cells, mouse and rat VSMCs (growing in culture), and many human cancer cell lines, including human breast cancer, nonsmall cell and small cell lung cancer, melanoma, mesothelioma, ovarian cancer, and prostate cancer cells (18, 50–53). Using both pharmacological inhibition and gene knockdown (siRNA and shRNA) techniques, we have found that the kinase activity of PKG-Iα is essential for preserving the survival in all of these cells. Furthermore, the proliferation (and, in some cases, the migration) of these cells is also dependent on the intracellular kinase activity of PKG-Iα (50–52).

The role of PKG-Iα in promoting cell survival, proliferation, and migration is included in the model of Fig. 2. Although there is currently limited information about the PKG-Iβ isoform in neural cells, we have included in the model the potential survival/proliferative effects of PKG-Iβ, based on studies of another mammalian cell type. For example, in colon cancer cells, which are reported to express predominantly the PKG-Iβ isoform, the high-level activation of PKG-Iβ or the forced overexpression of PKG-Iβ was shown to inhibit cell survival (specifically, inducing apoptotic cell death) and inhibit cell proliferation (54, 55). Thus, the two isoforms of PKG-I, when expressed and activated, may have very different biological effects (even opposite effects), likely mediated by the different subcellular localizations and different subsets of downstream target proteins that are phosphorylated. Further studies will be needed to determine the exact role of the two PKG-I isoforms in regulating cell survival, proliferation, and migration in the various types of neural cells.

Figure 2 illustrates how somewhat higher concentrations of NO (e.g., 1–50 nM, based on the data from the Garthwaite laboratory and the Marletta laboratory (40, 41)) would result in higher level activation of sGC, resulting in larger (likely 5- to 200-fold) increases in intracellular cGMP levels. These larger increases in cGMP levels would not only stimulate PKG-Iα kinase activity (likely causing nearly full activation of this isoform), but also stimulate the kinase activity of the other isoforms of PKG, i.e., PKG-Iβ and PKG-II, as well as potentially other target proteins affected by higher level cGMP, e.g., PKA (1, 3, 56), various phosphodiesterases (1), olfactory-type cyclic nucleotide-gated (CNG) cation channels, and rod-type CNG cation channels (57), if these alternative targets of cGMP are expressed. K_{act} values for each of these cGMP-target proteins are given in Fig. 2.

Figure 2 also shows the potential molecular and biological actions of NO at still higher levels, i.e., pathological and toxicological levels of NO (50 to >1,000 nM), which potentially could occur locally during high-level activation of the constitutive NO synthases (NOSs), neural-form NOS (nNOS, NOS1), and endothelial-form NOS (eNOS, NOS3). The higher levels of NO may also occur following the induced expression of the inducible-form NOS (iNOS, NOS2), which is capable of generating much higher levels of NO, typically as a result of ischemia and/or inflammatory conditions (e.g., increased production of pro-inflammatory cytokines, such as interleukin-1β, which, in an age-dependent manner, can stimulate the expression of iNOS) (58). Downstream, these higher levels of NO can result in (1) inhibition of cytochrome C oxidase in the mitochondria, inhibiting mitochondrial respiration and oxygen consumption (40), (2) S-nitrosylation of cysteine residues in many (>1,000) proteins, in some cases altering protein function (59–63), (3) nitration of tyrosine residues in various proteins, interfering with tyrosine-kinase-catalyzed phosphorylation of these tyrosine residues (64), and (4) oxidative stress, causing oxidizing damage to DNA, lipids, and proteins (42, 43, 64). Of particular interest is the recent discovery that higher/pathological levels of NO cause S-nitrosylation of dynamin-related protein 1 (Drp1, an important protein regulating mitochondrial fission), resulting in fragmentation, morphological abnormalities, and deficient production of energy in mitochondria within neural cells (61).

Overall, the many actions of NO in neural cells depend on the local concentration of NO as well as other concurrent events, such as the presence of elevated levels of ROSs. One ROS of particular importance is superoxide, which can combine with NO to form peroxynitrite ($ONOO^-$). At the low, physiological levels of NO, this reaction between NO and superoxide would result in the diversion of NO away from its many physiological functions (including neuroprotection, synaptogenesis, and synaptic plasticity) resulting in reduced activation of the sGC/cGMP/PKG-Iα and sGC/cGMP/PKG-Iβ signaling pathways and potentially increased risk of synaptic malfunction and neural cell death. At the higher pathological and toxicological levels of NO, the combination of superoxide with NO would result in the formation of sufficiently high levels of peroxynitrite to cause oxidative and nitrosative stress, which could further contribute to synaptic damage and neural cell death.

1.4. Use of Selective Pharmacological Inhibitors of the NO/sGC/cGMP/PKG-Iα Signaling Pathway and Gene Knockdown (siRNA and shRNA) to Genetically Silence the Expression of PKG-Iα

Figure 3 shows a model illustrating the use of pharmacological inhibitors that target key steps in the NO/sGC/cGMP/PKG-Iα signaling pathway as well as gene knockdown techniques, utilizing siRNA and shRNA constructs, to silence the expression of PKG-Iα. These inhibitors and siRNA/shRNA constructs are used to identify the downstream molecular targets and biological effects of the NO/sGC/cGMP/PKG-Iα pathway. Numerous inhibitors of the different NOS isoforms are available from commercial sources, and some of these inhibitors are capable of selectively inhibiting one isoform of NOS, as, for example, 1,400 W for inhibiting iNOS. Our previous studies have used 1,400 W and other (less selective) inhibitors of NOSs to show the critically important role of the NOSs in regulating apoptosis in human ovarian cancer cells (65).

Other studies from our laboratory have used pharmacological inhibitors of the endogenous-NO-induced heme-dependent activation of sGC (e.g., ODQ) and of PKG-Iα kinase activity (e.g., DT-2 and DT-3), as well as siRNA/shRNA constructs targeting PKG-Iα expression, to show the essential role of the NO/sGC/cGMP/PKG-Iα pathway in promoting cell survival (66) and cell proliferation/DNA synthesis in human ovarian cancer cells (50). Interestingly, the pro-growth effect of PKG-Iα was found to be dependent on a novel interaction between PKG-Iα and c-Src, a protein originally thought to be an oncogene but now recognized

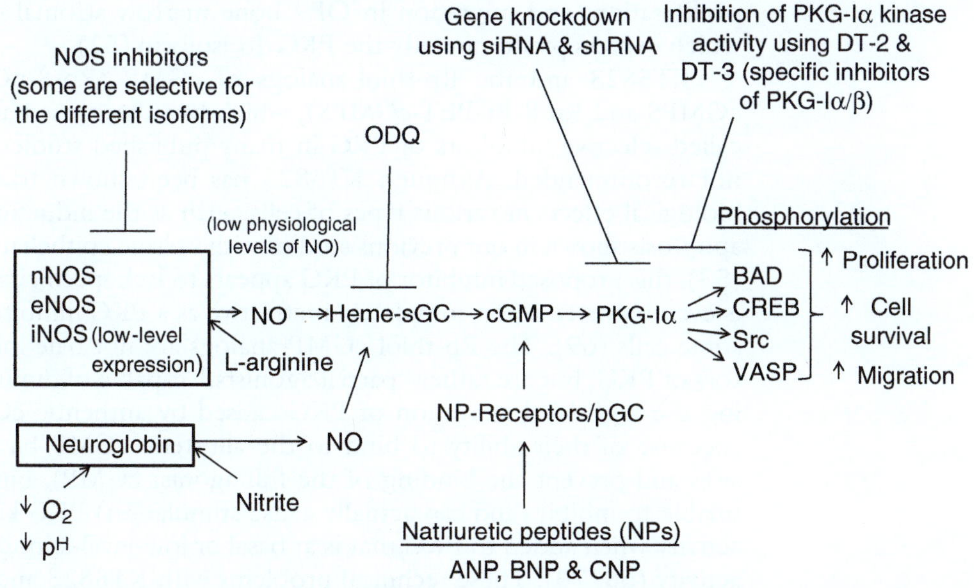

Fig. 3. Model showing the pharmacological inhibitors and gene knockdown techniques (e.g., siRNA and shRNA) used in the author's laboratory for determining the biological functions and downstream targets of endogenous PKG-Iα activity in mammalian cells. We have found that it is particularly important to use inhibitors and gene knockdown, rather than activators and forced overexpression of genes, when studying the involvement of the NO/sGC/cGMP/PKG-Iα and natriuretic peptide/pGC/cGMP/PKG-Iα signaling pathways, because these pathways are often already active under the experimental conditions (18, 38, 50–52, 65, 66). The use of pathway activators or force overexpression of a key pathway protein, in this situation, may not cause any effect. In most biological models (animal models, perfused tissues, and cultured cells), NO is endogenously produced by one or more of the NOSs, neural NOS (nNOS, NOS1), endothelial NOS (eNOS, NOS3), and inducible NOS (iNOS, NOS2), as well as via the conversion of endogenous nitrite into NO catalyzed by neuroglobin (and certain other heme-containing proteins). The cGMP/PKG-Iα pathway can also be activated by endogenously produced natriuretic peptides (as components of an autocrine/paracrine loop), as shown in our recent studies with bone marrow stromal cells (52), which would result in higher background activity, thus potentially interfering with any stimulatory effects that would be produced by added activators or forced overexpression.

as an important pro-growth/pro-survival signaling protein in both normal and cancerous mammalian cells (50).

DT-2 and the related DT-3 are cell permeable, highly specific inhibitors of PKG-Iα/β (respectively possessing 1,300-fold and 20,000-fold selectivity for PKG-Iα/β compared with the closest protein kinase, PKA), which have been developed by Wolfgang Dostmann and colleagues (67, 68); however, DT-2 and DT-3 are not able to discriminate between the two isoforms of PKG-I, because both of these inhibitors target the substrate-protein-binding site of the two PKG-I isoforms, a domain that is common for both PKG-Iα and PKG-Iβ. Nevertheless, DT-2 and DT-3 are very useful for identifying the molecular targets and biological roles of the PKG-Is in cells expressing only one of the isoforms, as, for example, in human ovarian cancer cells (50). Very recently, we have used DT-3, along with siRNA gene knockdown, to show a critically important role of PKG-Iα in promoting cell survival, proliferation, and migration in OP9 bone marrow stromal cells, which express predominantly the PKG-Iα isoform (52).

KT5823 and the Rp-thiol-analogs of cGMP (Rp-8-pCPT-cGMPS and Rp-8-Br-PET-cGMPS), which have been used as so-called selective inhibitors of PKG in many published studies, are not recommended. Although KT5823 has been shown to have biological effects in various types of cells, such as the induction of apoptosis shown in our previous studies with uterine epithelial cells (53), this proposed inhibitor of PKG appears to lack specificity and is even reported to be completely ineffective as a PKG inhibitor in some cells (69). The Rp-thiol-cGMP analogs are not true inhibitors of PKG, but are rather "partial agonists," capable of diminishing the high-level activation of PKG caused by authentic cGMP (because of their ability to bind to the allosteric cGMP binding sites and prevent the binding of the full agonist cGMP), but are unable to inhibit (and can actually cause stimulation) PKG kinase activity when added to PKG that is at basal or low-level-stimulated activity (68, 70). These technical problems with KT5823 and the Rp-thiol-cGMP analogs may explain, at least in part, the very different effects (sometimes opposite effects) observed in different studies using these agents.

Figure 3 also illustrates another potential source of endogenous NO, coming from the nitrite reductase activity of neuroglobin (and certain other heme-containing proteins, such as hemoglobin, myoglobin, and cytoglobin) (71, 72). Interestingly, the expression levels and nitrite reductase activity of neuroglobin are increased by low oxygen tension and low pH, as would occur during brain ischemia (71, 72), as well as via the stimulation of the sGC/cGMP/PKG signaling pathway in neural cells (HN33 cells, immortalized hippocampal neurons) (73), all potentially resulting in a NOS-independent production of NO from endogenous nitrite. Concentrations of nitrite are typically 1–10 μM in many tissues, including the brain, sufficient for the production of NO by the heme-containing globins (71, 72). Because of this alternative source of NO [potentially capable of generating picomolar (and possibly higher) concentrations of NO] and because of the exquisite sensitivity of the sGC/cGMP/PKG-Iα pathway, the total inhibition of endogenous NOS activity may not completely inhibit the activation of this signaling pathway.

Also shown in Fig. 3 is the activation of PKG-Iα by the natriuretic peptides, ANP, BNP, and CNP, potentially resulting in increased cell survival, proliferation, and migration. Early studies by Ronald Fiscus at Stanford University School of Medicine, in addition to showing ANP-induced activation of PKG (38), had also shown that ANP stimulates the production of cGMP in neural cells (PC12 pheochromocytoma and C6 glioma cells), resulting in the elevations of cGMP levels in both intracellular and extracellular compartments (74). Further studies in the Fiscus laboratory have shown that the cGMP elevations induced by ANP and BNP in PC12 cells dramatically protect these cells against the induction of apoptosis cause by trophic factor withdrawal (17) and that prior exposure of NG108-15 cells to ANP protected against the toxic/pro-apoptotic effects of high-level NO (19).

Four downstream target proteins, the apoptosis-regulating BAD (18), the transcription factor CREB (4, 11, 75), the pro-growth/pro-survival protein c-Src (50), and the actin-filament-/focal-adhesion-regulating protein VASP (50–52) are all directly phosphorylated by PKG-Iα, as shown in Fig. 3. Phosphorylation of BAD by PKG-Iα promotes cell survival in neural cells (18). PKG-catalyzed phosphorylation and activation of CREB is recognized to play an important role in memory consolidation in the lateral amygdala, specifically promoting late-phase LTP and fear memory (75). VASP and the related Ena protein are reported to be required for neuritogenesis in the developing cerebral cortex (76). The direct phosphorylation of VASP by the different PKG isoforms may contribute to the formation of filopodia and the regulation of growth cone guidance, potentially important for proper synaptogenesis during embryonic/fetal development as well as during recovery following neural injury in young and adults.

These studies have resulted in our current hypothesis that a primary physiological function of low-level NO and endogenous/locally generated natriuretic peptides (ANP, BNP, and CNP) within neural tissues may be to provide continuous, partial activation of PKG-Iα, which would result in downstream neuroprotection, specifically protecting the neural cells against the toxic effects of high-level ROSs and NO. Endogenous low-level NO and natriuretic peptides, via activation of PKG-Iα and other isoforms of PKG, may also play important roles in promoting the other neural functions shown in Fig. 1.

1.5. Previous Use of Capillary Electrophoresis (CE) for Measuring Apoptotic DNA Fragmentation, RT-PCR Products, and Protein Expression/Phosphorylation

Our laboratory has pioneered the use of CE technology coupled with the ultrasensitive laser-induced fluorescent detector (CE-LIF) as an ultrasensitive and highly quantitative method for determining levels of apoptosis (specifically the levels of apoptotic DNA fragmentation or apoptotic "DNA laddering") in tissue samples and cell culture samples. For example, we have used CE-LIF to determine the levels of apoptotic DNA fragmentation in various neural cell models, PC12 cells (11, 12, 17), NG108-15 cells (11, 12, 19), and N1E-115 cells (18), as well as in uterine epithelial cells (53, 77). Compared with agarose gels (i.e., agarose slab gel electrophoresis), which is the conventional technique for performing "DNA laddering" analysis, the new CE-LIF methodology is >1,000 times more sensitive (allowing accurate measurements of DNA fragmentation with fewer than 1,000 cells) and much more quantitative.

We have also used CE-LIF to measure the levels of RT-PCR products for determining gene expression levels (58). Our previous study showed that VSMCs from aging rats have dramatically (25-fold) higher levels of iNOS mRNA, compared with VSMCs from young controls. There are several advantages of using the CE-LIF methodology to analyze PCR products, including being able to clearly identify off-target products and the presence of primer-dimers, which may be a problem when using real-time PCR.

Although highly sensitive for measuring apoptotic DNA fragmentation and RT-PCR products, most CE-LIF systems have only a single capillary for the electrophoretic runs, thus resulting in a very low throughput, typically requiring about 1 h per sample for analysis. To partly solve this problem, we have begun using a 12-capillary CE system that uses an abbreviated electrophoretic run time described in Sect. 3.

A new methodology for ultrasensitive quantification of protein expression levels and protein phosphorylation levels, using CE technology coupled with in-capillary-Western blot-like analysis, has been developed by ProteinSimple (previously named Cell Biosciences, Inc., Santa Clara, CA, USA) (78). The equivalent of only 25 mammalian cells within each capillary (representing <1,000 cells in the total sample) was needed for determining ERK1/2 expression levels and phosphorylation levels. More recent studies by Alice Fan in the Dean Felsher Laboratory at Stanford University School of Medicine have shown that this new state-of-the-art methodology (now called NanoPro100 and NanoPro1000 systems) can be used for determining protein expression and phosphorylation levels of other signaling proteins (e.g., BCL2, MEK, STAT5, and JNK) in samples of lymphoma cells from human patients, again using <1,000 cells per sample (79). The Fiscus Laboratory at Roseman University of Health Sciences (Henderson/Las Vegas, Nevada, USA) has begun using this new methodology for determining the expression and phosphorylation levels of the two PKG-I isoforms (and many other proteins) in neural cell samples described in Sect. 3.

2. Materials

The materials used and their suppliers included purified PKG-Iα (Axorra, San Diego, CA, USA); ODQ (1H-[1,2,4]Oxadiazolo[4,3-a] quinoxalin-1-one), Triton X-100 (Sigma, St. Louis, MO, USA); anti-PKG-Iα/β antibody (Cell Signaling Technology, Beverly, MA, USA); anti-rabbit-CW800 secondary antibody (LI-COR, Lincoln, NE, USA); Trizol, Superscript® VILO™ cDNA synthesis kit, 4–12% Bis-Tris NuPAGE gel system and EZQ protein Quantitation Kit (Invitrogen, Carlsbad, CA, USA); FastStart universal SYBR green master (Roche, Basel, Switzerland); QIAamp DNA mini kit and QIAxcel System [Qiagen, Germantown, MD (originally distributed through eGene, Inc., Irvine, CA, USA)]; PKG-I-specific primers (Integrated DNA Technologies, Coralville, IA, USA); NG108-15, PC-12, SH-SY5Y, N1E-115, NCI-H2052 and RPMI 1640 (ATCC, Manassas, VA, USA); fetal bovine serum (Gemini Biosciences, Sacramento, CA, USA); M-Per (Pierce, Rockland, IL, USA); Lambda phosphatase (Upstate, Billerica, MA, USA); Premix G2 pH 5–8 (nested) separation gradient, pI Standard Ladder 1, NanoPro100 Master Kit and NanoPro100 System (ProteinSimple/Cell Biosciences, Santa Clara, CA, USA). An antibody that is specific for the PKG-Iα isoform has been developed recently in collaboration with Cell Signaling Technology (Beverly, MA, USA). This antibody has been used as the primary antibody in some of our traditional Western blots as well as in some NanoPro100 experiments (see Fig. 6), but because of its lower sensitivity (compared with the anti-PKG-Iα/β antibody targeting the C-terminal, common to both PKG-Iα and PKG-Iβ), the anti-PKG-Iα-specific antibody has been used only in a limited number of experiments.

3. Methods

3.1. Cell Cultures

All cells were cultured at 37°C in a humidified atmosphere containing 5% CO_2, passaged twice weekly and used between passages 5 and 20. The N1E-115 and NG108-15 murine neuroblastoma cells were cultured in DMEM with 4.5 g/L glucose and 10% fetal bovine serum. PC-12 rat pheochromocytoma cells and NCI-H2052 human mesothelioma cells were grown in RPMI-1640 with 4.5 g/L glucose and 10% fetal bovine serum. The SH-SY5Y human neuroblastoma cells were grown in Ham's F-12K with 4.5 g/L glucose and 10% fetal bovine serum.

3.2. Analysis of Apoptosis (Apoptotic DNA Fragmentation) by 12-Channel Capillary Electrophoresis with LED-Induced Fluorescence Detector (12-CE-LED-IF) System

For apoptotic DNA fragmentation studies, 10,000 N1E-115 cells were plated in a 48-well plate. After 24 h, cells were treated with 50 μM of ODQ (inhibitor of endogenous-NO-induced activation of heme-sGC) or control vehicle (DMSO) for 24 h, harvested for analysis, purified using a QIAamp DNA mini kit, and then analyzed on the QIAxcel system, a 12-channel capillary-electrophoresis-based RNA/DNA analysis system with light emitting diode (LED)-induced fluorescence detector (12-CE-LED-IF) system, marketed by Qiagen (originally called HDA-GT12 when produced and marketed by eGene, Inc., Irvine, CA, USA, which later combined with Qiagen). This system was originally designed as an automated high performance genetic analyzer that utilizes CE to resolve, identify, and quantify oligonucleotides.

Figure 4 shows the resulting electropherograms, illustrating the superior ability of the 12-CE-LED-IF instrument to resolve, identify,

Indentification and quantification of apoptotic DNA fragmentation using a new 12-channel capillary electrophoresis with LED-induced fluorescence detector (12-CE-LED-IF) methodology:

ODQ-induced depletion of cyclic GMP levels in N1E-115 mouse neuroblastoma cells causes induction of apoptosis (apoptotic DNA fragmentation, measured by 12-CE-LED-IF)

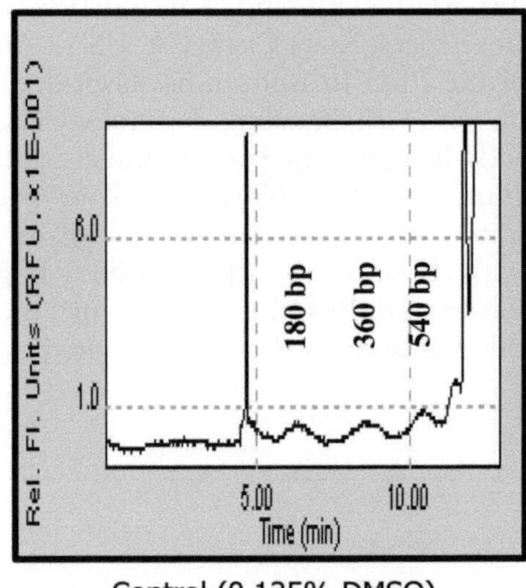

Control (0.125% DMSO)

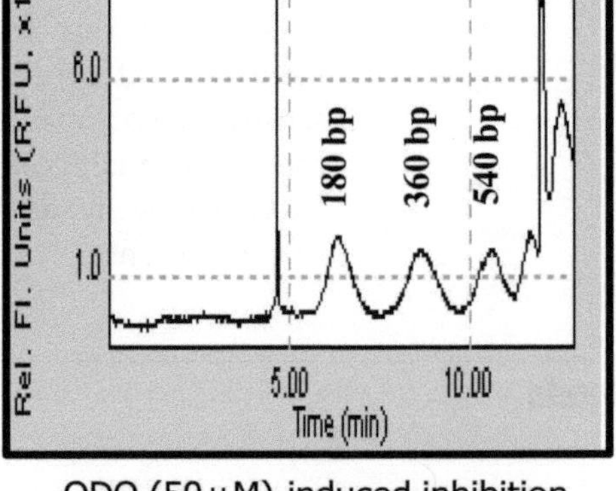

ODQ (50 μM)-induced inhibition of cyclic GMP biosynthesis (causing depletion of cyclic GMP levels)

Fig. 4. Data of apoptotic DNA fragmentation analysis obtained with a 12-channel capillary electrophoresis with LED-induced fluorescence detector (12-CE-LED-IF), showing the induction of apoptosis in N1E-115 neuroblastoma cells caused by ODQ-induced inhibition of endogenous NO-mediated activation of soluble guanylyl cyclase (and subsequent depletion of cGMP). The analysis was performed using the QIAxcel system by Qiagen (originally called HDA-GT12 when marketed by eGene, Inc.).

and quantify the apoptotic DNA fragments in the N1E-115 cells. Our previously published studies have shown direct comparisons between CE-based apoptotic DNA laddering and the laddering observed by conventional agarose gels (11, 12, 19, 53, 77).

3.3. Multiplexed Analysis of Transcript (mRNA) Levels Using a 12-Channel Capillary Electrophoresis System

PKG transcript levels were assessed using RT-PCR and subsequently analyzing the RT-PCR products on the 12-channel QIAxcel capillary electrophoresis system (same as in Sect. 3.2). RNA was Trizol extracted from 50,000 cells and DNAse treated. One microgram of purified total RNA was converted to cDNA using Superscript Vilo reverse transcriptase kit (Invitrogen) and was used in a PCR reaction using a GeneAmp PCR system 9700. Unique primers were designed for the PKG-Iα isoform (F: 5′ GGCGCAGGGCATCTCG 3′, R: 5′ ATCCACAATCTCCTGGA TCTG 3′) and the PKG-β isoform (F: 5′ GACCAGCCACCCAGCA 3′, R: 5′ TCCACAATCTCCTGGATCTG 3′). Expected product

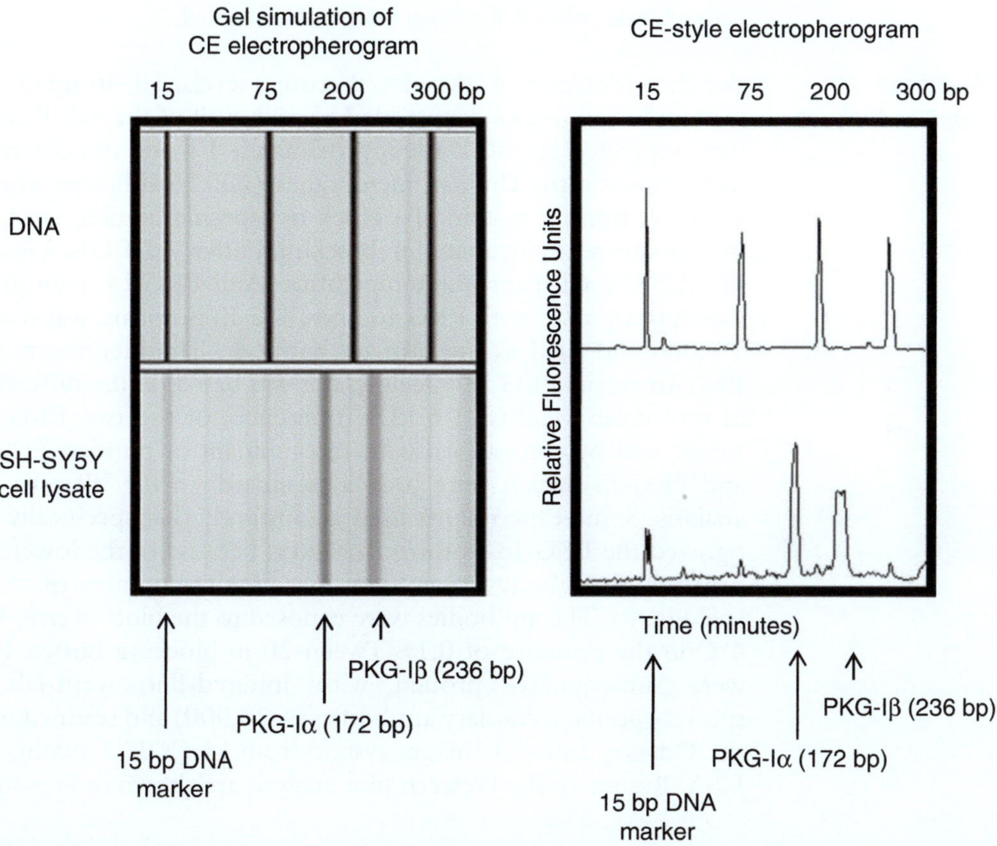

Fig. 5. Multiplexed analysis of RT-PCR products using the 12-CE-LED-IF system. Messenger RNA levels of both PKG-Iα and PKG-Iβ in human SH-SY5Y neuroblastoma cells could be simultaneously determined in a single electrophoretic run. As in Fig. 4, the QIAxcel system was used for analysis. This novel methodology allows clean separation, visualization, and quantification of the different RT-PCR products, thus avoiding some of the problems encountered with real-time PCR.

sizes were 172 and 236 base pairs, respectively. After 30 cycles, the PCR products were run on the QIAxcel system.

This 12-CE-LED-IF system has high resolution (separating and identifying oligonucleotides with only three base pairs differences) and is very sensitive, allowing for detection of low levels of PKG-I expression. Products were injected at 8 kV for 20 s and separated at 6 kV for 400 s (although only the relevant portions of the electropherograms are displayed in Fig. 5). Results were analyzed using BioCalculator 1.6 software. The RT-PCR products, which were generated using these primers, were previously run on 1.0% agarose gels, isolated, purified, and sequenced to verify their identity.

Figure 5 shows the electropherograms, displayed in two modes (gel simulation and CE-style), of the analysis for PKG-I mRNA levels in the SH-SY5Y human neuroblastoma cell line. DNA standards are used for determining the size of the PCR products. Multiplexed analysis allows the transcripts of both isoforms of PKG-I to be analyzed within the same electrophoretic run. Note that the SH-SY5Y cells express mRNA for both isoforms of PKG-I.

3.4. Traditional Western Blot Analyses

For determination of PKG-Iα/β protein levels, 10–40 μg of total protein in cell lysates were added to each well of the gel. Proteins were separated at 180 V for approximately 1 h and directly transferred onto nitrocellulose membranes (GE Healthcare) using a semi-wet transfer system. To block nonspecific binding sites, the membranes were incubated in blocking buffer (LI-COR, Lincoln, NE, USA) for 1 h at room temperature. Anti-PKG-Iα/β antibody, which recognizes both PKG-Iα and PKG-Iβ isoforms, was diluted 1:1,000 and used as the primary antibody. The identification of PKG-Iα versus PKG-Iβ in cell lysates was based on the differences in molecular weights (1.6 kDa difference) of the two PKG isoforms, and by comparison with recombinant or purified PKG-Iα and PKG-Iβ, which were used as standards in the Western blot analysis. Some experiments used an antibody that specifically recognized the PKG-Iα isoform, however, because of the lower sensitivity, this antibody was used in only a limited number of studies (see Fig. 6). The antibodies were exposed to the blots overnight at 4°C in the presence of 0.1% Tween-20 in blocking buffer. Blots were subsequently probed with infrared-fluorescent-labeled, species-specific secondary antibodies (1:20,000) and scanned using the Odyssey Infrared Imager system from LI-COR, Lincoln, NE, USA. Results of the Western blot analysis are shown in Figs. 6–8.

Fig. 6. Protein analysis using the new ultrasensitive NanoPro100 system (ProteinSimple, Santa Clara, CA, USA), based on separating proteins by isoelectric focusing using capillary electrophoresis, followed by in-capillary immunochemoluminescence detection. The NanoPro100 system is exceedingly sensitivity, has the capability of being highly quantitative and, because of using isoelectric focusing, can separate the different phosphorylated forms (including singly and doubly phosphorylated forms) of the detected proteins. (**a**) Ultrasensitive and linear detection of purified PKG-Iα using the NanoPro100 system and a primary antibody directed at the C-terminal region of PKG-Iα (a region shared by the PKG-Iβ isoform). The antibody recognizes a cluster of proteins, peaking at pI values of 5.7–5.8, representing different phosphoforms of PKG-Iα. If present in the sample, PKG-Iβ would be represented by an additional peak at a pI value of 5.5–5.6 (see Fig. 7 as example). The electropherogram in *panel* (**a**) illustrates the purity of the PKG-Iα preparation. (**b**) Western blot showing the specificity of a newly developed antibody, specific for the PKG-Iα isoform. Although specific for PKG-Iα, this new antibody is less sensitive compared with the other antibody (used in *panel* (**a**) and in most of our other studies, which recognizes both PKG-Iα and PKG-Iβ). (**c**) The detection of PKG-Iα in the NanoPro100 system was confirmed using the PKG-Iα-specific antibody.

a Detection of purified PKG-Iα protein using the NanoPro100 system: Proof of sensitivity and linearity

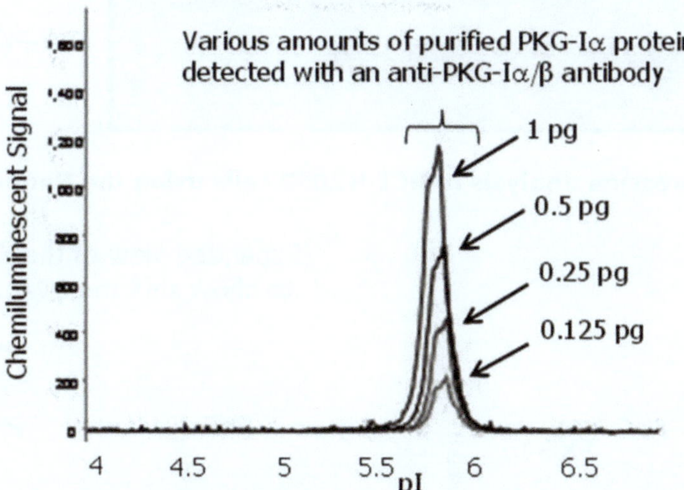

b Western Blot: Identification of a PKG-Iα-specific antibody

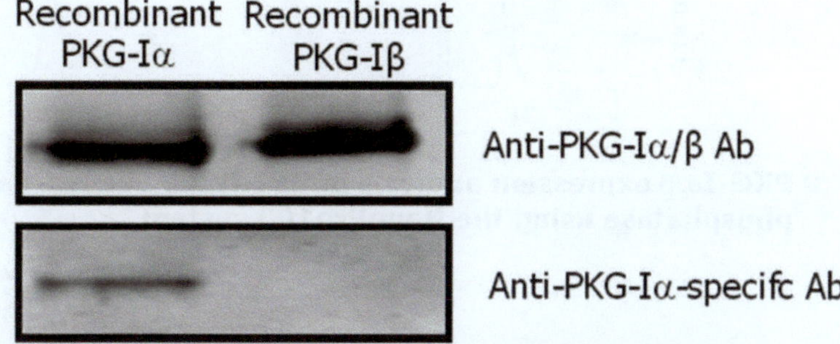

c Detection of purified PKG-Iα protein using the NanoPro100 system: Use of an isoform-specific antibody to verify detection of PKG-Iα

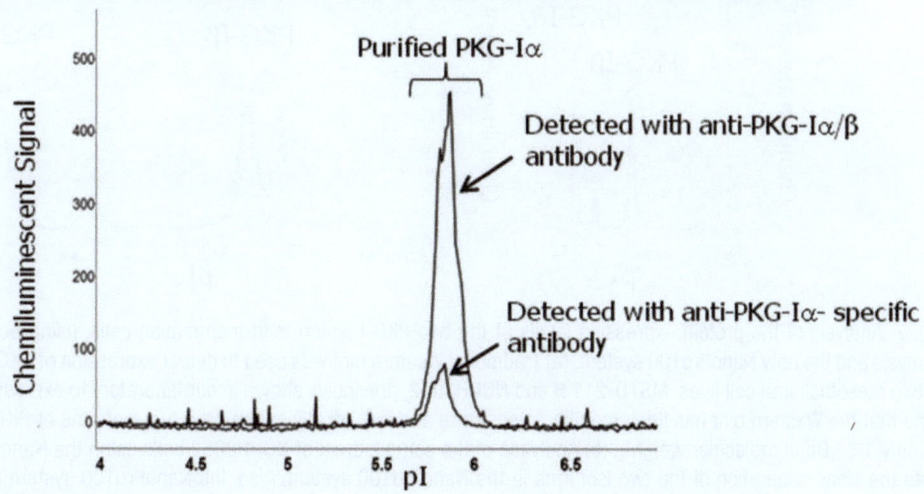

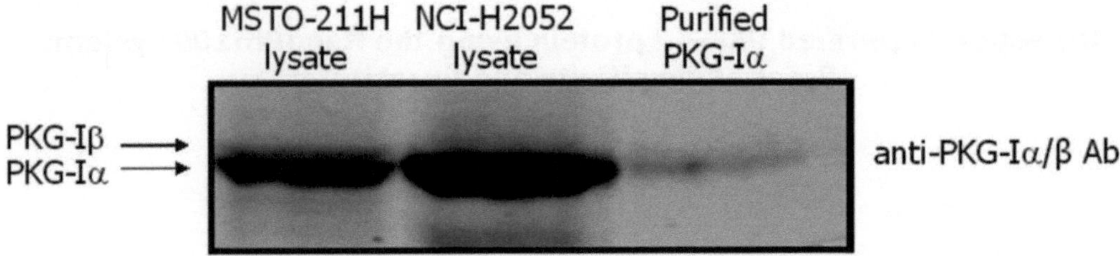

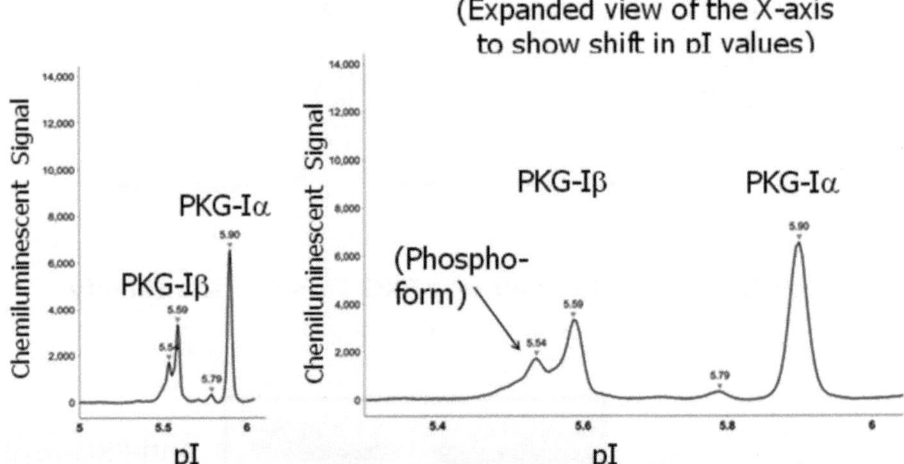

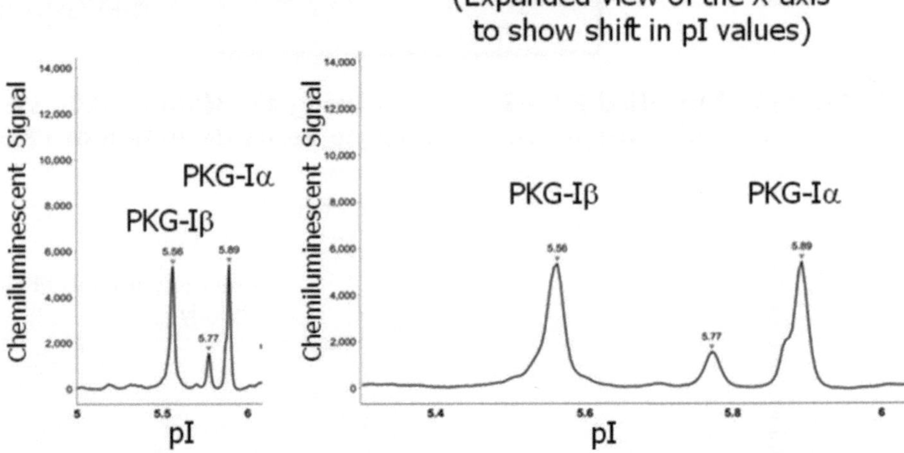

Fig. 7. Analysis of the protein expression levels of the two PKG-I isoforms in mammalian cells, using both Western blot analysis and the new NanoPro100 system. (**a**) Traditional Western blot was used to detect expression of PKG-Iα and PKG-Iβ in two mesothelioma cell lines, MSTO-211 H and NCI-H2052, previously shown in our laboratory to express both isoforms. Note that the Western blot has limited ability to separate and individually analyze the two isoforms of PKG-I (which differ by only 1.6 kDa in molecular weight). (**b**) Analyses of the same extract of NCI-H2052 cells using the NanoPro100 system. Note the clean separation of the two isoforms in the NanoPro100 system. Also, the NanoPro100 system detects a phosphorylated form of PKG-Iβ, represented by a second peak with somewhat lower pI value. (**c**) Analysis of the NCI-H2052 cell lysate following its incubation with lambda phosphatase, which removes the phosphorylation at serine, threonine, and tyrosine residues.

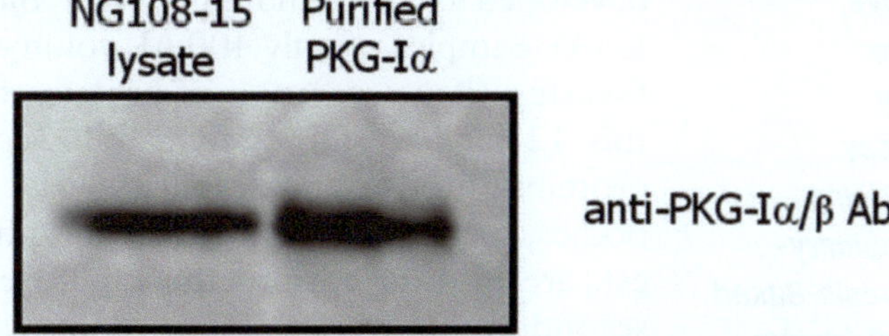

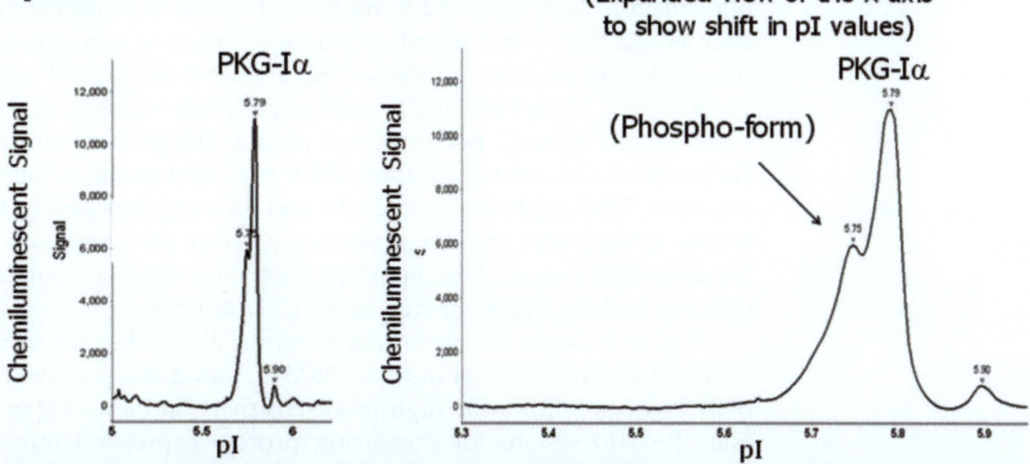

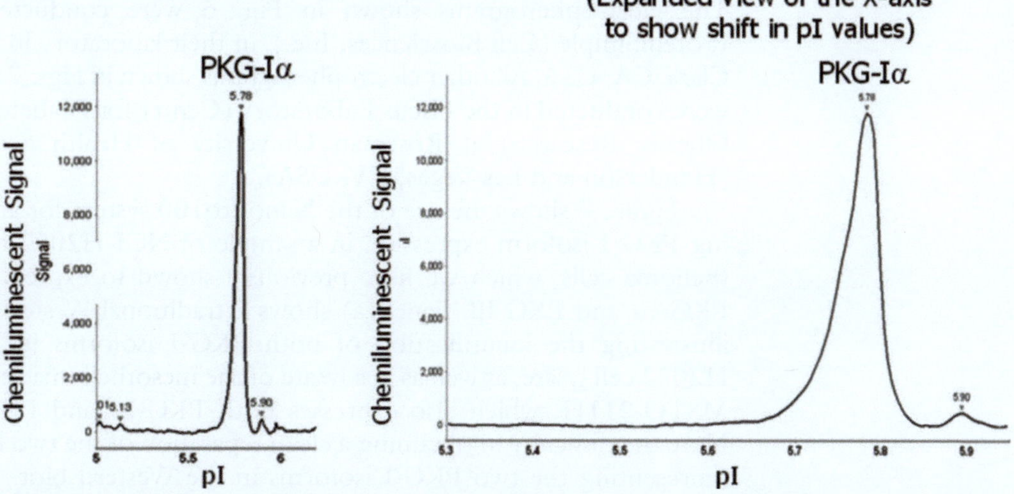

Fig. 8. Analysis of cell lysate of NG108-15 neuroblastoma-glioma hybrid cells using both Western blot and NanoPro100 system. (**a**) Western blot showing protein expression of PKG-Iα. (**b**) NanoPro100 analysis of NG108-15 cell lysate, showing expression of only the PKG-Iα isoform. A second peak at a somewhat lower pI value suggests the presence of an additional phosphorylated form of PKG-Iα. (**c**) Analysis of the NG108-15 cell lysate following incubation with lambda phosphatase to remove the phosphorylation. Note that the second, smaller peak of PKG-Iα is lost after treatment with lambda phosphatase.

3.5. New Ultrasensitive, Quantitative In-Capillary Immunoassay Technique Using the New Capillary-Electrophoresis-Based NanoPro100 System from ProteinSimple (Cell Biosciences, Inc.)

A new ultrasensitive in-capillary CE-based immunoassay has been developed by ProteinSimple (Cell Biosciences, Santa Clara, CA, USA). Samples of only 400 nL volume are separated by isoelectric focusing (i.e., separation of proteins based on their pI values) in this 12-capillary CE system, followed by immobilization of the proteins onto the inner capillary walls. Subsequently, primary antibodies, specific for the proteins and phosphorylation sites of interest, are introduced into the capillaries, followed by HRP-linked secondary antibodies, followed by chemoluminescence reagents and the detection based on chemiluminescence. Cells were first lysed with M-Per lysis buffer in the presence of protease and phosphatase inhibitors. 62.5 ng/μL of lysate was prepared in Premix G2 pH 5–8 (nested) separation gradient containing pI standards and protease inhibitors. Separation time was 50 min at 15,000 μW. Anti-PKG-Iα/β antibody, which recognizes both PKG-Iα and PKG-Iβ isoforms, was used as the primary antibody (in most studies) and anti-rabbit-HRP was used as the secondary antibody. The incubations were 110 and 55 min, respectively. Cell lysates, treated with lambda phosphatase before the analysis on the NanoPro100 system, were prepared according to the manufacturer's instructions and incubated at 37°C for 60 min.

Figure 6 shows the electropherograms, illustrating the separation and identification of purified PKG-Iα, used as a positive control. Panel (a) shows the high-level sensitivity and linearity of the NanoPro100 system for measuring protein expression levels of PKG-Iα. Panel (b) shows Western blots, indicating the specificity of the anti-PKG-Iα-specific antibody. Panel (c) shows the relative sensitivities of the NanoPro100 system using the two antibodies. The electropherograms shown in Fig. 6 were conducted by ProteinSimple (Cell Biosciences, Inc.), in their laboratory in Santa Clara, CA, USA. All other electropherograms shown in Figs. 7 and 8 were conducted in the Fiscus Laboratory (Center for Diabetes and Obesity Research) at Roseman University of Health Sciences (Henderson and Las Vegas, NV, USA).

Figure 7 shows the use of the NanoPro100 system for analyzing PKG-I isoform expression in a sample of NCI-H2052 mesothelioma cells, which we have previously shown to express both PKG-Iα and PKG-Iβ. Panel (a) shows a traditional Western blot illustrating the identification of both PKG-I isoforms in NCI-H2052 cell lysate, as well as in a lysate of the mesothelioma cell line MSTO-211H, which also expresses both PKG-Iα and PKG-Iβ. Note the difficulty in obtaining a clean separation of the two bands representing the two PKG-I isoforms in the Western blot. Panel (b) shows the NanoPro100 electropherograms, which illustrate the clean separation and the relative expression levels of the two PKG-I isoforms. Interestingly, the PKG-Iβ appeared to have a phosphorylated form that was present, represented by the shoulder peak. To confirm that this peak was a phosphorylated form, the cell lysate was first treated with lambda phosphatase to remove the phosphates. Panel (c) shows the loss of the PKG-Iβ shoulder peak after treating with lambda phosphatase, indicating that part of the PKG-Iβ expressed in NCI-H2052 cells is in a phosphorylated form.

Figure 8 shows the use of the NanoPro100 system for analyzing PKG-I isoform expression in a sample of NG108-15 neural cells. Previously, we have shown that NG108-15 cells express predominantly the PKG-Iα isoform, using traditional Western blot analysis (53). This was confirmed in Fig. 8, Panel (a) of the present study. Panel (b) shows the NanoPro100 electropherogram, illustrating the presence of only PKG-Iα. In this case, a phosphoform of the PKG-Iα appeared as a peak on the left shoulder of the major peak. Panel (c) confirms that the shoulder peak in Panel (b) was indeed a phosphorylated form of PKG-Iα, because it was removed by prior treatment with lambda phosphatase.

4. Notes/ Troubleshooting

One must take into consideration several factors when conducting experiments and analyzing data with the QIAxcel and NanoPro100 systems.

4.1. QIAxcel Experiments

1. Apoptotic DNA fragmentation analysis ("DNA laddering") using a traditional agarose slab gel electrophoresis system requires a relatively large sample size, typically more than 1,000,000 cells to get consistent, visually identifiable DNA laddering. The QIAxcel system is far more sensitive, allowing for analysis of just a fraction of the sample size needed with agarose gels, requiring just a few 1,000 cells for analysis by the QIAxcel system. Nevertheless, it is important to optimize the time point for the apoptotic DNA fragmentation analysis when working with different cell types, because DNA fragmentation can occur at very different time points in different types of cells. For example, our early studies with PC-12 pheochromocytoma cells showed that apoptotic DNA fragmentation after removal of trophic factors or exposure to neurotoxins occurs as early as 4 h after the stressful insult (11, 12, 17). In contrast, many other cell types, such as NG108-15 cells and uterine epithelial cells, require longer time points (e.g., 24 h) before showing the apoptotic DNA fragmentation (11, 12, 19, 53, 77). Therefore, a complete time course of the onset of apoptotic DNA fragmentation should be conducted with any new types of cells.
2. When analyzing gene expression, it is critical to effectively treat prepared RNA with DNase to avoid contamination of dsDNA

once cDNA is prepared. Most RNA kits provide a DNase treatment step in the protocol. If preparing RNA by conventional phenol extraction, treat with DNAse (1 U/ug of RNA for 20 min at 37°C), and then deactivate for 15 min at 65°C. Quantification can be achieved by extrapolation from a known standard curve. Limit of detection is at least 0.1 ng/μl.

4.2. NanoPro100 Experiments

1. Given the exquisitely high level of sensitivity, the choice of proper antibody is critical. The antibody chosen should be as specific for the target of interest as possible to avoid additional peaks from being detected, which could confuse interpretation of the results. Keep in mind that the cell lysates are prepared in a nondenaturing fashion, so antibodies that work in traditional Western blots could potentially not work, or recognize the target of interest differently in this assay, depending on epitope exposure. In other words, conformation of the native protein may render the epitope in an inaccessible position. Additionally, the cell lysates should be prepared in a low/no salt lysis buffer to prevent sample drift during the isoelectric focusing step. Since samples are in a nondenaturing environment, proteases and phosphatases present are free to cleave the target protein of interest, so be sure to add appropriate amounts of protease/phosphatase inhibitors and perform the NanoPro100 electrophoretic runs within minutes of lysing the cells.
2. If available, phospho-specific antibodies can aid in the identification of specific phosphoforms, which will usually migrate to a lower pI value, due to the acidic nature of the phosphate groups.
3. Theoretical pI values may differ from the observed pI values in this system due to posttranslational modifications that could modify the charge of the protein. For example, acetylation can add a negative charge to a protein. Phosphatase treatment, as shown in Figs. 7 and 8, will allow one to identify peaks that are true phosphoforms. If phosphorylated, a peak should shift to the right after treatment with the phosphatase.
4. Although a recombinant version of the target of interest may not migrate to the exact same pI value as the native protein, it can be used to generate a standard curve to extrapolate protein amount/concentration in the sample.

5. Conclusions

Although early studies had raised questions about the true endogenous activators of PKG in biological samples, evidence over the years now suggest that intracellular kinase activity of PKG is regulated by endogenous NO and natriuretic peptides (ANP, BNP and CNP), all of which stimulate cGMP synthesis, elevate intracellular levels of cGMP, and increase the PKG kinase activity in many types of mammalian cells. The PKG-Iα isoform, by comparison with the PKG-Iβ isoform, is 10–100 times more sensitive to cGMP-induced stimulation of PKG kinase activity, and thus likely represents the isoform of PKG responding to the lower end of the physiological levels (i.e., picomolar concentrations) of NO. The activation of PKG-Iα by the low-level NO promotes phosphorylation of multiple downstream target proteins, including BAD, CREB, c-Src, and VASP, which in turn mediate the neuroprotective effects of the low-level-NO/cGMP/PKG-Iα signaling pathway, helping to protect neural cells against the toxic effects of ROSs and high/toxic-level NO. These data suggest a key role of the low-level-NO/cGMP/PKG-Iα pathway in protecting against the onset of various neurodegenerative diseases, thus preventing or at least slowing down the progression of Alzheimer's disease, ALS, HIV-dementia, and Parkinson's disease.

The new developments in capillary electrophoresis technologies, resulting in the new QIAxcel system from Qiagen and the NanoPro100/NanoPro1000 system from ProteinSimple (Cell Biosciences, Inc.), can now be used for conducting ultrasensitive, highly quantitative measurements of apoptotic DNA fragmentation, multiplexed RT-PCR products, and protein expression and phosphorylation levels in very small sized biological samples (fewer than 1,000 mammalian cells).

Acknowledgments

Funding for this project was provided by startup funding for the Fiscus Laboratory in the Center for Diabetes & Obesity Prevention, Treatment, Research & Education at the Roseman University of Health Sciences (formerly the University of Southern Nevada), Henderson and Las Vegas, NV, USA.

References

1. Francis SH, Busch JL, Corbin JD, Sibley D (2010) cGMP-dependent protein kinases and cGMP phosphodiesterases in nitric oxide and cGMP action. Pharmacol Rev 62:525–563
2. Hofmann F, Feil R, Kleppisch T, Schlossmann J (2006) Function of cGMP-dependent protein kinases as revealed by gene deletion. Physiol Rev 86:1–23
3. Lincoln TM, Dey N, Sellak H (2001) Invited review: cGMP-dependent protein kinase signaling mechanisms in smooth muscle: from the regulation of tone to gene expression. J Appl Physiol 91:1421–1430
4. Pilz RB, Casteel DE (2003) Regulation of gene expression by cyclic GMP. Circ Res 93: 1034–1046
5. Barnstable CJ, Wei JY, Han MH (2004) Modulation of synaptic function by cGMP and cGMP-gated cation channels. Neurochem Int 45:875–884

6. Feil R, Hofmann F, Kleppisch T (2005) Function of cGMP-dependent protein kinases in the nervous system. Rev Neurosci 16:23–41
7. Feil R, Kleppisch T (2008) NO/cGMP-dependent modulation of synaptic transmission. Handb Exp Pharmacol 184:529–560
8. Kleppisch T, Feil R (2009) cGMP signalling in the mammalian brain: role in synaptic plasticity and behaviour. Handb Exp Pharmacol 191: 549–579
9. Wang X, Robinson PJ (1997) Cyclic GMP-dependent protein kinase and cellular signaling in the nervous system. J Neurochem 68:443–456
10. Fiscus RR (1988) Molecular mechanisms of endothelium-mediated vasodilation. Semin Thromb Hemost 14(Suppl):12–22
11. Fiscus RR (2002) Involvement of cyclic GMP and protein kinase G in the regulation of apoptosis and survival in neural cells. Neurosignals 11:175–190
12. Fiscus RR, Yuen JP, Chan SL, Kwong JH, Chew SB (2002) Nitric oxide and cyclic GMP as pro- and anti-apoptotic agents. J Card Surg 17:336–339
13. Fiscus RR, Murad F (1988) cGMP-dependent protein kinase activation in intact tissues. Methods Enzymol 159:150–159
14. Fung E, Fiscus RR, Yim AP, Angelini GD, Arifi AA (2005) The potential use of type-5 phosphodiesterase inhibitors in coronary artery bypass graft surgery. Chest 128:3065–3073
15. Barger SW, Fiscus RR, Ruth P, Hofmann F, Mattson MP (1995) Role of cyclic GMP in the regulation of neuronal calcium and survival by secreted forms of beta-amyloid precursor. J Neurochem 64:2087–2096
16. Barger SW, Mattson MP (1995) The secreted form of the Alzheimer's beta-amyloid precursor protein stimulates a membrane-associated guanylate cyclase. Biochem J 311(Pt 1):45–47
17. Fiscus RR, Tu AW, Chew SB (2001) Natriuretic peptides inhibit apoptosis and prolong the survival of serum-deprived PC12 cells. Neuroreport 12:185–189
18. Johlfs MG, Fiscus RR (2010) Protein kinase G type-Ialpha phosphorylates the apoptosis-regulating protein Bad at serine 155 and protects against apoptosis in N1E-115 cells. Neurochem Int 56:546–553
19. Cheng Chew SB, Leung PY, Fiscus RR (2003) Preincubation with atrial natriuretic peptide protects NG108-15 cells against the toxic/proapoptotic effects of the nitric oxide donor S-nitroso-N-acetyl. Histochem Cell Biol 120:163–171
20. Lohmann SM, Walter U, Miller PE, Greengard P, De Camilli P (1981) Immunohistochemical localization of cyclic GMP-dependent protein kinase in mammalian brain. Proc Natl Acad Sci U S A 78:653–657
21. Ariano MA (1983) Distribution of components of the guanosine 3′,5′-phosphate system in rat caudate-putamen. Neuroscience 10:707–723
22. Feil S, Zimmermann P, Knorn A, Brummer S, Schlossmann J, Hofmann F, Feil R (2005) Distribution of cGMP-dependent protein kinase type I and its isoforms in the mouse brain and retina. Neuroscience 135:863–868
23. Zhuo M, Hu Y, Schultz C, Kandel ER, Hawkins RD (1994) Role of guanylyl cyclase and cGMP-dependent protein kinase in long-term potentiation. Nature 368:635–639
24. Ota KT, Monsey MS, Wu MS, Schafe GE (2010) Synaptic plasticity and NO-cGMP-PKG signaling regulate pre- and postsynaptic alterations at rat lateral amygdala synapses following fear conditioning. PLoS One 5:e11236
25. Contestabile A, Monti B, Ciani E (2003) Brain nitric oxide and its dual role in neurodegeneration/neuroprotection: understanding molecular mechanisms to devise drug approaches. Curr Med Chem 10:2147–2174
26. Tegenge MA, Bicker G (2009) Nitric oxide and cGMP signal transduction positively regulates the motility of human neuronal precursor (NT2) cells. J Neurochem 110:1828–1841
27. Wang L, Gang Zhang Z, Lan Zhang R, Chopp M (2005) Activation of the PI3-K/Akt pathway mediates cGMP enhanced-neurogenesis in the adult progenitor cells derived from the subventricular zone. J Cereb Blood Flow Metab 25:1150–1158
28. Yoneyama M, Kawada K, Shiba T, Ogita K (2011) Endogenous nitric oxide generation linked to ryanodine receptors activates cyclic GMP/protein kinase G pathway for cell proliferation of neural stem/progenitor cells derived from embryonic hippocampus. J Pharmacol Sci 115:182–195
29. Firestein BL, Bredt DS (1998) Regulation of sensory neuron precursor proliferation by cyclic GMP-dependent protein kinase. J Neurochem 71:1846–1853
30. Andoh T, Chiueh CC, Chock PB (2003) Cyclic GMP-dependent protein kinase regulates the expression of thioredoxin and thioredoxin peroxidase-1 during hormesis in response to oxidative stress-induced apoptosis. J Biol Chem 278:885–890
31. Qian Y, Chao DS, Santillano DR, Cornwell TL, Nairn AC, Greengard P, Lincoln TM, Bredt DS (1996) cGMP-dependent protein kinase in dorsal root ganglion: relationship with nitric oxide synthase and nociceptive neurons. J Neurosci 16:3130–3138

32. Sung YJ, Chiu DT, Ambron RT (2006) Activation and retrograde transport of protein kinase G in rat nociceptive neurons after nerve injury and inflammation. Neuroscience 141: 697–709
33. Fiscus RR, Torphy TJ, Mayer SE (1984) Cyclic GMP-dependent protein kinase activation in canine tracheal smooth muscle by methacholine and sodium nitroprusside. Biochim Biophys Acta 805:382–392
34. Fiscus RR, Mayer SE (1982) Neural regulation of cyclic AMP, cyclic AMP-dependent protein kinase, and phosphorylase in bullfrog ventricular myocardium. Circ Res 51:551–559
35. Fiscus RR, Rapoport RM, Murad F (1983) Endothelium-dependent and nitrovasodilator-induced activation of cyclic GMP-dependent protein kinase in rat aorta. J Cyclic Nucleotide Protein Phosphor Res 9:415–425
36. Ignarro LJ, Buga GM, Wood KS, Byrns RE, Chaudhuri G (1987) Endothelium-derived relaxing factor produced and released from artery and vein is nitric oxide. Proc Natl Acad Sci U S A 84:9265–9269
37. Palmer RM, Ferrige AG, Moncada S (1987) Nitric oxide release accounts for the biological activity of endothelium-derived relaxing factor. Nature 327:524–526
38. Fiscus RR, Rapoport RM, Waldman SA, Murad F (1985) Atriopeptin II elevates cyclic GMP, activates cyclic GMP-dependent protein kinase and causes relaxation in rat thoracic aorta. Biochim Biophys Acta 846:179–184
39. Murad F, Rapoport RM, Fiscus R (1985) Role of cyclic-GMP in relaxations of vascular smooth muscle. J Cardiovasc Pharmacol 7(Suppl 3): S111–S118
40. Bellamy TC, Griffiths C, Garthwaite J (2002) Differential sensitivity of guanylyl cyclase and mitochondrial respiration to nitric oxide measured using clamped concentrations. J Biol Chem 277:31801–31807
41. Fernhoff NB, Derbyshire ER, Marletta MA (2009) A nitric oxide/cysteine interaction mediates the activation of soluble guanylate cyclase. Proc Natl Acad Sci U S A 106:21602–21607
42. Ridnour LA, Thomas DD, Switzer C, Flores-Santana W, Isenberg JS, Ambs S, Roberts DD, Wink DA (2008) Molecular mechanisms for discrete nitric oxide levels in cancer. Nitric Oxide 19:73–76
43. Thomas DD, Ridnour LA, Isenberg JS, Flores-Santana W, Switzer CH, Donzelli S, Hussain P, Vecoli C, Paolocci N, Ambs S, Colton CA, Harris CC, Roberts DD, Wink DA (2008) The chemical biology of nitric oxide: implications in cellular signaling. Free Radic Biol Med 45: 18–31
44. Sato M, Nakajima T, Goto M, Umezawa Y (2006) Cell-based indicator to visualize picomolar dynamics of nitric oxide release from living cells. Anal Chem 78:8175–8182
45. Batchelor AM, Bartus K, Reynell C, Constantinou S, Halvey EJ, Held KF, Dostmann WR, Vernon J, Garthwaite J (2010) Exquisite sensitivity to subsecond, picomolar nitric oxide transients conferred on cells by guanylyl cyclase-coupled receptors. Proc Natl Acad Sci U S A 107:22060–22065
46. Nausch LW, Ledoux J, Bonev AD, Nelson MT, Dostmann WR (2008) Differential patterning of cGMP in vascular smooth muscle cells revealed by single GFP-linked biosensors. Proc Natl Acad Sci U S A 105:365–370
47. Francis SH, Woodford TA, Wolfe L, Corbin JD (1988) Types I alpha and I beta isozymes of cGMP-dependent protein kinase: alternative mRNA splicing may produce different inhibitory domains. Second Messengers Phosphoproteins 12:301–310
48. Lincoln TM, Thompson M, Cornwell TL (1988) Purification and characterization of two forms of cyclic GMP-dependent protein kinase from bovine aorta. J Biol Chem 263:17632–17637
49. Ruth P, Landgraf W, Keilbach A, May B, Egleme C, Hofmann F (1991) The activation of expressed cGMP-dependent protein kinase isozymes I alpha and I beta is determined by the different amino-termini. Eur J Biochem 202:1339–1344
50. Leung EL, Wong JC, Johlfs MG, Tsang BK, Fiscus RR (2010) Protein kinase G type Ialpha activity in human ovarian cancer cells significantly contributes to enhanced Src activation and DNA synthesis/cell proliferation. Mol Cancer Res 8:578–591
51. Wong JC, Fiscus RR (2010) Protein kinase G activity prevents pathological-level nitric oxide-induced apoptosis and promotes DNA synthesis/cell proliferation in vascular smooth muscle cells. Cardiovasc Pathol 19:e221–e231
52. Wong JC, Fiscus RR (2011) Essential roles of the nitric oxide (NO)/cGMP/protein kinase G type-Ialpha (PKG-Ialpha) signaling pathway and the atrial natriuretic peptide (ANP)/cGMP/PKG-Ialpha autocrine loop in promoting proliferation and cell survival of OP9 bone marrow stromal cells. J Cell Biochem 112:829–839
53. Chan SL, Fiscus RR (2003) Guanylyl cyclase inhibitors NS2028 and ODQ and protein kinase G (PKG) inhibitor KT5823 trigger apoptotic DNA fragmentation in immortalized uterine epithelial cells: anti-apoptotic effects of basal cGMP/PKG. Mol Hum Reprod 9:775–783
54. Deguchi A, Thompson WJ, Weinstein IB (2004) Activation of protein kinase G is

sufficient to induce apoptosis and inhibit cell migration in colon cancer cells. Cancer Res 64:3966–3973

55. Liu L, Li H, Underwood T, Lloyd M, David M, Sperl G, Pamukcu R, Thompson WJ (2001) Cyclic GMP-dependent protein kinase activation and induction by exisulind and CP461 in colon tumor cells. J Pharmacol Exp Ther 299: 583–592
56. Cornwell TL, Arnold E, Boerth NJ, Lincoln TM (1994) Inhibition of smooth muscle cell growth by nitric oxide and activation of cAMP-dependent protein kinase by cGMP. Am J Physiol 267:C1405–C1413
57. Mottig H, Kusch J, Zimmer T, Scholle A, Benndorf K (2001) Molecular regions controlling the activity of CNG channels. J Gen Physiol 118:183–192
58. Chan GH, Fiscus RR (2004) Exaggerated production of nitric oxide (NO) and increases in inducible NO-synthase mRNA levels induced by the pro-inflammatory cytokine interleukin-1beta in vascular smooth muscle cells of elderly rats. Exp Gerontol 39:387–394
59. Ahern GP, Klyachko VA, Jackson MB (2002) cGMP and S-nitrosylation: two routes for modulation of neuronal excitability by NO. Trends Neurosci 25:510–517
60. Gow AJ, Chen Q, Hess DT, Day BJ, Ischiropoulos H, Stamler JS (2002) Basal and stimulated protein S-nitrosylation in multiple cell types and tissues. J Biol Chem 277: 9637–9640
61. Nakamura T, Lipton SA (2010) Redox regulation of mitochondrial fission, protein misfolding, synaptic damage, and neuronal cell death: potential implications for Alzheimer's and Parkinson's diseases. Apoptosis 15:1354–1363
62. Sayed N, Baskaran P, Ma X, van den Akker F, Beuve A (2007) Desensitization of soluble guanylyl cyclase, the NO receptor, by S-nitrosylation. Proc Natl Acad Sci U S A 104:12312–12317
63. Seth D, Stamler JS (2011) The SNO-proteome: causation and classifications. Curr Opin Chem Biol 15:129–136
64. Beckman JS, Koppenol WH (1996) Nitric oxide, superoxide, and peroxynitrite: the good, the bad, and the ugly. Am J Physiol 271:C1424–C1437
65. Leung EL, Fraser M, Fiscus RR, Tsang BK (2008) Cisplatin alters nitric oxide synthase levels in human ovarian cancer cells: involvement in p53 regulation and cisplatin resistance. Br J Cancer 98:1803–1809
66. Fraser M, Chan SL, Chan SS, Fiscus RR, Tsang BK (2006) Regulation of p53 and suppression of apoptosis by the soluble guanylyl cyclase/cGMP pathway in human ovarian cancer cells. Oncogene 25:2203–2212
67. Dostmann WR, Tegge W, Frank R, Nickl CK, Taylor MS, Brayden JE (2002) Exploring the mechanisms of vascular smooth muscle tone with highly specific, membrane-permeable inhibitors of cyclic GMP-dependent protein kinase Ialpha. Pharmacol Ther 93:203–215
68. Taylor MS, Okwuchukwuasanya C, Nickl CK, Tegge W, Brayden JE, Dostmann WR (2004) Inhibition of cGMP-dependent protein kinase by the cell-permeable peptide DT-2 reveals a novel mechanism of vasoregulation. Mol Pharmacol 65:1111–1119
69. Burkhardt M, Glazova M, Gambaryan S, Vollkommer T, Butt E, Bader B, Heermeier K, Lincoln TM, Walter U, Palmetshofer A (2000) KT5823 inhibits cGMP-dependent protein kinase activity in vitro but not in intact human platelets and rat mesangial cells. J Biol Chem 275:33536–33541
70. Valtcheva N, Nestorov P, Beck A, Russwurm M, Hillenbrand M, Weinmeister P, Feil R (2009) The commonly used cGMP-dependent protein kinase type I (cGKI) inhibitor Rp-8-Br-PET-cGMPS can activate cGKI in vitro and in intact cells. J Biol Chem 284:556–562
71. Gladwin MT, Kim-Shapiro DB (2008) The functional nitrite reductase activity of the heme-globins. Blood 112:2636–2647
72. Tiso M, Tejero J, Basu S, Azarov I, Wang X, Simplaceanu V, Frizzell S, Jayaraman T, Geary L, Shapiro C, Ho C, Shiva S, Kim-Shapiro DB, Gladwin MT (2011) Human neuroglobin functions as a redox regulated nitrite reductase. J Biol Chem 286(20):18277–18289
73. Zhu Y, Sun Y, Jin K, Greenberg DA (2002) Hemin induces neuroglobin expression in neural cells. Blood 100:2494–2498
74. Fiscus RR, Robles BT, Waldman SA, Murad F (1987) Atrial natriuretic factors stimulate accumulation and efflux of cyclic GMP in C6-2B rat glioma and PC12 rat pheochromocytoma cell cultures. J Neurochem 48:522–528
75. Paul C, Stratil C, Hofmann F, Kleppisch T (2010) cGMP-dependent protein kinase type I promotes CREB/CRE-mediated gene expression in neurons of the lateral amygdala. Neurosci Lett 473:82–86
76. Kwiatkowski AV, Rubinson DA, Dent EW, Edward van Veen J, Leslie JD, Zhang J, Mebane LM, Philippar U, Pinheiro EM, Burds AA, Bronson RT, Mori S, Fassler R, Gertler FB (2007) Ena/VASP is required for neuritogenesis in the developing cortex. Neuron 56: 441–455
77. Fiscus RR, Leung CP, Yuen JP, Chan HC (2001) Quantification of apoptotic DNA fragmentation in a transformed uterine epithelial cell line, HRE-H9, using capillary

electrophoresis with laser-induced fluorescence detector (CE-LIF). Cell Biol Int 25: 1007–1011

78. O'Neill RA, Bhamidipati A, Bi X, Deb-Basu D, Cahill L, Ferrante J, Gentalen E, Glazer M, Gossett J, Hacker K, Kirby C, Knittle J, Loder R, Mastroieni C, Maclaren M, Mills T, Nguyen U, Parker N, Rice A, Roach D, Suich D, Voehringer D, Voss K, Yang J, Yang T, Vander Horn PB (2006) Isoelectric focusing technology quantifies protein signaling in 25 cells. Proc Natl Acad Sci U S A 103:16153–16158

79. Fan AC, Deb-Basu D, Orban MW, Gotlib JR, Natkunam Y, O'Neill R, Padua RA, Xu L, Taketa D, Shirer AE, Beer S, Yee AX, Voehringer DW, Felsher DW (2009) Nanofluidic proteomic assay for serial analysis of oncoprotein activation in clinical specimens. Nat Med 15:566–571

Chapter 19

Electrophysiological Technique for Analysis of Synaptic Function of PKN1 in Hippocampus

Hiroki Yasuda and Hideyuki Mukai

Abstract

PKNs are serine/threonine protein kinases that have conserved catalytic domains homologous to those of protein kinase C (PKC) family members and regulatory regions containing antiparallel coiled-coil (ACC) domains and C2-like domains. PKN1 in particular is abundant in the brain, and the physiological role of this enzyme has been examined by generating PKN1 genetically modified mice and inhibitors of this enzyme. Here, we review electrophysiological techniques to analyze synaptic functions in the CA1 region of the hippocampus using these mice and inhibitors.

Key words: PKN, PRK, PAK, Rho, Hippocampus, Synaptic plasticity, Electrophysiology

1. Introduction

1.1. Protein Kinases and Phosphatases in Synaptic Plasticity

Phosphorylation and dephosphorylation in signaling cascades that underlie synaptic plasticity decide the direction of synaptic changes in the hippocampus (1). Strong inputs to synapses induce large excitatory postsynaptic potentials (EPSPs) with action potentials, fully relieve magnesium block of NMDA receptors (NMDARs) on postsynaptic sites, and then massive calcium influx is induced and this activates calcium/calmodulin-dependent protein kinase II (CaMKII) (2). CaMKII triggers signaling processes that result in the insertion of AMPA receptors (AMPARs) into postsynaptic sites; which is called long-term potentiation (LTP) of synaptic transmission (3). Although the key substrates for CaMKII that are highly essential for triggering LTP remain unclarified, CaMKII itself (4), AMPARs (5, 6), and stargazin (7) among others are the candidates. We have also reported that protein kinase A mediates LTP in the CA1 region of the immature hippocampus where CaMKII is not fully expressed (8). On the other hand, weak synaptic inputs cause lower calcium influx through NMDARs and

Hideyuki Mukai (ed.), *Protein Kinase Technologies*, Neuromethods, vol. 68,
DOI 10.1007/978-1-61779-824-5_19, © Springer Science+Business Media, LLC 2012

induce long-term depression (LTD) that is dependent on calcineurin and protein phosphatases (9). Other than these, there are many protein kinases and phosphatases in the brain and their roles in the central nervous system are being investigated.

1.2. PKN in CNS

PKN was first described as a fatty acid- or phospholipid-activated serine/threonine protein kinase and also as a protease-activated protein kinase (10–12). Subsequent molecular cloning and protein analyses revealed that PKN is a family consisting of multiple isoforms, and hence the original PKN was redesignated as "PKN1." (Since these studies have been carried out independently in several laboratories, different nomenclatures were proposed for the PKN isoforms derived from different genes: Regarding the nomenclature adopted herein, we use PKN1 (instead of PKNα (11), PRK1 (13), or PAK1 (12)), PKN2 (instead of PRK2 (13), or PAK2 (14)), and PKN3 (instead of PKNβ (15)) in accordance with the Human Kinome nomenclature (16).) Human and mouse PKN1s have at least two splicing variants having different first exons (E1a and E1b), designated as PKN1a and PKN1b (manuscript in preparation). PKN isoforms have conserved domains: ACC domains and the C2-like domain in the amino-terminal region, and the catalytic serine/threonine kinase domain in the carboxyl-terminal region. However, PKN isoforms show different enzymatic properties and tissue distributions, and have been implicated in various distinct cellular processes (reviewed in (17)).

PKN1 is expressed in almost every tissue and is particularly enriched in the spleen, thymus, testes, and brain (PKN1 is approximately 0.01% of total protein in the rat brain) (10). PKN1a is the major constituent of PKN1 in the brain, reaching approximately 90% of PKN1 (manuscript in preparation). PKN2 is also widely expressed in various tissues, and the amount of PKN2 is ~1/3 of that of PKN1 in the brain. Northern blotting of PKN3 showed that PKN3 mRNA is almost undetectable in normal tissues (15), but a subsequent experiment using a specific antibody against PKN3 showed that PKN3 is present in most of the tissues and its amount is ~1/10 of that of PKN1 in the brain (manuscript in preparation). In situ hybridization histochemistry revealed that PKN1 mRNA is heterogeneously localized in the rat brain, but most neurons from the forebrain to the hindbrain contain PKN1 mRNA, suggesting the involvement of this enzyme in processes that are common to multiple neuronal types and the non-exclusive association of PKN1 with any one neurotransmitter system (18). Immunohistochemical analysis revealed that PKN1 is concentrated in a subset of endoplasmic reticulum (ER) and ER-derived vesicles localized to the apical compartment of the juxtanuclear cytoplasm, as well as late endosomes, multivesicular bodies, Golgi bodies, secretory vesicles, and cell nuclei (19), suggesting that the enzyme is implicated in the regulation of vesicle movement in neurons.

PKN1 has been suggested to be involved in the organization of various cytoskeletal components. PKN1 associates with, phosphorylates and depolymerizes intermediate filament proteins, such as neurofilament and glial fibrillary acidic protein (GFAP) (20–22). PKN1 is also reported to be involved in the phosphorylation of the microtubule-associated protein tau, directly and indirectly through GSK-3β (23, 24), leading to the regulation of microtubule assembly. In Alzheimer disease-affected neurons, PKN1 is redistributed to the cortical cytoplasm and neurites and is closely associated with neurofibrillary tangles (NFTs) and their major constituent, abnormally modified tau. PKN1 is also found in degenerative neurites within senile plaques (19), suggesting a specific role of PKN1 in NFT formation and neurodegeneration in Alzheimer disease-damaged neurons. Loh et al. conducted a genome-wide RNA interference-based forward genetic screening and showed that PKN1 is involved in lysophosphatidic acid (LPA)-induced neurite retraction, which is known to be related to the activation of Rho GTPase (25). They also reported that the haploinsufficiency of a Drosophila orthologue of PKN1 is able to suppress light-induced retinal degeneration in the fly model of class III autosomal dominant retinitis pigmentosa (ADRP) (25), suggesting the close relationship of PKN1 with neurodegeneration.

PKN1 is cleaved at specific sites by caspase-3 or related proteases in apoptotic Jurkat and U937 cells, and it generates constitutively active kinase fragments (26). The intense 55-kDa band that corresponds to the active catalytic fragment of PKN1 appears 5 days after middle cerebral artery occlusion (MCAO) in the rat brain, and is sustained at an increased level until at least day 28 (27). A 55-kDa PKN1 cleavage fragment is also induced in an ischemia/reperfusion model of the rat retina generated by increasing intraocular pressure or by N-methyl-D-aspartate (NMDA) administration, and DEVD-CHO significantly inhibits the appearance of the 55-kDa fragment and protects against retinal cell loss (28). Fragmentation of PKN1 was also observed in samples collected from the parietal cortex, striatum, septal nucleus, hippocampus, and periaqueductal gray matter in a hydrocephalus model rat generated by kaolin treatment (29). Furthermore, a 55-kDa PKN1 fragment was reported to be induced by exposure of rat primary cortical neurons to excitotoxic glutamate and in the spinal cords of SOD1G93A transgenic mice, the most well-characterized model of amyotrophic lateral sclerosis (ALS) (20). Constitutive activity of PKN1 may be important for neuronal degenerative or regenerative processes.

Thus, there have been accumulating evidences suggesting that PKN1 has some roles in a pathological state of the nervous system. On the other hand, the physiological function of PKN1 in the nervous system has not been clarified yet.

1.3. Role of PKN1 in Synaptic Plasticity

We have been trying to investigate PKN functions in the central nervous system by developing PKN1 genetically modified mice and inhibitors of PKN. Here, we describe how to analyze the synaptic functions in PKN1a knockout (KO) mice using electrophysiological techniques.

2. Materials

2.1. Generation of PKN1a KO Mouse

To evaluate the functions of PKN1a in the brain, the gene encoding PKN1a was disrupted in 129SvJ mouse embryonic stem cells by replacing the first exon (E1a) with a neomycin-resistance cassette manuscript in preparation). No truncated PKN1 protein was detected in PKN1a$^{-/-}$ mouse cells by immunoblot analysis. The abundance of other PKN isoforms was similarly observed in the brain of PKN1a$^{+/+}$ and PKN1a$^{-/-}$ mice. All mice were backcrossed to a C57BL/6J genetic background for 15 generations.

2.2. Development of PKN Inhibitory Peptide

An inhibitory peptide selective for PKNs, designated as the PRL peptide (PRLRRQKKIFSKQQG), was prepared as previously described (30). The PRL peptide was derived from the portion of the autoinhibitory region of human PKN1. This peptide selectively inhibits the kinase activity of all isoforms of PKN toward a δpeptide substrate with a Ki of 0.7 μM. The PRL peptide itself is not cell permeable; however, it is useful for inhibiting a postsynaptic neuron only by loading it into such a neuron through a patch pipette in slice patch experiments (see below).

3. Methods

3.1. Hippocampal Slice Preparation

Animal use and all experimental procedures were approved by the Ethical Committee for Animal Experiments of Gunma University. Hippocampal slices were cut from littermates of KO and wild-type mice born from heterozygous parents.

3.1.1. Preparation of Extracellular Solution

The extracellular solution (artificial cerebrospinal fluid, ACSF) we usually use contains 119 mM NaCl, 2.5 mM KCl, 4.0 mM $CaCl_2$, 4.0 mM $MgSO_4$, 1.0 mM NaH_2PO_4, 26.2 mM $NaHCO_3$, and 11 mM glucose. Before use, the solution should be bubbled well with 95% O_2/5% CO_2 gas.

3.1.2. Anesthetizing

A mouse was anesthetized in a bell jar with isoflurane. The mouse should be unconscious before decapitation, but an overdose of the anesthetic should be avoided. We carefully checked the animal was breathing and make sure not to kill it.

3.1.3. Dissection of Hippocampus

We removed the mouse from the jar, decapitated, removed the skull, removed the forebrain, and kept it for 2 or 3 min in an ice-cold oxygenated (95% O_2/5% CO_2) ACSF. Then, we put the forebrain on a filter paper wet with the ice-cold ACSF on a glass petri dish. We cut the forebrain into left and right half hemispheres and kept one of the two hemispheres in an ice-cold oxygenated ACSF. Then (a) we cut the frontal part of one hemisphere, allowed it to stand and removed the brainstem and the thalamus with spatulas. (b) We inserted a spatula between the hippocampus and the cortex, flipped over the hippocampus and removed the rest. (c) The dissected hippocampus was kept in ice-cold oxygenated ACSF. Another hemisphere was placed on a petri dish, and steps (a)–(c) were repeated again.

3.1.4. Preparation of Slices

Two hippocampi were placed in an agar holder, and the holder was placed on a cutting chamber filled with ice-cold oxygenated ACSF. The hippocampi were cut into slices using a vibratome. The slices were kept in a submersion- or interface-type incubation chamber for at least 1.5–2 h before the start of recording.

3.2. Electrophysiological Recordings

3.2.1. Field Excitatory Postsynaptic Potential Recording

A hippocampal slice was taken from the incubation chamber using a transfer pipette, placed in a submersion-type recording chamber mounted on an upright microscope (e.g., BX51WI, Olympus, Tokyo, Japan) and perfused with oxygenated ACSF (at 30°C) including a GABAA receptor antagonist, picrotoxin at 100 μM. Field potential recordings were made from neurons in the CA1 area using an amplifier (e.g., Multiclamp 700B, Molecular Devices, Sunnyvale). Glass recording and stimulating electrodes were made from patch pipettes with broken tips, filled with ACSF, and located in the stratum radiatum in the CA1 region. Field excitatory postsynaptic potentials (fEPSPs) were evoked with a negative voltage pulse with a duration of 100 μs applied through the stimulating electrode. fEPSP sweeps were collected and stored using a software in a PC computer (e.g., Igor, Wavemetrics, Lake Oswego, USA).

3.2.2. Slice Patch Experiments

When an inhibitor of a kinase or a phosphatase, we are interested in, is available, we can load the inhibitor into a postsynaptic neuron through a patch pipette and test the function of the enzyme at postsynaptic sites even if the inhibitor is membrane impermeable.

A recording glass electrode for slice patch experiment has a resistance with a few or several Mohms depending on the targeted cell size and types of experiments. For voltage-clamp experiments, excitatory postsynaptic currents (EPSCs) are recorded through an electrode with a relatively lower resistance and filled with a cesium-based internal solution, for example, a solution containing (in mM) 135 $CsMeSO_4$, 10 HEPES, 0.2 EGTA, 8 NaCl, 5 QX-314 Cl, 4 Mg-ATP, and 0.3 Na_3GTP (pH 7.2 with CsOH; osmolality adjusted to 280–290 mOsm). When we block the activity of the enzyme we can include the inhibitor in the internal solution and passively load the neuron with it. For current-clamp experiments, EPSPs are recorded through a glass patch electrode filled with a potassium-based internal solution, for example, a solution containing (in mM) 135 K-gluconate, 10 HEPES, 0.2 EGTA, 4 Mg-ATP, and 0.5 Na_3GTP (pH 7.2 with KOH, osmolality adjusted to 280–290 mOsm).

We recorded from hippocampal neurons using the blind patch technique (31). Briefly, a patch pipette was inserted into the cell layer of the CA1 region applying a moderate positive pressure inside the pipette. The pipette was moved forward stepwise until the tip touched a neuron and pipette resistance was slightly elevated. After ensuring that the pipette caught a neuron by moving another few steps and checking that the resistance kept increasing, we released the internal positive pressure. When the increase in resistance was slow and not sufficient, we blew through the pipette to release the neuron, applied the positive pressure, move the electrode, and tried to obtain another neuron again. When the pipette resistance was elevated appropriately, we sucked in the neuron and made a giga seal. Before rupturing the cell membrane and form a whole-cell configuration, we usually waited for a few minutes because the neural tissue around the electrode was depolarized by the cesium- or potassium-based internal solution blown out while approaching the neuron.

3.3. Tests for Postsynaptic Function

3.3.1. Long-Term Potentiation

LTP is a major form of synaptic plasticity observed in vitro and in vivo, and a long-lasting enhancement of synaptic strength induced by a strong conditional stimulation. Especially, LTP in the hippocampus is very prominent and is hypothesized to underlie learning and memory; therefore, the mechanisms of induction and expression of LTP have been well investigated. LTP in the CA1 region is NMDAR-dependent (2). Strong inputs induce massive calcium influx through NMDARs, which activates calcium/CaMKII, resulting in the induction of LTP (2). In 1995, Liao et al. (32) and Isaac et al. (33) reported on "silent synapses," which have only NMDARs and do not have AMPARs. They also proved that LTP is the activation of AMPARs at silent synapses, meaning that the expression of LTP is postsynaptic in origin (2). Moreover, many studies proved that the expression of LTP in the CA1 region of the hippocampus involves the incorporation of AMPARs into postsynaptic sites on spines in pyramidal neurons (3). Thus, LTP in this area has a postsynaptic function.

The role of the enzyme of interest in LTP was analyzed by comparing the magnitude of LTP of fEPSPs or EPSP(C)s between wild-type mice and KO mice in which the gene encoding the target enzyme was deleted. Initially, stable baseline fEPSPs with an amplitude in the range of 0.2–0.4 mV were recorded at intervals of 20 s in the presence of a GABAA receptor antagonist, picrotoxin, at 100 μM. Next, one or a few bursts of 100 Hz 1 s tetanic stimula-

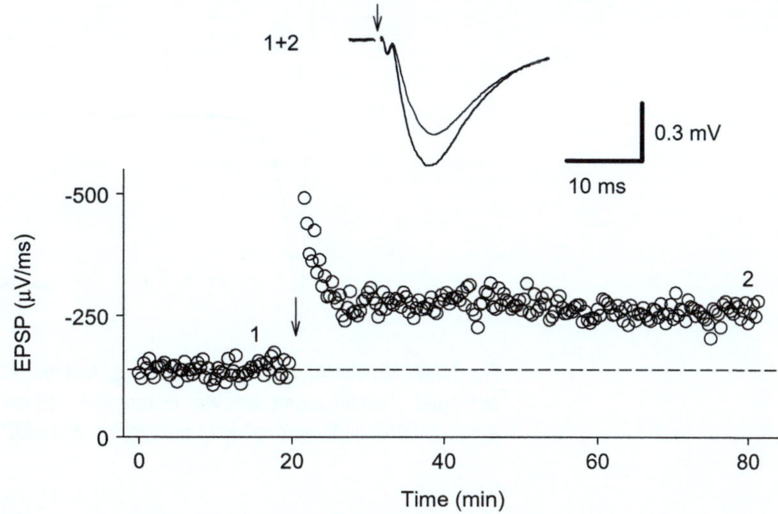

Fig. 1. LTP in CA1 region of mouse hippocampus. The *lower graph* shows an example of LTP in the hippocampus. Tetanic stimulation was delivered at the time indicated by the *arrow* after recording baseline fEPSP responses for 20 min. The inset shows average fEPSP sweeps taken at times indicated by numbers.

tion were delivered to induce LTP, and then fEPSPs were recorded at the same frequency as that of baseline responses (Fig. 1). To induce a robust LTP, we can increase the stimulus intensity during tetanus. We have analyzed PKN1a KO mice and we did not see any difference in LTP between wild type and KO mice (Yasuda and Mukai, in preparation).

3.3.2. Long-Term Depression (Optional)

LTD is also a major form of synaptic plasticity and is the down-regulation of synaptic transmission. LTD in the CA1 region of the hippocampus is also dependent on NMDARs. Weak synaptic inputs cause calcium influx through NMDARs that is not sufficient to activate CaMKII, but sufficient to activate calcineurin, resulting in LTD induction (1, 9). The expression of NMDAR-dependent LTD involves the internalization of AMPARs by endocytosis at postsynaptic sites (34); therefore, LTD is also a postsynaptic function. In fEPSP recordings, LTD can be induced by low-frequency stimulation (Ex. 1 Hz 15 min). However, the change in fEPSPs in LTD is smaller in wild-type mice hippocampus than in the rat hippocampus (Yasuda, personal communication); therefore, possibly it may be difficult to detect a difference in the magnitude of LTD between wild-type and KO mice.

3.3.3. AMPAR/NMDAR Ratio

If the target enzyme is involved in the regulation of synaptic plasticity mentioned above, the expression of AMPARs at postsynaptic sites may be up- or down-regulated. To determine the amount of synaptic AMPARs, we measured the ratio of AMPAR-mediated EPSCs to

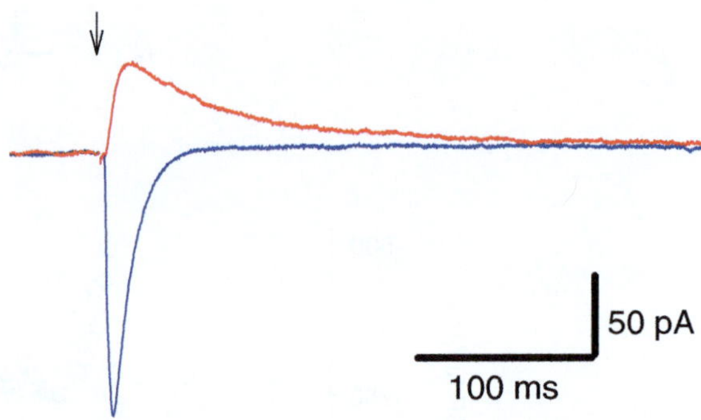

Fig. 2. AMPAR/NMDAR ratio in hippocampus. AMPAR EPSCs were recorded at −70 mV and averaged. Then the membrane was clamped at +40 mV in the presence of CNQX (10 μM) a non-NMDAR glutamate receptor antagonist, and NMDAR EPSCs were recorded (*red*).

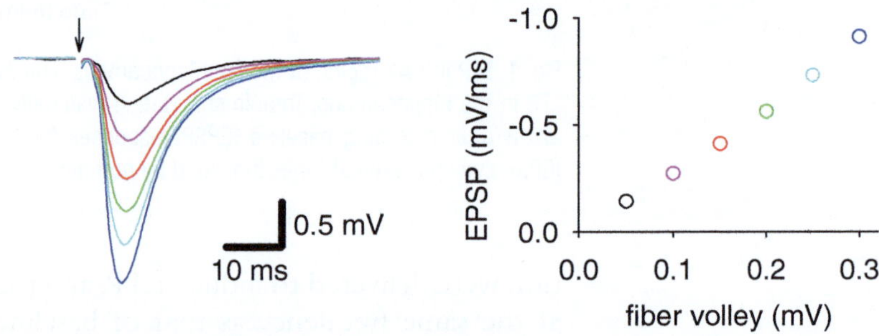

Fig. 3. Input–output relationship. Stimulus intensity was adjusted so as to obtain fEPSPs with fiber volley amplitudes of 0.05, 0.1, 0.15, 0.2, 0.25, and 0.3 mV (*right*) and the slope of fEPSPs was plotted against the fiber volley amplitudes (*left*).

NMDAR-mediated EPSCs (AMPAR/NMDAR ratio). Initially, AMPAR EPSCs were recorded from a CA1 pyramidal neuron at −70 mV in the presence of picrotoxin, a GABAA receptor antagonist, at 100 μM. Next, CNQX (10 μM), a AMPAR antagonist, was added to extracellular ACSF, the neuron was depolarized at +40 mV, and NMDAR EPSCs were recorded (35) (Fig. 2).

3.3.4. Input–Output Relationship

In order to investigate whether synaptic transmission is intact or altered in the KO mice, we examined the input–output relationship. In field potential recordings, fiber volley is the activity of presynaptic fibers and its amplitude reflects the number of fibers activated by the stimulus. The ratio of amplitudes or initial slopes of fEPSPs to amplitudes of fiber volley corresponds to a postsynaptic response per certain amount of presynaptic fibers (Fig. 3). When synaptic plasticity is altered in KO mice and the expression of postsynaptic AMPARs is not the same as that in wild-type mice, the input–output relationship might be different between them. Changes in transmitter release in KO mice can also affect the input–output relationship; therefore the input–output relationship is a gross estimation of both pre- and postsynaptic function.

3.4. Tests for Presynaptic Function

3.4.1. Paired-Pulse Ratio

When an electrical shock through a stimulating electrode causes an action potential in an axon, it travels throughout the axon, induces calcium influx through voltage-dependent calcium channels, and triggers transmitter release at synaptic boutons or presynaptic terminals. If the second shock is delivered immediately after the first shock, the transmitter release is enhanced as long as the transmitter is not depleted in the first release, and the second synaptic response is facilitated. This is called "paired-pulse facilitation (PPF)," which is prominent in synapses whose release probability is low, such as the hippocampus (36, 37) (Fig. 4). Facilitation is hypothesized to be caused by elevated calcium level in the synaptic terminals after the first action potential (38). In contrast, when release probability is elevated by increasing extracellular calcium or in synapses that has originally have a high release probability, for example, in the cerebral cortex, the transmitter is to some extent depleted after the first release and the paired-pulse ratio (PPR) is lowered or even paired-pulse depression (PPD) occurs (36–38). Thus, we can roughly estimate release probability by measuring "PPR," which is the ratio of the amplitude of the second synaptic response to that of the first one. We did not see any significant difference in PPR between wild-type and PKN1a KO mice (Yasuda and Mukai, in preparation).

3.4.2. Posttetanic Potentiation

In fEPSP recordings, LTP is induced by high-frequency tetanic stimulation, as mentioned above. Just after the tetanus, fEPSPs were robustly enhanced, then immediately started to decline and reached an elevated stable level. The enhancement of synaptic responses immediately after tetanus is called posttetanic potentiation (PTP). Similar to PPF, PTP is considered to be a temporal enhancement of transmitter release (39) caused by residual calcium in presynaptic terminals after tetanus (38, 40); therefore, it is a presynaptic function. In theory, similar to PPR, PTP is larger when

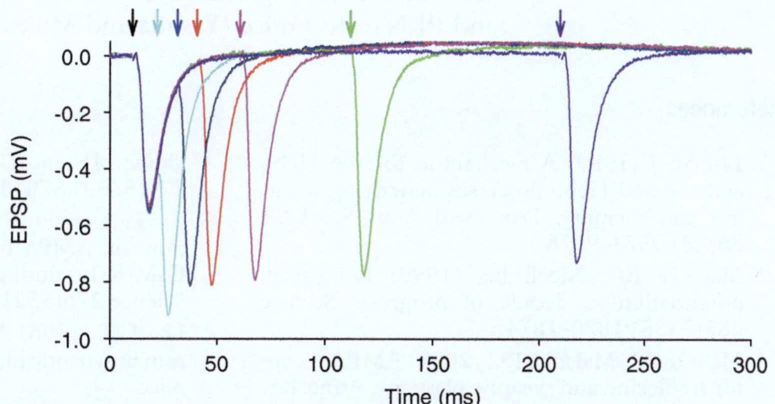

Fig. 4. Paired-pulse fascilitation in hippocampus. Paired-pulse stimuli with intervals of 10, 20, 30, 50, 100, 200 ms were delivered and fEPSPs were recorded.

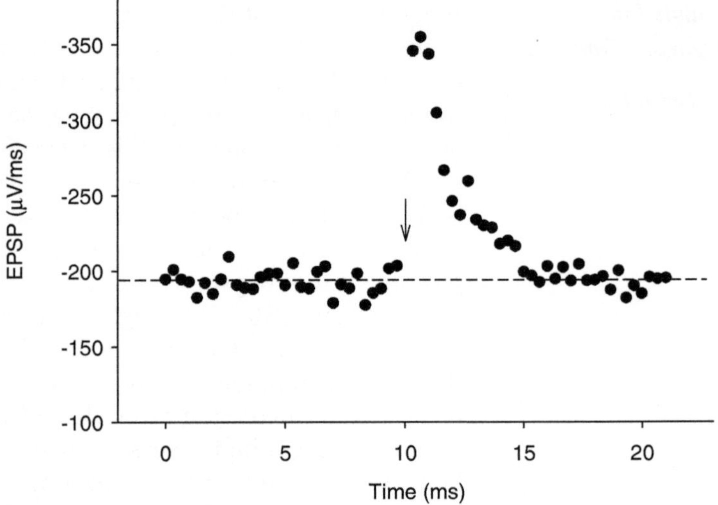

Fig. 5. Posttetanic potentiation in hippocampus. A 100 Hz 1 s tetanic stimulation was delivered in the presence of D-APV (50 μM), a NMDAR antagonist at the time indicated by the *arrow*.

the release probability is low because the transmitter is still available even after tetanus. In contrast, PTP is smaller when the release probability is high, because the transmitter is already depleted after a massive transmitter release happens during tetanus (for example, see Fig. 3d in (37) and Fig. 6 in (39)). The discrepancy in changes between PPR and PTP in the hippocampus in αCaMKII (41) and Rab3A (37) KO mice and in parallel fiber to Golgi cell (PF–GC) synapses in the cerebellum (39) has been reported, suggesting that PPR and PTP are independently regulated (37, 39).

When we measured PTP, we delivered high-frequency stimulation (for example, 100 Hz for 1 s) in the presence of a NMDAR antagonist D-APV at 50 μM to inhibit LTP induction (Fig. 5). We did not see any significant difference in PTP between wild-type and PKN1a KO mice (Yasuda and Mukai, in preparation).

References

1. Lisman J (1989) A mechanism for the Hebb and the anti-Hebb processes underlying learning and memory. Proc Natl Acad Sci USA 86(23):9574–9578
2. Malenka RC, Nicoll RA (1999) Long-term potentiation—a decade of progress? Science 285(5435):1870–1874
3. Malinow R, Malenka RC (2002) AMPA receptor trafficking and synaptic plasticity. Annu Rev Neurosci 25:103–126
4. Giese KP et al (1998) Autophosphorylation at Thr286 of the alpha calcium-calmodulin kinase II in LTP and learning. Science 279(5352):870–873
5. Barria A et al (1997) Regulatory phosphorylation of AMPA-type glutamate receptors by CaM-KII during long-term potentiation. Science 276(5321):2042–2045
6. Lee HK (2006) AMPA receptor phosphorylation in synaptic plasticity: insights from knockin mice.
7. Tomita S et al (2005) Bidirectional synaptic plasticity regulated by phosphorylation of stargazin-like TARPs. Neuron 45(2):269–277

8. Yasuda H et al (2003) A developmental switch in the signaling cascades for LTP induction. Nat Neurosci 6(1):15–16
9. Malenka RC, Bear MF (2004) LTP and LTD: an embarrassment of riches. Neuron 44(1):5–21
10. Kitagawa M et al (1995) Purification and characterization of a fatty acid-activated protein kinase (PKN) from rat testis. Biochem J 310 (Pt 2):657–664
11. Mukai H, Ono Y (1994) A novel protein kinase with leucine zipper-like sequences: its catalytic domain is highly homologous to that of protein kinase C. Biochem Biophys Res Commun 199(2):897–904
12. Peng B et al (1996) Phosphorylation events associated with different states of activation of a hepatic cardiolipin/protease-activated protein kinase. Structural identity to the protein kinase N-type protein kinases. J Biol Chem 271(50):32233–32240
13. Palmer RH, Ridden J, Parker PJ (1995) Cloning and expression patterns of two members of a novel protein-kinase-C-related kinase family. Eur J Biochem 227(1–2):344–351
14. Yu W et al (1997) Isolation and characterization of a structural homologue of human PRK2 from rat liver. Distinguishing substrate and lipid activator specificities. J Biol Chem 272(15): 10030–10034
15. Oishi K et al (1999) Identification and characterization of PKNbeta, a novel isoform of protein kinase PKN: expression and arachidonic acid dependency are different from those of PKNalpha. Biochem Biophys Res Commun 261(3):808–814
16. Manning G et al (2002) The protein kinase complement of the human genome. Science 298(5600):1912–1934
17. Mukai H (2003) The structure and function of PKN, a protein kinase having a catalytic domain homologous to that of PKC. J Biochem 133(1):17–27
18. Hashimoto T et al (1998) Localization of PKN mRNA in the rat brain. Brain Res Mol Brain Res 59(2):143–153
19. Kawamata T et al (1998) A protein kinase, PKN, accumulates in Alzheimer neurofibrillary tangles and associated endoplasmic reticulum-derived vesicles and phosphorylates tau protein. J Neurosci 18(18):7402–7410
20. Manser C et al (2008) Deregulation of PKN1 activity disrupts neurofilament organisation and axonal transport. FEBS Lett 582(15): 2303–2308
21. Matsuzawa K et al (1997) Domain-specific phosphorylation of vimentin and glial fibrillary acidic protein by PKN. Biochem Biophys Res Commun 234(3):621–625
22. Mukai H et al (1996) PKN associates and phosphorylates the head-rod domain of neurofilament protein. J Biol Chem 271(16): 9816–9822
23. Isagawa T et al (2000) Dual effects of PKNalpha and protein kinase C on phosphorylation of tau protein by glycogen synthase kinase-3beta. Biochem Biophys Res Commun 273(1): 209–212
24. Taniguchi T et al (2001) Phosphorylation of tau is regulated by PKN. J Biol Chem 276(13):10025–10031
25. Loh SH et al (2008) Identification of new kinase clusters required for neurite outgrowth and retraction by a loss-of-function RNA interference screen. Cell Death Differ 15(2):283–298
26. Takahashi M et al (1998) Proteolytic activation of PKN by caspase-3 or related protease during apoptosis. Proc Natl Acad Sci USA 95(20): 11566–11571
27. Ueyama T et al (2001) Generation of a constitutively active fragment of PKN in microglia/ macrophages after middle cerebral artery occlusion in rats. J Neurochem 79(4):903–913
28. Sumioka K et al (2000) Induction of a 55-kDa PKN cleavage product by ischemia/reperfusion model in the rat retina. Invest Ophthalmol Vis Sci 41(1):29–35
29. Okii N et al (2007) Fragmentation of protein kinase N (PKN) in the hydrocephalic rat brain. Acta Histochem Cytochem 40(4):113–121
30. Shiga K et al (2010) Development of an intracellularly acting inhibitory peptide selective for PKN. Biochem J 425(2):445–543
31. Blanton MG, Lo Turco JJ, Kriegstein AR (1989) Whole cell recording from neurons in slices of reptilian and mammalian cerebral cortex. J Neurosci Methods 30(3):203–210
32. Liao D, Hessler NA, Malinow R (1995) Activation of postsynaptically silent synapses during pairing-induced LTP in CA1 region of hippocampal slice. Nature 375(6530): 400–404
33. Isaac JT, Nicoll RA, Malenka RC (1995) Evidence for silent synapses: implications for the expression of LTP. Neuron 15(2): 427–434
34. Morishita W, Marie H, Malenka RC (2005) Distinct triggering and expression mechanisms underlie LTD of AMPA and NMDA synaptic responses. Nat Neurosci 8(8):1043–1050
35. Niisato K et al (2005) Age-dependent enhancement of hippocampal long-term potentiation and impairment of spatial learning through the Rho-associated kinase pathway in protein tyrosine phosphatase receptor type Z-deficient mice. J Neurosci 25(5):1081–1088

36. Castro-Alamancos MA, Connors BW (1997) Distinct forms of short-term plasticity at excitatory synapses of hippocampus and neocortex. Proc Natl Acad Sci USA 94(8): 4161–4166
37. Schoch S et al (2002) RIM1alpha Forms a protein scaffold for regulating neurotransmitter release at the active zone. Nature 415(6869): 321–326
38. Zucker RS, Regehr WG (2002) Short-term synaptic plasticity. Annu Rev Physiol 64:355–405
39. Beierlein M, Fioravante D, Regehr WG (2007) Differential expression of posttetanic potentiation and retrograde signaling mediate target-dependent short-term synaptic plasticity. Neuron 54(6):949–959
40. Greengard P et al (1993) Synaptic vesicle phosphoproteins and regulation of synaptic function. Science 259(5096):780–785
41. Chapman PF et al (1995) The alpha-Ca^{2+}/calmodulin kinase II: a bidirectional modulator of presynaptic plasticity. Neuron 14(3):591–597

Index

A

AChR pre-patterning ... 208
Action potential ... 349, 357
Adenomatous polyposis coli (APC) ... 160, 164
ADRP. *See* Autosomal dominant retinitis pigmentosa (ADRP)
Affinity electrophoresis ... 13–33
Agrin hypothesis ... 203
ALS. *See* Amyotrophic lateral sclerosis (ALS)
Alternative splicing ... 50, 51, 109, 205
Alzheimer disease ... 170–171, 175, 351
AMPA receptors (AMPARs) ... 349
AMPAR/NMDAR ratio ... 355–356
AMPARs. *See* AMPA receptors (AMPARs)
Amphetamine ... 172, 175
Amyotrophic lateral sclerosis (ALS) ... 169, 174, 320, 351
Antibody feeding ... 276, 277, 280–282, 284–286, 288
Antiparallel coiled-coil (ACC) domain ... 349, 350
Anti-phospho antibody ... 95, 96, 100
APC. *See* Adenomatous polyposis coli (APC)
Apoptosis ... 109, 110, 169, 191, 196, 239, 280, 320, 328, 330, 331, 333–334
Astrocyte ... 106, 116, 174
Autonomy ... 51
Autophosphorylation ... 50–52, 54, 62–63, 65, 67, 74, 80, 107–109, 158, 211, 221–223, 225–227, 232–234, 274, 287
Autosomal dominant retinitis pigmentosa (ADRP) ... 351
Axonal transport ... 273

B

Backsignaling ... 110, 113, 114
B-catenin ... 27–30, 160–162, 165, 167, 168, 212
Binding assay ... 49
Bio-pull down ... 291
Biotinylation ... 276, 277, 280, 282–284, 286, 288
Bipolar disorder ... 169, 173–175, 263
Blind patch technique ... 354
Brain slices ... 137, 144, 148, 149, 237–249
Butyrolactone I ... 95

C

CA1 ... 170, 171, 263, 267, 349, 353–356
Calcineurin ... 285, 350, 355
Calcium ... 285, 350, 355
Calcium/Calmodulin-dependent protein kinase II (CaMKII) ... 17, 349
Calmodulin ... 17, 49–69, 74, 75, 77, 78, 349
CaMKII. *See* Calcium/Calmodulin-dependent protein kinase II (CaMKII)
cAMP. *See* Cyclic AMP (cAMP)
Ca2+ oscillations ... 51, 253
Capillary electrophoresis ... 319–343
CASK ... 73–83
Caspase-3 ... 191, 192, 351
CD44 ... 109
Cdk5 ... 16, 74, 87, 163, 210
Cell permeable JNK inhibitor peptide ... 189
Cell proliferation ... 28, 134, 136, 169, 321, 327, 328
Cell signaling ... 29, 133–150, 167, 194, 222, 225, 227, 228, 230, 297, 298, 310, 323, 332, 333
Central nervous system (CNS) ... 31, 74, 107, 157, 191, 350, 352
cGMP. *See* Cyclic GMP (cGMP)
Chemokinetic ... 116, 123
Chemotactic ... 116, 123
Circadian rhythm ... 168, 293, 320–321
cJun N-terminal kinase (JNK) ... 134, 189–202, 332
CKAR ... 252–256
C2-like domain ... 350
CNS. *See* Central nervous system (CNS)
Compartmentalized cultures ... 275–288
Co-receptor ... 168, 213
Cortical neurons ... 81, 90, 169, 189–202, 240, 296, 298–301, 304, 308, 310–312, 351
CRD. *See* Cysteine rich domain (CRD)
Cyan fluorescent protein (CFP) ... 5
Cyclic AMP (cAMP) ... 4, 6, 154, 158, 159, 165, 168, 237–240, 321, 322
Cyclic AMP-dependent protein kinase (PKA) ... 154, 159
Cyclic GMP (cGMP) ... 320–324, 326–331, 334, 343
Cyclin-dependent kinase ... 16, 87–101
Cysteine rich domain (CRD) ... 112, 113, 205, 206

Hideyuki Mukai (ed.), *Protein Kinase Technologies*, Neuromethods, vol. 68,
DOI 10.1007/978-1-61779-824-5, © Springer Science+Business Media, LLC 2012

D

DAG. *See* Diacylglycerol kinase (DAG)
D-APV....358
Dendritic arbor....292, 293, 302, 305, 306
Dendritic spine....309
DG. *See* Diacylglycerol (DG)
Diacylglycerol (DG)....170, 171, 173, 251, 259, 261, 262
Diacylglycerol kinase (DAG)....259–268
Dinuclear zinc(II) complex....20
D-JNKI1....192, 193, 196, 199, 201, 202
Dok7....210, 211
Dopamine....171, 172, 175, 292
Doxycycline-inducible expression....118

E

EGF. *See* Epidermal growth factor (EGF)
Electrophoretic mobility shift....95, 98
Electrophysiology....267, 349–358
Endocytosis....209, 273–275, 277, 281, 282, 284, 285, 355
Epidermal growth factor (EGF)....14, 111
EPSCs. *See* Excitatory postsynaptic currents (EPSCs)
EPSP. *See* Excitatory postsynaptic potential (EPSP)
ErbB receptor....106, 110–116, 121, 124
ERK. *See* Extracellular-signal regulated kinase (ERK)
Excitatory postsynaptic currents (EPSCs)....353, 356
Excitatory postsynaptic potential (EPSP)....267, 349, 353, 354
Extracellular-signal regulated kinase (ERK)....133–151, 191

F

Fiber volley....356
Field excitatory postsynaptic potentials (fEPSPs)....267, 353–357
Fluorescence....4–6, 8–10, 14, 16, 56, 126, 136, 144–145, 150, 196, 238, 239, 242–244, 246, 249, 256, 266, 296, 303, 308, 309, 324, 333–335
Fluorescent microscopy....49
Fluorescent protein....3
Förster resonance energy transfer (FRET)....4–10, 238, 239, 252, 254–256, 324, 326
FRET. *See* Förster resonance energy transfer (FRET)

G

GABAA receptor antagonist....353, 354, 356
Gamma-secretase....109, 110, 113, 114
GFAP. *See* Glial fibrillary acidic protein (GFAP)
GFP. *See* Green fluorescent protein (GFP)
Glia....105–108, 156, 282
Glial fibrillary acidic protein (GFAP)....136, 138, 144, 351
GluN2B....52, 55, 56, 66–69
Glutamate....49, 52, 56, 69, 77, 78, 137, 140, 141, 148, 169, 171, 191, 239, 292, 295, 296, 313, 351, 356
Green fluorescent protein (GFP)....4, 5, 56, 68, 69, 126, 136–138, 146, 147, 150, 238, 264, 266, 301, 303, 313, 324
GSK3beta....16, 17, 24, 26, 28, 98, 155–158, 162, 165, 166, 168–173, 175

H

HAMMOC. *See* Hydroxy acid-modified metal oxide chromatography (HAMMOC)
Hedgehog....153
Hippocampus....156, 169, 261–263, 265, 300, 321, 349–358
Histone H1....89, 94, 95, 97, 100
Human genetics....153
Hydroxy acid....36–37
Hydroxy acid-modified metal oxide chromatography (HAMMOC)....36–39, 41, 43, 44

I

Imaging....4–10, 56, 64, 68–69, 237–249, 251–256, 298, 307, 308, 314
Immunofluorescence....138, 144–145, 148–150, 298, 299, 301, 306, 308, 309, 314
Immunofluorescence staining
Immunoprecipitation....89, 92–94, 96–100, 166, 234, 298, 310, 311, 314
In-cell Western....301, 308
Integrin....109, 116, 121
In vitro kinase assay....94–95, 98–99

J

JNK. *See* cJun N-terminal kinase (JNK)
Juxtamembrane....112–114, 209

K

KD. *See* Kinase dead (KD)
Kinase activity assay....58, 62, 221
Kinase activity reporter....252, 256
Kinase dead (KD)....89, 221, 234
Kinome....75, 350
Knock-down/re-expression....49
Knockout mouse....50, 81, 82, 165, 175, 211
Kringle domain....205

L

Lactic acid....36, 37, 39
LDH assay....193, 196, 199

Lentiviral vector 124, 135–138, 145–146, 150
Leucine-rich repeat kinase 2, 219–234
Lipid 226, 251, 259, 292
Lipofection 4, 7, 9, 93, 118, 303
Lithium 155, 169–175, 263
Live cell imaging 251
Long term depression (LTD) 52, 170, 252, 263, 265, 321, 350, 355
Long term potentiation (LTP) 51, 52, 106, 169–171, 263, 321, 331, 349, 354–355, 357
LPA. *See* Lysophosphatidic acid (LPA)
Lrp4 ....206, 209–211, 213
LTD. *See* Long term depression (LTD)
LTP. *See* Long term potentiation (LTP)
Lysophosphatidic acid (LPA) 117, 351

M

Magnesium........................74, 77, 78, 80, 81, 175, 261, 349
Magnetofection 303, 304, 313
MAGUK....................................73, 75, 77, 79–81, 83, 209
Mammalian target of rapamycin (mTOR) 159, 259, 291–315
immunoprecipitation 298, 310
inhibitors 302
Mammalian target of rapamycin complex 1 (mTORC1) 291–294, 306, 309
Mammalian target of rapamycin complex 2 (mTORC2) 291–294, 306
MAPK. *See* Mitogen-activated protein kinase 2 (MAPK)
MASC. *See* Myotube associated specificity component (MASC)
Mass spectrometry.......................................14, 35–44, 310
MCAO. *See* Middle cerebral artery occlusion (MCAO)
McMahan, Jack 203
MEK inhibitor 140, 141
Memory..............51, 52, 134, 135, 168–171, 237, 252, 263, 268, 293, 321, 331, 354
Metal complex...20
Metal oxide chromatography (MOC) 36
Microscopy...............................68, 146, 149, 151, 240, 244, 246, 247, 266, 300, 312
Microtubule...........................14, 88, 91, 92, 164, 166, 168, 170, 210, 274, 292, 351
Middle cerebral artery occlusion (MCAO) 192, 351
Migration 16, 21, 25–27, 30, 32, 88, 105–127, 160, 168, 205, 208, 326, 327, 330
Migration substrate 115, 124
Mitochondria 109, 110, 169, 219, 220, 252, 292, 327, 328
Mitogen-activated protein kinase 2 (MAPK) 17, 24, 134, 135, 143, 154, 159, 162, 190, 191, 285
MOC. *See* Metal oxide chromatography (MOC)
Morphometric analysis 306
mTOR. *See* Mammalian target of rapamycin (mTOR)
mTORC1. *See* Mammalian target of rapamycin complex 1 (mTORC1)
mTORC2. *See* Mammalian target of rapamycin complex 2 (mTORC2)
Mucin 109, 111, 164
MuSK substrates 211
Mutational analysis..53
Myotube associated specificity component (MASC) 206

N

Nerve growth factor (NGF) 157, 164, 273
Neuregulin (NRG)
Neurexin 74, 80–82
Neurite branching 259
Neurite extension...259
Neurofilament 88, 164, 351
Neurogenesis 167–169
Neuromuscular junction 203–213
Neuron 4, 88, 100, 106, 136, 138, 169, 192, 206, 208, 209, 247, 261, 263, 275, 282, 295, 296, 299, 300, 304, 305, 352–354, 356
Neuronal cell death.................................88, 169, 171, 325
Neuronal death...............................170, 191, 192, 199, 201
Neuronal migration 88, 168
Neuroprotection 193, 196, 323, 328, 331
Neurotrophin.....134, 169, 204, 208, 274, 275, 279, 285, 286
NGF. *See* Nerve growth factor (NGF)
Nitric oxide (NO)..320–323, 325
NMDA receptor (NMDAR) 52, 69, 106, 170, 191, 349, 354–356
NMDA-toxicity 189
NO. *See* Nitric oxide (NO)
Notch 165, 168
NPXY motif...209
NR2B 49

O

Olomoucine...95
Optical biosensor...240

P

P25 ....88–94, 96–98, 100
P35 ....16, 17, 24, 87–89, 91–100, 193
P39 ....87–89, 92, 94, 97–100
PA. *See* Phosphatidic acid (PA)
Paired-pulse depression (PPD).......................................357
Paired-pulse facilitation (PPF) 357
Paired-pulse ratio (PPR) 357, 358
PAK..91
Parkin 74, 220

Parkinson disease....199, 219–234, 320, 343
Pharmacologic inhibitors....252, 320, 329
Phorbol esters....262
Phosphatidic acid (PA)....259
Phosphatidylinositol 3′ kinase....159
Phosphopeptide enrichment....36, 38, 39, 41
Phosphoproteomics....14
Phosphorylation....4, 13, 35–44, 49, 74, 88, 134, 153, 189, 204, 221, 239, 256, 260, 292, 321, 349
Phosphoserine....42, 167
Phosphosite antibody....49
Phospho-specific antibodies....137, 167, 342
Phosphotyrosine....167, 210
Phos-tag....13–33, 89
Phos-tag SDS-PAGE....16–19, 21–33, 99, 100
Picrotoxin....353, 354, 356
PKA. *See* Cyclic AMP-dependent protein kinase (PKA); Protein kinase A (PKA)
PKC. *See* Protein kinase C (PKC)
PKG. *See* Protein kinase G (PKG)
PKN....349, 350, 352
PKN1....349–358
PKNalpha....350
Polymorphism....172–174
Posttetanic potentiation (PTP)....357–358
PPD. *See* Paired-pulse depression (PPD)
PPF. *See* Paired-pulse facilitation (PPF)
PPR. *See* Paired-pulse ratio (PPR)
Presenilin....164, 167, 168
Presynaptic scaffold....73
PRK....349
PRK1....350
Proline-directed protein kinase....88
Protein kinase....3–11, 13–33, 49–69, 74, 88, 109, 134, 153, 189, 219, 238, 291, 320, 349
Protein kinase A (PKA)....4, 16, 237, 321, 324, 349
Protein kinase C (PKC)....52, 109, 159, 251–256, 259–268
Protein kinase G (PKG)....319–343
Protein phosphorylation....13–16, 21, 27, 31, 35–44, 61, 66, 140, 308, 332
Protein purification....234
Proteolytic cleavage....109, 113
Proximity ligation assay....299, 311–312, 314–315
Pseudokinase....74
Pseudosubstrate....75, 155, 158, 159
PTB domain....209, 210, 212
Pten-induced novel kinase 1....219–234
PTP. *See* Posttetanic potentiation (PTP)

R

Rab3A....358
Rapamycin....159, 259, 291–315
Rapsyn....203, 206, 208, 210
RATL....208, 209
Receptor dimerization....107
Recombinant proteins....112, 229, 233, 239, 298
Rho....209, 211, 351
RNAi. *See* RNA interference (RNAi)
RNA interference (RNAi)....135, 293, 351
Roscovitine....89, 95

S

Scaffold proteins....156, 191, 192, 208, 209
Schizophrenia....169–172, 175, 263, 293
Schwann cell....106, 115–117
Scintillation....54, 61, 94, 95, 220, 223, 230
SDS-PAGE....13–15, 33, 49–69, 74, 88, 94, 109, 134, 153, 189, 194, 219, 220, 238, 283, 291, 307, 311, 320, 349
Serotonin....173
SIDIC. *See* Site-determining ion combination (SIDIC)
Signaling endosomes....274, 275
Signal transduction....16, 133, 134, 189, 221, 259, 273, 275, 323
Site-determining ion combination (SIDIC)....38, 42
Slice patch....352–354
Spine formation....263
Stereotaxic injection of mice....143
Striatum....136, 138, 147, 156, 172, 351
Subcellular fractionation....167
Synaptic activity....83, 263
Synaptic plasticity....49, 52, 73, 74, 168, 170, 171, 237, 293, 319–343, 349–350, 352, 354–356
Synaptic transmission....263, 349, 355, 356
T286....50–52, 54, 62, 63, 65, 67
Synaptogenesis....73, 74, 81, 167, 179, 205, 319–343

T

Tau....14, 16–18, 21, 24–28, 88, 95, 164, 166–168, 170, 171, 191, 192, 351
Time-lapse imaging....7–10
Titanium dioxide (TiO2)....36, 37, 39, 41, 43, 44
Torpedo californica....203, 204
Transcription factors....82, 83, 113, 114, 134, 161, 163, 165, 168, 191, 192, 331
Transcytosis....281, 282, 284, 285
Transfection of neurons....69, 281
Transgenic mouse....171
Translocation....14, 52, 56, 68-69, 113, 262
Transwell migration assay....115, 116, 120, 122, 123
Trk receptor....273–288
T305/T306....49

U

Ubiquitination....74
Ubiquitinylation....161, 165
Unplugged....205, 208

V

Voltage-dependent calcium channel 357

W

Western blot 16, 21, 54, 59, 64, 67, 137, 143, 193, 197, 198, 201, 223, 232, 286, 296–298, 301, 306–308, 310, 311, 314, 332, 333, 336, 340–342
Wnt 28, 154, 160–162, 166, 168, 206, 207, 211, 213

Y

Y553 ....209, 210
Yellow fluorescent protein (YFP) 5

MIX
Papier aus verantwortungsvollen Quellen
Paper from responsible sources
FSC® C105338

If you have any concerns about our products, you can contact us on
ProductSafety@springernature.com

In case Publisher is established outside the EU, the EU authorized representative is:
Springer Nature Customer Service Center GmbH
Europaplatz 3, 69115 Heidelberg, Germany

Printed by Libri Plureos GmbH
in Hamburg, Germany